LOOKING AHEAD
IN WORLD FOOD AND AGRICULTURE:
Perspectives to 2050

Edited by **Piero Conforti**

**FOOD AND AGRICULTURE ORGANIZATION OF THE
UNITED NATIONS, 2011**

The designations employed and the presentation of material in this information product do not imply the expression of any opinion whatsoever on the part of the Food and Agriculture Organization of the United Nations (FAO) concerning the legal or development status of any country, territory, city or area or of its authorities, or concerning the delimitation of its frontiers or boundaries. The mention of specific companies or products of manufacturers, whether or not these have been patented, does not imply that these have been endorsed or recommended by FAO in preference to others of a similar nature that are not mentioned.

The views expressed in this information product are those of the authors and do not necessarily reflect the views of FAO.

ISBN 978-92-5-106903-5

FOREWORD

Anticipating future developments in global agriculture is by no means a simple exercise. In the last few years, many of the acute phenomena observed have complicated further the formulation of long-term prospects. The turbulence of world agricultural markets, the price spikes of 2008 and 2011, the wide climate variability experienced in important production regions, and the enhanced linkage among agriculture and other markets such as the energy and the financial markets have propelled interest in revisiting the relations among agriculture, its natural resource basis, economic development, food security and population growth. Discussions of the relationships among these phenomena are lively, as are those on what can be done to prevent the onset of more frequent and more critical conditions in the coming decades.

Given its various fields of expertise, FAO is at the centre of the technical debate on these themes. In 2009, FAO organized an expert meeting and forum around the question of "How to feed to the world in 2050". This initiative was supported by papers authored by world-class experts. This work has been revisited, and is now presented in this volume.

Several aspects of the perspectives for global agriculture are analysed. FAO's last global projection exercise to 2050 indicates that agricultural and food demand is expected to slow over the next decades, following slowing population growth and raising incomes. However, population will still grow considerably in the coming decades, and require world agricultural production to increase substantially by 2050. The macroeconomic outlook indicates that economic growth may bring significant reductions in poverty in the 2050 horizon, but climate change may impose additional constraints, particularly through increased pressures on land and water resources. Biofuel development may be another source of stress for markets, depending on the ability of technology to reduce the overlap between energy feedstock and food products.

Other areas explored in the volume are natural resources – notably land and water – as well as capital, investment and technology. Regarding natural resources, the amounts of land and water available at the global level are most probably sufficient to support the projected production increases. By 2050, the FAO baseline points to some net expansion of arable land, all in developing countries. However, the bulk of production increase will need to be generated through increased yields.

Technically, there seems to be considerable scope for pushing the agricultural technology frontier outwards. But much could also be done by simply applying existing technologies. To this end, farmers and other stakeholders along value chains need to receive the correct scarcity signals from markets to be able to access appropriate inputs and to invest. Global fixed capital stock in agriculture has been growing steadily over the last three decades, although at declining rates. Research is shown to be among the most productive investments for supporting agriculture, together with education, infrastructure and input credits. The concluding part of this volume takes stock of areas where consensus seems to emerge and those where controversies loom large. It identifies areas in which more information and analytical work are required and, last but not least, it provides insights into the strategies and policies to be enacted in support of global agriculture in the 2050 perspective.

In publishing this volume, FAO aims to keep the debate alive. The various chapters bring to the fore not only what we know about the long-term future of global agriculture, but also what we do not know, and the weaknesses of the methodologies used to make projections.

Hafez Ghanem
Assistant Director-General
Economic and Social Development
Department
FAO

Kostas Stamoulis
Director
Agricultural Development
Economics Division
FAO

CONTENTS

vii

About the Authors

Nikos Alexandratos is a consultant to FAO, and was formerly Chief of FAO's Global Perspective Studies Unit.

Gustavo Anriquez is an economist at the Agricultural Development Economics Division of FAO.

John Baffes is a senior economist at the Development Prospects Group of the World Bank.

Nienke Beintema is the leader of the Agricultural Science and Technology Indicators (ASTI) initiative of the International Food Policy Research Institute (IFPRI).

Gerold Boedeker is a programme officer at the FAO Sub-regional Office for Central and Eastern Europe, and was formerly an economist at FAO's Global Perspective Studies Unit.

Jelle Bruinsma is a consultant to FAO, and was formerly Chief of FAO's Global Perspective Studies Unit.

Andrew Burns is lead economist at the Development Prospects Group of the World Bank.

Derek Byerlee is an independent scholar, and was formerly Director, World Development Report 2008 at the World Bank.

Piero Conforti is an economist at the Agricultural Development Economics Division of FAO.

Stephan von Cramon-Taubadel is a professor at the Department of Agricultural Economics and Rural Development of the University of Göttingen, Germany.

Gregory O. Edmeades is an independent consultant, 43 Hemans Street, Cambridge, New Zealand.

Howard Elliott is an independent consultant, and was formerly Deputy Director General of the International Service for National Agricultural Research in The Hague, Netherlands.

Tony R.A. Fischer is an honourary research fellow at CSIRO Plant Industry, Canberra, Australia.

Günther Fischer is leader of the Land Use Change and Agriculture Program at the International Institute for Applied Systems Analysis (IIASA).

Hartwig de Haen is professor emeritus at the Department of Agricultural Economics and Rural Development of the University of Göttingen, Germany, and was formerly Assistant Director-General of FAO.

Evan Hillebrand is associate professor of international economics at the Patterson School of Diplomacy and International Commerce, University of Kentucky, United States of America.

Siwa Msangi is a senior research Fellow within the Environmental Production and Technology Division of the International Food Policy Research Institute (IFPRI).

Oleg Nivyevskiy is a research associate at the Department of Agricultural Economics and Rural Development of the University of Göttingen, Germany.

Israel Osorio-Rodarte is a consultant at the Development Prospects Group of the World Bank.

Mark Rosegrant is director of the Environment and Production Technology Division of the International Food Policy Research Institute (IFPRI).

Alexander Sarris is professor of economics at the University of Athens, Greece, and was formerly Director of the Trade and Markets Division of FAO.

Josef Schmidhuber is principal officer at the Statistics Division of FAO, and was formerly senior economist at FAO's Global Perspective Studies Unit.

Dominique van der Mensbrugghe is lead economist at the Development Prospects Group of the World Bank.

ACKNOWLEDGEMENTS

Competent advice from Ira Matuschke and Michelle Kendrick is gratefully acknowledged. Special thanks are due to Jane Shaw, for editorial support and Anna Doria Antonazzo who did the layout.

Acronyms

ACIAR	Australian Centre for International Agricultural Research
ACS	agricultural capital stock
AEZ	Agro-Ecological Zone (model)
AGVP/PC	agricultural gross value of production per capita
AKST	agricultural knowledge, science and technology
ARI	agricultural research intensity ratio
AR4	Fourth Assessment Report (IPCC)
ASARECA	Association for Strengthening Agricultural Research in Eastern and Central Africa
ASTI	Agricultural Science and Technology Indicators
AY	attainable yield
AYa	attainable yield under current economics
AYb	attainable yield under efficient institutions and markets
CAADP	Comprehensive Africa Agriculture Development Programme
CARD	Centre for Agriculture and Rural Development
CARDI	Caribbean Agricultural Research and Development Institute
CATIE	Tropical Agricultural Research and Higher Education Center
CBS	country balance sheet(s)
CDM	Clean Development Mechanism
CGE	computable general equilibrium (model)
CGIAR	Consultative Group on International Agricultural Research
CIMMYT	International Maize and Wheat Improvement Center
CIRAD	International Cooperation Centre of Agricultural Research for Development
CO_2	carbon dioxide
CSIR0	Commonwealth Science and Industrial Organization
CV	coefficient of variation
DEA	data envelopment analysis
DSSAT	Decision Support System for Agrotechnology Transfer
EM	expert meeting

ENVISAGE	ENVironmental Impact and Sustainability Applied General Equilibrium (model)
EU	European Union
FAPRI	Food and Agricultural Policy Research Institute
FBS	food balance sheet(s)
FDI	foreign direct investment
FT	Fischer-Tropsch (diesel)
FTE	full-time equivalent
FY	farm yield
GAEZ	Global Agroecological Zones Study
GCM	general circulation model
GDP	gross domestic product
GEO	Global Environmental Outlook
GHG	greenhouse gas
GIDD	Global Income Distribution Dynamics
GIS	Geographic Information System
GMC	global middle class
GTAP	Global Trade Analysis Project
HGCA	Home Grown Cereal Authority
HI	harvest index
HIC	high-income country
IAASTD	International Assessment of Agricultural Science and Technology for Development
ICOR	incremental capital output ratio
ICP	International Comparison Project
ICRISAT	International Crop Research Institute for the Semi-Arid Tropics
IEA	International Energy Agency
IF	International Futures (model)
IFPRI	International Food Policy Research Institute
IIASA	International Institute for Applied Systems Analysis
IITA	International Institute of Tropical Agriculture
IMF	International Monetary Fund
IMPACT	International Model for Policy Analysis of Agricultural Commodities and Trade
IP	intellectual property
IPCC	Intergovernmental Panel on Climate Change

IR	Interim Report
IRD	Institute of Research for Development
IRRI	International Rice Research Institute
JIRCAS	Japan International Research Center for Agricultural Sciences
K	potassium
LMY	low and middle-income country
LUT	land utilization type
MARS	marker-assisted recurrent selection
MAS	marker-assisted selection
MDER	minimum daily/dietary energy requirement
MDG	Millennium Development Goal
MEA	Millennium Ecosystem Assessment
MIROC	Model for Interdisciplinary Research on Climate
MUV	Manufacturing Unit Value index
MV	modern varieties
N	nitrogen
NARS	National Agricultural Research Systems
NCAR	National Center for Atmospheric Research
NCD	non-communicable disease
NEPAD	New Partnership for Africa's Development
NUE	nitrogen use efficiency
ODA	official development assistance
OECD	Organisation for Economic Co-operation and Development
OFID	OPEC Fund for International Development
OPEC	Organization of the Petroleum Exporting Countries
P	phosphorus
PAR	photosynthetically active radiation
ppm	parts per million
PPP	purchasing power parity
PWT	Penn World Tables
PY	potential yield
PY_w	water-limited potential yield
R&D	research and development
ReSAKSS	Regional Strategic Analysis and Knowledge Support System
RUE	radiation use efficiency

SAP	standard accounting procedure
SAR	Special Administrative Region
SME	small and medium enterprise
SNS	sensitivity (scenario)
SOFI	The State of Food Insecurity in the World
SRES	Special Report on Emissions Scenarios
SUA	supply utilization account
TAR	target (scenario)
TDW	total above-ground dry weight
TE	transpiration efficiency
TFP	total factor productivity
TOE	tonne(s) of oil equivalent
TV	traditional varieties
UN	United Nations
UNEP	United Nations Environment Programme
USDA	United States Department of Agriculture
WEO	World Energy Outlook
WFS	World Food System (model)
WTO	World Trade Organization

Introduction

Piero Conforti

The last few years have witnessed a revamping of the debate on the relations among agriculture, natural resources, population growth and economic development. Various events, taking place within a short time span, have driven such renewed interest. Weather variability has seemed to accelerate, deeply affecting agricultural production around the world. Food and energy prices have shown large swings, while the biofuel industry has been expanding rapidly in some countries. Financial investment in agricultural commodity-based derivatives has grown at a fast pace, while large international investments in land, especially targeting developing countries, have gained the newspaper headlines.

These phenomena, and the relations among them, are complex in nature and enmeshed in a wide technical debate involving a host of different subjects and fields of expertise. Controversies loom large, but consensus seems to be emerging, at least around a few facts. For instance, the fact that agriculture has been neglected in many developing countries, and that this has at least accentuated the negative consequences of the events of recent years; and the fact that the phenomena observed – such as the 2008 spike in agricultural prices and the peak in 2011 – are acute symptoms generated by the combination of several causes. Beyond these facts there are no simple answers or solutions to the many questions and problems posed by current events. Some experts and commentators stress the finite dimension of natural resources available on the planet. Global population is growing at a decreasing pace, but is expected to continue growing over the next decades. And in certain regions the pressure on natural resources is already approaching critical levels.

Information and discussion on what is to be expected in the coming decades seem to be in high demand. How are the evolution of demand and supply in the next decades going to shape agricultural markets? How are long-term growth prospects and the expected evolution of per capita income going to affect agriculture and food production? Are the natural resources available, such as land and water, sufficient to feed a growing population? What role can economic incentives and technical change play in shaping supply? And what are the priority areas where investment and research should be directed? How may the use of agricultural products in biofuel production affect markets? How can climate change affect production possibilities, and hence markets? Projections and impact analyses of these phenomena are subject to a wide debate; some figures point to catastrophic outcomes, albeit with attached probability values that are, fortunately, smaller than one.

In such a context, it is imperative to attempt to gain a better understanding of how these perspective events may interact. All of them directly affect food security, and especially the ability of vulnerable population groups to express an effective demand on the market, to produce enough food sustainably, and to earn a viable income. Agriculture plays a notoriously vital role in the dynamic of poverty, given that a large share of the poor make their living directly or indirectly from this sector.

In 2009, FAO's Economic and Social Development Department organized an expert meeting and high-level expert forum around the question of "How to feed the world in 2050". The scientific basis of this initiative was provided by world-class experts, who analysed key aspects of the matter, including population, poverty and macroeconomic developments; investment and technology; and markets and price development, the connection between food and energy markets, and the impact of climate change.

This volume follows up on these events in two ways. First, the volume proposes a selection of revised and updated results – which in some cases resulted from the interaction of experts after the 2009 expert meeting – and takes stock of the conclusions reached, highlighting areas where there is fundamental consensus, and those where more information and analytical work are required. Second, the results form the basis of discussion of the strategies and policies that can be considered to prepare global agriculture for the 2050 perspective.

Needless to say, the volume is far from exhaustive on all of these issues, and some of them are almost absent. Environmental issues and related policies are hardly emphasized, but contributions are made to the debate on resources and their availability for agricultural production. The debate surrounding the volatility of food prices, its causes, consequences and related policies is also outside the domain of this volume, along with the debate on the relation between agriculture and financial markets, and discussion of the dynamics of foreign direct investment.

Many of the contributions to the 2009 expert meeting collected in this volume report on projection exercises. It is worth emphasizing that these are not meant to be speculations on the future; rather, they aim to fuel a discussion of how the phenomena observed today may evolve and combine in the future. Their ultimate goal is to offer insights into what can be done today to shape some underlying trends. In so doing, projections also help to identify which of the many phenomena observed today may be long-lasting and may produce significant impacts in the future.

The volume is divided into four main parts. The first part, The global agriculture outlook, reports the results of three fairly comprehensive projection exercises produced by FAO, the International Food Policy Research Institute (IFPRI) and the International Institute for Applied Systems Analysis (IIASA).

Chapter 1, by N. Alexandratos, analyses the perspectives of global agriculture to 2050, using as a starting point the base period 1999 to 2001. Alexandratos works around the latest global projection exercise produced by FAO (2006), known and quoted as the "Interim Report", whose main results are summarized in an Annex at the end of the chapter. Alexandratos validates the results of the 2006 exercise. First he analyses how projections compare with recent developments observed in world agricultural markets. Second, the chapter looks at how the projections in the FAO (2006) report compare with those formulated three years later for the Organisation for Economic Co-operation and Development (OECD)/ FAO Outlook. These are based on more recent information – the 2008 price spike had already occurred – and for the period from 2009 to 2018. This part of the validation also looks at the market impact of the increased use of crops by the biofuel industry. The chapter revisits food consumption projections formulated on the 1999/2001 base, and projections of the number of undernourished people in developing countries. Over the last years, food consumption data have been considerably revised, as have the parameters used to compute numbers of undernourished and population projections. The results presented by Alexandratos are not based on a single equilibrium model; rather, they are generated through an accounting system that projects demand and supply for 110 countries and 34 agricultural products using a wealth of econometric technical parameters and geo-referenced data.

A different approach is taken by the second projection exercise presented in this part of the volume, prepared by S. Msangi and M. Rosegrant. Contrary to Alexandratos, these authors base their projections on intensive modelling, using the International Model for Policy Analysis of Agricultural Commodities and Trade (IMPACT), developed by IFPRI. This model has been extensively used to project long-term global food supply and demand and food security. A set of counterfactual scenarios are analysed to gain insights into the impact that market drivers such as income and population have on world food prices, production and food security. Another set of simulations focus on three specific aspects:

the development of the biofuel industry, which is deemed to have significantly affected world prices in 2008; the potential role of technical change in world markets, particularly yield-increasing innovations, and their impact on food security; and climate change, which is assumed to have a significant impact on markets in the coming decades. Scenarios are run also to understand how increases in agricultural production and productivity may offset the increased use of feedstocks in the bioenergy industry, in terms of world prices and changes in the nutrition status of vulnerable population groups. The chapter concludes with a discussion of the prospects for food security, and some policy implications.

The last projection exercise included in the first part of this volume is authored by G. Fischer. This chapter focuses mainly on the development of biofuels and the impact of climate change. The analysis is deeply rooted in geo-referenced data, coupled with an economic equilibrium model. Geo-referenced data are drawn from the FAO/IIASA Agro-Ecological Zone (AEZ) model, while the economic equilibrium model used is the IIASA World Food System (WFS) model. Concerning climate change, Fischer starts from country-level assumptions of gross domestic product (GDP) growth and their likely impact on main environmental variables, particularly greenhouse gas emissions and atmospheric concentrations of carbon dioxide (CO_2). The impacts of the resulting climate change scenarios are analysed on the basis of geo-referenced data, to understand how the potential for agriculture and food production may change over the coming decades. This allows the computation of projected changes in yields. In building biofuel development scenarios, Fischer draws on the 2008 World Energy Outlook (WEO), which reports projections on energy uses. This same source is employed to derive assumptions on future shares of biofuels in total energy. Simulations also analyse the uncertainty about technical change, by specifying alternative patterns of diffusion and reduction in the use of traditional feedstocks, such as sugar and maize. The consequences of alternative scenarios are analysed mainly in terms of relative prices, production, consumption, trade and number of people at risk of hunger.

The second part of the volume, Growth, poverty and macroeconomic prospects, discusses the evolution of macro variables and income distribution. Growth plays a key role in shaping the economic environment in which agriculture is expected to operate. At the same time, in several countries, the macroeconomic framework is directly affected by what happens in agriculture. This is particularly the case for poorer countries, where agriculture is still the backbone of the economy, and the agriculture sector's performance is more closely related to food insecurity. This second part of the volume consists of two chapters.

The first, by E. Hillebrand, analyses the outlook for growth, poverty and inequality. The author starts from the observation that global poverty rates have fallen in the last 50 years, while income inequality has increased, as shown by the growing distance between OECD countries and the rest of the world. Hillebrand applies an accounting procedure to forecast poverty levels based on exogenous GDP projections, assuming constant income distribution and a constant ratio of consumption to income. Reduced-form equations are employed to compute an elasticity of the poverty headcount ratio to income growth. The chapter analyses two main scenarios. The first, called "market first", is optimistic, and implies fast growth and the tendency for poor countries to converge rapidly towards OECD ones. The second scenario, "trend growth", is more pessimistic. It assumes that countries continue to evolve along the paths observed over the last 25 years, which means continued fast growth in some countries, such as China and India, and more mixed outcomes in Latin America, the Near East and Africa. Projections of poverty turn out to be sensitive to assumptions on population growth and the resource outlook. Projections also turn out to be quite sensitive to the assumption, embedded in all scenarios, that technical change can address the limitations imposed by finite resource availability. The author correctly notes that a number of common assumptions in forecast exercises are highly questionable: technology is not necessarily a residual, as in Solow-type models; and catching up is not a necessary outcome, especially if institutions and investment are not adequate.

On related themes, the second contribution in this part of the volume is the chapter by D. van der Mensbrugghe, I. Osorio-Rodarte, A. Burns and J. Baffes. Focus here is on the linkage between poverty and commodity markets. The analysis is wide, and dwells on a number of aspects, starting from experience of the 2008 and 2010/2011 price spikes and their causes. The chapter analyses how commodity price formation can be affected in the future by growth prospects, and the related expansion of consumption, which is expected to take place mostly in developing countries and emerging economies. The impact of climate change, particularly on production and trade, is also analysed, along with the increased interdependence between energy and non-energy markets. The interaction among macro variables, the environment and commodity markets is studied through the World Bank general equilibrium global model called ENVironmental Impact and Sustainability Applied General Equilibrium Model (ENVISAGE). This is a so-called "integrated assessment model", which specifies the relations among economic activity, greenhouse gas emissions and temperatures up to 2100. Scenarios analysed concern climate change, the evolution of agricultural productivity, and its potential impact on agricultural commodity prices, with special reference to developing countries. Prospects for poverty to 2050 are analysed with the World Bank's Global Income Distribution Dynamics (GIDD) model.

The third part of the volume is titled Land, water and capital requirements, and is made up of three chapters that propose, respectively, quantifications of natural resources, human-made capital, and investment to satisfy agricultural and food demand projected in 2050.

Chapter 6, by J. Bruinsma, discusses the outlook for natural resources to 2050, and deals with the *vexata quaestio* of whether resources available on the planet are sufficient to support the population expected in 2050. As the starting point, the author takes the FAO baseline projections to 2050, partially revised with updated information for recent years, and analyses the implications of the expected size of demand and supply in terms of land use, water use and crop intensification. Consistent with the FAO approach to long-term projections, the analysis is not based on alternative scenarios. Similar to Fischer's work in Chapter 3, it relies on the FAO-IIASA AEZ database. For land, Bruinsma starts from overall potential, and works out an amount of surface that could potentially be converted into arable land, taking into account the constraints to expansion, such as those arising from environmental concerns, the need to protect fragile ecologies, and the lack of infrastructure. Projections on water use rely on the FAO ACQUASTAT database, and a wealth of scattered information for specific countries. The author projects the likely expansion of areas equipped for irrigation, water use efficiency and a measure of pressure on water resources.

The following two chapters of this third part discuss agricultural capital stock, its evolution over the last decades, and its likely evolution to 2050, given the requirements in terms of predicted agricultural supply. Chapter 7, by S. von Cramon-Taubadel, G. Anriquez, H. de Haen and O. Nivyevskyi, start by estimating the evolution of capital stocks and investment in agriculture over the last four decades. Two approaches to measuring agricultural capital stocks are proposed, implemented and compared. The first is based on national accounts; it is more comprehensive and allows the capture of a wide set of farming assets, but the necessary data are available for only a few countries. The second method is based on physical inventories reported in FAOSTAT. This information is available for a large set of countries over a long time series, but it covers only a small set of fixed assets in farming; hence this second method is less comprehensive. The relations among capital, investment, agricultural productivity and food insecurity are subsequently analysed in this chapter, and estimates of changes in agricultural TFP are presented. The chapter also explores the role that public expenditure on agriculture has played in encouraging the growth of agricultural capital stock and factor productivity.

Investment requirements to 2050 are analysed in Chapter 8, by J. Schmidhuber, J. Bruinsma and G. Boedeker, with reference to developing countries. The authors compute the amount of capital required to produce crops and livestock products projected in FAO's long-term outlook to 2050. The idea is that given the technical parameters, projected production quantities can be used to infer the required input use; in other word, given the amount of a certain crop expected to be produced in 2050, the requirements are computed for that production in terms of hectares of land, amount of irrigation, machinery, livestock herds, hand tools, etc. The authors also consider investment in production chains that strictly speaking go beyond primary production, such as those in storage, processing and marketing. No distinction is made between private and public investment.

The fourth and last main part of the volume, Raising productivity: research and technology, discusses how research and development (R&D) can contribute to increasing productivity in agriculture, the role of public expenditure and the specific types of technologies that should be emphasized. This part includes two chapters: the first analyses past trends and the current situation, while the second looks at the 2050 perspective.

Chapter 9 is by N. Beintema and H. Elliott and relies extensively on the Agricultural Science and Technology Indicators (ASTI) database, supported by IFPRI. The authors review past trends of investments in agricultural R&D, considering both public and private resources for agriculture. Underinvestment in agriculture is then analysed in three ways: by comparing the rate of return on agricultural research relative to the social rate of return; by considering trends in productivity growth; and by looking at the inability to attain political commitments, such as the Millennium Development Goals (MDGs). The agricultural research intensity ratio (ARI) – the ratio of agricultural R&D investments to agricultural GDP – is computed for a large sample of developing and middle-income countries. The authors analyse ARIs through four components: i) the share of agricultural research in total agricultural expenditure; ii) the share of public expenditure on agriculture in total public expenditure; iii) the share of public revenue and expenditure in GDP; and iv) the inverse of agriculture's share in GDP. The chapter also discusses emerging challenges for agricultural research, such as adaptation to climate change, increasing weather variability and water scarcity.

The following and last chapter in this part, authored by T.R.A. Fischer, D. Byerlee and G.O. Edmeades, discusses the possibility of technology achieving the level of agricultural intensification required to meet expected demand in 2050. Specifically, the chapter discusses yield prospects for three key staple crops – wheat, rice and maize – which are the basis for many consumption patterns around the world. Actual farm yields, attainable yields and potential yields are analysed, along with the constraints that can undermine intensification, ranging from inadequate infrastructure and institutions to farmers' skills and attitude towards technology. Incentives are also considered among the elements that can affect technology adoption by farmers as well as technology development. The technical possibilities for increasing potential yields are also discussed, including through the use of biotechnology.

The concluding part of the volume also contains two chapters. Chapter 11, by N. Alexandratos, presents a critical review of the projection exercises, highlighting the main messages that can be derived from the previous chapters, the information gaps and inconsistencies, and the methodological differences that lead to different results. Alexandratos organizes his review around five topics: i) the expected behaviour of world food prices; ii) the impact of climate change; iii) the impact of biofuel development on world agricultural markets; iv) the outlook for GDP and global inequality and poverty; and v) the projections for sub-Saharan Africa, which is undoubtedly the most sensitive area, both today and in perspective.

The insights provided by Alexandratos serve as a useful introduction to the following and last chapter, by P. Conforti and A. Sarris. Chapter 12 gives a broad overview on the results, the challenges that they pose in the 2050 perspective, and the policy directions that can be derived from them. A brief discussion of the major areas in which more information and analysis would be useful is also included. The need to shape policy action is one of the main reasons for taking an interest in long-term projections and for attempting to distinguish what are likely to be short-lived phenomena from what are likely to continue having an impact on the world agriculture and food system into the coming decades. The discussion of policy directions is organized around three pillars: measures affecting the demand side of the global food system; measures affecting its supply side; and measures aimed at producing global public goods related to agriculture.

The book seeks to contribute to maintaining the debate on the future of the global agricultural and food economy. Its content was designed to interest both a technical audience and a wider range of professionals working around the world in areas related to agriculture and food security, in both public and private institutions.

References

FAO. 2006. World agriculture: towards 2030/2050. *Prospects for food, nutrition, agriculture and major commodity groups. Interim report.* Rome, Global Perspective Studies Unit.

Fischer, G., van Velthuizen, H., Shah, M. & Nachtergaele, F.O. 2002. *Global agro-ecological assessment for agriculture in the 21st century: Methodology and results.* Laxenburg, Austria, IIASA and Rome, FAO.

UNDP. 2005. *Halving hunger: It can be done. London*, UN Millennium Project, Task Force on Hunger.

PART I

THE GLOBAL AGRICULTURE OUTLOOK

PART I

WORLD FOOD AND AGRICULTURE TO 2030/2050 REVISITED. HIGHLIGHTS AND VIEWS FOUR YEARS LATER[1]

Nikos Alexandratos

This chapter sketches out the possible evolution of world food and agriculture to 2050 in terms of key variables: production and consumption of the main commodity groups; and the implications for food and nutrition in developing countries. It presents a view of how these variables may evolve over time, not how they should evolve from the normative perspective of solving problems of nutrition and poverty. The chapter's contents are based on food and agriculture projections to 2015, 2030 and 2050, prepared in the years 2003 to 2005 and published in 2006 (FAO, 2006 – hereafter referred to as the Interim Report [IR]). The main findings from Chapter 1 of the IR (Overview) are attached as Annex 1.2. The reader is referred to the full IR for details.

The IR projections were based on historical data from the complete FAO (FAOSTAT) food balance sheets (FBS) available for all countries. The FBS data then available went up to 2001, so the base year for the projections was the three-year average for 1999 to 2001. The projected rainfed and irrigated land use and yield configurations underlying the production projections were not evaluated against the land and water potentials of each country at that time. The latest attempt in this area dates from work carried out in 2000 to 2002, with projections going to 2030 from base year 1997/1999 and published in 2003 (Bruinsma, 2003), using the land potential estimates from an older edition of the Global Agroecological Zones Study (GAEZ) of FAO and the International Institute for Applied Systems Analysis (IIASA) (Fischer *et al.*, 2002). The IR evaluation was delayed by waiting for fresh estimates of such potentials to be produced for the revised GAEZ. These estimates from the new GAEZ are currently being prepared for publication (Fischer, van Velthuizen and Nachtergaele, forthcoming), but are not yet available in the format required for use in analyses of the IR-type. In the

---

1. The author thanks colleagues in FAO's Markets and Trade Division for making available the preliminary results of their ongoing work on the projections for the 2009 OECD/FAO Agricultural Outlook.

meantime, an attempt has been made to unfold the land-use and yield growth implications of the production projections to 2030/2050 of the IR using the old GAEZ estimates of land potentials. These are presented in Chapter 6.

Naturally, presenting in mid-2009 projections completed in 2005 and based on historical data up to 2001 and on the outlook for key exogenous variables (population and GDP projections) as known at that time presents some problems. The last few years have witnessed upheavals, and these must be taken into account in passing judgement regarding the relevance today of views into the future from four years ago. In the first place, the energy markets have intruded into those for agricultural produce, via high energy prices and the boost these gave to the demand for crops as biofuel feedstocks, helped by government policies favouring such use of crops. It is now widely accepted that this was a key factor explaining the food price surges up to mid-2008. Second, the overall economic outlook is being severely affected by the ongoing economic crisis, although the issue of how important this may prove to be in the longer term is moot. In addition, the latest demographic assessments (from 2006) (UN, 2007) and the just-released assessment for 2008 suggest that projected populations to 2050 may be higher than those of the 2002 assessment (UN, 2003) used in the IR, particularly in several countries of sub-Saharan Africa.[2]

It would be desirable to account for these new circumstances by redoing the entire projections exercise. This proved practically impossible, however, given the great country and commodity detail involved (FAO, 2006: 66–68) and the delay in updating FAO's FBS data (Box 1.1). The second best option is to review the IR projections on the basis of the FAO data set used predominantly for current monitoring, and published (for major countries and aggregates only) in the six-monthly *Food Outlook,* and also for the annual OECD/FAO medium-term projections (hereafter referred to as the country balance sheets [CBS] data).[3] The current round of these medium-term projections for the ten years 2009 to 2018 has just been completed (OECD/FAO, 2009).[4] These projections *ex-hypothesi* incorporate all the information currently available concerning developments in the last few years and views of what may be in store up to the year 2018 in terms of the overall economy, the energy sector and prices. As such, the projections provide a valid benchmark for comparison with those of the IR to draw inferences

---

2. World population was projected to reach 8.9 billion by 2050 in the 2002 assessment, and 9.2 billion in the latest 2008 assessment. The IR projections for developing countries of sub-Saharan Africa are 1.5 billion and 1.7 billion, respectively.

3. CBS is a database maintained by the Trade and Markets Division of FAO. The CBS data used here were updated on 3 July 2009.

4. Data and projections available at www.agri-outlook.org/document/6/0,3343,en_36774715_36775671_40969158_1_1_1_1,00.html.

about the continued validity, or otherwise, of these IR projections. Comparability is limited by differences in commodity coverage/specifications and in the country groups distinguished (Box 1.1). However, some comparisons at the level of large country aggregates (developing, developed, world) can be made to provide a reality check of the IR projections. Regional-level projections are presented in the section on Food consumption and nutrition in developing countries.

Box 1.1 - The data situation

Before proceeding, a note on the data situation is in order. Projections published in the IR and previous work were based exclusively on FAO's FAOSTAT data sets of production and trade of all commodities, including non-food ones such as cotton and rubber, as they had been standardized and processed into the supply utilization accounts (SUAs) and the FBS. Revisiting these projections in mid-2009, to take account of recent developments, required inspecting them against SUA/FBS data updated to a more recent year, as many changes had occurred owing to the advent of biofuels and the surge in food prices. However, such data were not yet available: at the time of writing (May 2009) FAO's published SUA/FBS data go only to 2003, with provisional unpublished ones to 2005. These estimates include some radical revisions to the historical data, including those for 1999/2001, the base year for the IR, particularly as regards per capita food consumption, which is of key importance in diagnosing the nutrition situation (see the section on Food consumption and nutrition in developing countries). Non-SUA/FBS FAOSTAT data go to 2007 for production and 2006 for trade. It is obvious that the existing updates of the SUA/FBS data do not provide an adequate basis for revisiting the IR projections in light of the new circumstances.

The following analysis resorts to the CBS data set, which covers a more limited number of commodities than the SUA/FBS data; for example, it does not cover key food commodities such as roots and tubers or pulses, which are the mainstay of diets in several countries. It has data up to 2008 (for which year the data are often estimates) for production, trade and stocks (and hence also includes the implicit total domestic disappearance or consumption for all uses). It often, but not always, includes utilization categories (food, feed, etc.). The country coverage and detail in this data set are not always sufficient to generate the country groups used in the IR projections (FAO, 2006: 67). For example, the IR projections include Romania and Bulgaria in the group "Other Eastern Europe". Likewise, the ten countries that entered the European Union (EU) in 2004 are projected as a separate group from the older EU15 countries. For recent years, the CBS data do not generally show data for these countries individually, but only for the EU as a whole. This makes it impossible to generate data suitable for comparing the IR projections for many country groups with actual outcomes to 2007 and with estimates for 2008. In the following discussion, data are therefore compared between developing countries and the rest of the world or developed countries. The latter comprise the "industrial" and "transition" groups in the IR.

Even more serious problems arise from the non-comparability of data resulting from differences in commodity specifications. For example, in the IR the commodity "sugar" includes all sugar crops and derived products (including non-centrifugal sugar, which is important in countries such as India) converted into raw sugar-equivalent quantities.

> The CBS do not use the same coverage, so direct comparison is not possible. The same goes for the commodity "vegetable oil": in the IR specification it comprises all oilcrops, oils and derived products converted into oil equivalent. This means that consumption of oilseeds – directly as pulses (e.g., soybeans, groundnuts) or in other forms – is counted as consumption of the oil content equivalent in the IR data and projections but not in those of other databases and projection studies.

This chapter presents such a reality check, together with the IR projections, for a few commodity aggregates, focusing particularly on cereals (sum of wheat, rice and coarse grains) and meat (sum of bovine, pig meat, poultry and ovine, in carcass weight). There were two reasons for selecting these aggregates:

- They do not present major comparability problems with the commodity specifications of the IR.

- They have held centre stage in the debate on food price surges: at the early stages of the price surges, there was quasi-consensus around the view that spurts in food/feed demand, particularly in the fast-growing emerging economies (India, China) with their allegedly voracious appetite for meat, were a key determinant. This is no longer a proposition that many would defend, but it is an idea that is hard to die[5] (Alexandratos, 2008).

In addition, comparisons for the commodity "vegetable oils" are presented. These comparisons are of a more limited nature because of incompatibilities in the commodity specifications.

Interim report projections and reality checks

A major point made in the IR was that the growth of demand in developing countries and the world for both cereals (excluding their use for biofuels, which was not accounted for in the IR) and meat would gradually decelerate. However, as noted, in the debate on the recent food price surges up to mid-2008, it was often stated (or rather assumed, given that food consumption data were scarce) that the spurt in demand for meat and the associated demand for feed cereals in developing countries, particularly China and India, were a major factor explaining why cereals prices surged. The first question is therefore whether the predicted deceleration is actually happening. Attention should then turn to examining

---

5. For example, see a recent article in the *Economist* ("Green shoots", 21 March 2009) holding that the steady increase in demand from poorer countries is the single largest cause of rising prices. It would be more correct to state that increases in demand in the developing countries represent the major component of global demand growth, but this is nothing new. This phenomenon was present even when prices were not rising, and often when they were falling.

whether or not the OECD/FAO projections sketch out future trajectories that are close enough to those of the IR.

Cereals

Table 1.1 compares the IR's projections with the most recent data for 1999/2001 and the latest three-year average for 2006/2008, with and without cereal use for biofuels. Figure 1.1 illustrates the relevant trajectories.

Figure 1.1
World cereal consumption, with and without United States maize for ethanol

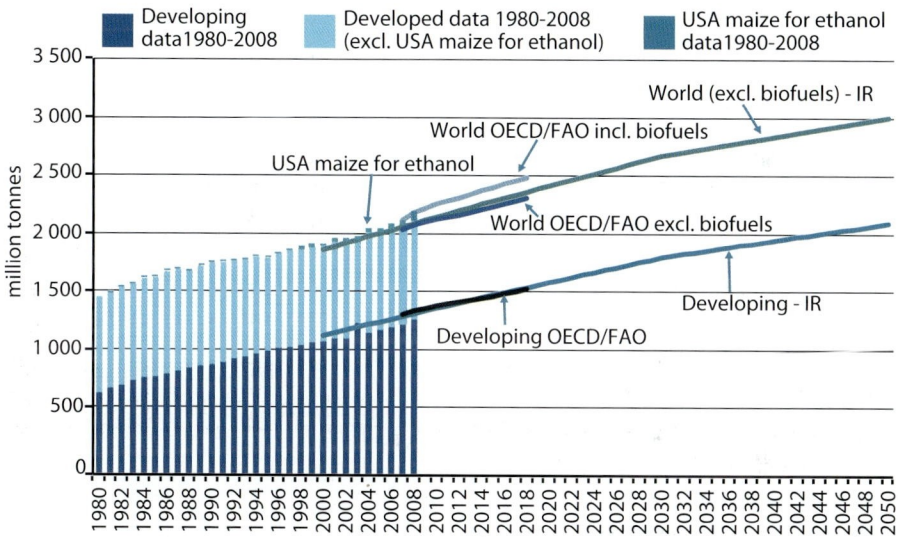

Sources: FAO, 2006; OECD/FAO, 2009.

Consumption, developing countries: A gradual slowdown in the growth of cereals consumption (all uses, not only food) in the developing countries was projected – to 1.8 percent per annum in the first sub-period 1999/2001 to 2015. This is happening. From 1999/2001 to 2006/2008 growth decelerated to 1.8 percent per annum, from 3.0 percent in the 1980s and 2.0 percent in the 1990s, while per capita consumption increased to 244 kg per annum in 2006/2008. Therefore, for this criterion, the IR projections seem to be on the right track. Will they continue to be so in the future? The OECD/FAO medium-term projections to 2018 foresee that aggregate consumption in developing countries will rise to 1 462 million tonnes in 2015 (close enough to the 1 472 million tonnes of the IR, Table 1.1) and on to 1 522 million tonnes by 2018. The IR projections seem to be on the right track for this criterion too.

Table 1.1
Cereals (wheat, rice milled, coarse grains): IR data to 2001 and projections, versus revised CBS data to 2008 and OECD/FAO projections to 2018

	Quantity *(million tonnes)*					
Consumption	1999/ 2001	2006/ 2008	2015	2018	2030	2050
World: IR data and projections (excl. biofuels)	1 866		2 287		2 677	3 010
World: CBS data	1 900	2 130				
USA: maize for ethanol (USDA data)[a]	16	74				
World: CBS data (excl. USA maize ethanol)	1 884	2 056				
World: OECD/FAO projections (incl. biofuels)		2 121	2 407	2 490		
World: OECD/FAO projections biofuels		84	172	175		
World: OECD/FAO projections (excl. biofuels)		2 037	2 235	2 314		
Developing countries: IR data and projections (excl. biofuels)	1 125		1 472		1 799	2 096
Developing: CBS data	1 148	1 301				
Developing: OECD/FAO projections		1 301	1 462	1 522		
Developed countries: IR data and projections (excl. biofuels)	741		815		877	914
Developed: CBS data	752	829				
Developed: CBS data (excl. USA maize ethanol)	736	755				
Developed: OECD/FAO projections (incl. biofuels)		820	945	967		
Developed: OECD/FAO projections biofuels		80	168	172		
Developed: OECD/FAO projections (excl. biofuels)		740	777	796		
Production						
World: AT data and projections	1 885		2 290		2 679	3 012
World: CBS data	1 887	2 147				
World: OECD/FAO projections		2 127	2 416	2 500		
World: OECD/FAO projections (excl. biofuels)		2 043	2 244	2 325		
Developing countries: AT data and projections	1 026		1 304		1 567	1 799
Developing: CBS data	1 026	1 205				
Developing: OECD/FAO projections		1 192	1 327	1 374		
Developed countries: AT data and projections	859		985		1 112	1 212
Developed: CBS data	861	942				
Developed: OECD/FAO projections		935	1 088	1 126		
Developed: OECD/FAO projections (excl. biofuels)		855	920	955		
Net imports						
Developing countries: AT data and projections	112		168		232	297
Developing: CBS data	110	121				
Developing: OECD/FAO projections		122	140	154		

[a] Historical data for cereals use for biofuels going back to 1980 exist for only the United States of America www.ers.usda.gov/data/feedgrains/feedgrainsqueriable.aspx.

USDA = United States Department of Agriculture.

Table 1.1 (*continued*)

	Growth rate *(% per annum)*						
Consumption	1980-1990	1990-2000	1999/2001-2006/2008	2006/2008-2018	1999/2001-2015	2015-2030	2030-2050
World: IR data and projections (excl. biofuels)					1.4	1.1	0.6
World: CBS data	1.9	1.0	1.6				
USA: maize for ethanol (USDA data)	20.2	4.5	24.4				
World: CBS data (excl. USA maize ethanol)	1.9	0.9	1.3				
World: OECD/FAO projections (incl. biofuels)				1.5			
World: OECD/FAO projections biofuels				6.9			
World: OECD/FAO projections (excl. biofuels)				1.2			
Developing countries: IR data and projections (excl. biofuels)					1.8	1.3	0.8
Developing: CBS data	3.0	2.0	1.8				
Developing: OECD/FAO projections				1.4			
Developed countries: IR data and projections (excl. biofuels)					0.6	0.5	0.2
Developed: CBS data	0.8	-0.4	1.4				
Developed: CBS data (excl. USA maize ethanol)	0.7	-0.5	0.4				
Developed: OECD/FAO projections (incl. biofuels)				1.5			
Developed: OECD/FAO projections biofuels				7.1			
Developed: OECD/FAO projections (excl. biofuels)				0.7			
Production							
World: AT data and projections					1.3	1.1	0.6
World: CBS data	1.6	0.9	1.9				
World: OECD/FAO projections				1.5			
World: OECD/FAO projections (excl. biofuels)				1.2			
Developing countries: AT data and projections					1.6	1.2	0.7
Developing: CBS data	2.8	1.8	2.3				
Developing: OECD/FAO projections				1.3			
Developed countries: AT data and projections					0.9	0.8	0.4
Developed: CBS data	0.6	0.0	1.3				
Developed: OECD/FAO projections				1.7			
Developed: OECD/FAO projections (excl. biofuels)				1.0			

Source: Data and projections from FAO, 2006.

Consumption, developed countries: The IR projected a rebound of growth in the early years of the projection period because of the expected recovery of transition countries after the deep declines of the 1990s. Growth did rebound, to 1.4 percent per annum in the period 1999/2001 to 2006/2008, which was more than projected in the IR (0.6 percent for 1999/2001 to 2015). However, much of the rebound was due to the growing use of grains for biofuels (overwhelmingly maize for ethanol in the United States of America)[6] and the associated price rises. Without these, the rebound was a much more modest 0.4 percent per annum – lower than the IR projections. That it was lower than projected can be interpreted as reflecting the fact that not all use of maize for ethanol represented additional consumption: part of it was matched by reductions in, mainly, the use of grain for livestock feed following the higher prices, hence the lower than projected growth of consumption for food and feed (see section on Biofuels: significance for the long-term outlook).[7]

What about the future? The OECD/FAO projections foresee faster growth in developed countries, at an annual 1.5 percent from 2006/2008 to 2018, than the IR projections do. However, the OECD/FAO projections for developed countries *include biofuels* (80 million tonnes in 2006/2008, 172 million tonnes in 2018). Excluding such use from the projections, the growth of consumption for all other uses from 2006/2008 to 2018 is reduced to 0.8 percent per annum. In the end, the IR projection for 2015 of 815 million tonnes compares with the 945 million tonnes (with biofuels) and the 777 million tonnes (without biofuels) of the OECD/FAO projections for the same year. Again, it is implicit that the growth of biofuels will squeeze out some of the IR-projected consumption for food and, predominantly, feed. Overall, therefore, the IR projections for developed countries (excluding biofuels use) seem to be on track.

Consumption, world totals: The sum of the two country groups – developed and developing – shows that for the world as a whole consumption growth was higher (1.6 percent per annum from 1999/2001 to 2006/2008) than the projected 1.4 percent for 1999/2001 to 2015. However, it was lower (1.3 percent per annum) than projected in the IR if the United States of America's maize use for biofuels

---

6. Use of maize for fuel alcohol in the United States of America had reached 91 million tonnes in 2008 (*www.ers.usda.gov/data/feedgrains/feedgrainsqueriable.aspx*). This is the only source with data of cereals use for biofuels extending back to 1980. Data for more recent years are available for some other countries in the data set used in the OECD/FAO projections: for 2008, they indicate, 6 million tonnes in the EU27, 2 million tonnes in Canada and 4 million tonnes in China.

7. Not all the maize used for biofuels should be subtracted from the supplies available for feed: some 30 percent is returned to the feed sector in the form of by-products (mainly distillers' dry grains).

is excluded from world consumption. The OECD/FAO projections for 2015 are 2 407 million tonnes (with biofuels) and 2 235 million tonnes (without biofuels) versus the 2 287 million tonnes in the IR projection for the same year.

Conclusion: By and large, the trajectories of actual consumption to 2008, for the world as a whole and separately for developing and developed countries (excluding the biofuels component), follow the IR projection paths fairly closely (Figure 1.1), which are for gradually decelerating growth. It would therefore be possible to use the existing IR projections (at least for these large country aggregates and the world as a whole) and add one or more alternative views of future use of cereals for biofuels (this topic is addressed later in this chapter, with deeper coverage in Chapters 2 and 3). This would generate a path of possible developments in the global demand for cereals over the projections' time horizon that would be compatible with the IR projections, the developments to date, the OECD/FAO medium-term outlook and at least one view of cereal use for biofuels. Obviously, updating the study's views of cereal futures for individual countries and small country groups would require considerable work to run similar reality checks at the country level, while also taking on board the drastic revisions of the FBS historical data on food consumption for all commodities (discussed in the section on Food consumption and nutrition in developing countries).

Production and net imports, developing countries: The interface between production historical data and the projections is not as neat as those for consumption, given fluctuations caused by both weather and policies. Data for developing countries production are plotted in Figure 1.2 (and shown in Table 1.1). Production was nearly stagnant over the period 1996 to 2002 (1 023 million tonnes in 1996/1998, 1 030 million tonnes in 2001/2003), while consumption kept growing and stocks were depleted. This was one of the factors that presaged the price spikes in subsequent years (Alexandratos, 2008). During this period, almost all the increases in consumption were met by stocks drawdown. The role of China was particularly important in developments during this period: China started running down the huge stocks it had accumulated in the 1990s, with closing stocks of 309 million tonnes in 1999 (84 percent of annual consumption), falling to 148 million tonnes by 2005 (40 percent of consumption).[8] From 2003 onwards there was a rebound in production (reaching 1 205 million tonnes in 2006/2008),

---

8. Problems associated with China's huge stocks accumulated by the late 1990s included overflowing granaries with losses due to quality deterioration, and large financial losses from sales (domestic and export) at below-cost prices. These problems prompted policy reforms to reduce stocks, including some relaxation of the policies that obliged farmers to produce cereals (OECD, 2005: 37; USDA, 2001).

---

and production increases were more than sufficient to meet the growth in consumption. Indeed, part of the increased production went to rebuilding stocks (Figure 1.3). China's role was also important in this second period; without China, the turnaround from stock depletion to stock rebuilding is much less pronounced, although still evident in the data.

Figure 1.2
Cereal production, consumption (left axis) and net imports (right axis), developing countries

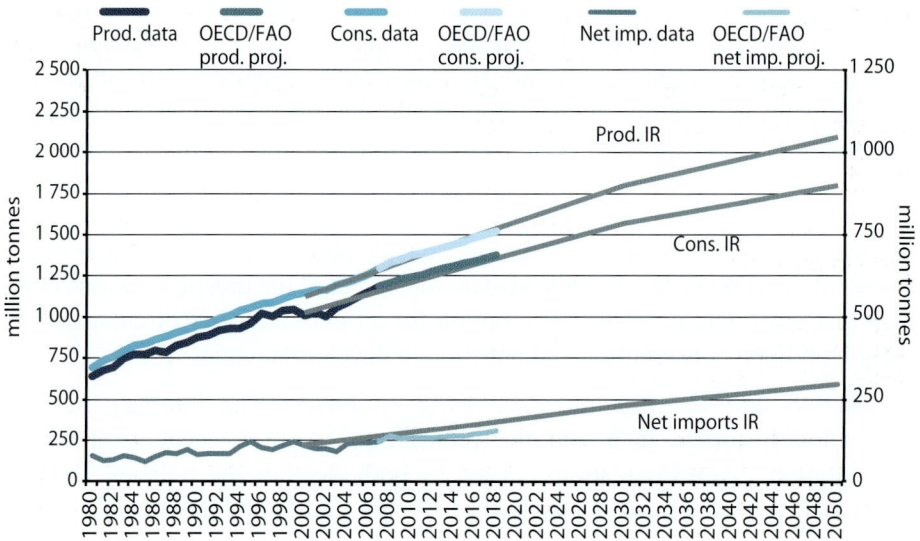

Sources: FAO, 2006; OECD/FAO, 2009.

It is important to note that in both periods, changes in net imports played a minor role as contributors to changes in aggregate consumption. They fluctuated in the range of 91 million tonnes (2003) to 136 million tonnes (2008). The IR had projected that net imports would play a larger role as contributors to the growth of consumption in developing countries. Net imports were projected to rise from the 112 million tonnes of 1999/2001 to 168 million tonnes in 2015, and on to 232 million tonnes in 2030 and 297 million tonnes in 2050. The OECD/FAO projections have 140 million tonnes in 2015 and 154 million tonnes in 2018. If developments in the first half of the current decade are a harbinger of things to come, there may be need for radical rethinking of how the future of developing countries is viewed in terms of growing dependence on imported cereals. Lower imports than projected in the IR mean lower projected consumption and/or higher projected production. As already noted, consumption growth in the developing countries is largely on the projected path. Therefore, if projected imports must be lower, the production projections must be revised upwards. This raises the question: Is the IR projection of 1.6 percent per annum from 1999/2001 to 2015 too low in light of the production growth rebound of recent years (2.3 percent from 1999/2001 to 2006/2008)?

Before jumping to conclusions, there is need to take a closer look at the production increases and examine whether the acceleration of growth is likely to prove durable or is the result of extraordinary circumstances. This requires looking at the data for individual countries. Annex 1.1 lists the 29 developing countries (accounting for 16 percent of developing countries' cereal production in 1999/2001) that in the period 1999/2001 to 2006/2008 achieved cereal production growth rates exceeding 4 percent per annum, with 5.7 percent per annum as a group, up from 1.7 percent in the 1990s. For several of these countries the spurt in growth of the last few years represented recoveries from troughs in the preceding years. Such growth rates are certainly not very informative for judging long-term growth prospects.

Figure 1.3
Growth in cereal demand met by changes in production, stocks and net imports, 1997/2002 versus 2002/2008

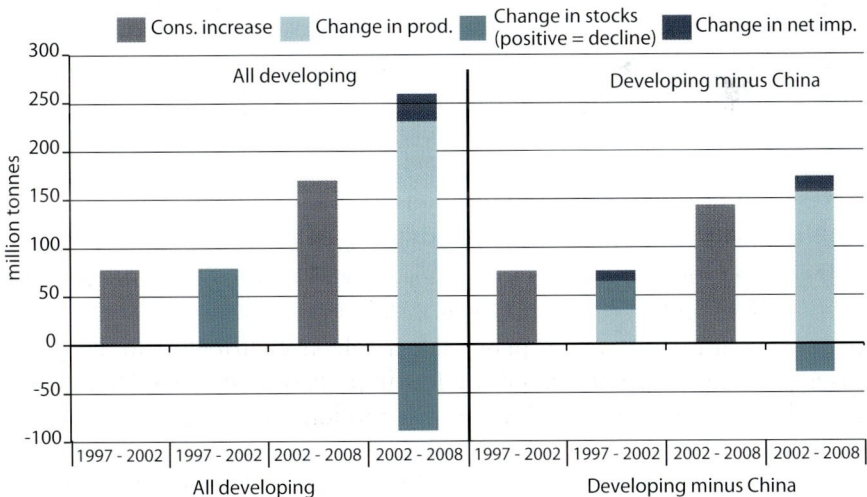

Source: FAO CBS data.

Were the IR projections to be redone today, it would certainly be necessary to revisit the production projections of individual countries in the light of developments in the last few years. The key issue is, of course, whether this would affect in any significant way the aggregates for all developing countries and the prospects for growth of their net cereal imports. For this the OECD/FAO projections to 2018 can be referred to. In these projections, the spurt in the growth of production for the period 1999/2001 to 2006/2008 (2.3 percent per annum) is not maintained;[9] they project a growth rate of 1.3 percent per annum from 2006/2008 to 2018, with projected production for 2015 of 1 327 million tonnes, versus 1 304 million tonnes in the IR (Table 1.1). It has also been noted that consumption in developing countries is somewhat lower than that in the IR. By implication, developing countries' projected net imports, being the difference between two much larger numbers, are lower than those of the IR for 2015, at 140 million tonnes versus 168 (Figure 1.3).

Much of the difference in net imports is due to India and China:[10] in the IR they turn into modest (for their size) net importers by 2015, while the OECD/FAO projections have them continuing as small net exporters (6.4 million tonnes in 2015 and 5.1 million tonnes in 2018). Excluding India and China, the two projections of net imports for 2015 are close – 143 million tonnes in the IR, 146 million tonnes in OECD/FAO. China and India have the potential to influence decisively the cereal trade prospects of the developing countries. The two countries together were net importers in the past, but became net exporters after 1999, reaching peak net exports of 26 million tonnes in 2002, which declined to 4 to 6 million tonnes per annum in the four years 2005 to 2008.

A few years ago, these two giants were seen as turning into net importers again over the medium term. Thus, the 2004 issue of the Food and Agricultural Policy Research Institute (FAPRI) projections to 2013 had them as net importers of 11 million tonnes in 2013. The latest edition (FAPRI, 2009) has them as net exporters of only 1 million tonnes in 2018. Similarly, the OECD *Agricultural*

---

9. The latest cereals production forecast for 2009 for the developing countries indicates virtually no increase over that of 2008 (FAO, *Food Outlook*, June 2009).

10. China's net trade position does not include the situation in Taiwan Province of China and China Hong Kong Special Administrative Region (SAR), both of which have been net importers, to the tune of 7 million tonnes per annum in the last ten years, and are projected to remain so in the future. Thus, all China is really a net importer of cereals, both at present and in the projections.

outlook of 2004 had China becoming a significant net importer.[11] A recent International Food Policy Research Institute (IFPRI) report (Rosegrant *et al.*, 2008: Figure 4.7) has, in its baseline scenario, China's net cereal imports exceeding 50 million tonnes in both 2025 and 2050, while India remains a net exporter in 2025 turning into a net importer in 2050. In conclusion, the net trade position of the developing countries, being the difference between the much larger numbers of production and consumption, remains sensitive to even small variations of these two larger numbers. Views about the future cereal trade positions of China and India can cause any outlook of developing countries' import needs to swing around. As noted, such views tend to change over time. Back in the mid-1990s, Brown (1995) considered the prospect of China's burgeoning cereals imports as a major threat to world food security, a clear exaggeration at the time (see critique in Alexandratos, 1996) and even more so today. Many people seem to be mesmerized by the hugeness and high economic growth rates of China and its apparently voracious appetite for livestock products and food in general. This perception may be accurate (for some time) for things such as energy and metals, but is much less so for food: the income elasticity of the demand for food tends to decline rather rapidly, being limited (as it were) by the elasticity of the human stomach. The IR projection of the status of China and India as modest net importers by 2015 reflected the dominant view of a few years ago. There is no compelling reason for changing the long-term projections just now, but the matter should certainly be kept under constant review.

Production, developed countries: The IR had projected an acceleration of developed country production (not accounting for the effects of biofuels) in the first projection sub-period (to 0.9 percent per annum between 1999/2001 and 2015, up from zero in the 1990s) because of the expected recovery in transition countries. The advent of the additional demand for biofuels led to production increases that were even faster than projected in the IR (1.3 percent per annum in 1999/2001 to 2006/2008; Table 1.1), and that were significantly influenced by a quantum jump of 13 percent in 2008, following the price spikes. The OECD/FAO projections foresee even higher growth in the future (1.7 percent per annum from 2006/2008 to 2018), largely because of growing use for biofuels. The latter

---

11. "From a net exporter of both wheat and coarse grains at the beginning of the Outlook, China could become a significant importer of cereals assuming that the Tariff Rate Quotas (TRQs), implemented by China under the World Trade Organization (WTO) accession agreement, will be used efficiently. By the end of this decade, China could import more than ten times as much wheat, coarse grains and rice as in the recent past. Both wheat and rice import quotas are projected to become filled at least in some years, and coarse grain imports, already by far the largest part of Chinese cereals imports, could reach levels equivalent to twice the import quota for maize" (OECD, 2004: 52).

Table 1.2
Meat (bovine, ovine, pig, poultry): IR data to 2001 and projections, versus revised CBS data to 2008 and OECD/FAO projections to 2018

	Carcass weight (million tonnes)						Growth rate (% per annum)						
	1999/2001	2006/2008	2015	2018	2030	2050	1981–1990	1990–2000	1999/2001–2006/2008	2006/2008–2018	1999/2001–2015	2015–2030	2030–2050
Consumption													
World: IR data and projections	228		305		380	463	3.3	3.3			2.0	1.5	1.0
World: CBS data	230	270							2.3				
World: OECD/FAO projections		267	312	328						1.9			
Developing countries: IR data and projections	127		191		258	334					2.8	2.0	1.3
Developing: CBS data	128	159					5.2	6.5	3.1				
Developing: OECD/FAO projections		157	195	208						2.6			
Developed countries: IR data and projections	101		113		123	130					0.8	0.5	0.3
Developed: CBS data	102	112					2.3	0.4	1.3				
Developed: OECD/FAO projections		110	117	120						0.8			
Production													
World: IR data and projections	230		306		382	465	3.3	3.2			1.9	1.5	1.0
World: CBS data	230	271							2.4				
World: OECD/FAO projections		268	312	329						1.9			
Developing countries: IR data and projections	125		190		255	332					2.8	2.0	1.3
Developing: CBS data	126	158					5.1	6.2	3.3				
Developing: OECD/FAO projections		156	192	205						2.5			
Developed countries: IR data and projections	104		116		126	133					0.7	0.6	0.3
Developed: CBS data	104	113					2.3	0.4	1.2				
Developed: OECD/FAO projections		112	120	124						0.9			

Source: Data and projections from FAO, 2006.

are projected to increase by slightly more than 100 percent, with almost all of the increase occurring in developed countries. If it is assumed that all cereals used for biofuels come from home production in the developed countries, when they are excluded, production in 2015 becomes 920 million tonnes versus the 985 million tonnes in the IR. This difference can be attributed in part to the possibility that the use of cereals as biofuels would squeeze out some of the demand for other uses (mostly feed), and in part to the lower net imports required by developing countries in the OECD/FAO projections, as discussed in previous paragraphs.

Production, world: For the world as a whole (the sum of developed and developing countries), the IR projection is 2 287 million tonnes for 2015. This compares with the 2 416 million tonnes (with biofuels) or 2 244 million tonnes (without biofuels) of the OECD/FAO projections for the same year. If it were not for biofuels, the IR projection of 3 012 million tonnes for 2050 would not be in need of major revision. However, the advent of biofuels requires at least speculation on possible upwards revisions, perhaps to some 3 150 million tonnes, as discussed in the section on Biofuels: significance for the long-term outlook.

Meat

Consumption, developing countries: The IR emphasized that the fast growth of meat consumption in the developing countries in the 1980s and 1990s reflected predominantly developments in China and a few other countries (e.g., Brazil) (FAO, 2006: Table 3.7). It projected that such growth was bound to slow as these countries reached medium to high levels of per capita consumption. Other developing countries would experience faster growth than in the past, but that would not be sufficient to sustain the growth of consumption in the developing countries and the world as a whole at the high rates of the preceding two decades. Is this forecast slowdown happening?

Table 1.2 shows that deceleration is taking place in developing countries: from growth of more than 5 percent per annum in the 1980s and 1990s to 3.1 percent in the first seven years of the projection period – more or less on target to meet the 2.8 percent per annum projected for the period to 2015. The OECD/FAO projections foresee a growth rate of 2.6 percent per annum from 2006/2008 to 2018, in line with the IR's 2.8 percent for 1999/2001 to 2015. OECD/FAO projects a slow rise in per capita consumption, from 29 kg in 2006/2008 to 33 kg in 2015, the same as the IR projects for that year.

Consumption, developed countries: In contrast, meat consumption in developed countries has been growing faster than anticipated in the IR. Per capita consumption rose from 75 kg in 1999/2001 to 80 kg in 2006/2008. In the OECD/FAO projections it rises further to 85 kg in 2015 and on to 87 kg in 2018. This contrasts with the IR projections of 83 kg for 2015 and 95 kg in 2050. The overshooting is wholly due to the strong rebound of consumption in transition countries (the former Soviet Union and Eastern Europe) in the early years of the projection period after the slump of the 1990s. Their per capita consumption rose from 46 kg per annum in 1999/2001 to 57 kg in 2006/2008, a level that the IR had projected would be reached in a later year. Clearly, this must be taken into account in any further discussion of livestock sector prospects. Needed revisions would raise the transition countries' consumption in the medium term, but the key issue is whether this would alter in any significant way the longer-term prospects as depicted in the IR: these countries' consumption of 23 million tonnes of meat in 2006/2008 accounts for 8.5 percent of the world total; their population is on the decline (from 404 million people in 2007 to 346 million in 2050); therefore, even if they continue their rapid growth of meat consumption to reach the developed countries' average (some 95 kg per capita in 2050), they will add only another 9 million tonnes (or 2 percent) to the 465 million tonnes world consumption the IR had projected for 2050 – not a significant change. The key issue therefore remains whether or not the developing countries, with their growing weight in world population and meat consumption, are likely to make faster progress than projected (from 26.7 kg per capita per annum in 1999/2001 to 44 kg in 2050). So far growth is on the projected trajectory, with per capita consumption reaching 29 kg per annum in 2006/2008. As noted, the OECD/FAO projections indicate 33 kg for 2015 and 34 kg in 2018.

Consumption, world: The growth of world meat consumption has been slowing, from 3.3 percent per annum in the 1980s and 1990s to 2.3 percent per annum for the period 1999/2001 to 2006/2008. The IR projects 2.0 percent per annum for 1999/2001 to 2015 and further declines in growth in subsequent projection periods. OECD/FAO projects 1.9 percent per annum from 2006/2008 to 2018, i.e., the acceleration caused by the rebound of consumption in the transition countries in recent years is not maintained. Overall, therefore, the IR projections of world meat consumption can be considered an acceptable longer-term outlook in the light of developments to date, at least in global totals.

Production: Production projections mirror those of consumption, given that net trade is a very small fraction (less than 1 percent) of production/consumption for the large country aggregates considered here. The commentary on consumption magnitudes therefore also applies to those of production.

Vegetable oils

The IR (FAO, 2006: 27, 52–58) highlighted the importance of vegetable oils as a fast-growing item in the food consumption growth of developing countries. It projected that such growth would continue for some time (FAO, 2006: Tables 2.7 and 3.9). It also highlighted the growing weight of non-food uses of oils in industry (paints, detergents, lubricants, oleochemicals in general and, increasingly, biodiesel). It projected that world consumption for both food and non-food uses would continue to grow at high rates, although not as high as those of the recent past. As the historical data on non-food uses included biodiesel, the IR projections must be considered as containing an allowance for biodiesel, albeit of unknown magnitude. How do the IR projections compare with developments in the current decade and with the OECD/FAO projections?

Straightforward comparisons of quantities such as those shown earlier for cereals cannot be made for vegetable oils. This is because the CBS data are not of the same specification as those used in the IR analyses (Box 1.1). In addition, the OECD/FAO projections treat the oilseeds-oils complex as two commodities: "vegetable oil" (the sum of only the four major oils – soybean, rapeseed, sunflower seed and palm) and "oilseeds" (the sum of rapeseed, soybeans and sunflower seed). They do not cover other oils and oilseeds (coconut, groundnut, sesame, cottonseed, olive, etc.), some of which are important in several countries. Therefore, the IR data and projections cannot be compared directly (in terms of quantities of production and consumption) with the data in CBS or with the OECD/FAO projections. At best, comparisons can be made between the IR's projected growth rates of consumption of vegetable oil only (not oilseeds) and the OECD/FAO projections, which are not affected significantly by the differences in commodity coverage and specification.

Comparisons of the consumption growth rates are shown in Figure 1.4. The growth rates in the IR projections for the period 1999/2001 to 2015 are generally lower than those of the OECD/FAO projections for 2006/2008 to 2018. However, the latter include an allowance for biodiesel. Without this, the OECD/FAO growth rates of consumption are lower than the IR's. In practice, the IR growth rates are halfway between the OECD/FAO projections of growth rates with biodiesel and those without biodiesel; for example, the IR world growth rate of 2.7 percent per annum is halfway between OECD/FAO's 3.4 percent with biodiesel and 2.2 percent without biodiesel.

As noted, the IR projections contain a component for biodiesel, which is unknown but must be small: the use of oils for biodiesel really shot up in the last few years, from less than 1 million tonnes in 1999/2001 to 10 million tonnes in 2006/2008 (mostly in the EU and to a lesser extent the United States of America and several developing countries – Argentina, Brazil, Malaysia, Indonesia and Thailand), according to the data used in OECD/FAO projections. It is noted that the four oils included in the OECD/FAO definition of vegetable oils are the fastest growing ones. It is therefore to be expected that the growth rate is higher in the OECD/FAO projections than in those of the IR, which also include the slower-growing oils. By and large, therefore, the IR projections can be considered an acceptable basis for generating a long-term outlook for the sector after adding one or more alternatives for biodiesel use of vegetable oils.

Figure 1.4
Growth of vegetable oil consumption

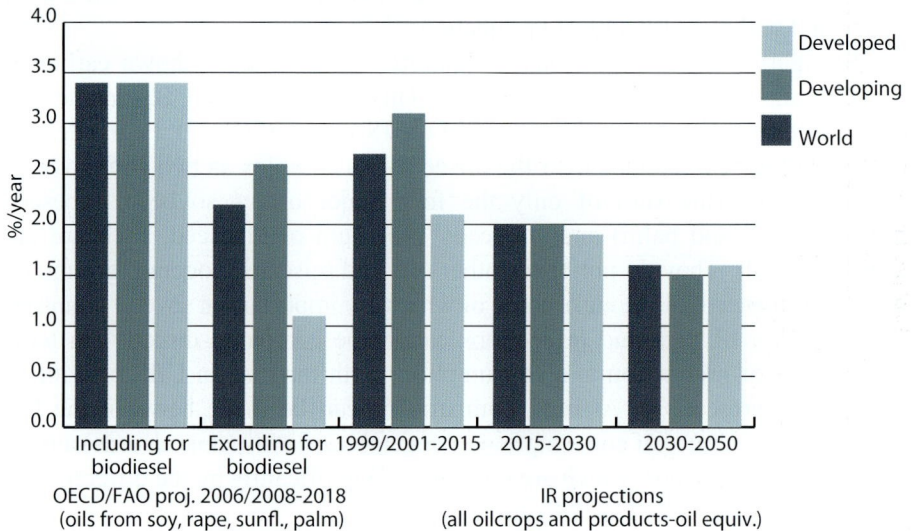

Sources: FAO, 2006; OECD/FAO, 2009.

The IR projections indicate a growing export orientation for the vegetable oil sector in developing countries (a growing share of total production going to exports), and a growing import dependence in developed countries (a growing share of consumption coming from net imports from developing countries, Figure 1.5). The OECD/FAO projections confirm these prospects, although direct comparability of quantities is not possible. Developed countries are increasing their net imports of oils from 8.1 million tonnes (20.4 percent of consumption) in 2006/2008 to 16 million tonnes (28.2 percent of consumption) in 2018. At the same time, they continue to be net exporters of oilseeds, predominantly soybeans from the United States of America, to the tune of 20.5 million tonnes in 2018, up from 15.5 million tonnes in 2006/2008. These net oilseed exports correspond to roughly 4 to 5 million tonnes of oil equivalent,[12] so developed countries' net imports of all oils and oilseeds (in oil equivalent) would be some 11 to 12 million tonnes in 2018 (16 million tonnes minus 4 to 5 million tonnes). This is higher than the IR projection for 2015, of 7.2 million tonnes. The difference can be attributed to higher oil and oilseed imports following growth of the biodiesel industry in developed countries.

Figure 1.5
Oilseeds, vegetable oils and products (oil equivalent), IR projections

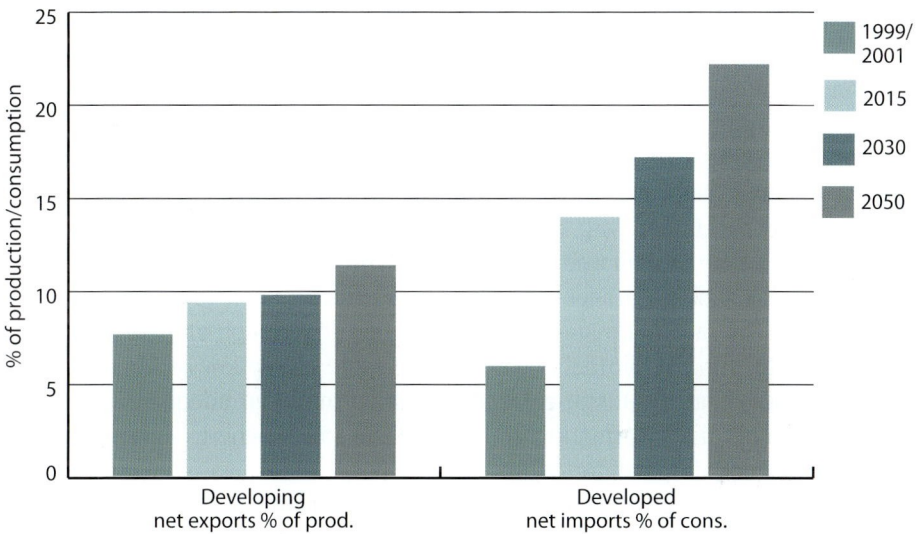

Source: FAO, 2006.

Biofuels: significance for the long-term outlook

The potential for using crops to produce biofuels had its moment of glory during the recent price surges of both energy and food commodities. At one extreme, biofuels were vilified as causing the food price surges and, occasionally, as destroying the environment and land and water resources. At the other extreme,

---

12. The bulk of developed countries' oilseed exports are soybeans from the United States of America, and rapeseed and sunflower seed (mainly from Canada, Eastern Europe and Ukraine). Therefore, if developed countries' net oilseed exports were converted into oil equivalent (to obtain a number that can be compared with the definition used in the IR), they would correspond to some 4 to 5 million tonnes of oil (using an average oil extraction rate near that of soybean's 18 to 19 percent, increased to 20 to 25 percent to account for the higher extraction rates of rapeseed and sunflower seed, of 41 to 43 percent).

they were seen as offering great opportunities for boosting farm incomes and energy independence, and for mitigating adverse environmental effects by reducing the burning of fossil fuels.

The debate subsided with the collapse of oil prices. These days, headlines are usually concerned with the woes of the biofuels industry following its rapid expansion during the boom years. The industry is now largely kept alive by mandates and subsidies, with the possible exception of sugar ethanol, mainly in Brazil.

However, the issue is not dead. High energy prices are likely to return (IEA, 2008; Stevens, 2008; McKinsey Global Institute, 2009), and the geopolitical causes driving the quest for energy security are not going away. Add the strength of the farm and biofuel industry lobbies, the continuing relevance of environmental concerns and the prospects for technological change in converting biomass to liquid fuels, and the debate can be expected to reignite. It follows that any assessment of long-term food prospects cannot ignore the possibility that the expected "normal" slowdown in the growth of demand for agricultural produce (and the underlying claims on agricultural resources and technology development) may not materialize. There is therefore need for one or more projection alternatives to account for biofuel effects. Such projections are not easily made. The fact that mandates and subsidies drive much of the use of grains and vegetable oils for biofuels (with the possible exception of Brazil's sugar ethanol) means that the historical data do not provide an adequate basis from which to glean valid information concerning the role of energy-to-crop relative prices as triggers of demand growth.

Currently, biofuel projections are commonly an integral part of most food and agriculture projections. In this area, the latest attempts that contain (to varying degrees) sufficient detail of the biofuels modules are all medium-term (ten years), not long-term.[13] They include the latest annual issues of ten-year outlooks by USDA (2009), FAPRI (2009) and OECD/FAO (2009). The last of these provides the most detail, so it is used to illustrate the orders of magnitude involved; Figures 1.6 and 1.7 show the volumes of biofuels (ethanol and biodiesel, respectively) projected to be produced by 2018.

World production of ethanol is projected to increase by slightly more than 100 percent from 2008 to 2018, with the United States of America, Brazil and the EU27 as the major players. Both Brazil and the EU are projected to increase their shares in the world total. The United States of America's share will be somewhat reduced (from 43 to 37 percent) and will lose its top position to Brazil's

---

13. The IIASA work (Chapter 3), which contains long-term biofuels projections, was not available at the time of writing this chapter.

Figure 1.6
Ethanol production, OECD/FAO projections

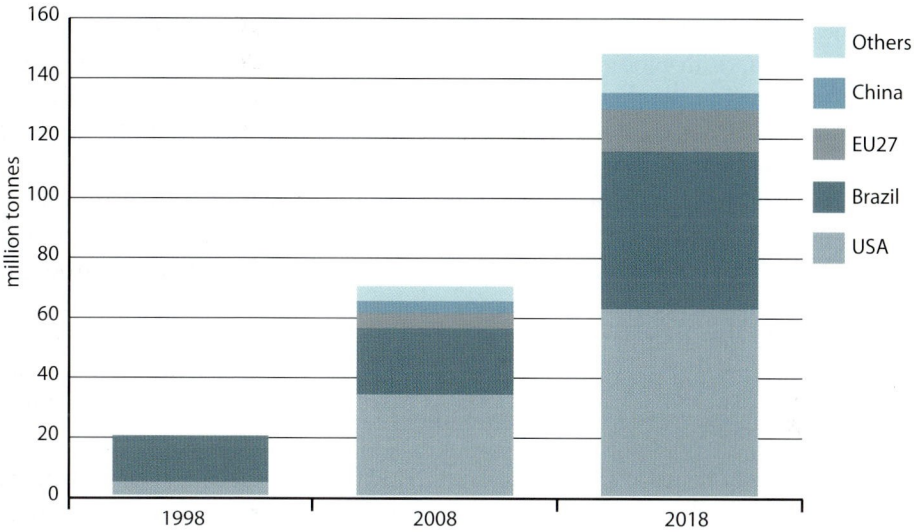

Legend:
- Others
- China
- EU27
- Brazil
- USA

Source: OECD/FAO, 2009.

Figure 1.7
Biodiesel production, OECD/FAO projections

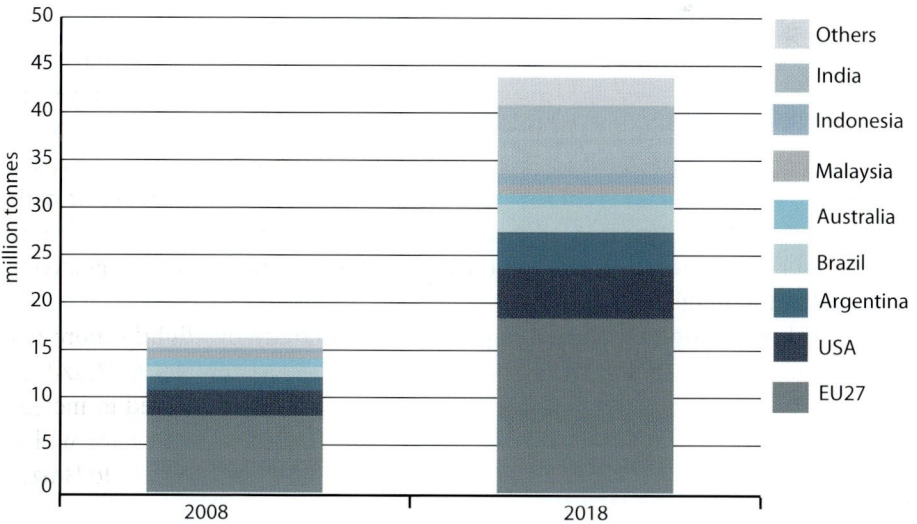

Legend:
- Others
- India
- Indonesia
- Malaysia
- Australia
- Brazil
- Argentina
- USA
- EU27

Source: OECD/FAO, 2009.

share, which will increase from 34 to 39 percent. Biodiesel production is seen as growing even faster than ethanol, by 170 percent in the ten-year period. The EU will continue to hold top place, with 42 percent of world production (down from the current 50 percent). The great revelation (according to these projections) could be India, with biodiesel production going from very little today to some 7 million tonnes in ten years, all of it from *jatropha*, making India the world's second largest producer, with a 16 percent share. This reflects the mandate for having a 20 percent biofuels blend in gasoline and diesel by 2017.

The key issue is what all this may imply for food security and nutrition. Would food consumption be lower with the use of food crops for biofuel production than it would be without it? It is difficult to provide a concrete answer to this question without running counterfactual scenarios, which is not practicable at the moment. It is not just a question of whether world food and feed consumption would be lower because of the price rises caused by, mainly, biofuels. This can be taken for granted, as the diversion of grain to biofuels most directly affects the feed/livestock sector in developed countries, which is more sensitive to price changes than other components of the food system. However, food security and nutrition issues are related to the food consumption in countries where large proportions of the population are undernourished or just above the threshold for undernourished (in terms of minimum daily energy requirements [MDERs] – see section on Food consumption and nutrition in developing countries). In such countries, food price rises could aggravate the situation of those below the threshold and push some of those above it into the class of undernourished.

None of the ten-year projection studies offers scenarios with and without biofuels.[14] Developments in the last few years of price surges embody information that can help to answer the question: Do the ten-year projections provide information about the impact of biofuels on per capita food consumption, and in which country groups? As seen in Table 1.1, some 84 million tonnes of cereals were used for biofuels in the three-year average for 2006/2008, and 105 million tonnes in 2008 alone. Has this led to a reduction in per capita consumption? In the absence of FBS data beyond 2005, it is not known what happened to the per capita food consumption of all commodities expressed in kilocalories per person per day (kcal/person/day). The CBS data can be used to figure out only how per capita consumption of cereals for all uses evolved over the last few years. Figure 1.8

---

14. A recent IIASA study for the Organization of the Petroleum Exporting Countries (OPEC) Fund for International Development (OFID) (IIASA, 2009: Part III, Box 3.4-1) indicates that in 2020, 66 percent of the additional demand for cereals generated by scenarios with growing biofuels use (over and above use in 2008) would be met by increased production, and the rest by reduced consumption of feed (24 percent) and food (10 percent).

plots consumption in kilograms per capita (all uses, with and without cereals use for biofuels). It is seen that:

- there have been no declines, but rather small increases, in per capita consumption in developing countries (cereals use for biofuels in these countries – some 4 million tonnes in China – is very small, so the entire change can be attributed to non-biofuel uses);

- the only declines occurred in developed countries, from 2006 to 2007 in food and feed consumption (i.e., all consumption minus the part going to biofuels); with biofuels, developed countries' per capita consumption rose significantly.

Does this mean that the diversion of cereals to biofuels, and the associated price increases did not lead to reduced per capita food consumption and/or increases in the numbers undernourished in countries with nutrition problems?[15]   In the

Figure 1.8
Cereal consumption, all uses including and excluding use for biofuels

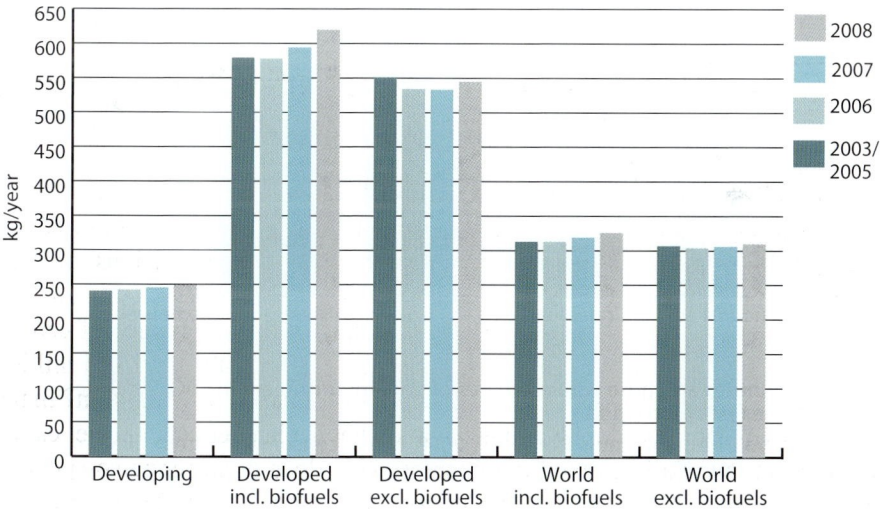

Developing countries are those with estimates of undernourishment in FAO, 2008; developed countries are the rest of the world

Sources: Cereals consumption from CBS data, 3 July 2009; population from UN, 2007.

---

15. The relevant question is whether or not per capita consumption is less than it would have been in the absence of the price surges. It is also noted that the number (but not the percentage of population) undernourished may increase even when per capita consumption does not decline, or even increases a little. This can happen because of population growth.

absence of updated FBS data covering all food products, this question cannot be answered.

As noted, the risk of the nutritional situation deteriorating in the wake of price surges is highest and most relevant in countries with low food consumption levels and significant proportions of their populations undernourished. To shed light on this, Figure 1.9 unfolds developments in the per capita consumption of cereals by developing country sub-groups, according to their nutrition status in 2003/2005 (as given in FAO, 2008). Again, it is seen that no country group suffered a decline. On the contrary, per capita consumption increased in all groups.

Figure 1.9
Developing countries cereal consumption: all uses, by proportion of population undernourished in 2003/2005

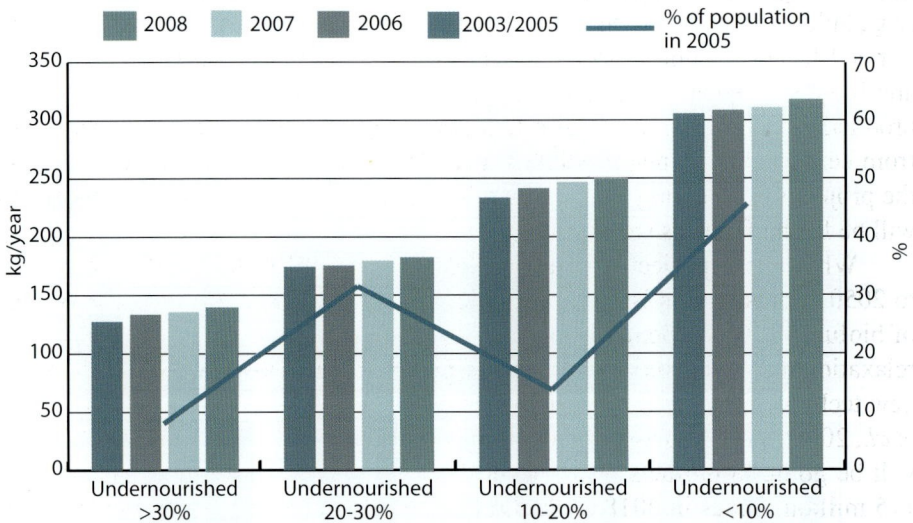

Developing countries are those with estimates of undernourishment in FAO, 2008; developed countries are the rest of the world.

Sources: Undernourishment status from FAO, 2008; cereal consumption from CBS data 3 July 2009; population from UN, 2007.

This is not equivalent to saying that the diversion of grain to biofuels and the associated price rises had no impact on the numbers undernourished: it is possible that were it not for biofuels, the per capita consumption of cereals would have improved by more than shown in Figures 1.8 and 1.9. Naturally, not all the amount devoted to biofuels would have been available for food and feed: part would simply not have been produced, as the high prices were to a large measure responsible for the rebound in world cereals production in both 2007 (+ 5.4 percent) and 2008 (+ 7.3 percent). As noted, the IIASA analysis (2009) suggests that in the projection period to 2020, some two-thirds of the cereals going to biofuels could come from increased production, and the balance from reduced consumption of food and, mainly, feed.

The increases of biofuel production in the OECD/FAO ten-year projections (Figures 1.6 and 1.7) imply further increases in the demand for feedstock crops (presented earlier in the sections on Cereals and Vegetable oils). Naturally, not all additional ethanol will be produced from cereals, and not all biodiesel will come from the four major vegetable oils covered in the OECD/FAO analyses. Even without resorting to feedstocks of non-food crop biomass (second-generation biofuels), by-products (e.g., molasses), food crops other than cereals (mainly sugar cane, sugar beet, cassava, etc.) and fats other than the major oils (e.g., tallow, coconut oil) will contribute a share to biofuels, which may grow – as implied by, for example, Brazil's increasing share in world ethanol production and India's ascendancy in biodiesel based on *jatropha*. The increases in biofuel production will therefore require less than proportional increases in feedstocks from cereals and the major edible vegetable oils (Figure 1.10). Nevertheless, in the projections, growing shares of world cereals and vegetable oils consumption will be for biofuels, as shown in Figure 1.11.

What about projections beyond 2018? The IFPRI study with projections to 2050 addresses this issue in the following assumption: "We hold the volume of biofuel feedstock demand constant starting in 2025, in order to represent the relaxation in the demand for food-based feedstock crops created by the rise of the new technologies that convert non-food grasses and forest products" (Rosegrant *et al.*, 2008:, 11). This assumption implies that some 200 million tonnes of cereals will be going to biofuels by 2050 (up from the 105 million tonnes in 2008 and 175 million tonnes in 2018 of the OECD/FAO projections). Assuming that two-thirds of this additional demand would come from increased production (as in the estimate in IIASA, 2009), the original projection of 3 010 million tonnes in 2050 (Table 1.1) would need to be raised to some 3 150 million tonnes, and food/feed consumption lowered by some 60 million tonnes, to 2 950 million tonnes.

These are all speculative ballpark numbers and are offered to provide some orders of magnitude. If they turn out to be approximately correct, world agriculture could perhaps cope with the problem without incurring significantly higher stress over that implied by the need to increase cereals production by some 900 million tonnes projected in the IR, in terms of the required land-irrigation-yield configurations shown in Chapter 6.

Figure 1.10
Increases in biofuel production versus increases in cereal and vegetable oil feedstocks, 2008 to 2018

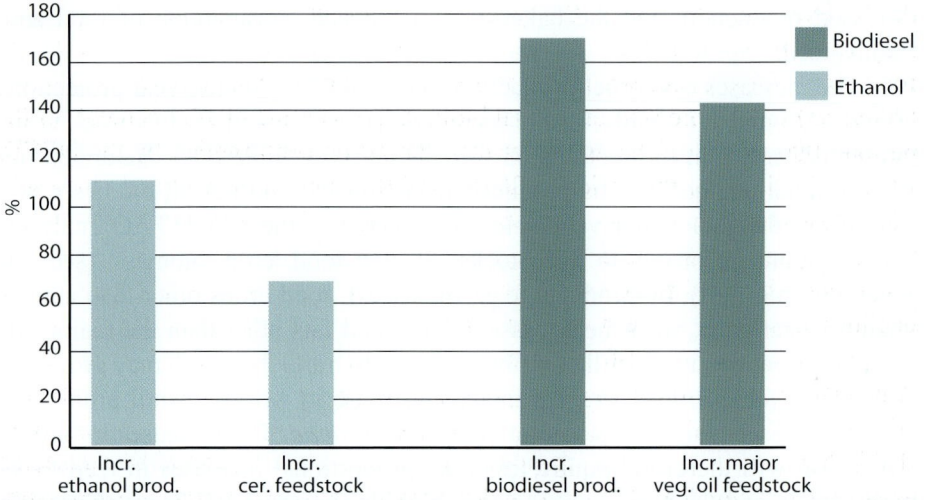

Source: OECD/FAO, 2009.

Figure 1.11
Cereal and vegetable oil biofuel feedstocks as shares of aggregate world consumption

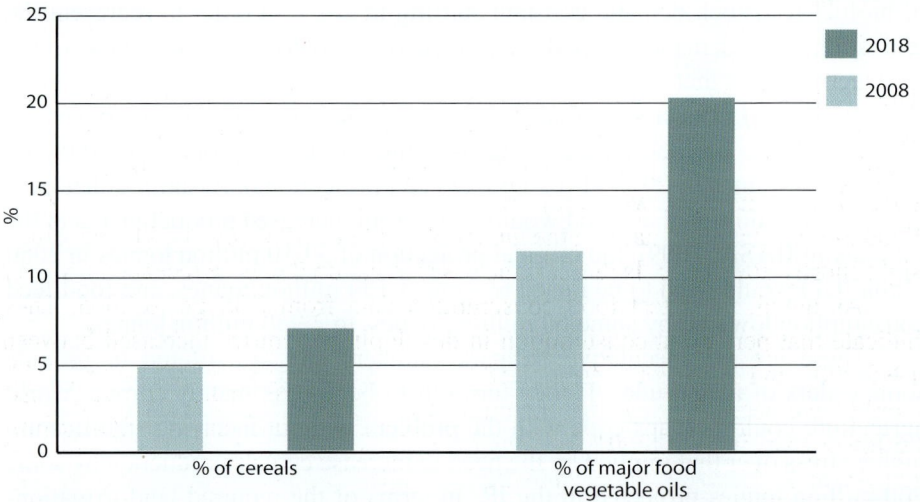

Vegetable oils include those from soybeans, rapeseed, sunflower seed and oil-palm, but not *jatropha*.
Source: OECD/FAO, 2009.

However, things may turn out quite differently if energy prices were to explode and make the conversion of food crops to biofuels profitable, even without subsidies and mandates. The investment frenzy that underpinned expansion of the biofuels industry during the recent price surges for petroleum is telling. It may happen again, and the energy sector must be seen as competing with the food sector for supplies when it is profitable for it to do so. The latest McKinsey (2009: 63) report forecasts an annual biofuels growth rate of 14.4 percent for the period 2006 to 2020. This is higher than the 10 percent implied by the OECD/FAO projections for the period 2006 to 2018 (the sum of ethanol and biodiesel). The latest United States Government energy outlook to 2030 (EIA, 2009) has world biofuels growth rates in the range of 10 percent (low oil price case) to 14 percent (high oil price case) per annum from 2006 to 2020. Annual growth declines drastically in the subsequent decade, to between 3.7 and 4.6 percent.

In conclusion, food-fuel competition is likely to continue into the future. Any analysis must address the eventuality of such competition intensifying, with adverse effects on the food security of some countries and population segments: if this happens, the purchasing power of those demanding more energy could easily overwhelm that of the poor demanding food (see further discussion in Schmidhuber, 2006; Alexandratos, 2008). Among the major tasks of any remake of FAO's long-term projections would be addressing this issue, unfolding the implications for food security and exploring the alternatives.

Food consumption and nutrition in developing countries

Revisiting current estimates and possible future outcomes

The IR (FAO, 2006: Table 2.3) projected a gradual rise in per capita food consumption in developing countries. As a result, the number of undernourished people would gradually fall, from 811 million in 1999/2001 to 582 million in 2015. Further declines were projected for 2030 and 2050, with the 1996 World Food Summit target of halving the numbers undernourished by 2015 being within sight shortly after 2030. Is this still the case? What do the more recent data show?

As noted, the latest food consumption data from FBS go to 2005. They indicate that per capita consumption in developing countries increased between 1999/2001 and 2003/2005, from 2 580 to 2 620 kcal/person/day (Table 1.3). It would have been expected that the number undernourished in 2003/2005 would be lower than in 1999/2001. However, FAO's *The State of Food Insecurity in the World 2008* (FAO, 2008, hereafter referred to as SOFI 2008) estimates the number undernourished in developing countries at 823 million in 2003/2005, i.e., an increase, although food consumption per capita also increased. This seems to be going against the grain of the arguments made in the IR: that rising per capita consumption and the accompanying slight improvement in the equality of distribution would lead to declining undernourishment. Of course, it is quite possible for the number undernourished to increase because of population growth, if the increase in per capita calorie intake is small, as is the case here (see previous paragraph and Table 1.3). However, the question arises: Do the most recent estimates indicate a real reversal of the trend towards gradual and slow declines in the numbers of undernourished, or is this just data noise?

Table 1.3
Per capita food consumption (kcal/person/day)

	IR Table 2.1				New data and adjusted projections				
Country group	1999/2001	2015	2030	2050	1999/2001 new	2003/2005	2015	2030	2050
World	2 789	2 950	3 040	3 130	2 725	2 771	2 884	2 963	3 047
Developing countries	2 654	2 860	2 960	3 070	2 579	2 622	2 770	2 864	2 966
Sub-Saharan Africa	2 194	2 420	2 600	2 830	2 128	2 167	2 319	2 494	2 708
- excluding Nigeria	2 072	2 285	2 490	2 740	2 016	2 061	2 206	2 406	2 643
Near East and North Africa	2 974	3 080	3 130	3 190	2 991	2 995	3 072	3 134	3 197
Latin America and Caribbean	2 836	2 990	3 120	3 200	2 798	2 899	2 953	3 084	3 151
South Asia	2 392	2 660	2 790	2 980	2 334	2 344	2 532	2 656	2 843
East Asia	2 872	3 110	3 190	3 230	2 764	2 839	3 034	3 112	3 144
- excluding China	2 698	2 835	2 965	3 100	2 475	2 538	2 614	2 740	2 870
Industrial countries	3 446	3 480	3 520	3 540	3 429	3 462	3 501	3 548	3 569
Transition countries	2 900	3 030	3 150	3 270	2 884	3 045	3 043	3 159	3 283

Source: Author.

To understand what is happening, it should be noted that the data for per capita consumption, population, MDERs (the threshold for classifying people as undernourished) and the measure of inequality (the coefficient of variation [CV]) have all been revised rather drastically. These are the key data and parameters used to estimate undernourishment. They are now different from those in SOFI 2004, which were used to prepare the IR.

For example, the average kcal/person/day in developing countries for 1999/2001 is now 2 580 kcal, down from the 2 654 used in the IR and SOFI 2004. The decline is particularly sharp for some countries, such as Myanmar (from 2 840 to 2 160), Ecuador (from 2 720 to 2 220), Indonesia (from 2 910 to 2 420) and Benin (from 2 500 to 2 190). Such declines cannot but take a heavy toll of the estimates of undernourished, *ceteris paribus*.

If these revised kcal figures had been used in the computation on the IR estimates, with all the other data (population, MDERs and CVs) as they were known then, the undernourished would have been 920 million, not 811 million, in the starting three-year average 1999/2001 of the IR. Again, using the new kcal for 2003/2005 with the population data from the United Nations (UN) 2002 assessment (UN, 2003) – to be compatible with that of the IR for 1999/2001 – and unchanged MDERs and CVs, the undernourished for 2003/2005 would have been 910 million, a small decline from the estimate for 1999/2001, not an increase.

The revisions of the other data (population, MDERs and CVs) explain why the estimates of undernourishment in SOFI 2008 imply a small increase rather than a small decline between 1999/2001 and 2003/2005. To appreciate what is involved, it is necessary to examine how and why the data were so drastically revised:

- Regarding population data, the UN assessment of 2006 (UN, 2007) had revised estimates for several countries, which had to be taken on board. This concerned particularly several African countries, such as Togo (the new estimate for 2000 is 18 percent higher than the old one), Benin (16 percent higher), Angola (12 percent higher), Senegal (10 percent higher), Nigeria (9 percent higher) and Mali (16 percent lower).

- The reasons why the MDERs and the CVs were revised are explained elsewhere (FAO, 2004). The IR (FAO, 2006: Box 2.2) provides a more general discussion of the estimation of numbers undernourished.

- Concerning data on food consumption per capita, for some countries the change was predominantly the direct consequence of the population revisions: approximately the same amount of food was now divided by a larger population (e.g., in Togo, Benin, Angola, Senegal). In other countries, changes in both population and total food supplies were responsible for the changes in per capita consumption (e.g., in Nigeria, Mali). At the other extreme, for some countries the change in per capita consumption was almost entirely due to revised estimates of total food consumption in the FBS (e.g., in Indonesia, Myanmar, Ecuador). Such changes in the total food consumption data are not necessarily (or only) the result of changed total national availabilities of food commodities (production + imports – exports + stock changes). They also reflect changes introduced in the final-use allocations of total availabilities in the course of preparing the revised FBS (allocations among food, feed, stock changes, etc.).

Generating a revised base year data set

The preceding discussion suggests that the IR projections of per capita food consumption (kcal/person/day, given in FAO, 2006: Table 2.1) and the derived projections of undernourishment (FAO, 2006: Table 2.3) need to be adjusted before they can be compared with the latest situation presented in SOFI 2008 and provide a basis for making statements about the future course of undernourishment in relation to the present. This assumes that the SOFI 2008 depiction is accepted as representing the reality for the latest year with estimates, 2003/2005. The SOFI 2008 estimates are based on: i) the revised kcal/person/day data; ii) the new population data from the UN 2006 assessment; and iii) the new MDERs and CVs. These are used to create new estimates of undernourishment in the starting situation of the IR (the base year 1999/2001) and are shown in Table 1.4, columns 3 and 9. It is seen that use of the revised data generates a total estimate of 810 million undernourished people in the developing countries in 1999/2001.

Table 1.4
Incidence of undernourishment, developing countries, from SOFI 2008

Country group	Share of population (%)					
	1990/ 1992	2003/ 2005	1999/ 2001	2015	2030	2050
	(1)	(2)	(3)	(4)	(5)	(6)
Developing countries	n/a	16.3	17.0	11.3	8.1	4.8
Sub-Saharan Africa	n/a	30.5	32.0	22.3	13.9	7.0
- excluding Nigeria	n/a	35.8	37.6	26.5	16.1	7.9
Near East and North Africa	n/a	7.9	8.1	6.1	5.0	3.2
Latin America and the Caribbean	n/a	8.3	9.7	6.9	4.4	3.1
South Asia	n/a	21.3	21.1	13.8	10.2	5.2
East Asia	n/a	11.3	12.7	7.1	5.1	3.9
- excluding China	n/a	15.0	17.0	12.8	8.8	5.3
	Number of people[a] (millions)					
	(7)	(8)	(9)	(10)	(11)	(12)
Developing countries	813	823	810	664	556	370
Sub-Saharan Africa	169	213	202	204	174	118
- excluding Nigeria	154	200	190	196	165	110
Near East and North Africa	19	33	31	31	31	24
Latin America and Caribbean	53	45	50	43	31	24
South Asia	283	313	289	238	206	118
East Asia	290	219	237	149	115	87
- excluding China	112	97	105	93	72	46

[a] The absolute numbers differ from those published in SOFI 2008 because the latter includes Central and Western Asian countries of the former Soviet Union in the developing countries.
Source: FAO, 2008.

This is practically identical to that of the IR (based on SOFI 2004 estimates), although regional estimates are somewhat different, even though the underlying kcal and parameters have been revised – some of them drastically. Obviously, the impacts of the revisions of the key data and parameters used in the estimation have cancelled one another.

Adjusting the projected food consumption

Obviously, if the SOFI 2008 estimates for 2003/2005 are taken as representing the actual undernourishment situation, the projections must be adjusted before statements can be made about how the situation may evolve in the future compared with the present, i.e., 2003/2005. Adjustments must be made to the projected values of: i) kcal/person/day (to take into account the new starting data for 2003/2005); ii) the population in the projection years from the UN 2006 assessment of population prospects, which was used to generate the SOFI 2008 estimates (the IR population projections used those of the UN 2002 assessment); and iii) the revised MDERs and CVs.

Ideally, the new historical FBS data (available in unpublished form up to 2005) would be used, and the whole projections exercise redone by country and commodity to generate the new projected values for kcal/person/day. This is not practically possible at this stage, so shortcuts have to be devised to make adjustments. Box 1.2 describes the rules used to make the adjustments. These rules were applied directly for each country at the level of kcal/person/day (not by commodity). It is noted again that these adjustments are necessary to account for the fact that food consumption levels and the estimates of undernourishment depicted in SOFI 2008 differ from those that formed the basis for the IR food consumption and undernourishment projections.

The revised projections of kcal/person/day resulting from these adjustments are shown in Table 1.3 (reproducing Table 2.1 of the IR [FAO, 2006] for comparison). The following comments apply:

- With the exception of the Near East and North Africa region, the revised base year data for kcal/person/day in all other developing regions are lower than the data used in the IR. The difference is very marked in the East Asia region, particularly when China is excluded from the regional totals (see earlier discussion on the data for Myanmar and Indonesia).

- These lower starting levels have an impact on the projected values when the latter are adjusted as indicated. Although projected per capita levels in developing countries are lower than in the IR (by an average of 3.4 percent in 2050), the aggregate projected consumption in 2050 is virtually the same as that of the IR. This is because the new projected population of

UN 2006 is higher than that of UN 2002 (used in the IR), by 3.2 percent. That the new projections of per capita consumption combined with the new population values generate aggregate food demand equal to that of the IR is important: the aggregate food demand of the IR was derived as an integral component of the entire configuration of production, consumption (all uses, not only food) and trade. However, although this feature applies to the developing countries aggregate, it may not apply at the level of individual countries.

- The result is that in 2050 fewer developing countries than reported in the IR will have reached medium to high levels of per capita food consumption (more than 2 700 kcal/person/day): 73 countries in these revised estimates account for 80 percent of the developing country population in 2050, versus 85 countries and 90 percent of the population in the IR (Table 1.5).

Table 1.5
Developing country populations with given per capita food consumption[a]

kcal/person/day		IR Table 2.2			Revised		
		1999/2001	2030	2050	2003/2005	2030	2050
< 2 200	Population (million)	584	29		515	217	
< 2 200	Average kcal	2 001	2 060		1 928	2 087	
< 2 200	Countries (number)	32	2		32	6	
2 200–2 500	Population (million)	1 537	785	128	2 087	785	381
2 200–2 500	Average kcal	2 403	2 380	2 460	2 365	2 368	2 367
2 200–2 500	Countries (number)	26	17	3	26	20	9
2 500–2 700	Population (million)	201	510	618	368	2 575	1 148
2 500–2 700	Average kcal	2 547	2 605	2 625	2 616	2 653	2 632
2 500–2 700	Countries (number)	14	23	12	13	26	20
2 700–3 000	Population (million)	1 925	2 336	1 622	1 372	801	3 035
2 700–3 000	Average kcal	2 933	2 835	2 870	2 987	2 854	2 856
2 700–3 000	Countries (number)	16	31	42	14	25	35
> 3 000	Population (million)	484	3 049	5 140	735	2 495	3 185
> 3 000	Average kcal	3 174	3 280	3 200	3 163	3 309	3 262
> 3 000	Countries (number)	14	29	45	17	25	38
All developing	Population (million)	4 731	6 709	7 509	5 077	6 873	7 748
All developing	Average kcal	2 654	2 960	3 070	2 622	2 864	2 966
All developing	Countries (number)	102	102	102	102	102	102

[a] Only countries with FBS.
Source: Author.

Box 1.2 - Rules for adjusting the IR food (kcal/person/day) projections

If the FBS kcal for 2003/2005 is lower than that of the IR base year (i.e., $kcal_{2003/2005}$ < $IR\text{-}kcal_{1999/2001}$), the $kcal_{2003/2005}$ is taken as the base year and the projected values are derived by applying the growth rates of kcal in the IR projections (49 of the 97 developing countries in the IR are in this category). Thus, for these countries:

$Revkcal_{2015} = kcal_{2003/2005} \times (2 + g)11$

where g is the annual growth rate between 1999/2001 and 2015 of the kcal in the IR. $Revkcal_{2030}$ and $Revkcal_{2050}$ are derived applying the same rule, i.e., applying the respective growth rates of the IR projections to the $Revkcal_{2015}$.

If the FBS kcal for 2003/2005 is higher than that of the IR base year, but lower than the kcal for 2015 in the IR projections (i.e., $IRkcal_{2015}$ > $Newkcal_{2003/2005}$ > $IRkcal_{1999/2001}$), the $IRkcal_{2015}$ remains unchanged and so do the IR projected kcals for 2030 and 2050 (38 countries in this category).

If the FBS kcal for 2003/2005 is higher than the IR kcal projected for 2015, then $Revkcal_{2015} = Newkcal_{2003/2005}$ (ten countries in this category). An upper limit of 3 500 kcal is imposed to prevent countries with very drastic upward revisions in their kcal from exploding towards unrealistically high levels of consumption in the projection years. Cuba is an example: it had 2 833 kcal in the IR base year 1999/2001; in the revised data used in SOFI 2008 it has 3 022 kcal for the same year and 3 276 kcal for 2003/2005.

Revised estimates of undernourishment in the future

The implications of the changes for undernourishment in the future are unfolded in Table 1.4. The following comments apply:

- SOFI 2008 indicates that the numbers undernourished in the developing countries increased between 1990/1992 and 2003/2005, although the percentages of the population affected declined. As described previously, the same applies to changes in the period 1999/2001 to 2003/2005. However, it was noted that revisions in the kcal/person/day data alone would have produced a small decline, not an increase. It is the application of the whole package of data and parameter revisions that generates a small increase. Does this indicate that the problem is getting worse, rather than improving towards the World Food Summit target of halving absolute numbers by 2015 (from those of 1990/1992)? It can only be noted that the increase in the estimate of absolute numbers is small and may not be significant, given the data noise.

- Comparing the new projected numbers of undernourished in Table 1.4 with those in Table 2.3 of the IR (FAO, 2006), it is evident that projected undernourishment is now higher, both in absolute numbers and as percentages of the population. The higher percentages of the population result from lower projected per capita kcal (Table 1.4). This impact is reinforced in the absolute numbers because the projected population of developing countries is higher (UN, 2006) (Table 1.5).

• The revised projections indicate a slow decline in undernourishment. However, the rate of decline in the IR was such that the World Food Summit target could be within reach shortly after 2030. In the revised estimates, achievement of this target is shifting further into the future – to just before 2050.

Conclusions

This chapter examined whether the long-term projections to 2050 in the IR study (FAO, 2006, prepared between 2003 and 2005 from historical data to 2001 and base year 1999/2001) were still valid as predictions (for selected broad country and commodity aggregates) of what may be in store in world food and agriculture to mid-century. The projections were tested against: i) actual outcomes, as far as data permitted, in the first eight years of the projection period (to 2008); and ii) the ten-year projections for 2009 to 2018 that OECD/FAO had just completed, both with and without the quantities used as biofuel feedstocks. It was concluded that on both counts, and disregarding biofuels, the IR's projections are still broadly valid at the level of the aggregates considered.

The advent of biofuels requires a fresh look at the long-term picture. The existing medium-term projections of biofuel production and, in some cases, of the corresponding crop quantities to be used as feedstocks indicate that further growth is in prospect, although not at the very high rates of the last few years. In these projections, the quantities of cereals by which world aggregate consumption would be higher because of biofuels are still relatively modest (7 percent of world consumption in 2018, up from the current 4.8 percent; Figure 1.11), and much of the increase would likely come from increased production over and above what it would have been without biofuels. However, biofuels have the potential to be a major disruptive force, conditioning agricultural futures because of growing integration of the energy and agriculture markets. This theme, together with the possible impact of climate change, must inform all future attempts to speculate about long-term futures of world food and agriculture.

The chapter also examined the IR's projections of food consumption and numbers undernourished in developing countries in the light of some drastic revisions of the historical data and parameters used to compute such numbers, as well as in the projected populations. The projected food consumption levels had to be adjusted to account for these revisions and make it possible to compare the projections with the latest published estimates (in SOFI 2008) of per capita consumption and numbers undernourished. These adjustments indicate that the rate at which the numbers undernourished were projected to decline – slow and inadequate though it was in the IR projections – may turn out to be even slower. Achievement of the 1996 World Food Summit target of halving the number undernourished in the developing countries by 2015 (from that of 1990/1992) may well recede further into the future.

References

Alexandratos, N. 1996. China's cereals deficits in a world context. *Agricultural Economics*, 15–16.

Alexandratos, N. 2008. Food price surges: Possible causes, past experience, and longer term relevance. *Population and Development Review*, 34(4): 663–697. www.fao.org/es/esd/foodpricesurges-alexandratos.pdf.

Brown, L. 1995. *Who will feed China, wake-up call for a small planet.* New York, W.W. Norton.

Bruinsma, J., ed. 2003. *World agriculture: Towards 2015/30*, an FAO perspective. London, Earthscan and Rome, FAO. www.fao.org/es/esd/gstudies.htm.

EIA. 2009. *International energy outlook 2009.* Washington, DC, United States Department of Energy, Energy Information Administration (EIA).

FAO. 2004. *Human energy requirements. Report of a Joint FAO/WHO/UNU Expert Consultation*, Rome, 17–24 October 2001. Food and Nutrition Technical Report Series No. 1. Rome.

FAO. 2006. *World agriculture: towards 2030/2050, interim report.* Rome. www.fao.org/es/esd/gstudies.htm.

FAO. 2008. *The State of Food Insecurity in the World 2008.* Rome.

FAO. 2009. *Crop prospects and food situation,* No. 2 (April 2009). Rome.

FAPRI. 2009. *US and world agricultural outlook.* FAPRI Staff Report No. 09-FSR 1. Ames, Iowa, USA, Iowa State University.

Fischer, G., van Velthuizen, H. & Nachtergaele, F. (forthcoming), *Global agro-ecological assessment – the 2009 revision.* Laxenburg, Austria, IIASA.

Fischer, G., van Velthuizen, H., Shah, M. & Nachtergaele, F. 2002. *Global agro-ecological assessment for agriculture in the 21st century: Methodology and results.* RR-02-002., Laxenburg, Austria, IIASA.

IEA. 2008. *World energy outlook 2008.* Paris, OECD and International Energy Agency (IEA).

IIASA. 2009. *Biofuels and food security: Implications of an accelerated biofuels production.* A study prepared by G. Fischer, E. Hizsnyik, S. Prieler, M. Shah and H. van Velthuizen for OPEC/OFID). Laxenburg, Austria, IIASA.

McKinsey Global Institute. 2009. *Averting the next energy crisis: the demand challenge*. New York.

OECD. 2004. *Agricultural outlook 2004–2013*. Paris.

OECD. 2005. *Review of agricultural policies in China, main report*. Document No. AGR/CA(2005) 6. Paris.

OECD/FAO. 2009. *Agricultural outlook, 2009–2018: Highlights*. Paris.

Rosegrant, M.W., Huang, J., Sinha, A., Ahammad, H., Ringler, C., Zhu, T., Sulser, T.B., Msangi, S. & Batka, M. 2008. *Exploring alternative futures for agricultural knowledge, science, and technology (AKST)*. ACIAR Project Report No. ADP/2004/045. Washington, DC, IFPRI.

Schmidhuber, J. 2006. Impact of an increased biomass use on agricultural markets, prices and food security: a longer-term perspective. Paper for International Symposium of Notre Europe, Paris, 27–29 November 2006. www.fao.org/es/esd/pastgstudies.html.

Stevens, P. 2008. *The coming oil supply crunch. A Chatham House Report*, London, Royal Institute of International Affairs.

UN. 2003. *World population prospects: the 2002 revision*. New York.

UN. 2007. *World population prospects: the 2006 revision*. New York.

USDA. 2001. China's grain policy at a crossroads. *Agricultural Outlook* (September). Washington, DC.

USDA. 2009. *USDA agricultural projections to 2018*. Long-term Projections Report No. OCE-2009-1. Washington, DC.

CEREAL PRODUCTION IN DEVELOPING COUNTRIES

Developing countries with cereal production growth > 4.0 percent per annum[a]

	Amount ('000 tonnes)			Growth (% per annum)		
	1989/ 1991	1999/ 2001	2006/ 2008	1980-1990	1990-2000	1999/2001-2006/2008
Sierra Leone	353	181	913	0.2	-7.2	26.0
Iraq	2 456	1 379	3 471	2.8	-9.0	14.1
Paraguay	787	1 236	2 531	6.8	4.1	10.8
Guinea	521	955	1 888	0.0	7.9	10.2
Afghanistan	2 645	2 311	4 493	-3.8	0.8	10.0
Algeria	2 481	1 871	3 337	-1.2	-4.4	8.6
Chad	647	1 103	1 952	4.6	5.0	8.5
Ethiopia and Eritrea	6 370	8 858	15 378	1.7	4.0	8.2
Cambodia	1 649	2 740	4 663	8.6	6.3	7.9
Madagascar	1 779	1 910	3 183	1.9	0.4	7.6
Uruguay	1 101	1 568	2 600	2.1	5.8	7.5
Morocco	7 452	3 478	5 649	8.7	-5.1	7.2
Mali	1 999	2 350	3 578	7.1	2.3	6.2
Niger	1 898	2 690	4 093	0.4	2.9	6.2
Myanmar	9 110	14 002	20 934	-0.2	3.8	5.9
Sudan	2 918	3 988	5 818	-1.3	2.0	5.5
Venezuela	1 834	2 565	3 722	5.6	2.6	5.5
Iran, Islamic Republic	12 248	13 224	19 139	3.8	0.0	5.4
Zambia	1 461	1 137	1 615	6.6	0.3	5.1
Brazil	34 910	46 873	65 483	2.8	2.5	4.9
Yemen	700	689	958	0.4	0.3	4.8
Tanzania, United Republic	3 897	3 826	5 311	3.8	0.8	4.8
Angola	297	546	754	-2.5	7.1	4.7
Philippines	10 781	12 732	17 312	3.3	1.0	4.5
Burkina Faso	1 961	2 660	3 613	6.2	1.9	4.5
Nigeria	16 896	20 045	27 223	10.4	2.1	4.5
El Salvador	764	781	1 055	2.1	-0.4	4.4
Bolivia, Plurinational State	801	1 142	1 541	2.4	3.3	4.4
Malawi	1 543	2 347	3 096	1.1	7.1	4.0
Total	132 257	159 187	235 303	3.4	1.7	5.7
Other developing	737 239	866 952	969 742	2.7	1.8	1.6
All developing	869 496	1 026 138	1 205 045	2.8	1.8	2.3

[a] Only countries with 2006/2008 production > 500 000 tonnes.

Main findings

Continued growth of world agriculture even after the end of world population growth

The main reason is that zero population growth at the global level will be the net result of continuing increases in some countries (e.g., by some 31 million annually in 2050 in Africa and South and Western Asia together) compensated by declines in others (e.g., by some 10 million annually in China, Japan and Europe together). Nearly all the further population increases will be occurring in countries several of which even in 2050 may still have inadequate food consumption levels, hence significant scope for further increases in demand. The pressures for further increases of food supplies in these countries will continue. Much of it will have to be met by growing local production or, as it happened in the past and is still happening currently, it may not be fully met – a typical case of production-constrained food insecurity. The creation of slack in some countries with declining population (e.g., the transition economies, when growth of aggregate demand will have been reduced to a trickle -.01 percent per annum in the final two decades 2030 to 2050) will not necessarily be made available to meet the still growing demand in countries with rising population, e.g., demand growth at 2.0 percent per annum in sub-Saharan Africa.

In conclusion, zero population growth at the global level will not automatically translate into zero growth in demand and cessation of the building-up of pressures on resources and the wider environment. The need for production to keep growing in several countries will continue to condition their prospects for improved nutrition. In those among them that have limited agricultural potential, the problem of production-constrained food insecurity and significant incidence of undernourishment may persist, even in a world with stationary population and plentiful food supplies (or potential to increase production) at the global level. Nothing new here: this situation prevails at present and it will not go away simply because population stops growing at the global level. Projections to 2050 provide a basis for thinking about this possible outcome.

Food and nutrition

The historical trend towards increased food consumption per capita as a world average and particularly in the developing countries will likely continue, but at slower rates than in the past as more and more countries approach medium-high levels. The average of the developing countries that rose from 2 110 kcal/person/day 30 years ago to the present 2 650 kcal, may rise further to 2 960 kcal in the next 30 years and on to 3 070 kcal by 2050. By the middle of the century the great bulk of their population (90 percent) may live in countries with over 2 700 kcal, up from 51 percent at present and only 4 percent three decades ago. As in the past, the great improvements in China and a few other populous countries will continue to carry a significant weight in these developments.

However, not all countries may achieve food consumption levels consonant with requirements for good nutrition. This may be the case of some of the countries which start with very low consumption (under 2 200 kcal/person/day in 1999/2001), high rates of undernourishment, high population growth rates, poor prospects for rapid economic growth and often meagre agricultural resources. There are 32 countries in this category, with rates of undernourishment between 29 percent and 72 percent, an average of 42 percent, Yemen and Niger among them. Their present population of 580 million is projected to grow to 1.39 billion by 2050, that of Yemen from 18 million to 84 million and that of Niger from 11 million to 53 million. Their current average food consumption of 2 000 kcal/person/day is actually a little below that of 30 years ago. Despite the dismal historical record, the potential exists for several of these countries to make gains by assigning priority to the development of local food production, as other countries have done in the past. Under this fairly optimistic assumption, the average of the group may grow to 2 450 kcal in the next 30 years, although this would still not be sufficient for good nutrition in several of them. Hence the conclusion that reducing undernourishment may be a very slow process in these countries.

Notwithstanding the several countries with poor prospects for making sufficient progress, the developing countries as a whole would record significant reductions in the relative prevalence of undernourishment (percent of population affected). However, these will not be translated into commensurate declines in the numbers undernourished because of population growth. Reduction in the absolute numbers is likely to be a slow process. Numbers could decline from the 810 million in 1999/2001 to 580 million in 2015, to 460 million in 2030 and to just over 290 million by 2050. This means that the number of undernourished in developing countries, which stood at 823 million in 1990/1992 (the three-year average used as the basis for defining the World Food Summit target), is not likely to be halved by 2015. However, the proportion of the population undernourished could be halved by 2015 – from 20.3 percent in 1990/1992 to 10.1 percent in 2015 and on to 6.9 in 2030 and to 3.9 by 2050. It is noted that the UN Millennium Development Goals (MDGs) refer not to halving the numbers undernourished but rather to a target to "halve, between 1990 and 2015, the proportion of people who suffer from hunger". In this sense, the MDG goal may be achieved.

Despite this slow pace of progress in reducing the prevalence of undernourishment, the projections do imply considerable overall improvement. In the developing countries the numbers well-fed (i.e., not classified as undernourished according to the criteria used here) could increase from 3.9 billion in 1999/2001 (83 percent of their population) to 5.2 billion in 2015 (90 percent of the population), to 6.2 billion (93 percent) in 2030 and to 7.2 billion (96 percent) by 2050. That would be no mean achievement. Fewer countries than at present will have high incidence of undernourishment, none of them in the most populous class. The problem of undernourishment will tend to become smaller both in terms of absolute numbers affected and, even more, in relative terms (proportion of the population), hence it will become more tractable through policy interventions, both national and international.

The progress in raising per capita food consumption to 3 000+ kcal/person/ day in several developing countries is not always an unmixed blessing. The related diet transitions often imply changes towards energy-dense diets high in fat, particularly saturated fat, sugar and salt, and low in unrefined carbohydrates. In combination with lifestyle changes, largely associated with rapid urbanization, such transitions, while beneficent in many countries with still inadequate diets, are often accompanied by a corresponding increase in diet-related chronic non-communicable diseases (NCDs). In many countries undergoing this transition, obesity-related NCDs tend to appear when health problems related to undernutrition of significant parts of their populations are still widely prevalent. The two problems coexist and these countries are confronted with a "double burden of malnutrition" resulting in novel challenges and strains in their health systems.

Growth of agriculture and main commodity sectors

Aggregate agriculture: World agriculture (aggregate value of production, all food and non-food crop and livestock commodities) has been growing at rates of 2.1 to 2.3 percent per annum in the last four decades, with much of the growth originating in the developing countries (3.4 to 3.8 percent per annum). The high growth rates of the latter reflected, among other things, developments in some large countries – foremost among them China. Without China, the rest of the developing countries grew at 2.8 to 3.0 percent per annum. They also reflected the rising share of high-value commodities like livestock products in the total value of production: in terms of quantities (whether measured in tonnage or calorie content), the growth rates have been lower.

The future may see some drastic decline in the growth of aggregate world production, to 1.5 percent per annum in the next three decades and down to 0.9 percent in the subsequent 20 years to 2050. The slowdown reflects the lower population growth and the gradual attainment of medium-high levels of per capita consumption in a growing number of countries. The latter factor restricts the scope for further growth in demand per capita in several countries that had very high growth in the past, foremost among them China. In contrast, developing countries that experienced slow growth in the past (and as a result still have low per capita consumption – less than 2 700 kcal/person/day) and potential for further growth should not experience any slowdown but rather some acceleration. Increasingly, world agriculture will have to depend on non-food uses of commodities if growth rates are not to be sharply lower compared with the past. As noted, the biofuels sector may provide some scope, perhaps a significant one, for relaxing the demand constraints represented by the declining rates of increase in human consumption.

Cereals: All the major commodity sectors should participate in the deceleration of agricultural growth. The cereals sector (sum of wheat, milled rice and coarse grains) has already been in such downward trend for some time now, with the growth rate having fallen from 3.7 percent per annum in the 1960s, to 2.5 percent, 1.4 percent and 1.1 percent in the subsequent three decades to 2001. In this latter year world production stood at just under 1.9 billion tonnes. It has grown further since then to some 2 billion tonnes in 2005 (preliminary estimate). We project increases to some 3 billion tonnes by 2050 and this would afford some increase in world per capita availability to around 340 kg (for all food and non-food uses), some 10 percent over present levels. It is noted that the current level of per capita consumption (309 kg in 1999/2001) is lower than what was achieved in the past mainly due to the sharp declines in the transition economies (the former socialist countries of the Soviet Union and Eastern Europe) in the 1990s. Recovery in their consumption as well as continued growth in the developing countries should raise the world average to levels it had attained in the past (in the mid-1980s). A good part of the increase in world cereals consumption should be for animal feed (mostly coarse grains), with the bulk of such consumption increases originating in the developing countries to support the expansion of their livestock production.

The decline in the growth rate notwithstanding, the absolute increases involved should not be underestimated: an increase of world production by another 1.1 billion tonnes annually will be required by 2050 over the 1.9 billion tonnes of

1999/2001 (or 1 billion tonnes over the 2 billion of 2005). Achieving it should not be taken for granted, as land and water resources are now more stretched than in the past and the potential for continued growth of yield is more limited.

Not all countries will be able to increase cereals production *pari passu* with their consumption. Therefore, past trends of ever-growing net cereals imports of the developing countries should continue and grow to some 300 million tonnes by 2050, a 2.7-fold increase over the 112 million tonnes of 1999/2001. This is a much lower rate of increase compared with the past when they had grown more than fivefold in 40 years. The novel element in the projections is that transition economies are transforming themselves from the large net importers of cereals they were up to the early 1990s (net imports of 43 million tonnes in 1993) to net exporters (18 million tonnes net exports annual average in 2002 to 2004). Such net exports could increase further in the future and, therefore, the traditional cereals exporters (North America, Australia, the EU and the developing exporters) would not have to produce the full surplus needed to cover this growing deficit.

Livestock: Production and consumption of meat will also experience a growth deceleration compared with the high growth rates of the past, although the milk sector should accelerate, mainly because of growth in the developing countries demand. The growth of the meat sector had been decisively influenced upwards by the rapid growth of production and consumption in China, and to a smaller extent also Brazil. This upward influence on the world totals was counterbalanced in the 1990s by the drastic shrinkage of the livestock sector in the transition economies, leading to a growth rate in the decade of 2.1 percent per annum versus 3.1 percent if the transition economies data are excluded from the world totals. These influences will not be present with the same force in the future – with the exception of continued rapid growth of production in Brazil (mainly for export). The decline in the transition economies has already been reversed, while the growth of meat consumption in China, which grew from 9 kg per capita to more than 50 kg in the last three decades, cannot obviously continue at the same high rates for much longer.

The rest of the developing countries still have significant scope for growth, given that their annual per capita meat consumption is still a modest 16 kg. Some of this growth potential will materialize as effective demand, and their per capita consumption could double by 2050, i.e., faster than in the past. It is unlikely that other major developing countries will replicate the role played by China in the past in boosting the world meat sector. In particular, India's meat consumption growth may not exert anything like the impact China had in the past, notwithstanding its huge population and good income growth prospects. The country may still have low levels of consumption (although significantly above the current 5 kg) for the foreseeable future.

Vegetable oils: The sector has been in rapid expansion, fuelled by the growth of food consumption and imports of the developing countries. The growth of the non-food uses (including in recent years for the production of biofuels in some countries) was also a major factor in the buoyancy of the sector, as was the availability of ample expansion potential of land suitable for the major oil crops – mainly soybeans in South America and the oilpalm in Southeast Asia. Indeed, oil crops have been responsible for a good part of the increases in total cultivated land in the developing countries and the world as a whole. These trends are likely to continue as the food consumption levels of the developing countries are still fairly low and the income elasticity of demand for vegetable oils is still high in most countries. In parallel, the growing interest in using vegetable oils in the production of biofuels may provide a significant boost. In this respect, concerns have been expressed that the rapid expansion of land areas under oil crops can have significant adverse impacts on the environment, mainly by favouring deforestation. This is just another example of the trade-offs between different aspects of sustainability that often accompany development: benefits in terms of reduced emissions of greenhouse gases when biofuels substitute petroleum-based fuels in transport versus the adverse impacts of land expansion.

Sugar: There are a number of features that characterize the evolution of the sector and determine future prospects: (a) rapidly rising food consumption in the developing countries (3.2 percent per annum in the last 30 years); (b) the emergence of several of them as major net importers (net imports of the deficit developing countries rose from 10 million tonnes to 29 million tonnes over the same period); (c) the growing dominance of Brazil as the major low-cost producer and exporter (production rose from 7.5 million tonnes to 32 million tonnes and net exports from 1 million tonnes to 11 million tonnes over the same period); (d) the growing use of sugar cane as feedstock for the production of biofuels (ethanol, mainly in Brazil, which now uses some 50 percent of cane production for this purpose); and (e) the prospect that after many years of heavy protectionism of the sugar sector and declining net imports in the industrial countries (which turned into net exporters from the mid-1980s, mainly owing to the protection of the sector in the EU and the substitution of maize-based sweeteners for sugar in the United States of America), the stage may be set for a reversal of such trends and the resumption of growth in their imports.

Many developing countries, including China, still have low or very low sugar consumption per capita (28 countries have less than 10 kg per annum and another 18 have 10 to 20 kg). Therefore, the potential exists for further growth in consumption, although it will not be as vigorous as in the past, when 60 developing countries had less than 20 kg in 1969/1971. Depending on the evolution of petroleum prices, sugar cane use as feedstock for the production of biofuels may keep growing in several producing countries (or those that have the resource potential to become major producers). Already several countries have plans to do so. It is possible that this development would contribute to keeping the growth rate of world aggregate demand (for all uses) and production from declining in line with the deceleration in the demand for food uses.

Roots, tubers and plantains: These products play an important role in sustaining food consumption levels in the many countries that have a high dependence on them and low food consumption levels overall. Many of these countries are in sub-Saharan Africa. In some countries (e.g., Nigeria, Ghana, Benin, Malawi) gains in production following the introduction of improved cultivars have been instrumental in raising the per capita food consumption levels. There is scope for other countries in similar conditions to replicate this experience. This prospect, together with the growing consumption of potatoes in many developing countries, should lead to a reversal of the trend for per capita food consumption of these products to decline – a trend that reflected largely the decline of food consumption of sweet potatoes in China. In addition, the potential use of cassava in the production of biofuels (actively pursued in Thailand) would further sustain the demand growth for this sector.

Agricultural trade of the developing countries

The growing imports of, mainly, cereals, livestock products, vegetable oils and sugar of many developing countries have resulted in the group of the developing countries as a whole turning from net agricultural exporters to net importers in most years after the early 1990s, reaching a deficit of USD 12 billion in 2000, before recovering in subsequent years to 2004. The recovery of recent years reflected above all the explosive growth of Brazil's agricultural exports, including oilseeds and products, meat, sugar, etc. Without Brazil, the deficit of the rest of the developing countries, already present from the late 1980s onwards, grew further from USD 20 billion in 2000 to USD 27 billion in 2004. Their traditional export commodities (tropical beverages, bananas, natural rubber, etc.) did not exhibit similar dynamism and for long periods stagnated or outright declined (in value terms), with the exception of the group fruit and vegetables.

The structural factors underlying these trends are likely to continue. The growing food demand in the developing countries will continue to fuel the growth of import requirements of basic foods in many of them, while the scope is limited for growth of consumption and imports of their traditional exportables to the developed countries. If anything, the growing competition among the developing exporters to supply those nearly saturated markets will continue to put pressure on prices (levels and instability) and lead to shifts in market shares at the expense of the weakest exporters among them, as happened with coffee in recent years. It may happen with sugar if the preferences protecting the weakest developing exporters were to be diminished or outright removed under the thrust of trade reforms. What will be somewhat different from the past is that the traditional dichotomy between developed (net importers) and developing (net exporters) will be further blurred: the markets facing the major developing exporters will be increasingly those of the importer developing countries, as is already happening with commodities such as sugar and vegetable oils.

Conclusions

The slowdown in world population growth and the attainment of a peak of total population shortly after the middle of this century will certainly contribute to easing the rate at which pressures are mounting on resources and the broader environment from the expansion and intensification of agriculture. However, getting from here to there still involves quantum jumps in the production of several commodities. Moreover, the mounting pressures will be increasingly concentrated in countries with persisting low food consumption levels, high population growth rates and often poor agricultural resource endowments. The result could well be enhanced risk of persistent food insecurity for a long time to come in a number of countries, in the midst of a world with adequate food supplies and the potential to produce more.

The slowdown in the growth of world agriculture may be mitigated if the use of crop biomass for biofuels were to be further increased and consolidated. Were this to happen, the implications for agriculture and development could be significant for countries with abundant land and climate resources that are suitable for the feedstock crops – assuming, of course, that impediments to biofuels trade do not stand on the way. Several countries in Latin America, Southeast Asia and sub-Saharan Africa, including some of the most needy and food-insecure ones, could benefit. Whether and to what extent this will happen is very uncertain, but the issue deserves serious analysis and evaluation. Of particular interest are: (a) possible adverse effects on the food security of the poor and the food-insecure if food prices were to rise because of resource diversion towards the production of feedstock crops for biofuels; and (b) the environmental implications of cultivated land expansion into pasturelands and forested areas. As noted, this is a typical case of possible trade-offs between different aspects of the environment and sustainability: benefits from the reduction in greenhouse gas emissions when biofuels substitute fossil fuels in transport and adverse effects from the expansion and intensification of agriculture.

WORLD AGRICULTURE IN A DYNAMICALLY CHANGING ENVIRONMENT: IFPRI'S LONG-TERM OUTLOOK FOR FOOD AND AGRICULTURE[1]

Siwa Msangi
Mark Rosegrant

The sharp increases in food prices that occurred in global and national markets over the 2006 to 2008 period sharpened the awareness of policy-makers and agricultural economic analysts of the stresses facing global food systems and the ecosystems that support them. The rapid increases in prices of key food commodities such as maize, wheat, rice and soybeans have mirrored the increases in prices of energy products, and strengthened the perception that energy and agricultural markets are becoming more closely linked (Schmidhuber, 2006). In the period 2002 to 2008, the international market prices of basic grain commodities more than doubled, while the prices of wheat and rice tripled. Although this might present different impacts on the consumer price indices in different countries – owing to the different shares of these commodities in total consumption – it represents a significant and sharp change in market conditions. While many see the reversal of historically declining real prices of agricultural commodities as an opportunity for agricultural producers in both developed and developing countries, others remain concerned about the implications of high food prices and increased volatility in food markets for the welfare and well-being of vulnerable populations, who consist mostly of net consumers of these products and who largely reside in the poorest regions of the developing world (Evans, 2008; FAO, 2008).

The nearly fourfold increase in oil prices over the same period led to second-round price effects on the wide range of goods and services that depend significantly on fossil fuels as inputs to production, including agricultural ones. Looking into the future, a number of researchers project the continued elevation of world prices for agricultural goods to above historical trends, despite a levelling off in the short term from the current highs. The medium-term projections generated by

---

1. The authors gratefully acknowledge the invaluable assistance of Miroslav Batka and Leonard Gwanmesia in the preparation and revision of this paper, and the valuable contribution of Simla Tokgoz.

the joint Organisation for Economic Co-operation and Development (OECD)/ FAO modelling effort show that a prevailing tightness remains in most major agricultural markets, keeping price levels significantly above historical trends (OECD/FAO, 2008). The world market price projections of the International Food Policy Research Institute (IFPRI) show that world grain prices will increase a further 30 to 50 percent over the period 2005 to 2050, while meat prices in the same period will increase an additional 20 to 30 percent beyond the levels seen in 2007/2008 (von Braun, 2008).

The underlying factors that led to the rapid increases in food prices up to 2008 have been widely discussed in the policy literature and are varied – in both nature and relative strength in driving market dynamics across various commodities. In both the published literature and the press, a number factors have been attributed to the rapid increase in food prices, ranging from the rapid increase in production of first-generation, food-based biofuels (Oxfam International, 2008; Runge and Senauer, 2008), to the increase of cereal and meat demand in East and South Asia and the increase in speculative activity in food markets. Several comprehensive discussions of this issue have appeared in recent literature (Headey and Fan, 2010; Headey, 2010), which seeks to assess the relative merit of each of these factors while providing an overview of the global macroeconomic picture and the relative decline of the United States dollar in relation to other currencies (Abbot, Hurt and Tyner, 2008). The steady decline in the global level of cereal stocks, resulting from the private sector taking over the operation of cereal stocks from governments and adopting a more "just-in-time" management orientation (Trostle, 2008), has also been cited as a factor that reduced the ability of national governments to stabilize consumer and producer prices (OECD, 2008). Most authors, however, do not isolate a single cause as being to blame for the current world food situation, but cite a complex interaction among several coincident factors.

The challenges and increased stresses that face global food production and distribution systems in the decade starting in 2010 are particularly acute and pressing for sub-Saharan Africa, where persistent levels of food insecurity already exist. For example, roughly 33 percent of the population of sub-Saharan Africa lives with insufficient food supplies (FAO, 2005) and an even greater proportion – 43 percent – lives below the international dollar poverty line (Dixon, Gulliver and Gibbon, 2001). Myriad constraints lie in the way of Africa's benefiting from higher producer prices of agricultural commodities on the world market, and include the fact that most of sub-Saharan Africa's agricultural production relies on rainfed cultivation and receives lower input levels of improved seed technology and fertilizer applications than agriculture in other regions. Additionally, the area affected by land degradation within the region is expanding, thereby causing a decline in soil fertility that reduces yield levels and increases the difficulty in maintaining sufficient production levels, especially given the lack of technological innovation and fertilizer use (FAO, 2005).

This chapter examines the key environmental, technological and socio-economic drivers that underpin the global world food situation, and evaluates the potential role of alternative policy interventions that might address these. It discusses these policy interventions in terms of the role they can play in enhancing market stability, food security and human well-being in the face of the increasing stresses that continue to face global agricultural markets and food systems. Specifically, it looks at the role that biofuels might play in raising food prices, and the role that agricultural technology investments might have in counteracting these effects. Based on this analysis, the chapter concludes with some final recommendations for both policy intervention and further research.

Drivers of change in food systems

The upward pressure on key commodity prices mentioned in the previous section can be accounted for by a number of underlying factors or drivers of change that are diverse in nature. These drivers range from environmental to socio-economic and from slow- to fast-moving, and affect outcomes differently in the short and long terms. In addition, underlying factors driving the long-term trends in food supply and demand have also contributed towards a tightening of global food markets during the past decade. These trends are driven by both environmental and socio-economic changes, as well as by agricultural and energy policies, including those that encourage biofuel production from agricultural feedstocks. Figure 2.1 illustrates the interactions among the various key drivers of change in global food systems, and their linkages to other components of the food economy and to important outcomes of human well-being, such as nutrition. Although Figure 2.1 does not include all the factors of importance, it incorporates the main elements of global environmental and economic change in food production and consumption systems that are addressed in this chapter.

Socio-economic change in the form of increasing growth in population numbers and total income, is among the major drivers that change the economic behaviour of consumers in terms of their demand for food and energy products. Urbanization, which is related to these demographic changes, also has an impact on consumption patterns and the transformation of consumer preferences for food, fibre and energy products. These changes in consumption and consumption preferences introduce increased stresses into the demand side of food and energy systems, while other environmental factors might restrain the supply side of food systems from responding readily – as a result of either resource scarcity or degraded land and water quality. Reduced investments in crop and energy technology, over time, can also lead to a longer-term slowdown in the expansion of supply, which eventually leads to higher prices as demand begins to grow faster.

Figure 2.1
Interrelationships among key drivers of change in food systems, and their connection to human well-being

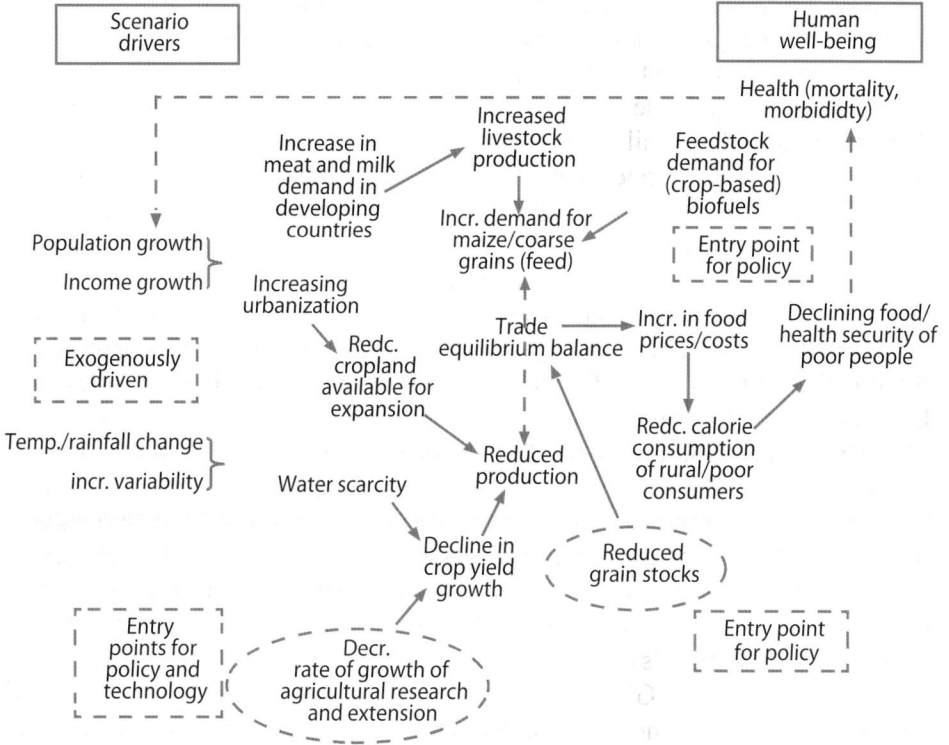

Source: Authors.

Taking these factors into account, as illustrated in Figure 2.1, a variety of entry points for policy or technological intervention present themselves. These offer a menu of options for policy-makers to consider when deciding how best to cope with the current stresses on food or energy systems, or how to mitigate the severity of such stresses in the future. The following subsections discuss some of these components and drivers of the food system in more detail, putting them into the context of food and energy supply and demand systems.

Socio-economic factors

Both demographic growth and socio-economic change – in the form of overall income growth, rates of urbanization or changes in the incidence of poverty in the population over time – are key factors that determine observed patterns of food consumption and nutrition outcomes. Since the oil crisis in the 1970s, there has been notable socio-economic progress and growth in various regions of the world, in terms of human welfare. Despite population growth, the number of malnourished people in developing countries has declined over time – albeit at various rates. According to *The State of Food Insecurity in the World 2006* report (FAO, 2006a), a decrease of 37 million from 1970 to 1980 was followed by a decrease of almost 100 million between 1980 and 1990, but then by a decrease of only 3 million in the period 1990 to 1992, which was set as the baseline for the 1996 World Food Summit. Food has become more affordable, as it is now less than half as expensive in real terms as it was in 1960. This decline in the cost of food can be attributed to a large increase in food production – even in per capita terms, the world now produces 40 percent more food than it did 40 years ago (MEA, 2005). Nonetheless, these positive trends might be reversed in the future, if the major tipping points of climate change and accompanying degradation of land and water resources intensify.

The main socio-economic factors that drive increasing food demand are population increases, rising incomes and increasing urbanization. Global population is set to increase from approximately 6 billion in 1995 to 8 billion in 2025, with more than 98 percent of this increase occurring in developing countries, according to the United Nation's (UN) medium-variant projections (UN, 2004). In addition, 84 percent of the population increase from 1995 to 2025 in developing countries is expected to occur in urban areas. Incomes, measured by gross domestic product (GDP) per capita, are expected to grow strongly in recently industrialized nations, and most rapidly in East Asia and the Pacific, according to the projections of growth used by a number of key policy centres (World Bank, 2007a; UNEP, 2007). Based on the rates used in IFPRI's International Model for Policy Analysis of Agricultural Commodities and Trade (IMPACT) projections in its most recent world food situation report (von Braun, 2008), per capita GDP in China is expected to increase by 5.2 percent per year from 1995 to 2025, while those in the Republic of Korea, Thailand and India grow at approximately 4.5 percent per year. In general, growth rates in Asia will be the highest, ranging from 2.1 to 5.2 percent per year, while Eastern European incomes will rise by 4.1 percent per year. On the other hand, rapid population growth in sub-Saharan Africa is projected to depress per capita growth rates to approximately 0.8 to 1.7 percent per year.

The combination of rising income and increasing urbanization is also changing the nature of diets. Rapidly rising incomes in the developing world have led to increased demand for livestock products. In addition, it has been shown that urbanized populations consume fewer basic staples and more processed foods and livestock products (Rosegrant *et al.*, 2001). Diets with a higher meat content put additional pressure on land resources for pasture and coarse grain markets for feed, including maize. As a result of these trends, it is predicted that by 2020 more than 60 percent of meat and milk consumption will occur in the developing world, and the production of beef, meat, poultry, pork and milk will at least double from their 1993 levels (Delgado *et al.*, 1999).

Increasing urbanization compounds the pressure on adjacent areas to meet the demand of large, concentrated populations. While urbanized areas themselves do not require large portions of land, the terrestrial and water resources necessary to support their populations can overwhelm existing rural-urban linkages. Many developing countries with large land endowments find it easier to convert forest and other land cover to agricultural production than to disseminate yield-enhancing technologies, especially where extension services are limited or non-existent. It is estimated that an additional 120 million ha of cropland will need to be converted to agriculture to meet food demands in developing countries over the next 30 years, with seven countries in Latin America and sub-Saharan Africa providing most of the additional land potential (FAO, 2006b).

These agricultural land requirement projections assume that 70 percent of food needs will be met through yield enhancements (FAO, 2006b). However, agricultural research dedicated to productivity enhancement of staple crops has declined over the years. As the United States of America and other developed regions shift their research focus to reflect consumer preferences for processed, organic and humane products, the diffusion of more relevant yield-enhancing technology in developing countries has slowed (Alston and Pardey, 2006). Only one-third of global, public agricultural research in the 1990s was in developing countries, more than 50 percent of it in Brazil, China, India and South Africa (Alston and Pardey, 2006). Therefore, better technology diffusion and more public money dedicated to developing country research programmes are critical to meeting growing food needs.

Environmental drivers

Population and income growth increase the pressure on natural resources to meet domestic, agricultural and industrial demand. Many large water basins, including the Yellow River and the Ganges, are expected to pump relatively less water for irrigation over the next 20 years, owing to unfavourable competition from other sectors. As a result, compared with 1995 levels, irrigated cereal yields in water-scarce basins are expected to decline by 11 to 22 percent by 2025 (Rosegrant, Cai and Cline, 2005).

Climate change and increasing demand for water resources will have an impact on growing conditions, significantly affecting food production in the future. Integrated assessment models have shown that climate change effects on temperature and rainfall will have positive yield effects in cooler climates, while decreasing cereal yields in low-latitude regions, where most developing countries are located (Easterling *et al.*, 2007). Specifically, owing to global warming, developing countries face declines of 9 to 21 percent in overall agricultural productivity, while the effects on industrialized countries will range from a 6 percent decline to an 8 percent increase, depending on the offsetting effects that additional atmospheric carbon could have on rates of photosynthesis (Cline, 2007). As a result of these differentials in predicted production capabilities, some regions will benefit from increased yields, while others will be forced to import increasing amounts of food to meet demand. Fischer *et al.* (2005) estimate that cereal imports will increase in developing countries by 10 to 40 percent by 2080. Although this prediction covers a large variation, the combined effects of rapid population growth, lower yields and increasing reliance on trade policy for food imports could leave an additional 5 to 170 million people malnourished in 2080 – depending on the projection scenario – with up to 75 percent of them in Africa (Schmidhuber and Tubiello, 2007). Parry, Rosenzweig and Livermore (2005) have shown that the regional variation in the numbers of food-insecure people is better explained by population changes than by climate impacts on food availability. A recent report released by IFPRI (Nelson *et al.*, 2010) looks at a wide range of scenarios illustrating the complex interplay between climate and socio-economic outcomes that leads to future outcomes for food and agriculture to 2050, and the critical role that productivity growth plays in offsetting the negative impacts. As a result, economic and other development policy, especially that pertaining to agricultural research and technology, will be critical in influencing future human well-being.

Policy-based drivers

In addition to the socio-economic and environmental processes described in the previous subsections, other factors can help create the kind of tight market environment that was observed in 2006 to 2008. These include the decline in cereal stocks, and unilateral trade actions by individual countries (such as India), as they both restrict supply in the market. For example, world wheat stocks-to-use ratios have declined from more than 40 percent in 1970 to 20 percent today – below the oil crisis level. Maize stocks-to-use ratios have declined from their 45 percent peak in the 1980s to about 12 percent, a level also previously seen only during the world oil crisis (Abbot, Hurt and Tyner, 2008). There have also been increasing levels of private capital invested in grain (and other commodity) markets, in search of portfolio diversification and in response to the recent poor performance of the stock market. In addition, unfavourable macroeconomic developments (such as the United States dollar devaluation) can further complicate the situation for some consumers. The thorough overview of these issues given by Headey and Fan (2010) illustrates the importance of various country-level policy decisions (over grain reserve and trade policies, for example) in creating the conditions that led to the spike in food prices, and their implications for institutional design. In this assessment, production shocks and productivity trends played less of a role in explaining the surge in food prices seen in 2006 to 2008 than other important policy drivers did. Looking to 2050 and beyond, however, the role of yield growth and productivity improvements in enabling production to meet future demands while "saving land" in the process becomes more critical.

A closer look at productivity growth finds that yield growth rates for major grains have been declining in recent decades (World Bank, 2007b) and have dropped by roughly 50 percent since their highs during the 1960s and late 1970s. One of the causes of this decline is no doubt a fall in the growth of public agricultural research and development (R&D) spending, in both the developing and the developed world (World Bank, 2007b). At the global level, R&D spending growth has declined by 51 percent in real terms in the two decades since the 1980s, with the decline occurring mainly in the developed world while the developing world has taken a larger share of the world's agricultural research spending than the developed world since the 1990s (Alston and Pardey, 2006). This is especially troubling as both FAO and IFPRI project that future production growth will depend more heavily on yield improvements than area expansion, as has been found in past assessments of global agricultural futures, such as the Millennium Ecosystem Assessment (MEA, 2005). In fact, some regions, such as East Asia, Europe and North America, will need to increase production as the agricultural area shrinks.

Characterizing the drivers of change

Given the rather complex interplay of factors described in this chapter and the wider literature, it is useful to try and separate the slower-acting, long-term drivers of change from the faster-moving ones that might have more of an impact in the short term. Population and income growth both tend to act relatively slowly and steadily over time, evolving in a rather predictable fashion – given the nature of the drivers that underlie demographic and economic growth, and past experience. There are also long-term shifts in climatic conditions at play, which also tend to unfold more gradually over time than do the shorter-term manifestations of climatic variability such as weather events that occur within the cyclical progression of seasons. Another slow-moving change is the gradual slowing of crop yield growth relative to the rate of food demand growth, which is driven by socio-economic changes.

In contrast to these slow-moving drivers of change are the faster-moving ones, which can take the form of sudden climatic and environmental shocks that cause seasonal losses of harvest. Although food demand tends not to surge upwards over short periods, there have been relatively rapid increases in the demand for energy, especially for transportation, which manifest themselves in the increasing demand for fossil-based fuels and renewable substitutes such as biofuels. The demand for biofuels, such as ethanol and biodiesel, tends to be strong when fossil-based fuel prices are high and national fuel policies push for increased levels of blending to reduce the cost of fuel imports. This has been the case in a number of countries around that world, and is a major determinant in the rapid expansion of biofuel production observed over the past six years.

It is worthwhile to consider the characteristics of these various drivers of change, to develop a better understanding of their relative importance in explaining the tightening of market conditions observed in global food markets in recent times. Despite some of the fairly comprehensive overviews and discussions of high food prices – in terms of their causes and consequences – relatively little effort has been made to distinguish their dynamic characteristics of change, to identify their relative importance in explaining short-term versus long-term phenomena. Such a distinction is helpful, not only in allowing identification of the most urgent issues to be addressed from a policy point of view, but also in identifying which issues are more temporary in nature and which might persist into the future, preventing market and food system characteristics from returning to a stable equilibrium, or causing prices to rise even further later on.

While Figure 2.1 shows how the various drivers of change interact with each other and where the critical feedback loops might be, it does not identify the type of distinguishing characteristics that can explain short- and longer-lived effects on food systems. Figure 2.2 does more to make this distinction. It shows where some key drivers of change lie in relation to each other and in terms of their dynamic characteristics, which are a combination of the speed with which they act and the degree to which they explain short- or long-term phenomena. At the end of the spectrum containing the fast-acting drivers that help to explain short-term effects, market speculation stands out as a factor that might explain the "bubbles" that can form in markets as a result of expectations about short- to medium-term trends, which can reverse themselves fairly rapidly on the basis of economic conditions and fast-changing market information. This type of activity has been cited as a factor in the spikes that developed in some markets and were contrary to the indicators provided by the supply and demand fundamentals that usually determine price formation (von Braun *et al.*, 2008). Other authors (Headey and Fan, 2010) are more cautious of attributing the influence of speculative activity to the rise in food prices up to 2008, given the lack of econometric evidence from the available data.

Figure 2.2
Characteristics of the drivers of change in food systems

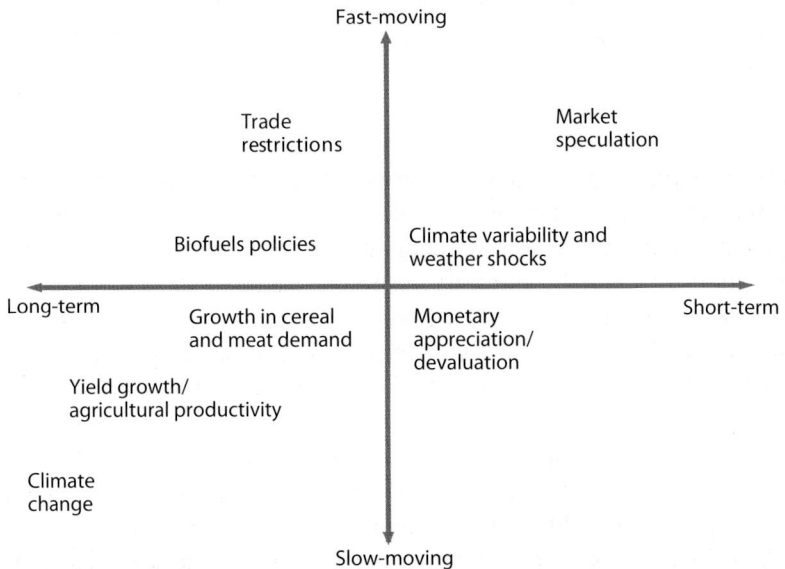

Source: Authors.

At the other end of the spectrum, among the relatively slow-moving phenomena that play a part in determining the long-term evolution of food systems and the performance of the underlying ecosystems that support them, is climate change, which encapsulates the changes in long-term means of temperature, precipitation and even atmospheric carbon content that affect crop growth potential and the characteristics of key agro-ecological systems. Climate change as a phenomenon should be distinguished from the effects of climate variability and extreme weather incidents that are currently occurring in many regions and that act over a far quicker time scale. These types of weather shocks drive the supply side of the food equation and lead to sudden drops in output that can push up market prices, whereas sudden surges from the demand side of the equation (such as those due to growth in crop-based biofuel production) might tighten market conditions and contribute to similar price increases.

Other drivers of supply and demand change that operate on a slower-moving trajectory are growth in demand for key consumer food products such as cereals and meat, which also have implications for feed demand, and trends for crop yield growth, which determine how well the supply side can adjust to increases in demand. Changes in demand for food and fibre products tend not to surge as rapidly as those for energy-intensive products, such as petroleum for transport, but represent a component of food system change that will continue to keep prices at an elevated level into the future, as cited by OECD in its projections of agricultural production and prices to 2017 (OECD/FAO, 2008), and in longer-term projections (IAASTD, 2009).

Entry points for policy

Given these various drivers, several possible entry points for policy intervention that might address the current global food situation can be considered. As shown in Figure 2.1, these entry points are on both the supply and the demand sides. On the demand side, policies that govern the use of food-based feedstocks for biofuel production could be altered, so that the overall quantities from food and feed sources are substituted by other non-food feedstocks or feedstock conversion technologies. Other policies that might affect direct food and feed use of grains would rely on the alteration of consumer preferences for food products (including meat), and are not as straightforward to address within the analytical framework discussed in this chapter. Therefore, attention should focus on the use of food crops in first-generation biofuel production.

From the supply side, a number of interventions can be considered. The first is boosting the output of cereals by raising yields over time, through policies that accelerate the improvement of crop technologies so that higher growth rates of yield are realized. This can be done directly through improved seed technologies, which might enhance the productivity and hardiness of plant varieties, or through the expansion of area under irrigated production, which typically has a higher yield than rainfed alternatives. Improved seed technologies can even reduce the loss in productivity that occurs when irrigated crops become water-logged or subject to increased salinity and submergence, thus allowing the expansion of irrigated area to raise overall production levels further. Through analytical work supported by IFPRI (Rosegrant *et al.*, 2009), the comprehensive Strategy and Results Framework of the Consultative Group on International Agricultural Research (CGIAR) demonstrates that these kinds of intervention across the range of mandate commodities supported by CGIAR research could have profound and multiplicative effects on future food outcomes.

Another supply-side intervention would be improving the management of grain storage, so that there are sufficient quantities of grain on hand to provide an adequate buffer when shocks in either production or supply cause prices to spike. This has been discussed at length in the recent literature, without a great deal of analysis. Considerable attention is paid to this aspect of policy in the analytical framework presented in the following section.

Quantitative outlook to 2050

This section presents some forward-looking outlooks for food production and consumption that are based on IFPRI's IMPACT model (Rosegrant *et al.*, 2001; 2005; Rosegrant, Cai and Cline, 2002), and outlines the implications observed for long-term food security. These simulations will help identify the impact of policy-based and socio-economic drivers on the evolution of agricultural prices, and the role that technological interventions and investments can play. They will also help to illustrate the types of entry point that are possible for helping to stabilize food prices and improve human well-being outcomes in the face of the various drivers of change discussed so far.

The model

To examine the potential impact of biofuel production growth on country-level and domestic agricultural markets, a partial equilibrium modelling framework is adopted to capture the interactions between agricultural commodity supply and demand, and trade, at the global level. The model used is IMPACT, which was developed by IFPRI for projecting global food supply, food demand and food security to 2020 and beyond (Rosegrant *et al.*, 2001). IMPACT is a partial equilibrium agricultural model for crop and livestock commodities, including cereals, soybeans, roots and tubers, meats, milk, eggs, oilseeds, oilcakes/meals, sugar/sweeteners, and fruits and vegetables. It is specified as a set of 115 country and regional sub-models, within each of which supply, demand and prices for agricultural commodities are determined. The model links the various countries and regions through international trade, using a series of linear and non-linear equations to approximate the underlying production and demand functions. World agricultural commodity prices are determined annually at levels that clear international markets. Growth in crop production in each country is determined by crop and input prices, the rate of productivity growth, investment in irrigation, and water availability. Demand is a function of prices, income, and population growth. IMPACT contains four categories of commodity demand: food, feed, biofuel feedstocks, and other uses.

Baseline model projections

Production growth: The profile of cereal production over time is presented in Figure 2.3, which shows steady trends of output growth to 2050. Cereal production is projected to grow steadily across all seven regions, with North America and Europe leading in production volume. When looked at on a per capita basis, however, the trends present a somewhat more static picture in terms of how the various regions are projected to maintain production levels relative to their populations (Figure 2.4). North American, European and Central Asian regions make significant increases in production relative to their own population growth, and are able to provide the surpluses needed to supply the food and feed needs of the rest of the world. The Near East and North African region is able to increase its per capita production levels over the production period, as is Latin America and the Caribbean. In contrast, the South and East Asian regions decrease their per capita production over time, as does sub-Saharan Africa.

Figure 2.3
Total cereal production to 2050

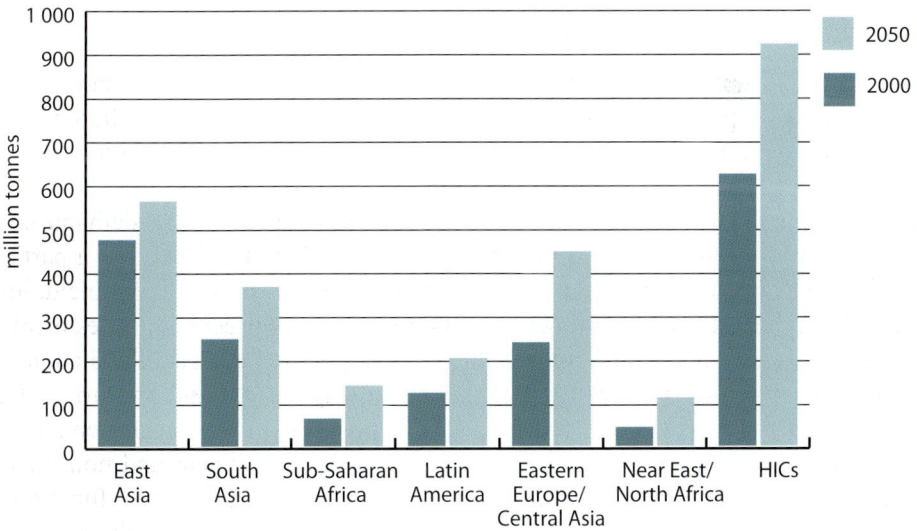

HIC = high-income country.
Source: Projections from von Braun, 2008.

Figure 2.4
Per capita cereal production to 2050

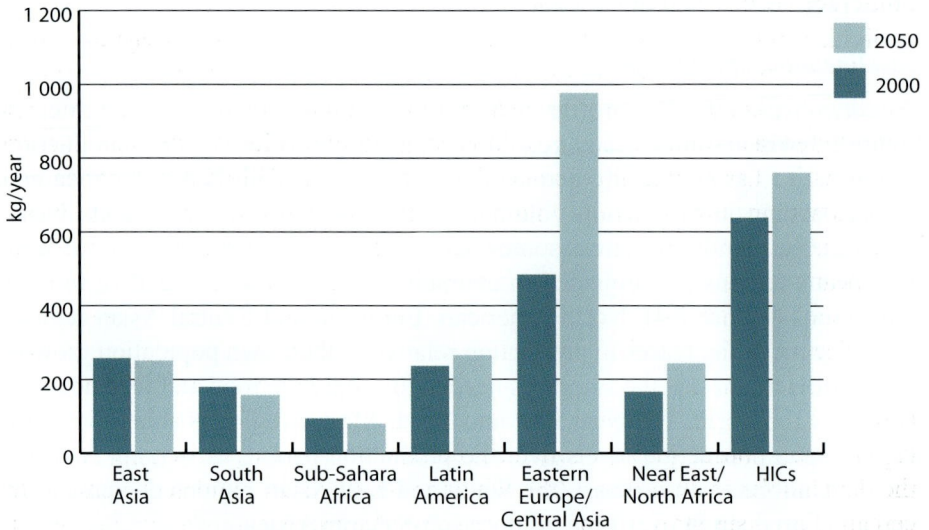

Source: Projections from von Braun, 2008.

Table 2.1
Total feed and food demand for cereals

Region	Total demand 2000 (million tonnes)	Total demand 2050 (million tonnes)	Change (%)	Food demand 2000 (million tonnes)	Food demand 2050 (million tonnes)	Change (%)	Feed demand 2000 (million tonnes)	Feed demand 2050 (million tonnes)	Change (%)
East Asia and Pacific	493	662	34	335	366	9	101	204	102
South Asia	244	421	73	217	359	66	3	12	266
Sub-Saharan Africa	83	239	189	64	185	188	7	18	156
Latin America and Caribbean	132	237	79	63	88	40	49	110	126
Eastern Europe and Central Asia	233	264	13	79	80	1	108	123	14
Near East and North Africa	88	178	102	56	102	83	23	58	147
High-income countries[a]	545	737	35	112	146	30	322	398	24

[a] United States of America, Canada, EU15, Switzerland, Norway, Cyprus, Israel, Japan, Republic of Korea, Singapore, Australia, New Zealand, Persian Gulf Region.

Source: Projections from von Braun, 2008.

Demand growth: Over the 50-year period, total food demand for cereals is projected to increase in all regions, with North America, Europe and East Asia leading all other regions in total volume. Table 2.1 shows how the total demand for cereals is divided into its largest two components: food and feed uses. Regarding food use, the region that shows the strongest demand growth for cereals is sub-Saharan Africa, although other regions such as South Asia, East Asia and the Pacific, and Latin America exceed it in terms of food consumption volume. The Near East and North Africa has similar food demand growth for cereals to South Asia, and the regions with the lowest levels of growth are Eastern Europe and Central Asia, and East Asia and the Pacific. Regarding feed uses of cereals, the North American and European regions lead the world in total volume of feed consumption, followed by East Asia, Latin America and the Caribbean, and the Near East and North Africa.

The patterns of food demand in per capita terms provide a more comparable basis for examining the changes in consumption patterns across regions (Figure 2.5). Regarding the demand for cereals, East and South Asia fall in per capita cereal consumption, while most of the rest of the world rises. In terms of the demand for meat (Figure 2.6), which is the main driver of feed demand for cereals, East Asia far outstrips other regions, in keeping with its rapid growth in per capita income compared with other developing and developed regions. Other regions that show large increases in per capita consumption of meat are North America and Europe; these have far higher levels of consumption compared with South Asia and sub-Saharan Africa, which grow steadily from relatively low levels owing to their steady income growth over the period.

Long-term trends in malnutrition: Given the patterns of supply and demand that have been highlighted, the IMPACT model infers a trend in levels of malnourished among the most vulnerable demographic of the population – those aged zero to five years. The determinants of malnutrition are derived primarily from four key indicators: per capita calorie availability; access to clean drinking-water; rates of secondary schooling among females; and the ratio of female-to-male life expectancy. The links between malnutrition and these determinants were established by Smith and Haddad (2000), who used the determinants as explanatory variables to account for changes in levels of child malnutrition across the developing world between 1975 and 1995. According to their work, a greater share of the reduction in child malnutrition levels over this period can be attributed to improvements in female schooling and access to clean water than to calorie availability alone. This finding is in line with the four-pillar concept of food security that underlies FAO's conceptual framework, in which availability is only one of the factors that accounts for food security status among vulnerable populations and must be evaluated along with access, utilization and stability.

Figure 2.5
Per capita cereal food demand to 2050

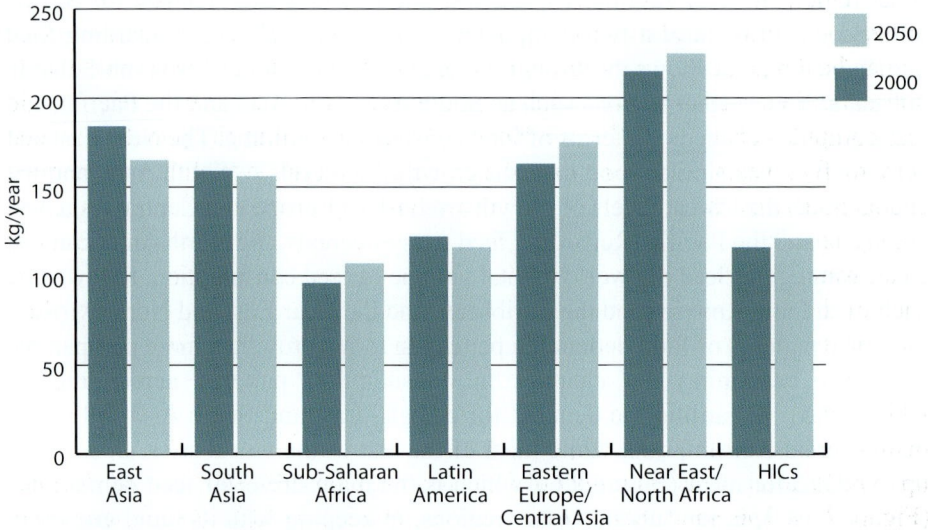

Source: Projections from von Braun, 2008.

Figure 2.6
Per capita meat demand to 2050

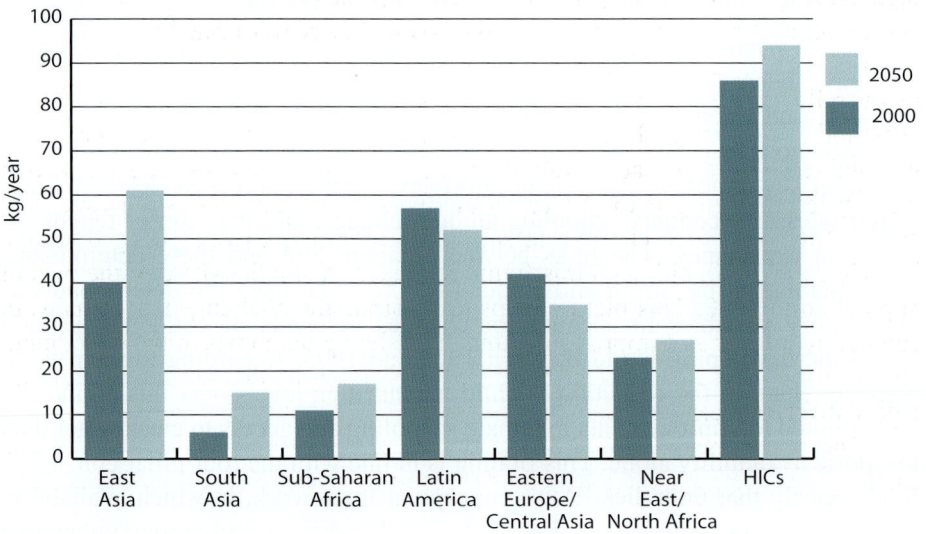

Source: Projections from von Braun, 2008.

The methodology used for tracking child malnutrition in IMPACT therefore covers aspects of availability, access and utilization, where the concept of access is grounded in the price response of consumption to market conditions, and utilization is influenced by access to clean water, which is a major determinant of human health and the body's ability to absorb and utilize available and accessible nutrients. This methodology is implemented through an analytical relationship that computes changes in the prevalence of malnutrition in the population aged zero to five years as a function of per capita calorie availability (generated endogenously by the model), as well as exogenous projections of schooling rates among females of secondary school age, the share of population with access to clean water, and the ratio of female to male life expectancies. The influence of each of the four explanatory factors on under-five malnutrition is determined by the statistical coefficients derived by Haddad and Smith's work (2000).

The baseline trends for malnutrition are illustrated in Figures 2.7 and 2.8, which show variation in the rates of change in malnutrition. The decline in malnutrition prevalence is steeper in Asia than sub-Saharan Africa in the period up to 2025, after which a number of African subregions also show steady declines (Figure 2.7). The South Asia region has the highest overall levels of prevalence, but is able to make significant reductions by 2050, compared with Southeast Asia and western sub-Saharan Africa, which are able to decrease the overall levels of prevalence only slightly. East Asia, which begins with the lowest levels, is able to draw these levels even further down in the longer term, to achieve single-digit prevalence rates, which no other region can match. The complete picture of child malnutrition emerges when total numbers of malnourished are examined. Figure 2.8 shows the Asian region as a whole to be the most aggressive in reducing its overall levels of malnutrition, which remain the highest in the world, even in 2050 and even compared with sub-Saharan Africa, which sees on overall increase in numbers before the acceleration of increases in production and per capita income levels allows it to reduce its numbers. In total numbers, however, the count of malnourished children in sub-Saharan Africa remains nearly the same in 2050 as in 2000, although this figure represents a smaller share of the overall population in 2050. This picture helps to illustrate the challenge that remains in combating hunger and improving human well-being outcomes in the developing world in the long term, given the impending pressures that environmental and policy-driven shocks will have on the world food system.

The following sections provide greater details about the nature of these challenges and their implications for future food security.

Figure 2.7
Prevalence of preschool child malnutrition in Asia and Africa (children aged 0 to 5 years)

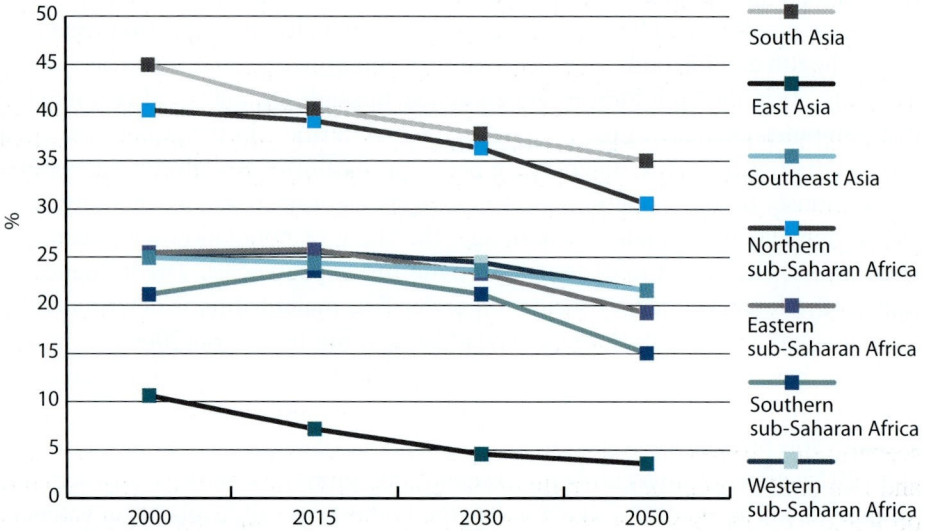

Source: Projections from von Braun, 2008.

Figure 2.8
Total numbers of malnourished preschool children in the developing world (children aged 0 to 5 years)

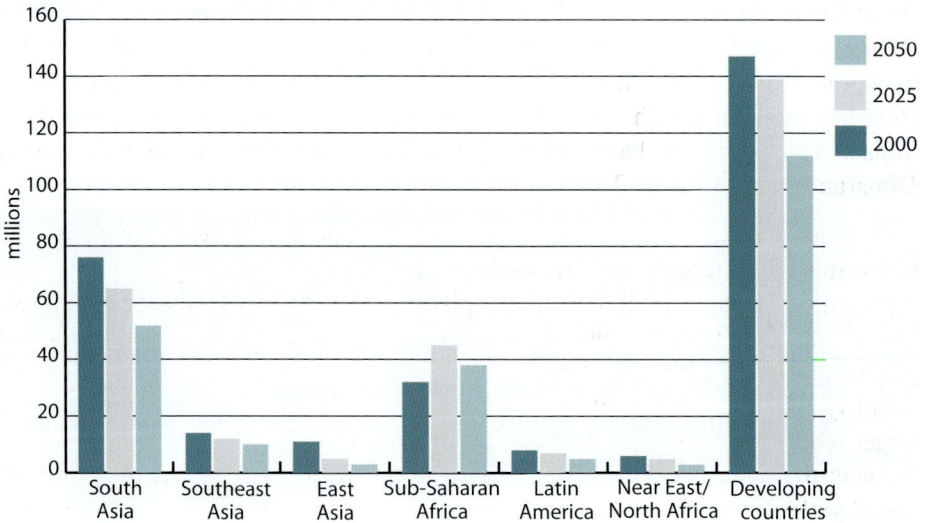

Source: Projections from von Braun, 2008.

The role of biofuels

Given the complex nature of the various drivers of change, and the way in which they interact within global agricultural and non-agricultural markets, it is not easy to isolate the effect of biofuels from that of other important factors. Nonetheless, using IMPACT, Rosegrant *et al.* (2008) set up a simple counterfactual experiment designed to show the contrasting impact on cereal prices that the observed historical trends of biofuels growth would exert if the global growth in biofuel production levels were reduced for the period 2000 to 2007, when most of this rapid growth was realized. The objective of this experiment was to see how much global cereal prices would have deviated from their observed baseline levels if biofuel production levels between 2000 and 2007 had remained on the same trajectory as in 1990 to 2000. The simulation results show a growth rate in average grain prices that is 30 percent lower than the actual rate of increase in world prices for 2000 to 2007. Other authors have carried out similar experiments to measure the effect of biofuels on market prices, although the choice of methodology (and scenario design) has a considerable impact on the measured impacts. As Headey and Fan (2010) point out, a study of the effects of United States maize policies on market prices (based on the Food and Agricultural Policy Research Institute [FAPRI] model) show that the combined effects of biofuel subsidies and tax credits amount to a level of support of 20 percent for maize prices in the United States of America, while the support to soybeans is an even more significant 73 percent (Meyers and Meyer, 2008). The analysis of McPhail and Babcock (2008) (using the international agricultural markets model from the Centre for Agriculture and Rural Development [CARD]) shows that the combined effect of subsidies provide a level of support to maize prices of 16 percent, although this study (and that using the FAPRI model) does not fully outline the effect of the tariff imposed on Brazilian ethanol, which other studies of trade liberalization's effect on biofuel impacts have shown to have had a significant influence on outcomes (Al-Riffai, Dimaranan and Laborde, 2010).

The implications of renewable fuel targets: Specific policies, such as the renewable fuel targets set by various countries for meeting blending and replacement rates of fossil fuels over a given time horizon, can also be examined. For example, the United States of America sets a target for first-generation biofuel production of 15 billion gallons by 2022, under the Energy Independence and Security Act. The additional production of maize feedstock needed to meet this target requires a higher level of yield growth, shown in Figure 2.9, to offset the impacts that it would otherwise have on food security; the average growth in cereal yields would have to increase from 1.3 to 1.8 percent a year (for the period 2000 to 2030) to counteract the implied trends in malnutrition. This translates into

Figure 2.9
Additional global cereal yield growth needed to offset impact of United States biofuels target

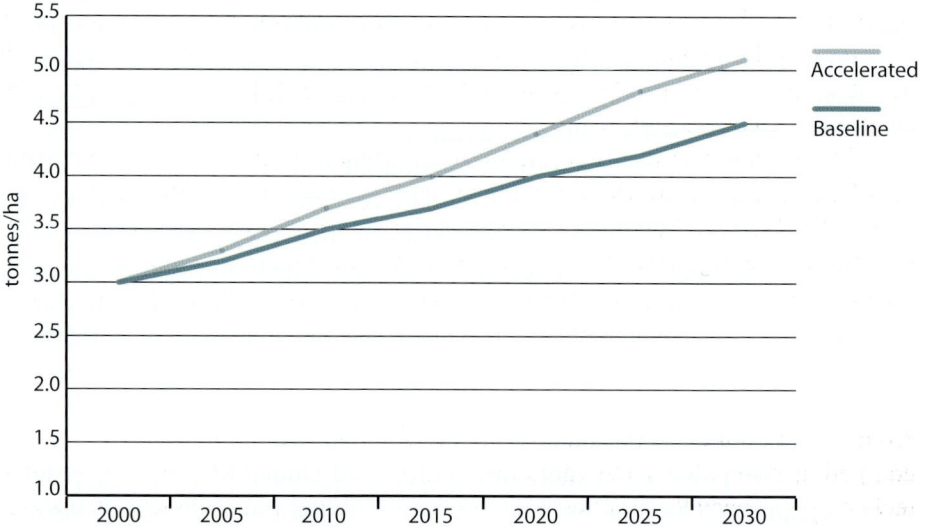

Source: Projections from von Braun, 2008.

Figure 2.10
Trends in child malnutrition to 2025 under the baseline case

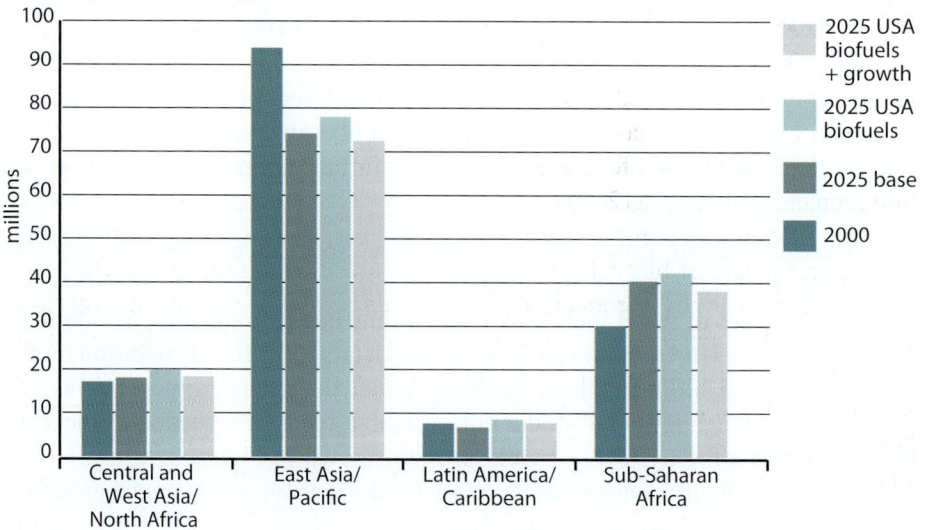

Source: Projections from von Braun, 2008.

an additional 1 percent of yield growth in the developing world and 0.5 percent in the developed world for 2000 to 2030 – presuming that higher yield gains can be made in less developed countries, where there might still be significant opportunities for closing yield gaps that could be exploited. The impact of this offsetting yield increase is shown in Figure 2.10, which shows the increases in malnutrition in 2025 resulting from the United States policy being offset by the additional cereal yield growth.

These scenarios illustrate the biofuels impact on global food prices fairly clearly, and lead to immediate implications for food security and human well-being. To illustrate how specific technological innovations can ameliorate the situation and reduce the pressure that crop-based biofuel production growth places on global food systems, further simulation-based experiments can be carried out, as described in the next subsection.

Yield-enhancing technologies and policies

An important policy intervention for alleviating the trade-offs that arise from the competing demands for land area to produce food, feed, fibre and fuel needs is technology, especially productivity-boosting technologies. Enhancing the yields of food, feed or fibre products per unit area of land has the effect of not only increasing the overall availability of these products (and lowering their market prices as a result), but also increasing the availability of land for non-agricultural uses, such as forestry, wildlife habitat or the provision of fuel from plantation-style biofuel systems. Increasing the yields of biofuel production systems, through improvements in the productivity and energy yield of the underlying conversion technologies, could also have a land-saving effect, increasing the area available for growing food and feed products, or for non-agricultural uses.

Some of these effects have been noted in recent global assessments of future trade-offs between food, feed and energy needs and the health of the environment and ecosystems. In MEA (2005), the scenario with the highest levels of technology adoption and high income growth (the "global orchestration" scenario) also had the highest levels of biofuel production. This arose from greater investments in increasing agricultural productivity, which reduced the competition for food-producing land, thereby making more land available for biofuel plantations and resulting in lower prices for both food and biofuel products. Conversely, the scenario with the lowest levels of income growth and technology adoption (the "Order from strength" scenario) also had the greatest competition for land under food production – owing to lower agricultural productivity and investments – and lower biofuels production, resulting in higher food and energy prices. The assessment scenario results also showed forest land decreasing because of higher levels of biofuel production, and more extensive agricultural land-use patterns also resulting in a similar encroachment on forest land. Both of these results underscore the persistent trade-offs between maintaining ecosystem health and meeting the demands for food, feed and fuel that exist in all of the scenarios considered. Although there are differences in the ways in which various drivers of change evolve under these scenarios – through increased demand for food, feed, fibre or fuel – they all involve competition for land uses and some encroachment on land that would otherwise remain unmanaged.

The fourth Global Environmental Outlook (GEO4) of the United Nations Environment Programme (UNEP) was similar to the MEA global assessment, and showed that increased emphasis on meeting targets for greenhouse gas reductions (under the "sustainability first" or "policy first" scenarios) could lead to increased biofuel production and decreases in area under forest (UNEP, 2007). In parallel with the global orchestration case in MEA, these GEO scenarios also embodied higher rates of income growth and technology adoption, thereby making agricultural growth more intensive and less extensive in nature, and allowing more land for non-agricultural uses (including biofuel production). In a similar way, both food and energy prices tended to be lower under these high-growth scenarios, owing to the higher production of food and energy products. At the same time, the area of land vulnerable to erosion risk also increases as a result of biofuel production, particularly under the policy first scenario, which pays less attention to soil conservation and improved land management than the sustainability first scenario does.

Other, more biofuel-specific scenarios examine the impact of biofuel production growth on food prices, through demand-side effects and the land-saving impact of increased technology growth, which affects the supply side of the agricultural market equation. As in the IMPACT-based simulations (Rosegrant *et al.*, 2001), the "business-as-usual" or "reference" scenario describes slowly declining rates of growth in agricultural research (and extension), following the same trends as observed in the past. As an alternative to the reference scenario, a case in which levels of agricultural knowledge, science and technology (AKST) are enhanced can be used. In this "high AKST"[2] variant, levels of investments in agriculture for the period 2005 to 2050 are elevated. These accelerated investments in agricultural technologies lead to increased growth in crop yields and livestock numbers. A further variant of this considers the implications of even more aggressive growth in agricultural R&D together with advances in other complementary sectors that

---

2. AKST refers to the broad conceptualization of agricultural technology and capital used in the recent International Assessment of Agricultural Science and Technology for Development (IAASTD) global assessment. Various scenarios embodying differing levels of AKST were quantified, using a number of models including IMPACT. The high AKST case described in this chapter was one of these scenarios.

provide key infrastructure and social services. Such sectors include investments in irrigation infrastructure (represented by accelerated growth in irrigated area and efficiency of irrigation water use, and accelerated or reduced growth in access to drinking-water) and changes in investments in secondary education for females, which is an important indicator for human well-being.

Implications for malnutrition

In the scenarios described in the previous subsection, the increase in crop prices resulting from expanded biofuel production is accompanied by a net decrease in availability of and access to food. Under the two biofuel scenarios, calorie consumption is estimated to decrease across regions, compared with baseline levels.

In the high AKST scenario, food security status and human well-being levels increase significantly owing to reductions in the prices of important tropical staple crops such as cassava and maize. Figure 2.11 shows how calorie availability is greatly enhanced over time by the acceleration in yield and production growth realized under high AKST levels. The effect is particularly strong in sub-Saharan Africa, where improvements in maize and cassava yields have a large impact on calorie availability, given the compositions of diets in the region and the fact that maize and cassava are important starch foods.

Figure 2.11
Increases in calorie availability under high AKST scenario compared with biofuel expansion under baseline technology levels

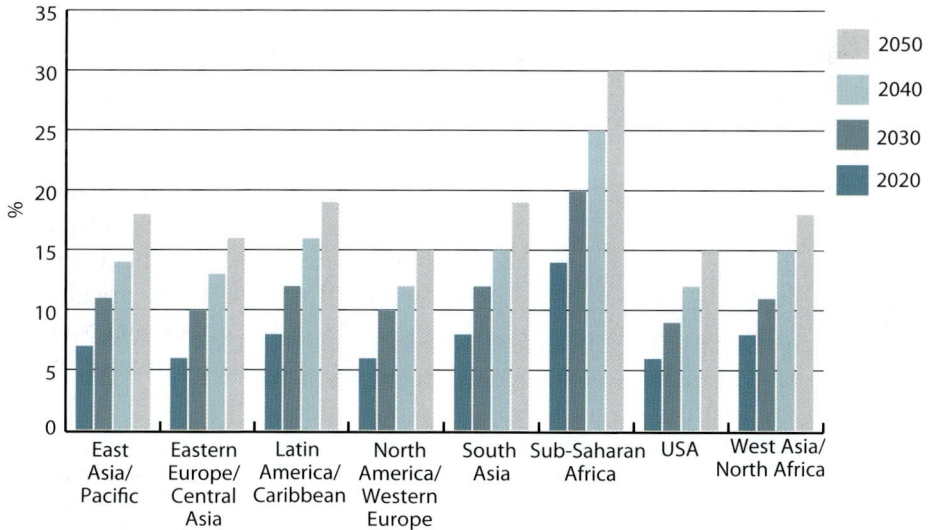

Source: Projections from von Braun, 2008.

Figure 2.12
Decreases in numbers of malnourished children under high AKST scenario compared with biofuel expansion under baseline technology levels

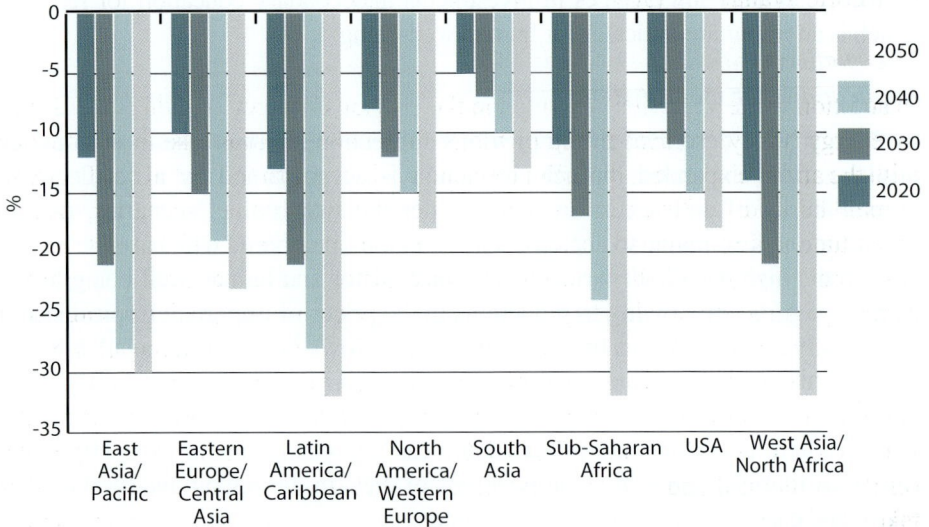

Source: Projections from von Braun, 2008.

Under high AKST, there is a significant reduction of malnourishment in small children over time, as a result of increased calorie availability in various regions (Figure 2.11) and other improvements in socio-economic conditions embedded in the high AKST scenario assumptions. Figure 2.12 shows that the level of malnourishment among small children drops strongly over time, in sub-Saharan Africa, North Africa and Latin America. The poorer regions of West Asia and North Africa benefit from enhanced access to water, better female schooling rates and lower food prices as much as the tropical regions do, owing to the poor state of social services in some of these regions. The rates of change are much faster in these regions, even compared with East Asia and the Pacific or South Asia, and the benefits appear to progress more strongly than even the improvement in calorie availability. This illustrates the importance of socio-economic factors other than food availability alone in determining malnutrition rates, and how the pillars of food security – availability, access, utilization and stability – interact to produce an effect that may be greater than the sum of the individual components. Although not all of the components of food security can be captured within the modelling framework described here, availability and access (which are closely connected to food prices) are well captured. Some elements of utilization are captured through the relationship between access to clean water and level of malnutrition, according to the empirical work of Smith and Haddad (2000), who found that 43 percent of the decrease in child malnutrition between 1975 and 1995 was due to female schooling, which was the leading determinant, followed by calorie availability (accounting for 26 percent).

The challenge of climate change

In addition to the scenarios presented in the previous subsection, which are driven by energy policy, the accounting of future food balances must also be reconciled with the added challenges that climate change will bring to the global food system. It must be said that the ultimate impacts of climate change – in terms of both magnitude and regional specificity – remain somewhat uncertain, and there is a wide spectrum of modelling results showing different degrees of impact for the same regions of the world. A great part of the uncertainty regarding results from different global circulation models results from the fact that each model presents different interactions among the atmosphere, the ocean and terrestrial systems, and the divergences in model results increase with time. The Intergovernmental Panel on Climate Change (IPCC) tried to portray the wide variance in model results in its third and fourth assessment reports, while many authors choose to take the more extreme of the examples to illustrate the types of impact that are plausible from climate change. Deviations in reported climate change impacts also result from different ways of translating the outputs of coupled (atmospheric-terrestrial) climate models into impacts on crop yield potential, and of translating these shifts in productivity into total production, consumption, price and trade effects in the various economic equilibrium models used.

At the heart of the challenge lies the reconciliation of biophysical modelling results, which are run at a relatively microlevel-scale of resolution, with the workings of an aggregate-level, market equilibrium-driven policy model such as IMPACT, which has to take the average of crop level effects across space. The marriage of these two elements – biophysical process-driven elements and economic equilibrium-driven mechanisms – is complex, and is the subject of continuing research among a number of research groups employing both partial and general equilibrium methods of economic market modelling. IFPRI's work has not fully attributed the possible effects that carbon fertilization could have on future crop yields, owing to the uncertainty that still exists in quantifying this result for various agronomic zones where on-the-ground reality could differ significantly from carbon fertilization experiments in the laboratory. Therefore, in this chapter's discussion of IFPRI results for climate change impacts, the reader should be aware that the methodology employed to account for climate change shocks within the modelling framework is still under revision.

Notwithstanding these difficulties, some results showing the overall magnitude of climate change impacts on global agricultural markets can be used to begin discussions of the implications for both national and household-level economic effects.[3] For this, the results from the more extreme "A2" climate scenario are used. This is the socio-economic scenario with greater emphasis on fossil-based fuels and less cooperation and (clean) technology sharing across the globe. This type of outcome is similar to those of the less favourable MEA and GEO4 scenarios in terms of portraying a less harmonious, cooperative and purely growth-driven kind of geopolitical atmosphere. More recent work by IFPRI (Nelson *et al.*, 2010) shows results from a scenario with more convergent socio-economic characteristics and balanced energy consumption patterns, the "A1b" scenario, which is used in this chapter to show contrast in the outcomes.

Table 2.2
Simulated impacts on yield in 2050 from various climate change scenarios

	Change from yields with climate as in 2000 *(%)*					
	CSIRO		NCAR		MIROC	
Crop and region	Irrigated	Rainfed	Irrigated	Rainfed	Irrigated	Rainfed
Maize						
Developing regions	-2.0	+0.2	-2.8	-2.9	-5.3	-3.5
Developed regions	-1.2	+0.6	-8.7	-5.7	-12.3	-29.9
Rice						
Developing regions	-14.4	-1.3	-18.5	-1.4	-11.9	+0.1
Developed regions	-3.5	+17.3	-5.5	+10.3	-13.3	-12.8
Wheat						
Developing regions	-28.3	-1.4	-34.3	-1.1	-13.4	-10.4
Developed regions	-5.7	+3.1	-4.9	+2.4	-11.6	-9.0

Sources: CSIRO and NCAR results from Nelson *et al.*, 2009, based on the A2 Special Report on Emissions Scenarios (SRES) scenario; MIROC results from Nelson *et al.*, 2010, based on A1b SRES scenario-based climate outputs.

Using recent IFPRI studies based on a variety of climate models, Table 2.2 shows the simulated effects of climate outcomes on crop yields for three major cereal commodities of key importance for food and feed uses. The model results from the Australian modelling group in the Commonwealth Science and Industrial Organization (CSIRO) tend to give "drier" outcomes than the United States model based at the National Center for Atmospheric Research (NCAR). This contrast shows up in the results (reported in Nelson *et al.*, 2009) in Table 2.2, where the

---

3. This subsection reflects improvements in modelling made since the Expert Meeting by presenting a wider range of scenario results to illustrate the dependence on climate model outputs.

yield impacts on maize, rice and wheat are more negative (or less positive) for the NCAR-based simulations than for the CSIRO outcomes, across both irrigated and rainfed systems in developing and more developed regions. To illustrate additional contrast, Table 2.2 includes the yield outcomes from a more recent IFPRI study (Nelson *et al.*, 2010), which simulates grain yields based on outputs from the Model for Interdisciplinary Research on Climate (MIROC). This gives precipitation patterns that are higher, on average, than those from CSIRO, although they also include stronger decreases of precipitation in some important food-growing regions. The differences between the NCAR and the CSIRO results arise mainly from differences in modelling approaches, while the differences from the MIROC model result from differences in the underlying socio-economic assumptions on which the climate models were run. This helps to explain why the differences in outcomes between MIROC and the other two climate models are not as systematic as those observed when comparing the A2-based outcomes of NCAR and CSIRO.

Table 2.3
World prices of selected grains (USD per metric tonne)

	Baseline		NCAR-based scenario		CSIRO-based scenario	
Cereal	2000	2050 no climate change *(million tonnes)*	2050 with climate change *(million tonnes)*	Change *(%)*	2050 with climate change *(million tonnes)*	Change *(%)*
Maize	95	155	235	51.9	240	55.1
Rice	190	307	421	36.8	406	32.0
Wheat	113	158	334	111.3	307	94.2

Source: Nelson *et al.*, 2009.

Table 2.3 shows the projected impact of climate change on global prices for the same three cereal commodities, limited to the two models that share the same underlying climate scenario. The doubling (NCAR) or near doubling (CSIRO) of the global market price for wheat in 2050, due to climate change, implies strong effects for consumers of wheat products in many developed and developing regions, especially for the more urbanized populations of Asia and sub-Saharan Africa. The strong increases in maize prices (which exceed 50 percent in both models) imply significant impacts on the livestock industry, which relies on maize for feed, and for regions where large numbers of consumers utilize maize as food, such as sub-Saharan Africa. Although the simulated impacts on rice are less pronounced, they are still important for regions that rely on rice as a key food staple and have experienced civil unrest due to increases in the rice price, as has been witnessed recently. Despite the differences in the underlying climate models, these results show that the impacts of climate change on food market outcomes to 2050 are non-trivial and significant. These increases do not necessarily represent sudden price spikes in 2050, but a gradual accumulation of price pressures building over time in response to the steady and constant tightening of supplies, as the area suitable for crop cultivation declines in various key cereal growing regions of the world. Nonetheless, the differences demonstrate that the pressure on global food supplies would be significantly increased if the environmental drivers embedded in these climate change scenarios were realized, and that responsive policy action and adaptation would have to occur to offset these effects. Such adaptive actions are not embedded in the results presented here, as agents' endogenous technology choices are not fully represented in the model. Adaptations and technology choices would have to be introduced into each scenario to account for the possibility of improved seed variety and other on-farm improvements, which are not endogenous within the framework. Such adaptation-focused scenarios will be included in further work.

Table 2.4
Total numbers of malnourished children, 2000 and 2050

	Baseline		NCAR-based scenario		CSIRO-based scenario	
Region	2000	2050 no climate change (millions)	2050 with climate change (millions)	Change (%)	2050 with climate change (millions)	Change (%)
South Asia	76	52	59	13	59	13
East Asia and Pacific	24	10	15	50	14	40
Eastern Europe and Central Asia	4	3	4	33	4	33
Latin America and Caribbean	8	5	6	20	6	20
Near East and North Africa	3	1	2	100	2	100
Sub-Saharan Africa	33	42	52	24	52	24
All developing countries	148	113	139	23	137	21

Source: Nelson *et al.*, 2009.

As already shown for the yield growth scenarios, the implications of these climate-driven scenarios for child malnutrition outcomes are presented in Table 2.4, which shows the impacts on malnourishment reported by Nelson *et al.* (2009) for the two A2 scenario-based climate outputs from NCAR and CSIRO models. The magnitudes of impacts on the headcount of malnourished children in 2050 are somewhat similar in both models, given that the price impacts of these scenarios were not vastly different. Although the percentage change in malnutrition is largest in the Near East and North Africa region, the sheer numbers of malnourished children in South Asia and sub-Saharan Africa imply greater overall changes in the headcount of malnourished children in these regions. While South Asia is projected to decrease its number of malnourished children by 24 million under a no-climate-change case for 2050, sub-Saharan Africa is likely to undergo an increase of 9 million to 2050, which increases by a further 10 million with climate change. So climate change represents a reversal of trends for South Asia and a deepening of an existing negative trend for sub-Saharan Africa. However, as already described, IMPACT's method for calculating malnutrition changes resulting from climate outcomes uses only changes in calorie availability, which is only one component of the food security determinants used elsewhere. This demonstrates how important it is to keep the other important socio-economic components of household food security on track (education, and access to water, sanitation and health services), if developing regions are to avoid being seriously derailed by the additional stress that global climate change poses to food futures. The effects of climate on these other non-calorie-based outcomes are not modelled here, but are worth further scrutiny, as the provision of services by climate-strained economies could be a significant factor in determining future welfare outcomes for vulnerable populations.

Implications for food security

This section discusses the implications of the scenario results in the context of the current global food situation, particularly the implications for household-level welfare.

Price changes in food and energy markets influence households, directly through market prices, or indirectly via the costs of production or transportation of other marketed goods. Net sellers and net buyers are affected differently, and although net sellers gain from price increases, their gains may not be enough to offset the negative impacts that net buyers undergo. FAO data show that in some of the poorest countries, a relatively small share of households are net sellers of the staple foods that are experiencing the strongest price effects. For example, slightly less than 16 percent of all the households in Bangladesh are net sellers of staples, according to data for 2000, compared with slightly more than 40 percent in Viet Nam in 1998 (FAO, 2008). Developing countries such as Madagascar, where almost 51 percent of all households were net sellers in 1993, are unusual, compared with countries such as Guatemala and Malawi, with slightly more than 10 percent in 2000 and almost 12 percent in 2004, respectively.

A recent paper by Ivanic and Martin (2008) shows that the impacts of high food prices had a differential effect on poverty rates and incidence, depending on the net seller or net buyer position of households. This analysis found that a country such as Viet Nam could (and probably did) experience a net reduction in poverty rates, because increased rice prices put rural households that were net sellers into a much better position than before. Peru too might experience poverty reductions, because increased maize prices would favour rural households that were net sellers. The benefits in Madagascar would arise from maize and dairy prices, and those in Pakistan from rice, dairy and wheat. The impacts therefore vary according to region and commodity, depending on the structure of the national economy concerned, particularly the agricultural economy. Most of the positive benefits that Ivanic and Martin document are in rural areas, while urban households tend to bear the negative impacts of higher prices, across the board. The Ivanic and Martin study also accounts for wage effects, which will be more pronounced (and positive) for rural households that sell their labour within the agriculture sector.

The means by which households adjust their production and consumption in response to economic shocks are shown in Figure 2.13, which illustrates the various dimensions that can be adjusted. Given that a number of expenses may be quasi-fixed, such as rent (especially for urban dwellers), more adjustment has to come from the food consumption side, often leading to poorer diets and lower levels of essential nutrient intake. Households with other assets can disinvest, to the extent possible, to smooth consumption in the short term. Often, however, these disinvestments are not reversed when economic conditions ease, resulting in reduced endowments and enhanced vulnerability to future shocks. The tendency to pull children, especially girls (Schultz, 2002), out of school in times of hardship leads to longer-term effects arising from decreased investments in human capital and reduced earning capacity and productivity in the future.

It might be argued that although biofuels cause increases in food prices, they could lower the costs of energy to households, thereby generating some benefits that would not otherwise occur. The specific outcome depends on the shares of household income going to food and energy purchases, which vary by income level. The available data on household-level expenditure patterns show that households on or below the poverty line tend to spend more than 50 percent of their incomes on food, and a far smaller share on energy (Ahmed, Hill and Wiesmann, 2007).

The evidence and experimental results presented in this chapter give rise to a number of policy recommendations for addressing the world food situation and its implications for current and future levels of human welfare. Some of these recommendations are of a technological nature, while others pertain more to policy-level interventions, at both the national and global levels.

Regarding specific technological interventions for addressing the declines in productivity of key staple crops that have been observed, a wide range of improved crop varieties can be adopted in regions that rely mostly on traditional lower-yielding varieties. Some varietal improvement is necessary, just to maintain yields at their current levels in the face of increasingly adverse environmental conditions, such as those brought on by elevated temperature levels, decreased rainfall or increased incidence of crop pests and diseases (which often move over space as a result of changes in temperature and rainfall conditions). One agricultural technology that was instrumental in allowing the South Asian green revolution to take off was irrigation, which faces drastic underinvestment in some regions, such as sub-Saharan Africa. However, increases in irrigation would have to be accompanied by corresponding investments in installing adequate drainage facilities, to avoid problems of salinity. In regions with (increasing) levels of soil salinity, improved drainage might also have to be accompanied by the adoption of more salt-tolerant crop varieties, to maintain yields at the levels needed for future supply growth.

Figure 2.13
Elements of household income and expenditure that can be adjusted in times of hardship

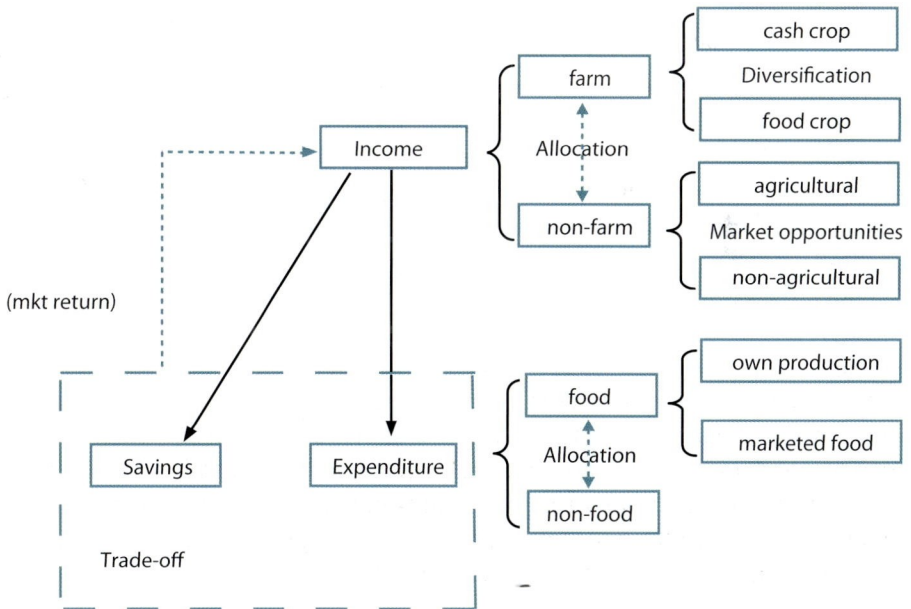

Source: Authors.

Policy interventions related to the use of agricultural feedstocks for first-generation, conventional biofuel production include limiting or even avoiding the use of food crops to produce biofuels such as ethanol and biodiesel. A variety of policy instruments support biofuel production, including direct support to biofuel producers and blenders, the setting of national blending targets or mandates, and trade instruments, which might raise the barriers for biofuel imports from some regions (or encourage exports from others). Technology adoption will largely continue to be driven by private industry, but can be helped from the policy level by increased spending on R&D aimed at pushing forward the next generation of conversion technologies and feedstocks. While a number of trade-related policy instruments need to be addressed at the country level, there is also a need for policy (and political) coordination at the global level, to effect multilateral agreements leading towards the liberalization of international trade. Trade policy has a large influence on biofuel trade and prices, through feedstocks and, even more so, the trade of biofuels themselves. In practice, allowing freer trade in ethanol makes it easier to replace gasoline with renewable fuels whenever energy prices rise. In addition, poorly designed tariffs, tax credits, subsidies and mandates can lead to perverse effects, such as the possibility of actually increasing fossil fuel consumption, as noted by De Gorter and Just (2007).

Regarding social protection of the most vulnerable sections of the population, much can be accomplished through policy-driven strengthening of national social safety net programmes that provide relief for those who are most threatened by escalating food prices, while avoiding blanket policies such as price controls, which are easier (and cheaper) for governments to enact but which have the perverse effect of reducing producer responses that could soften the price rises through increased outputs. In this case, the main challenge facing policy is to keep a balance between maintaining producer incentives and avoiding the distortions that could dampen the necessary self-correcting responses, while supporting human welfare through protecting the most vulnerable. The directing of interventions to those most in need requires deliberate and careful policy design, and this is often lacking in indiscriminate food subsidy schemes, which although they might benefit a lot of the poor (especially when they are the main consumers of the targeted staples), may also benefit better-off households that have other degrees of adjustment (or assets) to exploit.

Conclusions

This chapter has explored several key drivers of change in food systems and examined some possible entry points for policy interventions, determining their effects on food prices and other market-driven outcomes. Among the drivers of change discussed are policy-driven growth in biofuel production, which has played a role in the rapid increase in food prices, along with other factors such as global climate change. The chapter has demonstrated the offsetting impact that supply growth could have on the socio-economic impacts of biofuels, in terms of both price changes and changes in nutrition status. The chapter has also emphasized the need to be aware of all the components of food security – and not merely to focus on food production and output – to maintain progress towards reduced levels of malnutrition and improved human well-being.

Certain policy responses should be avoided when dealing with high prices. These include export bans (akin to a "starve-your-neighbour" policy), import subsidies, restoration of production subsidies, subsidies for the vocal middle class, policing and threatening traders, and attempting to curb food price inflation with macroeconomic policies. On the other hand, three broad policy areas represent desirable and effective tools in fighting the challenges and negative side-effects of high food prices: trade, agricultural growth, and protection of the vulnerable.

The pressures of high food prices can be alleviated by eliminating trade barriers and export bans, and making it easier for international institutions to raise the financing and mobilize the resources needed to effect emergency food imports for the neediest countries. Agricultural growth can be revitalized by expanding aid for rural infrastructure, services, agricultural research and technology. The vulnerable can be shielded from the worst effects of high food prices by expanding food and nutrition-related aid, including safety nets, child nutrition and employment programmes.

In summary, a two-track approach is needed in developing countries. It should include global and national food, health and nutrition security initiatives focusing on the vulnerable, and an agricultural productivity initiative focusing on small farmers.

Combining quantitative experiments with evidence from other studies, the chapter suggests a range of policy interventions that could be instrumental in offsetting the negative impacts of food prices and helping to promote benefits in situations where they exist, to encourage increased investments in the agriculture sector, and to reverse the steadily declining trend of R&D spending and decades of counterproductive agricultural trade and national-level sector policy.

References

Abbot, P.C., Hurt C. & Tyner, W.E. 2008. *What's driving food prices?* Issue Report. Oak Brook, Illinois, USA, Farm Foundation.

Ahmed, A.U., Hill R.V. & Wiesmann, D.M. 2007. *The poorest and hungry: Looking below the line.* 2020 Focus Brief on the World's Poor and Hungry People. Washington, DC, IFPRI.

Al-Riffai, P., Dimaranan, B. & Laborde, D. 2010. *Global trade and environmental impact study of the EU biofuels mandate.* Specific Contract No. SI2.537.787 implementing Framework Contract No TRADE/07/A2. (Final Report March 2010). Washington, DC, IFPRI.

Alston, J. & Pardey, P.G. 2006. Developing-country perspectives on agricultural R&D: New pressures for self-reliance? *In* P.G. Pardey, J.M. Alston and R.R. Piggott, eds. *Agricultural R&D in the developing world: Too little, too late?* Washington, DC, IFPRI.

Cline, W. 2007. *Global warming and agriculture: Impact estimates by country.* Washington, DC, Center for Global Development. 250 pp.

de Gorter, H. & Just, D.R. 2007. *The economics of US ethanol import tariffs with a consumption mandate and tax credit.* Department of Applied Economics and Management Working Paper No. 2007-21. Ithaca, New York, Cornell University. http://papers.ssrn.com/sol3/papers.cfm?abstract_id=1024532.

Delgado, C.L., Rosegrant, M.W., Steinfeld, H., Ehui, S. & Courbois, C. 1999. *Livestock to 2020. The next food revolution.* 2020 Vision for Food, Agriculture, and the Environment Discussion Paper No. 28. Washington, DC, IFPRI.

Dixon, J., Gulliver, A. & Gibbon, D. 2001. *Farming systems and poverty: Improving farmers' livelihoods in a changing world.* Washington, DC, World Bank, and Rome, FAO.

Easterling, W.E., Aggarwal, P.K., Batima, P., Brander, K.M., Erda, L., Howden, S.M., Kirilenko, A., Morton, J., Soussana, J.-F., Schmidhuber, J. & Tubiello, F.N. 2007. Food, fibre and forest products. Contribution of Working Group II to the Fourth Assessment Report of IPCC. *In* M.L. Parry, O.F. Canziani, J.P. Palutikof, P.J. van der Linden and C.E. Hanson, eds. *Climate change 2007: Impacts, adaptation and vulnerability*, pp. 273–313. Cambridge, UK, Cambridge University Press.

Evans, A. 2008. *Rising food prices: Drivers and implications for development.* Briefing Paper No. 08/02. London, Chatham House.

FAO. 2005. *Impact of climate change, pest and disease on food security and poverty reduction.* Background Paper for 31st Session of the Committee on World Food Security, 23–26 May 2005. Rome.

FAO. 2006a. *The State of Food Insecurity in the World 2006: Eradicating world hunger – taking stock ten years after the World Food Summit.* Rome.

FAO. 2006b. *World agriculture: towards 2030/2050.* Rome.

FAO. 2008. Soaring food prices: Facts, perspectives, impacts and actions required. Paper prepared for the High-Level Conference on World Food Security: The Challenges of Climate Change and Bioenergy. HLC/08/INF/1. Rome.

Fischer, G., Shah, M., Tubiello, F.N. & van Velhuizen, H. 2005. Socio-economic and climate change impacts on agriculture: an integrated assessment, 1990–2080. *Phil. Trans. R. Soc., B* 360: 2067–2083.

Headey, D.C. 2010. *Rethinking the global food crisis: The role of trade shocks.* IFPRI Discussion Paper No. 958. Washington, DC, IFPRI.

Headey, D.C. & Fan, S. 2010. *Reflections on the global food crisis: How did it happen? How did it hurt? And how can we prevent the next one?* IFPRI Research Monograph No. 165. Washington, DC, IFPRI.

IAASTD. 2009. *Agriculture at a crossroads.* Washington, DC, Island Press.

Ivanic, M. & Martin, W. 2008. *Implications of higher global food prices for poverty in low-income countries.* Policy Research Working Paper No. 4594. Washington, DC, Development Research Group of the World Bank.

McPhail, L.L. & Babcock, B.A. 2008. *Ethanol, mandates, and drought: Insights from a stochastic equilibrium model of the US corn market.* Working Paper No. 08-WP 464. Ames, Iowa, USA, Center for Agricultural and Rural Development, Iowa State University.

Meyers, W.H. & Meyer, S. 2008. *Causes and implications of the food price surge.* Columbia, Missouri, USA, Food and Agricultural Policy Research Institute, University of Missouri.

MEA. 2005. *Ecosystems and human well-being: Scenarios. Findings of the scenarios working group.* Washington, DC, Island Press.

Nelson, G.C., Rosegrant, M.W., Koo, J., Robertson, R., Sulser, T.B., Zhu, T., Ringler, C., Msangi, S., Palazzo, A., Batka, M., Magalhaes, M., Valmonte-Santos, R., Ewing, M. & Lee, D. 2009. *Climate change: Impact on agriculture and costs of adaptation.* IFPRI Food Policy Report. Washington, DC, IFPRI.

Nelson, G.C., Rosegrant, M.W., Palazzo, A., Gray, I., Ingersoll, C., Robertson, R., Tokgoz, S., Zhu, T., Sulser, T.B., Ringler, C., Msangi. S. & You, L. 2010. *Food security, farming and climate change to 2050: Scenarios, results, policy options.* IFPRI Research Monograph. Washington, DC, IFPRI.

OECD. 2008. *Rising food prices: Causes and consequences.* Paris.

OECD/FAO. 2008. *Agricultural outlook 2008–2017.* Paris.

Oxfam International. 2008. *Another inconvenient truth: How biofuels policies are deepening poverty and accelerating climate change.* Briefing Paper No. 114. Oxford, UK.

Parry, M., Rosenzweig, C. & Livermore, M. 2005. Climate change, global food supply and risk of hunger. *Phil. Trans. R. Soc.*, 360: 2125–2138.

Pinckney, T.C. 1989. *The demand for public storage of wheat in Pakistan.* Research Report No. 77. Washington, DC, IFPRI.

Rosegrant, M.W., Cai, X. & Cline, S. 2002. *World water and food to 2025: Dealing with scarcity.* Washington, DC, IFPRI.

Rosegrant, M.W., Paisner, M.S., Meijer, S. & Witcover, J. 2001. *Global food projections to 2020: Emerging trends and alternative futures.* Washington, DC, IFPRI.

Rosegrant, M.W., Cline, S.A., Li, W., Sulser, T.B. & Valmonte-Santos, R. 2005. *Looking ahead: Long-term prospects for Africa's agricultural development and food security.* 2020 Discussion Paper No. 41. Washington, DC, IFPRI.

Rosegrant, M.W., Zhu, T., Msangi, S. & Sulser, T.B. 2008. *The impact of biofuels on world cereal prices.* Background brief in support of testimony to United States Congressional briefing on the world food situation. Washington, DC, IFPRI.

Rosegrant, M.W., Ringler, C., Sulser, T.B., Ewing, M., Palazzo, A., Zhu, T., Nelson, G.C., Koo, J., Robertson, R., Msangi, S. & Batka, M. 2009. *Agriculture and food security under global change: Prospects for 2025/2050.* Background paper written in support of the CGIAR Strategy and Results Framework. Washington, DC, IFPRI.

Runge, C.F. & Senauer, B. 2008. How ethanol fuels the food crisis. Update 28 May 2008, *Foreign Affairs.* www.foreignaffairs.org/20080528faupdate87376/c-ford-runge-benjamin-senauer/how-ethanol-fuels-the-food-crisis.html.

Schmidhuber, J. 2006. Impact of an increased biomass use on agricultural markets, prices and food security: A longer-term perspective. Paper prepared for the International Symposium of Notre Europe, 27–29 November 2006. Paris.

Schmidhuber, J. & Tubiello, F. 2007. Global food security under climate change. *Proceedings of the National Academy of Sciences*, 104(5): 19703–19708.

Schultz, T.P. 2002. Why governments should invest more to educate girls. *World Development*, 30(2): 207–225.

Smith, L. & Haddad, L. 2000. *Explaining child malnutrition in developing countries: A cross-country analysis.* IFPRI Research Report. Washington, DC, IFPRI.

Trostle, R. 2008. *Global agricultural supply and demand: Factors contributing to the recent increase in food commodity prices.* WRS-0801. Washington, DC, Economic Research Service, USDA.

UN. 2004. *World population prospects: 2004 revisions.* New York.

UNEP. 2007. *Global Environmental Outlook (GEO4): Environment for Development.* Valletta, Malta, Progress Press.

von Braun, J. 2008. *The world food situation: New driving forces and required actions.* Food Policy Report. Washington, DC, IFPRI.

von Braun, J., Ahmed, A., Asenso-Okyere, K., Fan, S., Gulati, A., Hoddinott, J., Pandya-Lorch, R., Rosegrant, M.W., Ruel, M., Torero, M., van Rheenen, T. & von Grebmer, K. 2008. *High food prices: The what, who and how of proposed policy actions.* Policy Brief. Washington, DC, IFPRI.

World Bank. 2005. *Managing food price risks and instability in an environment of market liberalization.* Report No. 32727-GLB. Washington, DC, World Bank Agriculture and Rural Development Department.

World Bank. 2007a. *Global economic prospects: Managing the next wave of globalization.* Washington, DC, International Bank for Reconstruction and Development/World Bank.

World Bank. 2007b. *World Development Report 2008: Agriculture for development.* Washington, DC, International Bank for Reconstruction and Development/World Bank.

HOW CAN CLIMATE CHANGE AND THE DEVELOPMENT OF BIOENERGY ALTER THE LONG-TERM OUTLOOK FOR FOOD AND AGRICULTURE?

Günther Fischer[1]

Accumulating scientific evidence has alerted international and national awareness to the urgent need to mitigate climate change. Meanwhile, increasing and reoccurring extreme weather events devastate more and more harvests and livelihoods around the world.

Biofuels development has recently received increased attention as a means to mitigate climate change, alleviate global energy concerns and foster rural development. Its perceived importance in these three areas has made biofuels feature prominently on the international agenda. Nevertheless, the rapid growth of biofuel production has raised many concerns among experts worldwide, particularly regarding sustainability issues and the threat posed to food security (FAO, 2008a).

As recent events have shown, a number of factors – including the adoption of mandatory biofuel policies, high crude oil prices, increasing global food import demand, below average harvests in some countries and low levels of world food stocks – have resulted in sudden and substantial increases in world food prices. The consequences have been food riots around the world, from Mexico and Haiti to Mauritania, Egypt and Bangladesh. Estimates indicate that high food prices increased the number of food-insecure people by about 100 million.

This chapter presents an integrated agro-ecological and socio-economic spatial global assessment of the interlinkages among emerging biofuel developments, food security and climate change. Its purpose is to quantify the extent to which climate change and expansion of biofuel production may alter the long-term outlook for food, agriculture and resource availability, based on work

1. The work summarized in this chapter uses the modelling tools and databases developed by the Land Use Change and Agriculture Programme at IIASA. In particular, the chapter benefited from the model and data development and analysis carried out in the frame of a major global study on biofuels and food security (Fischer et al., 2009), commissioned by the Organization of the Petroleum Exporting Countries (OPEC) Fund for International Development (OFID). The author is grateful to colleagues Sylvia Prieler, Eva Hizsnyik, Mahendra Shah and Harrij van Velthuizen for their contributions and comments.

by FAO in its *World agriculture towards 2030/2050* assessment (Chapters 1 and 6 in this volume; FAO, 2006).

The International Institute for Applied Systems Analysis (IIASA) has developed a modelling framework and models to analyse the world food and agriculture system spatially and to evaluate the impacts and implications of agricultural policies. The modelling framework has recently been extended and adapted to incorporate biofuel development issues. A brief summary of the methods and models applied in this study is presented in the following section.

Methodology and data

The modelling framework

The analysis is based on a state-of-the-art ecological-economic modelling approach. The quantified findings of the scenario-based study rely on a modelling framework that includes the FAO/IIASA Agro-Ecological Zone (AEZ) model and the IIASA World Food System (WFS) model. The modelling framework encompasses climate scenarios, agro-ecological zoning information, demographic and socio-economic drivers, and production, consumption and world food trade dynamics (Fischer *et al.*, 2009; 2005). A summary of the main model components is provided in Annex 3.1.

This modelling framework comprises six main elements, as shown in Figure 3.1:

- A storyline and quantified development scenario is selected (usually from the extensive integrated assessment literature) to inform the WFS model about demographic changes in each region and projected economic growth in non-agricultural sectors. It also provides assumptions broadly characterizing the international setting (e.g., trade liberalization, international migration) and the priorities for technological progress. It quantifies selected environmental variables, such as greenhouse gas emissions and atmospheric concentrations of carbon dioxide (CO_2). In this study, it also defines scenarios of demand for first- and second-generation biofuels.

- The emissions pathway associated with the chosen development scenario is used to select from among the available matching published outputs from simulation experiments with general circulation models (GCMs). The climate change signals derived from the GCM results are combined with the observed reference climate to define future climate scenarios.

- The AEZ method is based on a selected climate scenario, estimates the likely agronomic impacts of climate change using a spatial grid of 5′ by 5′ latitude/longitude, and identifies adaptation options.

- Estimated spatial climate change impacts on yields of all crops are aggregated and incorporated into the parameterization of the national crop production modules of a regionalized WFS model.

- The global general equilibrium WFS model – informed by the development storyline and estimated climate change yield impacts – is used to evaluate internally consistent WFS scenarios.

In a final step, the results of the world food system simulations are downscaled to the resource database's spatial grid for quantification of land cover changes and further analysis of the environmental implications of biofuel feedstock production.

Figure 3.1
Framework for ecological-economic world food system analysis

Agro-ecological suitability and land productivity

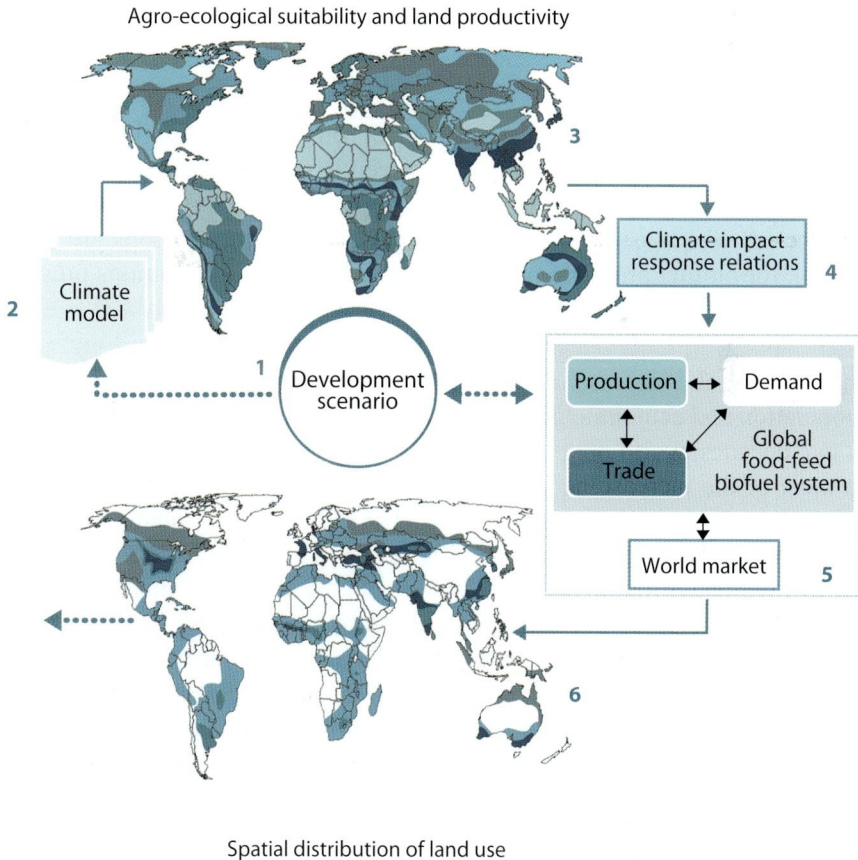

Spatial distribution of land use

Source: Author.

Potential impacts on the production, consumption and trade of agricultural commodities resulting from climate change and/or a rapid expansion of global biofuel use were evaluated in two steps. First, simulations were developed to represent possible futures where biofuel production was abandoned or frozen at current levels (i.e., as in 2008) and kept constant throughout the simulation period. Second, climate change impacts and alternative levels of biofuel demand, derived from different energy scenarios, were simulated with the WFS model and compared with the respective outcomes without additional biofuel demand or climate change.

The primary role of a reference scenario is to provide a neutral point of departure from which various scenarios take off as variants, with the impact of climate change and/or biofuel expansion being defined by the deviation of the simulation run from the outcomes of the reference scenario. The simulations were carried out yearly for 1990 to 2080.

Baseline assessment

Before turning to the impacts simulated for different assumptions regarding biofuel expansion and climate change, this section summarizes the results from a baseline projection. For this neutral point of departure, the FAO-REF-00 scenario[2] was selected, which assumes a system where no agricultural crops are used as feedstock for biofuel production and where current climate conditions prevail.

Population increase and economic growth

In the long run, the increase in demand for agricultural products is largely driven by population and economic growth, especially in developing countries. Over the next two decades, world population growth is projected at about 1 percent a year, with most of the increase being in developing countries. Population increase is an exogenous input in the model analysis. The most recent United Nations (UN) population projections available (UN, 2009), summarized in Table 3.1, were used. Details of regional groupings in the WFS model are shown in Annex 3.2.

Economic performance in the baseline projection for FAO-REF-00 is shown in Table 3.2. For the analysis reported here, the economic growth characteristics were calibrated by country or regional group to match basic assumptions in the FAO perspective study based on information provided by the World Agriculture Towards 2030/2050 study group at FAO (J. Bruinsma, May 2009, personal communication).

---

2. Details of the various scenarios are given in Table 3.26.

Table 3.1
Population development

Region	Total population *(million people)*					
	2000	2010	2020	2030	2040	2050
North America	306	337	367	392	413	430
Europe and Russian Federation	752	762	766	761	748	729
Pacific OECD	150	153	152	148	142	135
Sub-Saharan Africa	655	842	1 056	1 281	1 509	1 723
Latin America	505	574	638	689	725	744
Near East and North Africa	303	370	442	511	575	629
East Asia	1 402	1 500	1 584	1 633	1 630	1 596
South and Southeast Asia	1 765	2 056	2 328	2 553	2 723	2 839
Rest of world[a]	210	233	249	262	272	280
Developed	1 141	1 177	1 202	1 211	1 210	1 198
Developing	4 696	5 417	6 132	6 758	7 257	7 627
World	6 047	6 827	7 582	8 231	8 739	9 105

[a] The regionalization used in the WFS model is described in Annex 3.2.
Source: UN, 2009.

Table 3.2
GDP at constant 1990 prices, baseline projection FAO-REF-00

Region	GDP *(billion USD at constant 1990 prices)*					
	2000	2010	2020	2030	2040	2050
North America	8 286	10 582	12 427	13 817	15 480	17 050
Europe and Russian Federation	7 502	9 487	11 621	14 037	16 860	19 832
Pacific OECD	3 795	4 304	4 781	5 173	5 534	5 888
Sub-Saharan Africa	238	350	531	808	1 236	1 894
Latin America	1 450	2 014	2 822	4 267	6 284	8 828
Near East and North Africa	597	850	1 212	1 772	2 623	3 845
East Asia	1 596	4 165	8 037	13 106	18 373	24 625
South and Southeast Asia	1 255	2 020	3 136	4 840	7 293	10 139
Rest of world	2 418	3 000	3 640	4 343	5 103	5 913
Developed	19 583	24 372	28 830	33 028	37 875	42 770
Developing	5 135	9 399	15 738	24 795	35 810	49 331
World	27 136	36 771	48 207	62 165	78 788	98 014

Sources: IIASA WFS simulations; FAO-REF-00 scenario, May 2009.

While the recent economic growth rates of more than 8 percent a year in China and India may have been dented by the recent world financial crisis, relatively robust economic growth in China, India and other middle-income developing countries is expected in the next two decades.

Agricultural demand and production

Crop production is driven by developments in yields and crop areas. In many developing countries, the crop yields for most commodities are lower than those attained in developed countries (Table 3.3). At the global level, grain yields increased by an average of about 2 percent a year in the period 1970 to 1990, but this rate has halved since then.

Table 3.3
Total cereal production and consumption: baseline simulation without climate change and biofuel expansion, FAO-REF-00

Region	Cereal production (million tonnes)				Cereal consumption (million tonnes)			
	2000	2020	2030	2050	2000	2020	2030	2050
North America	474	588	645	707	304	354	376	404
Europe and Russian Federation	526	552	575	650	545	590	621	684
Pacific OECD	40	48	49	55	46	50	52	52
Sub-Saharan Africa	76	133	172	265	106	179	233	347
Latin America	130	197	221	269	139	196	227	272
Near East and North Africa	55	82	94	122	99	148	179	234
East Asia	423	525	568	636	461	570	620	677
South and Southeast Asia	345	450	496	573	341	453	494	573
Rest of world	75	94	103	125	103	120	128	146
Developed	1 008	1 149	1 229	1 363	858	945	993	1 072
Developing	1 060	1 425	1 590	1 914	1 183	1 596	1 808	2 171
World	2 143	2 668	2 923	3 402	2 144	2 661	2 928	3 388

Sources: IIASA WFS simulations; FAO-REF-00 scenario, May 2009.

With still considerable population growth in the reference projections of the FAO-REF-00 scenario, total production of cereals increases from 2.1 billion tonnes in 2000 to 2.9 billion tonnes in 2030, and to 3.4 billion tonnes in 2050. While developing countries produced about half the global cereal harvest in 2000, their share in total production increases steadily, reaching 57 percent by 2050. As developing countries' share in global consumption increases from 55 to 64 percent in the reference projection, their net imports of cereals grow over time, from 120 million tonnes in 2000 to about 220 million tonnes in 2030, and to 250 million tonnes by 2050.

Agricultural prices

Real prices of agricultural crops declined by a factor of more than two between the late 1970s and the early 1990s, and then stagnated until about 2002, when food prices started to rise. The long-term trend in declining food prices has had several drivers: population development and slowing demographic growth; technological development and growing input use in agriculture, notably a substantial increase in productivity since the green revolution of the early 1970s; and support policies maintaining relatively inelastic agricultural supply in developed countries.

The index of world food prices increased by 140 percent between 2002 and 2007, primarily as a result of increased demand for cereals and oilseeds for biofuels, low world food stocks, reduced harvests owing to drought conditions in locations such as Australia and Europe, record oil and fertilizer prices, and world market speculation. Since the second half of 2008, agricultural prices have again been decreasing substantially.

The baseline projection of the FAO-REF-00 scenario is characterized by modest increases in world market prices between 2000 and 2050. Table 3.4 shows the projected price indices for crops and livestock products in comparison with 1990 levels for a reference simulation without climate change or the expansion of biofuel production. This is also partly the outcome of an assumed further reduction of agricultural support and protection measures.[3]

Table 3.4
Agricultural prices, baseline projection FAO-REF-00

	Price index *(1990 = 100)*			
Commodity group	2020	2030	2040	2050
Crops	94	99	107	113
Cereals	104	106	114	123
Other crops	90	95	103	108
Livestock products	107	110	115	119
Agriculture	98	102	109	115

Sources: IIASA WFS simulations; FAO-REF-00 scenario, May 2009.

Risk of hunger

In 1970, 940 million people in developing countries – a third of their total population – were regarded as chronically undernourished. Over the following two decades, the number of undernourished people declined by some 120 million, to an estimated 815 million in 1990. The largest reduction occurred in East Asia, where the decline was from 500 million in 1970 to about 250 million in 1990. The numbers of undernourished people increased slightly in South Asia and almost

---

3. Price dynamics depend critically on assumed long-term rates of technological progress in agriculture. The price trends presented here should therefore not be interpreted as predictions of future price development but as a characteristic of the chosen reference simulation.

doubled in sub-Saharan Africa. The total number of undernourished people in developing countries declined further from 815 million in 1990 to 776 million in 2000. During this period, the number of undernourished in sub-Saharan Africa increased from 168 to 194 million. Africa has the highest proportion of undernourished people, at about 35 percent of the total population, compared with about 14 percent in the rest of the developing world (Figure 3.2).

Figure 3.2
Historical trends in numbers of undernourished people, developing countries

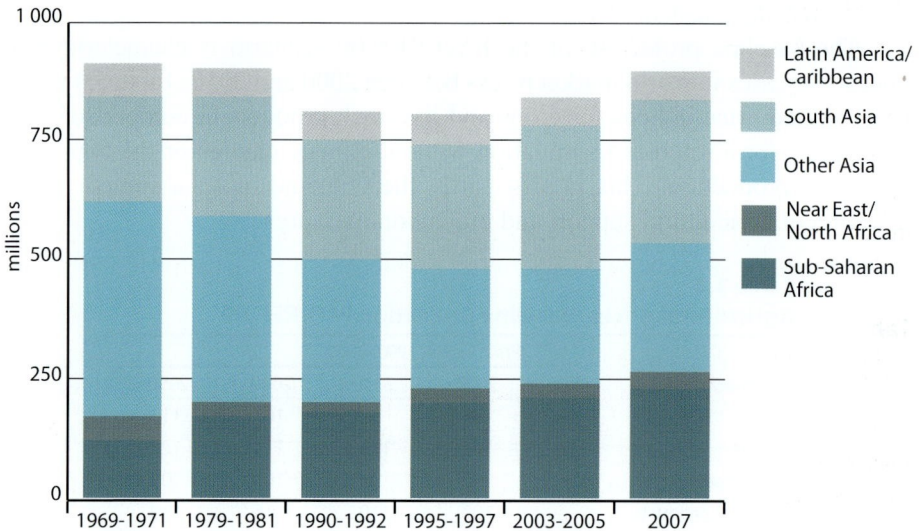

The estimate for 2007 is based on partial data for 2006 to 2008 and uses a simplified methodology, so should be regarded as provisional.
Sources: FAO, 2008b; 2001.

The FAO-REF-00 scenario projects a globally decreasing number of people at risk of hunger (Table 3.5). The projected decrease is most pronounced in East and South Asia. A projected further increase in the number of people at risk of hunger in Africa is expected to result in Africa accounting for 35 percent of the total in 2020 and 40 percent in 2030. Although representing some progress in mitigating hunger, the projected development in scenario FAO-REF-00 is far from sufficient to meet the reductions necessary for achieving the Millennium Development Goal (MDG).

Table 3.5
People at risk of hunger, baseline projection FAO-REF-00

	People at risk of hunger *(millions)*					
Region	2000	2010	2020	2030	2040	2050
Sub-Saharan Africa	196	252	286	271	258	239
Latin America	56	43	31	20	14	10
Near East and North Africa	42	51	57	53	52	47
East Asia	173	139	104	68	42	26
South and Southeast Asia	364	378	362	278	192	136
Developing countries	833	864	839	691	557	458

Sources: IIASA WFS simulations; FAO-REF-00 scenario, May 2009.

Crop and livestock production value added

In the FAO-REF-00 scenario, the global value added of crop and livestock production in 2000 amounts to USD 1 260 billion in 1990 dollars (Table 3.6). This is projected to increase by 30 percent in the 20 years to 2020. In 2030 and 2050 the projected values added amount to respectively USD 1 836 billion and USD 2 192 billion (in 1990 dollars).

Table 3.6
Value added of crop and livestock sector, baseline projection FAO-REF-00

	Value added *(billion USD)*					
Region	2000	2010	2020	2030	2040	2050
North America	166	179	192	203	214	226
Europe and Russian Federation	206	220	235	245	255	264
Pacific OECD	47	52	57	62	67	71
Sub-Saharan Africa	65	82	105	133	165	198
Latin America	155	190	227	262	289	308
Near East and North Africa	55	70	86	104	122	141
East Asia	249	282	314	342	365	384
South and Southeast Asia	252	299	348	400	450	498
Rest of world	65	71	78	85	93	101
Developed	419	451	483	510	535	561
Developing	775	923	1 081	1 241	1 391	1 530
World	1 259	1 445	1 642	1 836	2 019	2 192

Sources: IIASA WFS simulations; FAO-REF-00 scenario, May 2009.

Cultivated land

Some 1.6 billion ha of land is currently used for crop production, with nearly 1 billion ha under cultivation in the developing countries. During the last 30 years, the global crop area expanded by some 5 million ha a year, with Latin America accounting for 35 percent of this increase. The potential for arable land expansion exists predominantly in South America and Africa, where just seven countries account for 70 percent of this potential. There is relatively little scope for arable land expansion in Asia, which is home to some 60 percent of the world's population.

Table 3.7
Cultivated area, baseline projection FAO-REF-00

Region	Cultivated area *(million ha)*					
	2000	2010	2020	2030	2040	2050
North America	234	235	236	237	241	244
Europe and Russian Federation	339	337	336	334	334	334
Pacific OECD	57	57	57	57	60	61
Sub-Saharan Africa	226	245	265	284	301	315
Latin America	175	193	208	217	223	224
Near East and North Africa	67	69	70	72	73	74
East Asia	147	146	146	146	145	145
South and Southeast Asia	274	281	286	289	292	293
Rest of world	42	41	40	38	38	37
Developed	604	602	601	602	606	610
Developing	915	960	1 002	1 035	1 063	1 081
World	1 561	1 603	1 643	1 676	1 707	1 727

Sources: IIASA WFS simulations; FAO-REF-00 scenario, May 2009.

Table 3.8
Harvested area, baseline projection FAO-REF-00

Region	Harvested area *(million ha)*					
	2000	2010	2020	2030	2040	2050
North America	196	203	210	215	223	231
Europe and Russian Federation	215	216	218	219	221	223
Pacific OECD	25	26	27	28	30	31
Sub-Saharan Africa	134	152	174	194	214	231
Latin America	126	143	160	171	179	180
Near East and North Africa	42	46	50	53	56	59
East Asia	220	224	228	231	233	234
South and Southeast Asia	312	327	341	350	356	359
Rest of world	35	35	35	35	35	35
Developed	421	429	438	446	457	468
Developing	850	909	968	1 016	1 055	1 080
World	1 306	1 373	1 441	1 497	1 547	1 583

Sources: IIASA WFS simulations; FAO-REF-00 scenario, May 2009.

Projected global use of cultivated land in the FAO-REF-00 baseline scenario increases by about 165 million ha from 2000 to 2050. While aggregate arable land use in developed countries remains fairly stable, practically all of the net increases occur in developing countries. Africa and South America together account for 85 percent of the expansion of cultivated land (Table 3.7).

Cultivated land represents the physical amount of land used for crop production. In practice, part of this land is left idle or fallow, and part is used to produce more than one crop within a year. The total harvested area in the FAO-REF-00 scenario is shown in Table 3.8. The implied cropping intensity in the baseline projection increases from about 84 percent in 2000 to 89 percent in 2030, and to 92 percent in 2050.

Climate change impacts on crop suitability and production potential

Climate change and variability affect thermal and hydrological regimes, which in turn influence the structure and functionality of ecosystems and human livelihoods.

Scenarios of climate change have been developed to estimate the effects on crop yields, land areas with cultivation potential, and the numbers and types of crop combination that can be cultivated. A climate change scenario is defined as a physically consistent set of changes in meteorological variables, based on generally accepted projections of CO_2 (and other trace gas) levels.

For the spatial assessment of agronomic impacts of climate change on crop yields under the AEZ family of crop models, climate change parameters are computed for each grid point of the resource inventory by comparing the GCM monthly mean prediction for the given decade with those corresponding to the GCM baseline climate of 1960 to 1990. Such changes (e.g., differences in temperature, ratios of precipitation) are then applied to the observed climate for 1960 to 1990 used in AEZ, to generate future climate data as a plausible range of outcomes in terms of likely future temperatures, rainfall, incoming sunlight, etc. for the nominal years 2025 (termed the 2020s), 2055 (the 2050s) and 2085 (the 2080s).

The range of results computed in AEZ refers to different assumptions concerning autonomous adaptation in cropping and the effects of CO_2 fertilization on crop yields (Table 3.9). The first variant is quantified without considering the effects of CO_2 fertilization, and assumes that farmers would be able to change cropping dates and crop types but would be limited to local crop varieties, i.e., those with the same temperature characteristics and moisture requirements as the land utilization types (LUTs) used in the current climate. The second variant refers to results when CO_2 fertilization is still not considered, but best adapted plant types, such as those available elsewhere and adapted to higher temperatures, are available to maximize production potential. Variants 3 and 4 take into account the effects of CO_2 fertilization and quantify the outcomes with limited and full adaptation of crop types, respectively.

The results presented in Table 3.9 are based on a spatial climate change scenario derived from outputs of the United Kingdom's HadCM3 model (Gordon *et al.*, 2000; Pope *et al.*, 2000) for the IPCC SRES A2 emissions pathway (Nakicenovic and Swart, 2000).

Table 3.9
Impacts of climate change on production potential of rainfed wheat on current cultivated land, Hadley A2 (2050s)

		Change with respect to reference climate *(%)*			
Region	Cultivated land *(million ha)*	Without CO_2 fertilization: current crop types	Without CO_2 fertilization: adapted crop types	With CO_2 fertilization: current crop types	With CO_2 fertilization: adapted crop types
North America	230	-9	-9	-3	-3
Europe	179	-4	-4	3	3
Russian Federation	126	-1	-1	5	5
Central America and Caribbean	43	-48	-57	-45	-54
South America	129	-24	-26	-20	-22
Oceania and Polynesia	53	11	12	16	18
North Africa and West Asia	59	-8	-7	-2	-1
North Africa	19	-16	-14	-11	-9
West Asia	40	-4	-4	2	2
Sub-Saharan Africa	225	-56	-61	-54	-59
Eastern Africa	83	-59	-65	-57	-63
Central Africa	38	-76	-80	-75	-80
Southern Africa	17	-44	-47	-41	-44
Western Africa	86	-98	-99	-98	-98
Asia	519	-16	-17	-11	-13
Southeast Asia	98	-55	-58	-53	-56
South Asia	229	-40	-43	-37	-40
East Asia and Japan	151	-8	-9	-3	-5
Central Asia	41	15	15	21	21
Developed	591	-5	-5	1	2
Developing	972	-22	-24	-18	-20
World	1 563	-10	-11	-5	-5

Source: GAEZ 2009 simulations, May 2009.

Except for countries in Central Asia, the impact of climate change on wheat production in developing countries is generally negative. In contrast, the rainfed wheat production potential of current cultivated land in Europe, the Russian Federation and Oceania is increasing. The net global balance is projected to be a reduction of production potential of 5 to 10 percent by the 2050s.

Table 3.10
Impacts of climate change on production potential of rainfed maize on current cultivated land, Hadley A2 2050s

Region	Cultivated land *(million ha)*	Change with respect to reference climate *(%)*			
		Without CO_2 fertilization: current crop types	Without CO_2 fertilization: adapted crop types	With CO_2 fertilization: current crop types	With CO_2 fertilization: adapted crop types
North America	230	-5	-1	-2	2
Europe	179	23	23	28	27
Russian Federation	126	61	61	66	67
Central America and Caribbean	43	1	5	5	9
South America	129	-3	2	0	6
Oceania and Polynesia	53	27	30	31	34
North Africa and West Asia	59	31	30	34	34
North Africa	19	51	52	55	56
West Asia	40	23	22	26	25
Sub-Saharan Africa	225	-6	-3	-3	1
Eastern Africa	83	1	5	5	9
Central Africa	38	-4	1	-1	5
Southern Africa	17	-45	-44	-43	-43
Western Africa	86	-8	-5	-5	-1
Asia	519	-2	2	2	6
Southeast Asia	98	2	6	5	9
South Asia	229	-7	-3	-3	1
East Asia and Japan	151	3	7	7	11
Central Asia	41	23	26	26	30
Developed	591	13	15	17	19
Developing	972	-3	1	1	5
World	1 563	2	5	6	9

Source: GAEZ 2009 simulations, May 2009.

Table 3.10 summarizes the simulated AEZ results for rainfed grain maize. The global production potential of current cultivated land under projected HadCM3 climate conditions for the 2050s increases in all four variants, owing to a modest increase (or an only slight aggregated decrease) in the grain maize potential in developing countries and a significant increase in developed regions. Despite this improvement at the global level, there are several regions where maize production potential decreases, including sub-Saharan Africa.

Table 3.11
Impacts of climate change on production potential of rainfed cereals on current cultivated land, Hadley A2 2050s

Region	Cultivated land (million ha)	Change with respect to reference climate (%)			
		Without CO_2 fertilization: current crop types	Without CO_2 fertilization: adapted crop types	With CO_2 fertilization: current crop types	With CO_2 fertilization: adapted crop types
North America	230	-7	-6	-1	0
Europe	179	-4	-4	3	3
Russian Federation	126	3	3	9	9
Central America and Caribbean	43	-10	-6	-6	-2
South America	129	-8	-3	-4	1
Oceania and Polynesia	53	2	4	6	8
North Africa and West Asia	59	-8	-7	-2	-1
North Africa	19	-15	-13	-10	-8
West Asia	40	-4	-4	1	1
Sub-Saharan Africa	225	-7	-3	-3	1
Eastern Africa	83	-3	2	2	6
Central Africa	38	-7	-2	-3	3
Southern Africa	17	-32	-31	-29	-28
Western Africa	86	-7	-4	-3	1
Asia	519	-3	1	2	5
Southeast Asia	98	-5	-1	-1	4
South Asia	229	-6	-2	-2	2
East Asia and Japan	151	2	6	7	10
Central Asia	41	14	14	19	19
Developed	591	-3	-3	2	3
Developing	972	-5	-2	-1	3
World	1 563	-5	-2	0	3

Source: GAEZ 2009 simulations, May 2009.

The results compiled in Table 3.11 go beyond climate change impacts for single crops. The computations look at all cereal types represented in AEZ (some 118 LUTs covering wheat, rice, maize, barley, sorghum, millet, rye, oats and buckwheat) and determine separately for current and future climate conditions the most productive cereal type in each grid-cell of the spatial resource inventory. Results indicate a somewhat increasing global rainfed production potential, provided that CO_2 fertilization is effective and full adaptation of crop types is achieved; climate change could result in a reduction of about 5 percent of global production if these two conditions are not met. In the latter case, most regions would experience a reduction. At the regional level, results for Southern Africa, North Africa and Central America show the largest negative climate change impacts on rainfed cereal production potential.

Table 3.12 presents results for the temporal dimension of climate change impacts by summarizing simulated results based on HadCM3 for three periods – the 2020s, the 2050s and the 2080s. Numbers shown in the table are the best outcomes of the four variants assuming or not assuming effective CO_2 fertilization and full agronomic crop adaptation.

Table 3.12

Impacts of climate change on production potential of rainfed cereals on current cultivated land with CO_2 fertilization, Hadley A2

| | Change with respect to reference climate *(%)* | | | | | | | | |
| | Rainfed wheat | | | Rainfed maize | | | Rainfed cereals | | |
Region	2020s	2050s	2080s	2020s	2050s	2080s	2020s	2050s	2080s
North America	-1	-3	-2	7	2	-1	1	0	0
Europe	1	3	-1	22	27	21	1	3	-1
Russian Federation	3	5	-1	54	67	63	5	9	6
Central America	-33	-54	-76	6	9	-1	-1	-2	-15
South America	-14	-22	-33	2	6	5	1	1	-1
Oceania and Polynesia	-8	18	9	12	34	58	-7	8	2
North Africa and West Asia	2	-1	-12	19	34	39	2	-1	-11
North Africa	2	-9	-28	38	56	60	2	-8	-23
West Asia	2	2	-6	12	25	31	2	1	-5
Sub-Saharan Africa	-36	-59	-76	1	1	1	1	1	0
Eastern Africa	-38	-63	-81	6	9	11	3	6	9
Central Africa	-53	-80	-95	5	5	5	2	3	2
Southern Africa	-27	-44	-61	-29	-43	-32	-20	-28	-24
Western Africa	-77	-98	-100	1	-1	-6	1	1	-5
Asia	-7	-13	-31	2	6	4	3	5	3
Southeast Asia	-27	-56	-89	4	9	11	2	4	-1
South Asia	-10	-40	-71	1	1	-2	2	2	-1
East Asia and Japan	-9	-5	-16	1	11	12	1	10	12
Central Asia	10	21	9	25	30	16	16	19	11
Developed	0	2	-1	18	19	16	2	3	1
Developing	-11	-20	-36	2	5	3	2	3	0
World	-3	-5	-12	7	9	7	2	3	0

Source: GAEZ 2009 simulations, May 2009.

Results suggest that for the next decades global rainfed cereal production potential is not threatened by a gradual change of climate as projected by the HadCM3 model for the IPCC SRES A2 emissions scenario, provided that CO_2 fertilization effects materialize and farmers are prepared and empowered to adapt fully to a changing climate. It should also be noted that the results in Table 3.12 do not account for the impacts of possibly increased climatic variability.

Table 3.13 presents results for AEZ-estimated rainfed crop potentials of wheat, maize and sorghum (relative to the reference climate), based on the CSIRO GCM climate projections for IPCC A2 emission pathways. Estimates assume full adaptation of crop types and include effects of CO_2 fertilization due to increased atmospheric CO_2 concentrations. Table 3.14 summarizes changes relative to the crop potentials under current climate and excluding CO_2 fertilization effects on crop yield.

Table 3.13

Impacts of climate change on production potential of major rainfed cereals on current cultivated land with CO_2 fertilization, CSIRO A2

| | Change with respect to reference climate (%) | | | | | | | | |
| | Rainfed wheat | | | Rainfed maize | | | Rainfed sorghum | | |
Region	2020s	2050s	2080s	2020s	2050s	2080s	2020s	2050s	2080s
North America	3	10	7	3	9	7	15	25	28
Europe	2	3	-1	40	47	47	31	41	37
Russian Federation	4	4	-15	64	79	69	60	75	70
Central America	-19	-36	-53	2	7	13	3	10	17
South America	-12	-19	-30	2	3	4	8	10	15
Oceania and Polynesia	4	11	4	19	31	57	4	9	7
North Africa and West Asia	2	-1	-12	42	71	69	11	17	13
North Africa	1	4	-18	66	160	183	12	31	20
West Asia	3	-3	-9	33	38	26	11	12	9
Sub-Saharan Africa	-27	-45	-69	0	-2	-7	1	0	-4
Eastern Africa	-30	-48	-72	3	4	-1	4	4	-2
Central Africa	-34	-58	-84	2	2	-1	5	6	7
Southern Africa	-18	-34	-58	-26	-47	-51	-24	-41	-45
Western Africa	-76	-98	-100	0	-1	-7	1	2	-1
Asia	-8	-23	-45	0	1	0	3	5	4
Southeast Asia	-35	-48	-79	0	0	1	-2	-5	-5
South Asia	-22	-45	-70	-1	-3	-5	1	0	0
East Asia and Japan	-7	-21	-38	2	5	2	5	11	11
Central Asia	19	18	-7	34	87	110	27	35	31
Developed	3	7	0	23	30	29	27	38	37
Developing	-10	-23	-42	1	1	0	4	5	4
World	-1	-3	-13	8	10	9	12	16	16

Source: GAEZ 2009 simulations, May 2009.

Table 3.14

Impacts of climate change on production potential of major rainfed cereals on current cultivated land without CO_2 fertilization, CSIRO A2

Region	Change with respect to reference climate (%)								
	Rainfed wheat			Rainfed maize			Rainfed sorghum		
	2020s	2050s	2080s	2020s	2050s	2080s	2020s	2050s	2080s
North America	0	4	-3	2	5	2	12	20	21
Europe	-1	-3	-11	37	42	40	29	35	30
Russian Federation	1	-2	-23	61	73	62	57	68	62
Central America	-21	-39	-57	0	3	7	1	6	11
South America	-14	-23	-36	0	-1	-1	6	6	10
Oceania and Polynesia	2	6	-4	17	27	50	2	6	3
North Africa and West Asia	0	-7	-19	41	66	62	9	14	8
North Africa	-2	-2	-25	64	153	171	10	27	15
West Asia	0	-9	-17	32	34	21	9	8	5
Sub-Saharan Africa	-28	-47	-72	-2	-5	-12	-1	-3	-8
Eastern Africa	-31	-50	-74	1	0	-7	2	0	-7
Central Africa	-35	-60	-85	1	-1	-6	3	3	2
Southern Africa	-20	-37	-61	-27	-49	-54	-25	-43	-48
Western Africa	-76	-98	-100	-1	-5	-12	0	-2	-6
Asia	-10	-27	-49	-1	-2	-5	2	1	-1
Southeast Asia	-36	-51	-80	-2	-4	-4	-4	-7	-9
South Asia	-23	-47	-72	-3	-7	-10	-1	-4	-5
East Asia and Japan	-9	-24	-43	0	1	-3	3	7	5
Central Asia	17	12	-14	33	81	101	25	31	26
Developed	0	0	-10	21	25	22	25	32	30
Developing	-13	-27	-46	-1	-2	-5	2	1	-1
World	-4	-8	-21	6	6	4	10	12	10

Source: GAEZ 2009 simulations, May 2009.

The results of the AEZ analysis using HadCM3 and CSIRO climate projections for IPCC A2 emission pathways suggest three conclusions: i) in a number of regions, climate change poses a significant threat for food production; ii) the global balance of the food production potential from rainfed cereal production on current cultivated land may improve slightly in the short term – farmers' adaptation to a changing climate, and the strength of the CO_2 fertilization effect on crop yields will be decisive factors in realizing a positive global balance of food production potential; and iii) beyond 2050, negative impacts of warming will dominate and cause a rapid decrease of the crop production potential in most regions and for the global aggregate.

In the short term, policy-makers need to strengthen farmers' adaptation capacity and support strategies for coping with climate variability and extreme events that may severely affect the welfare of the most vulnerable populations. In the long run, if climate change is not halted, it will result in irreparable damage to arable land, water and biodiversity resources, with eventually serious consequences for food production and food security.

Impacts of climate change on world food system indicators

The potential impacts of climate change on production and trade of agricultural commodities, particularly cereals, were evaluated in two steps. First, simulations were undertaken in which current climate and atmospheric conditions prevailed. Second, yield impacts due to temperature and CO_2 changes, derived from the agro-ecological assessment, were simulated using the WFS model and compared with the respective outcomes without climate change. Assumptions and results for the reference projection were presented in the section on Baseline assessment.

Data on crop yield changes were estimated with AEZ for different climate change scenarios, and were compiled to provide yield impact parameterizations for the countries or regions covered in the WFS model. Yield variations caused by climate change were introduced into the yield response functions by means of a multiplicative factor with an impact on the relevant parameters in the mathematical representation (i.e., the crop yield functions).

Exogenous variables, population growth and technical progress were left at the levels specified in the respective reference projections. No specific adjustment policies for counteracting altered agricultural performance were assumed beyond the farm-level adaptations resulting from economic adjustments by individual actors in the national models The adjustment processes in the different scenarios are the outcome of the imposed yield changes triggering changes in national production levels and costs, leading to changes in agricultural prices in international and national markets; these in turn affect investment allocation and labour migration among sectors and within agriculture.

Agricultural prices

Table 3.15 summarizes the outcomes of scenario simulations with regard to agricultural prices. It shows the price index deviations, as percentages, relative to the equilibrium prices calculated in the reference projection without climate change. Price indices were calculated for cereals, overall crops and the aggregate of crop and livestock production. Climate scenarios were constructed for both the HadCM3 (Gordon *et al.*, 2000; Pope *et al.*, 2000) and the CSIRO (Gordon and O'Farrell, 1997; Hirst, Gordon and O'Farrell, 1997) GCM model outputs of

IPCC SRES A2 simulations. Table 3.15 gives the results for simulations using the Hadley Centre climate model outputs with and without the effects of CO_2 fertilization on crop yields. Again, the climate scenarios do not take into account the possibility of increased climate variability, and the results assume successful and full agronomic adaptation by farmers.

Table 3.15
Impacts of climate change on agricultural prices

		Change in price index relative to reference climate (%)			
Scenario	CO_2 *fertilization*	2020	2030	2050	2080
Cereals					
Hadley A2	with	-4	-1	-1	23
Hadley A2	without	1	6	10	44
CSIRO A2	with	1	3	2	21
Crops					
Hadley A2	with	-4	-3	-3	11
Hadley A2	without	0	4	7	27
CSIRO A2	with	-1	0	0	9
Agriculture					
Hadley A2	with	-3	-2	-2	8
Hadley A2	without	0	3	5	20
CSIRO A2	with	-1	1	0	7

Source: IIASA WFS simulations, May 2009.

Overall, climate change yield impacts have only a small impact on world market prices in the decades until about mid-century. In fact, the CO_2 fertilization effect and assumed autonomous adaptation to climate change more than compensate for negative yield impacts. Beyond 2050, negative yield impacts dominate and cause price increases, simulated at about 20 percent for cereals in the 2080s. When CO_2 fertilization effects are disregarded, prices start to increase gradually in the early decades, and increases are projected to accelerate after 2050. In this case, medium-term effects on cereal prices are in the order of 10 percent; in the long term – by 2080 – simulated price increases approach 50 percent.

Cereal production and consumption

The impacts of climate change on the production of cereals, resulting from both changes in land productivity and the economic responses of actors in the system, are summarized in Table 3.16.

The model results present a fairly consistent response to climate change and geographical patterns in regional cereal production. At the global level, taking into account economic adjustment of actors and markets, cereal production until 2050 is about 1 percentage point lower than it is when both CO_2 fertilization and agronomic adaptation are considered. For the 2080s the differences exceed 2 percentage points in both the HadCM3 and the CSIRO climate scenarios. When CO_2 fertilization effects are not considered, simulated global cereal production is 1.4 percent lower than the baseline in 2050 and more than 4.3 percent lower in 2080 (representing about 165 million tonnes).

Table 3.16
Impacts of climate change on cereal production

| | Change relative to reference scenario (%) | | | | | | | | | | | |
| | Hadley A2 | | | | CSIRO A2 | | | | Hadley A2, without CO_2 fertilization | | | |
Region	2020	2030	2050	2080	2020	2030	2050	2080	2020	2030	2050	2080
North America	1.9	-2.9	-2.9	-0.8	2.8	0.1	5.8	7.1	0.9	-3.9	-4.6	-4.8
Europe and Russian Federation	0.8	2.0	1.8	1.5	0.5	1.7	1.0	3.1	0.1	1.0	0.1	-1.1
Pacific OECD	-2.2	2.4	9.5	14.0	2.5	6.9	7.0	18.2	-1.8	2.8	9.3	13.6
Sub-Saharan Africa	-1.3	0.3	-2.0	-2.5	-0.6	0.4	-2.9	-7.2	-0.9	0.6	-2.0	-2.2
Latin America	0.9	4.7	5.5	6.0	1.3	3.5	-0.7	0.9	1.3	5.0	6.4	8.0
Near East and North Africa	-0.5	0.7	1.1	-1.0	5.2	7.7	7.4	-1.0	-0.7	0.3	0.3	-2.2
East Asia	0.1	0.7	2.0	-2.8	-2.2	-2.8	-3.4	-7.2	-0.6	-0.4	0.2	-5.3
South and Southeast Asia	-1.3	-1.3	-3.7	-12.2	-4.8	-5.9	-8.9	-12.8	-1.6	-1.9	-4.6	-13.2
Rest of world	-1.6	-1.7	-3.1	-4.6	-2.4	-2.8	-3.4	-4.6	-2.6	-3.4	-6.1	-9.0
Developed	1.2	-0.7	-0.3	0.5	1.7	1.1	4.2	5.9	0.3	-1.7	-2.0	-2.8
Developing	-0.3	0.7	0.2	-3.9	-1.8	-1.8	-4.2	-7.3	-0.6	0.2	-0.6	-4.9
World	0.3	0.1	-0.2	-2.2	-0.4	-0.6	-0.8	-2.1	-0.3	-0.7	-1.4	-4.3

Source: IIASA WFS simulations, May 2009.

Developing countries consistently experience significant reductions of cereal production in all climate scenarios in the long term up to the 2080s. Among the most severely affected regions are South Asia and sub-Saharan Africa.

In the world of the 2050s and 2080s, consumers are assumed to be much richer than they are today and to be largely separated from agricultural production processes. They earn their incomes mainly in the non-agricultural sectors. Therefore, aggregate changes in consumption depend mainly on food prices and income levels rather than on local production conditions. Table 3.17 summarizes the changes in total cereal consumption (food, feed, industrial and seed use, and waste) occurring in the WFS simulations in response to climate change.

Table 3.17 shows a fairly uniform decline in cereal consumption in the 2080s, of about 2 percent globally (representing a reduction of about 80 million tonnes compared with the 3.8 billion tonnes of consumption in the reference simulation) and about 2.5 percent in developing countries, for both climate model scenarios with CO_2 fertilization. In the HadCM3 simulation without CO_2 fertilization effects, the reduction is about 4 percent compared with a reference scenario without climate change.

Table 3.17
Impacts of climate change on cereal consumption

| | Change relative to reference scenario *(%)* | | | | | | | | | | | |
| | Hadley A2 | | | | CSIRO A2 | | | | Hadley A2, without CO_2 fertilization | | | |
Region	2020	2030	2050	2080	2020	2030	2050	2080	2020	2030	2050	2080
North America	0.7	0.3	0.5	-0.4	0.1	-0.3	1.2	1.0	-0.1	-0.8	-1.2	-3.6
Europe and Russian Federation	0.8	0.3	0.1	-1.2	0.1	-0.4	-0.7	-1.4	0.1	-0.6	-1.4	-3.6
Pacific OECD	2.2	0.3	1.5	-4.5	0.3	-1.5	-0.4	-5.0	0.3	-2.1	-3.2	-12.4
Sub-Saharan Africa	0.4	0.1	-0.1	-4.2	-0.2	-0.5	-0.6	-4.0	-0.2	-0.7	-1.4	-6.8
Latin America	0.8	0.3	-0.1	-2.6	0.1	-0.3	-0.5	-2.3	0.1	-0.4	-0.6	-3.4
Near East and North Africa	0.2	0.0	-0.1	-2.6	-0.2	-0.3	-0.2	-2.4	-0.3	-0.7	-1.1	-4.4
East Asia	0.0	-0.1	0.1	-1.0	-0.4	-0.8	-1.4	-0.8	-0.2	-0.4	-0.7	-1.6
South and Southeast Asia	0.0	-1.1	-1.0	-3.9	-0.9	-1.9	-1.5	-3.6	-0.7	-1.9	-2.0	-5.3
Rest of world	0.3	0.0	-0.1	-0.9	-0.1	-0.4	-0.4	-0.9	-0.1	-0.4	-0.7	-1.7
Developed	0.7	0.2	0.2	-1.6	0.0	-0.5	0.1	-0.8	0.0	-0.9	-1.7	-4.7
Developing	0.2	-0.2	-0.2	-2.5	-0.4	-0.9	-1.1	-2.5	-0.3	-0.8	-1.1	-3.8
World	0.4	-0.1	-0.1	-2.1	-0.2	-0.7	-0.7	-2.0	-0.2	-0.8	-1.2	-4.0

Source: IIASA WFS simulations, May 2009.

Risk of hunger

Estimates of the number of people at risk of hunger vary greatly, depending on socio-economic development trajectories (particularly assumed income levels and income distribution) and population numbers. Assumptions and results for the reference simulation were presented in the section on Baseline assessment. Under this reference projection, the estimated number of undernourished would slowly decrease from 2010, to about 900 million in 2020, 760 million by 2030, 530 million by 2050, and 150 million by 2080. For comparison, changes in the estimated numbers of people at risk of hunger, at different time points and under three different climate scenarios, are summarized in Table 3.18. It is worth noting that in these simulations the recorded climate change impacts on undernourishment are relatively small. In the early periods, this is owing to relatively small global yield impacts and small resulting price effects; in the long term, yield impacts become substantial owing to the improved socio-economic conditions and small absolute number of undernourished.

In summary, climate change impacts on agriculture will increase the number of people at risk of hunger. This impact will be of global significance if imposed on an already high level of undernourishment. In the socio-economic development scenario underlying the projections of *World agriculture: towards 2030/2050* (FAO, 2006), with solid economic growth and a transition to stable population levels after 2050, poverty and hunger, although negatively affected by climate change, are a much less ubiquitous phenomenon than they are today.

Cultivated land

The results for changes in cultivated land use are summarized in Table 3.19, and results for impacts on the harvested area are shown in Table 3.20. As for other food system indicators discussed previously, the changes in net cultivated area simulated in response to climate change scenarios up to 2050 are relatively small. Even when CO_2 fertilization effects are not taken into account, the additional land under cultivation globally is less than 10 million ha. Only after 2050, when climate change impacts become increasingly negative for crop yields, the additional land put into production increases compared with the reference climate simulations. In 2080, the estimated increase is 10 to 13 million ha in simulations with CO_2 fertilization effects accounted for, and 26 million ha in those without CO_2 fertilization. It should be noted that these estimated changes are net global effects and should not be confused with gross land conversion, which can be expected to be a lot higher in response to climate change impacts and adaptation efforts.

Impacts of biofuel expansion on world food system indicators

Biofuels, mainly ethanol and biodiesel, are produced from a number of agricultural crops that are also important for the provision of food and feed. At present, biofuel production is spreading around the world in a growing number of countries.

Several developed countries have embraced the apparent win-win opportunity to foster the development of biofuels in response to the threats of climate change, to lessen their dependency on oil and contribute to enhancing agriculture and rural development. Of course, these issues are also of concern to developing countries, where more than 70 percent of the poor reside in rural areas. Countries such as the United States of America, European Union (EU) countries, China, India, Indonesia, South Africa and Thailand have all adopted policy measures and set targets for the development of biofuels.

Table 3.18
Impacts of climate change on risk of hunger

	Change in number of people at risk of hunger relative to reference scenario *(millions)*								
	Hadley A2			CSIRO A2			Hadley A2, without CO_2 fertilization		
Region	2030	2050	2080	2030	2050	2080	2030	2050	2080
Sub-Saharan Africa	0	1	17	1	0	10	4	9	28
Asia	4	-2	5	22	4	3	27	18	14
Rest of world	-2	-2	6	1	0	5	5	9	16
World	1	-3	28	24	4	19	35	36	57

Source: IIASA WFS simulations, May 2009.

Table 3.19
Impacts of climate change on net use of cultivated land

	Change in cultivated area relative to reference scenario *(million ha)*								
	Hadley A2			CSIRO A2			Hadley A2, without CO_2 fertilization		
Region	2030	2050	2080	2030	2050	2080	2030	2050	2080
Sub-Saharan Africa	0	-1	3	1	0	2	1	2	7
Latin America	-1	-2	1	1	1	3	1	3	8
Other developing	0	0	1	0	0	1	1	1	4
Developed	1	1	5	3	3	6	2	2	6
Developing	-2	-4	5	2	1	7	3	5	19
World	-1	-3	10	4	4	13	5	8	26

Source: IIASA WFS, May 2009.

Table 3.20
Impacts of climate change on harvested area

	Change relative to reference scenario *(million ha)*								
	Hadley A2			CSIRO A2			Hadley A2, without CO_2 fertilization		
Region	2030	2050	2080	2030	2050	2080	2030	2050	2080
Sub-Saharan Africa	-1	-2	4	1	0	2	2	2	10
Latin America	-1	-2	1	1	1	4	1	4	10
Other developing	-1	-2	3	-1	-1	1	1	2	9
Developed	-1	0	6	3	3	6	2	4	9
Developing	-3	-5	8	1	0	7	4	8	29
World	-3	-6	14	4	2	14	6	12	39

Source: IIASA WFS, May 2009.

The main driving forces of biofuel expansion have been huge subsidies and the mandates and targets set by national governments. While the justification for biofuel targets as enhancing fuel energy security and contributing to climate change mitigation and agricultural rural development is appealing, the reality is complex as the consequences of biofuel developments result in local, national, regional and global impacts across interlinked social, environmental and economic domains well beyond the national setting of domestic biofuel targets.

The conditioning factors of biofuel development at the national level include the technical capabilities of biofuels as blending agents; agro-ecological conditions and the availability of land resources; the suitability, productivity and production potential of various biofuel feedstocks; the prospects for regional and international trade of biofuels; and the potential greenhouse gas emission savings and climate change mitigation.

Overview of biofuel scenarios

The biofuel scenarios used in the model simulations were designed to cover a wide and plausible range of possible future demand for biofuels. Scenario specification consisted of three steps: first, an overall energy scenario was selected, including the regional and global use of transport fuels as one of its components; second, pathways were identified based on biofuels' role in the total use of transport fuels; and third, assumptions were defined regarding the role and dynamics of second-generation biofuel production technologies, or conversely the fraction of total biofuel production expected to be supplied by first-generation feedstocks based on conventional agricultural crops (maize, sugar cane, cassava, oilseeds, palm oil, etc.). Data on current biofuel feedstock use, and the assumptions and biofuel scenarios used for the scenario analysis are described in detail in Fischer *et al.* (2009).

Future projections of transport fuel use

The World Energy Outlook (WEO 2008) reference scenario published by the International Energy Agency (IEA, 2008b) was used for describing regional energy futures. In this reference scenario, world primary energy demand grows by an average of 1.6 percent per year from 2006 to 2030, rising from 11 730 million tonnes of oil equivalent (TOE) to slightly more than 17 000 million TOE (or by about 45 percent). This projection embodies the effects of government policies and measures enacted or adopted up to mid-2008. The IEA World Energy Model – a large-scale mathematical system designed to replicate how energy markets function – was the principal tool used to generate sector-by-sector and fuel-by-fuel projections by region or country (IEA, 2008b).

World primary oil demand in the WEO reference scenario increases by about 40 percent, from 76.3 million barrels per day in 2000 to 106.4 million barrels per day in 2030. The transport sector consumes about three-quarters of the projected increase in world oil demand (IEA, 2008b).

In terms of total final consumption of transport fuel, the scenario projects an increase from 1 962 million TOE in 2000 to 3 171 million TOE in 2030. Regional totals of transport fuel consumption, derived from the WEO reference scenario for the period 1990 to 2030 and extrapolated to 2050 for use in the WFS simulations, are summarized in Table 3.21.

Table 3.21
Final consumption of transport fuels, WEO scenario

Region	2000	2020	2030	2050
	(million TOE)			
North America	655	773	773	781
Europe and Russian Federation	519	658	652	609
Pacific OECD	105	110	99	93
Africa	45	69	80	122
East Asia	114	337	495	625
South Asia	111	224	322	544
Latin America	149	253	285	332
Near East and North Africa	108	214	259	342
Rest of world	6	16	24	36
Developed	1 236	1 480	1 460	1 417
Developing	576	1 174	1 529	2 068
World[a]	1 962	2 830	3 171	3 750

[a] World totals include international marine bunkers and international aviation.
Source: IEA, 2008b.

Biofuels use and share in total final consumption of transport fuels

The level and regional pattern of total transport fuel consumption presented in the previous subsection has been applied in all the biofuel simulations with the WFS model discussed in this paper. Regarding biofuel use, two alternative scenarios were implemented: i) biofuel expansion based on the WEO 2008 projections; and ii) fast expansion of biofuel production in accordance with the mandates and targets announced by several developed and developing countries. In addition, a number of sensitivity scenarios were specified to gain understanding over a wide range of possible biofuel production levels to 2050.

Biofuel consumption in the WEO scenario: Final demand for biofuels in 1990 was about 6 million TOE, of which two-thirds were produced in Brazil. In 2006, world biofuel consumption reached 24.4 million TOE, with the United States of America as the largest producer and consumer. In the simulation for 2020, final consumption of biofuels in the developed countries is projected at 63 million TOE, with the United States and the EU27 accounting for 90 percent of this use. In 2030, the final consumption of biofuels reaches 79 million TOE in the developed world. For 2030 and 2050, the projections of biofuel consumption in developed countries amount to 79 and 124 million TOE respectively.

Among the developing countries, Brazil has been the pioneer, producing about 5 million TOE in 1990; this is projected to increase to 18 million TOE in 2020. Total biofuel consumption in developing countries starts from about 5.5 million TOE in 2000, increases to 31 million TOE by 2020, and reaches 46 million TOE in 2030. Biofuel use in developing countries in this scenario is dominated by Brazil throughout the projection period. Brazil, China and India together account for about 80 percent of biofuel use in developing countries, a combined share that decreases slightly, to about 75 percent, in 2050. Figure 3.3 shows the dynamics of projected biofuel consumption in the WEO-based scenario; panel A indicates the fuel split, panel B shows the distribution by region.

Figure 3.3
Final consumption of biofuels, WEO scenario

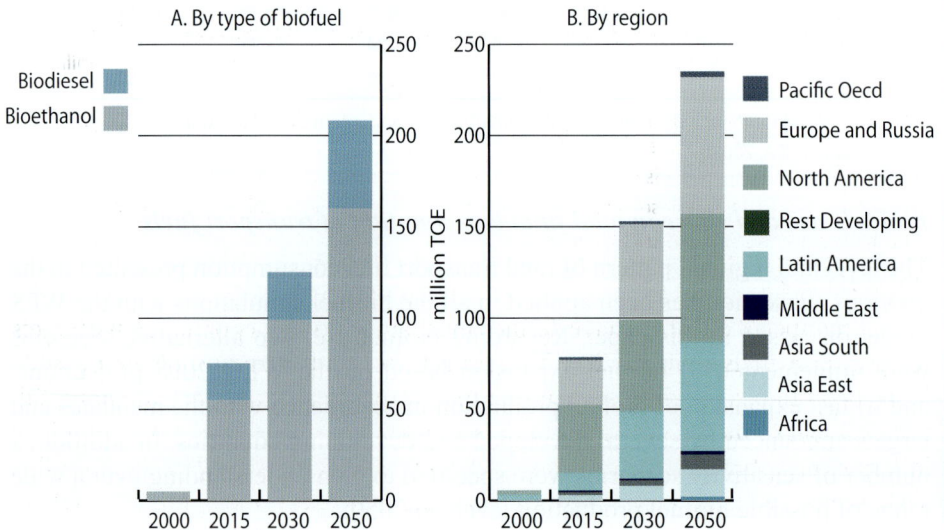

Source: Fischer *et al.*, 2009.

Biofuels consumption in the target (TAR) scenario: The WEO 2008 report states "… assume in the Reference Scenario that the biofuel mandates in China and the European Union will be met after a lag of a few years but that biofuels in the United States in 2030 will attain only about 40 percent of the very ambitious target in the 2007 Energy Independence and Security Act. Asia and OECD Europe experience faster rates of growth, but in absolute terms these increases trail those in the larger North American market. Biofuels demand in the OECD Pacific region remains modest. Growth in Latin America is moderate, a consequence of the sizeable share of the market in Brazil already held by biofuels" (IEA, 2008b: 172).

Table 3.22
Voluntary and mandatory targets for transport fuels in major countries

Country/region	Mandatory, voluntary or indicative target
Australia	At least 350 million litres of biofuels by 2010
Canada	5% renewable content in gasoline by 2010
EU	5.75% of biofuel by 2010, 10% by 2020
Germany	6.25% of biofuel by 2010, 10% by 2020
France	7% of biofuel by 2010, 10% by 2015, 10% by 2020
Japan	0.6% of auto fuel by 2010; fossil oil dependence in the transport sector reduced from 98% to 80% by 2030
New Zealand	3.4% of both gasoline and diesel by 2012
United States	12 billion gallons (55 billion litres) by 2010, 20.5 billion gallons (91 billion litres) by 2015, 36 billion gallons (164 billion litres) by 2022; 16 billion gallons (73 billion litres) from advanced cellulosic ethanol
Brazil	Mandatory 25% ethanol blend with gasoline; 5% biodiesel blend by 2010
China	2 million tonnes of ethanol by 2010, 10 million tonnes by 2020; 0.2 million tonnes biodiesel by 2010, 2 million tonnes by 2020
India	5% ethanol blending in gasoline in 2008, 10% in 2009; indicative target of 20% ethanol blending in gasoline and 20% biodiesel blending by 2017
Indonesia	2% biofuels in energy mix by 2010, 3% by 2015, 5% by 2020
Thailand	2% biodiesel blend by 2008, 10% by 2012; 10% ethanol blend by 2012
South Africa	2% of biofuels by 2013

Source: Fischer *et al.*, 2009.

A number of countries have defined mandatory, voluntary or indicative targets for transport fuels (Table 3.22). To gain a better understanding of the possible impacts on the world food system that may result from implementation and full achievement of the specified targets, a second biofuels scenario, more ambitious in terms of biofuel expansion than the WEO outlook, was implemented. In this TAR scenario, final consumption of biofuels increases to 189 million TOE in 2020 (about twice the value achieved in WEO) and climbs to 295 million TOE in 2030 and 424 million TOE in 2050. As hardly any country has announced biofuel targets beyond about 2020, this scenario should be interpreted as the extension of a rapid and ambitious biofuel development pathway based on the targets announced up to 2020. It approximately doubles biofuel consumption compared with the WEO projections. Figure 3.4 shows the distribution of biofuel consumption by type and region under the TAR scenario.

Figure 3.4
Final consumption of biofuels, TAR scenario

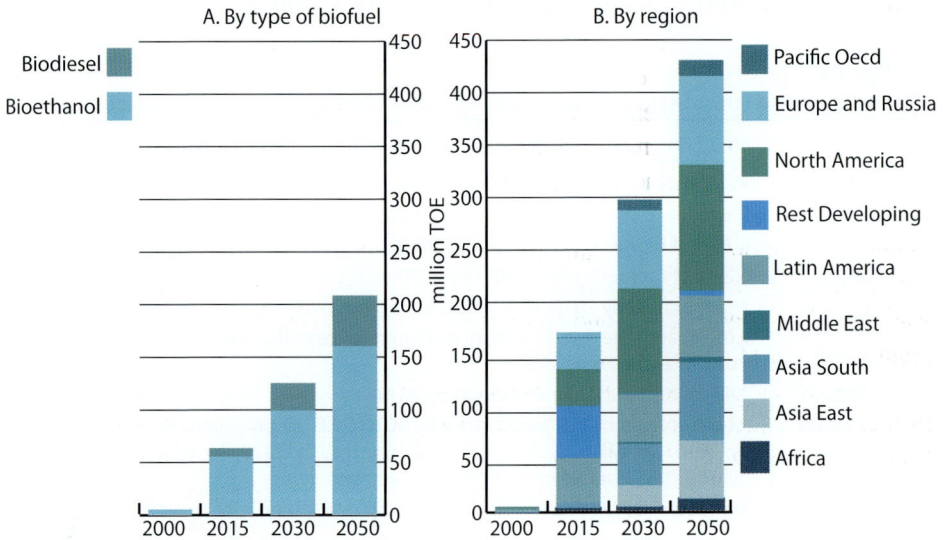

Source: Fischer *et al.*, 2009.

It is worth noting that in the TAR scenario the share of developing countries in total biofuel consumption is higher than in the WEO scenario, owing to the fairly ambitious proposed or announced targets for China, India, Indonesia and Thailand. This change in the regional distribution means that biodiesel's share in total biofuels increases somewhat compared with WEO.

Share of biofuel consumption in total transport fuels: In the developed world, the projected share of biofuel consumption in total transport fuels use in 2020 amounts to 4.3 percent in the WEO scenario. By 2030, this share increases to 5.5 percent. For the developing world, the WEO scenario projects biofuel shares in total transport fuel use of 2.7 percent in 2020 and 3.0 percent in 2030. At the global level, the shares come to 3.5 percent in 2020, 4.2 percent in 2030, and 6.0 percent in 2050. With a road transport share of 70 to 75 percent in total transport fuel use, biofuels would account for 4.5 percent of road transport in 2020, 5.4 percent in 2030, and 7.6 percent in 2050.

Share of second-generation biofuels in total biofuel consumption

In recent years, second-generation biofuels using woody or herbaceous non-food plant materials as feedstocks have attracted great attention because they are seen as superior to conventional feedstocks in terms of their greenhouse gas saving potential, while their potential for production on "non-food" land is seen as being even more valuable.

It is widely acknowledged that major technological breakthroughs will be required to improve feedstock materials and the efficiency of the conversion process before second-generation biofuels can make a significant contribution.

To complete the definition of biofuel scenarios in this assessment, three variants for both the WEO and the TAR biofuel scenarios were specified. These represent alternative views/expectations regarding the dynamics of technology deployment for second-generation fuels. The variants are defined on the basis of different pathways for the share of second-generation fuels in total biofuel consumption. Specification was done by broad regions and follows simple and transparent assumptions. The assumptions used for ethanol are summarized in Table 3.23.

The variant V1 (of both WEO and TAR) assumes that second-generation biofuel technologies will be available for commercial deployment in the United States of America by 2015. Lignocellulose conversion will contribute 7.5 percent of total bioethanol by 2020, rising to 25 percent by 2030. In other OECD countries, it is assumed that second-generation conversion plants will take off from 2020, occupying a share of 12.5 percent by 2030. The biofuel champions among developing countries (Brazil, China and India) will also start using second-generation technologies in 2020, but deployment follows a somewhat slower path to reach only 5 percent of total ethanol in 2030. The V2 variant portrays a delayed development of second-generation technologies. Conversion plants are assumed to become available only by 2030, implying that all transport biofuel production up to 2030 relies on conventional feedstocks.

Scenario variant V3 assumes an early and accelerated deployment of second-generation technologies. In TAR-V3, biochemical ethanol processing and Fischer-Tropsch (FT) diesel plants are already available in 2010, and in OECD countries contribute 10 percent of biofuels by 2015, increasing to more than 30 percent in 2020. In developed countries, second-generation biofuels account for about 50 percent of total biofuels in 2030, and more than two-thirds in 2050. China and India follow this development with a short delay. The shares of second-generation biofuels in these two countries are set at 10 percent in 2020, one-third in 2030, and half in 2050. Other developing countries start deploying second-generation plants in 2020, reaching shares of 10 percent in 2030, and 33 percent in 2050.

At the aggregate global level, second-generation biofuel shares in scenario variant WEO-V1 are 3 percent in 2020, 13 percent in 2030, and 30 percent in 2050. In scenario variant TAR-V1, these shares are respectively 2, 12 and 26 percent – somewhat lower than in the WEO scenario owing to the higher shares achieved by developing countries in the TAR scenario. For variant TAR-V3, with an assumed accelerated second-generation development and deployment path, the respective shares are 22, 38 and 55 percent.

Table 3.23
Shares of second-generation biofuels in total biofuels

Scenario	Region	Share of second-generation biofuels (%)			
		2015	2020	2030	2050
WEO-V1, TAR-V1	United States	Starts	7.5	25	50
	Other OECD	None	Starts	12.5	33
	Russian Federation	None	Starts	5	20
	Brazil/China/India	None	Starts	5	20
	Other developing	None	None	None	None
WEO-V2, TAR-V2	All countries	None	None	Starts	10
WEO-V3	United States	10	24	40	66
	EU27	None	10	33	50
	Other OECD	None	10	33	50
	Russian Federation	None	5	20	40
	China/India	Starts	5	20	40
	Other developing	0	0	10	20
TAR-V3	United States	10	35	55	70
	EU27	10	31	47	67
	Other OECD	10	31	47	67
	Russian Federation	Starts	10	33	50
	China/India	Starts	10	33	50
	Other developing	0	Starts	10	33

Source: Fischer *et al.*, 2009.

Sensitivity analysis of biofuels share in total transport fuels

In addition to the WEO and TAR biofuel scenarios, four sensitivity (SNS) scenarios were computed to scan the WFS model outcomes for a broad range of imposed first-generation biofuel production levels, from 2 to 8 percent in 2020, 2.5 to 10 percent in 2030, and 3 to 12 percent in 2050. Table 3.24 summarizes the assumed shares of first-generation biofuels in total transport fuel use for different scenarios and time points.

Table 3.24
Shares and total amounts of first-generation biofuels

Scenario	Share in total transport fuels (%)			First-generation biofuels consumption (million TOE)		
	2020	2030	2050	2020	2030	2050
SNS-V1	2	2.5	3	54	76	106
SNS-V2	4	5	6	107	151	211
SNS-V3	6	7.5	9	161	227	317
SNS-V4	8	10	12	214	302	423

Source: Fischer *et al.*, 2009.

First-generation biofuel feedstocks demanded in selected biofuel scenarios

Estimates for 2008 indicate that about 80 to 85 million tonnes of cereals were used for ethanol production, mainly maize in the United States of America, and about 10 million tonnes of vegetable oil for biodiesel production, dominated by the EU. In the reference scenario FAO-REF-01, these amounts are kept constant for the simulation period to 2050. The amounts increase in both the WEO and TAR scenario variants. The time path in each scenario variant depends on the level and geographical distribution of biofuel production and on assumptions regarding the availability of second-generation technologies. The amounts of cereals and vegetable oils required for transport biofuel production in 2020, 2030 and 2050 in selected biofuel scenarios are shown in Table 3.25.

Table 3.25
Use of cereals and vegetable oils for biofuel production

Scenario	Cereals (million tonnes)			Vegetable oils (million tonnes)		
	2020	2030	2050	2020	2030	2050
FAO-REF-01	83	83	83	10	10	10
WEO-V1	181	206	246	26	30	44
WEO-V2	192	258	376	26	33	48
TAR-V1	327	437	446	58	85	112
TAR-V3	238	272	262	46	59	61

Source: IIASA WFS simulations, May 2009.

Impacts of first-generation biofuel expansion on food system indicators

This section presents the results of an integrated spatial ecological and economic assessment of the impacts of accelerated expansion of biofuel production, evaluated in the context of the world food economy and global resource base.

Previous sections presented the analysis framework used in this study, and key assumptions regarding economic development and transport energy demand, particularly use of first- and second-generation biofuels. Internally consistent sets of assumptions were formulated as model scenarios and used to quantify the impacts of expanding biofuel use on agriculture and world food system outcomes. A total of ten scenarios were analysed; the acronyms used and a brief description of each are given in Table 3.26.

Table 3.26
Biofuel scenarios analysed in this study

Acronym	Description
FAO-REF-00	Starting in 1990, assumes a world with no agricultural crops used for biofuel production.
FAO-REF-01	Assumes historical biofuel development until 2008; biofuels feedstock demand kept constant after 2008; used as a reference simulation with which alternative biofuel scenarios are compared to identify their impacts.
WEO-V1	Assumes transport energy demand and regional biofuel use as projected by IEA in the WEO 2008 reference scenario. Second-generation conversion technologies become commercially available after 2015; deployment is gradual (Table 3.23)
WEO-V2	Assumes transport energy demand and regional biofuel use as projected by IEA in the WEO 2008 reference scenario. Assumes that delayed arrival of second-generation conversion technologies results in all biofuel production until 2030 being based on first-generation feedstocks.
TAR-V1	Assumes transport energy demand as projected by IEA in the WEO 2008 reference scenario. Assumes that mandatory, voluntary or indicative targets for biofuel use announced by major developed and developing countries will be implemented by 2020, resulting in about twice the biofuel consumption projected in WEO 2008. Second-generation conversion technologies become commercially available after 2015; deployment is gradual (percentages as in WEO-V1).
TAR-V3	Assumes transport energy demand as projected by IEA in the WEO 2008 reference scenario. Assumes that mandatory, voluntary or indicative targets for biofuel use announced by major developed and developing countries will be implemented by 2020. Accelerated development of second-generation conversion technologies permits rapid deployment; 33% and 50% of biofuel use in developed countries from second-generation in 2020 and 2030 respectively.
SNS-V1, V2, V3, V4	Sensitivity scenarios assuming low (V1), intermediate (V2), high (V3) and very high (V4) shares of first-generation biofuels in total transport fuels (Table 3.24).

The impacts of additional demand for first-generation biofuels on production, consumption and trade of agricultural commodities – particularly food staples – were evaluated by comparing the results of a range of biofuel expansion scenarios with a reference projection of the WFS simulated without imposing additional biofuel demand. Results of the reference projection are presented in the section on Baseline assessment.

The biofuel expansion scenarios analysed involved several simulation experiments related to two aspects:

- share of transport energy supplied from biofuels;

- sensitivity of results to development speed of second-generation technologies.

As in climate change analysis, all exogenous variables, such as population growth, technical progress and growth of non-agricultural sectors, are left at the levels in the reference projection. No specific adjustment policies to counteract altered performance of agriculture are assumed beyond the farm-level adaptations resulting from economic adjustments by individual actors in the national models. The adjustment processes in the different scenarios are the outcome of the imposed additional biofuel demand causing changes in agricultural prices in international and national markets; these in turn affect investment allocation and labour migration among sectors and within agriculture. Time is an important aspect in this adjustment process.

Agricultural prices

As expected in a general equilibrium WFS model, when simulating scenarios with increased demand for food staples due to the production of first-generation biofuels, the resulting market imbalances at prevailing prices push international prices upwards.

Table 3.27 shows the results for selected scenarios: biofuel demand according to projections in scenario variants WEO-V1 and WEO-V2 (the latter assuming delayed introduction of second-generation technologies); and high biofuel consumption levels under scenario variants TAR-V1 and TAR-V3 (assuming accelerated introduction of second-generation biofuels).

Table 3.27
Impacts of biofuel expansion on agricultural prices

	Change in price index relative to reference scenario FAO-REF-01 *(%)*								
	Cereals			Crops			Agriculture		
Scenario	2020	2030	2050	2020	2030	2050	2020	2030	2050
WEO-V1	11	5	10	10	7	10	8	5	7
WEO-V2	14	13	21	12	11	15	9	8	11
TAR-V1	38	38	27	35	34	27	27	26	20
TAR-V3	19	17	12	22	18	13	17	12	9
SNS-V1	5	5	7	4	5	6	3	3	4
SNS-V2	21	15	21	17	15	18	13	11	13
SNS-V3	37	35	40	30	29	31	24	22	23
SNS-V4	55	58	60	47	47	47	36	36	35

Sources: IIASA WFS simulations; FAO-REF-01 scenario, May 2009.

For 2020, the price increases for both cereals and other crops under the WEO scenario are in the order of 10 percent. As the contribution of second-generation biofuels is still small in WEO-V1, the further delay assumed in WEO-V2 causes only moderate additional crop price increases. For biofuel demand specified in the TAR scenario (which is about twice that projected in the WEO scenario), the impact on crop prices in 2020 is fairly substantial, at about 35 percent. With accelerated introduction of cellulosic ethanol, as assumed in TAR-V3, the price impact on cereals is halved to about 19 percent.

For 2030, the pattern of price impacts remains similar to that of 2020. As second-generation biofuels gain importance towards 2030, the differences in price impacts between WEO-V1 and WEO-V2 become more visible. With accelerated deployment of second-generation fuels, even the large volumes of biofuels produced in TAR-V3 can be achieved with price increases of only about 15 percent.

Figure 3.5
Cereal price index compared with share of first-generation biofuels in transport fuels, 2020

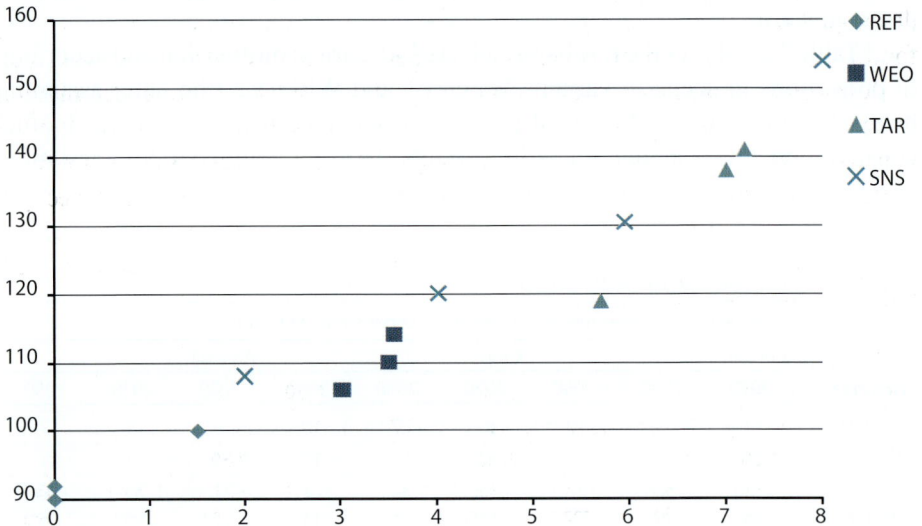

SNS = sensitivity scenarios.
TAR = scenario simulations based on mandates and indicative voluntary targets.
WEO = simulations based on WEO 2008 projections of biofuel demand.
REF = reference projections with constant, decreasing or no biofuel demand beyond 2008.
Source: IIASA WFS simulations, May 2009.

Summarizing these scenario experiments, it emerges that agricultural prices depend considerably on the aggregate share that first-generation biofuels are mandated to contribute to total transport fuel consumption. This is shown in Figure 3.5.

Cereal demand and production

The rising agricultural prices in the biofuel scenarios provide incentives on the supply side for intensifying production and augmenting and reallocating land, capital and labour. At the same time, consumers react to price increases and adjust their patterns of consumption. Figure 3.6 shows the producer responses in cereal sectors under different biofuel scenarios in 2020 and 2030, i.e., the amount of additional cereal production realized in each scenario compared with FAO-REF-01.

In 2020, compared with 83 million tonnes in 2008 under the reference scenario, the additional global demand for cereal commodities for ethanol production is about 100 million tonnes in WEO-V1 and WEO-V2, 240 million tonnes in TAR-V1 and 155 million tonnes in scenario TAR-V3. Figure 3.6 highlights that the production increases in response to higher agricultural prices are greater in developed countries, as are the reductions in feed use (Figure 3.7). Regarding food use, however, consumption in developed countries is less responsive than in developing countries, which account for 75 percent of the forced reduction in cereal food consumption. Rising food commodity prices tend to have greater negative effects on lower-income than higher-income consumers. First, lower-income consumers spend a larger share of their income on food, and second, staple food commodities such as maize, wheat, rice and soybeans account for a larger share of their food expenditure. Responses on the consumer side, with reduced food and feed use of cereals, are shown in Table 3.28.

Table 3.28
Impacts of biofuel expansion on cereal production and demand

	Change relative to reference scenario FAO-REF-00 (*million tonnes*)								
	2020			2030			2050		
Scenario	Biofuel use	Produc-tion	Food/feed	Biofuel use	Produc-tion	Food/feed	Biofuel use	Produc-tion	Food/feed
FAO-REF-00	83	64	-19	83	66	-17	83	68	-15
WEO-V1	181	134	-46	206	167	-45	246	180	-62
WEO-V2	192	140	-48	258	194	-68	376	271	-102
TAR-V1	327	229	-96	437	308	-133	446	313	-127
TAR-V3	238	174	-59	272	201	-69	262	198	-62

Sources: IIASA WFS simulations; FAO-REF-00 scenario, May 2009.

Figure 3.6
Changes in cereal production relative to baseline FAO-REF-01

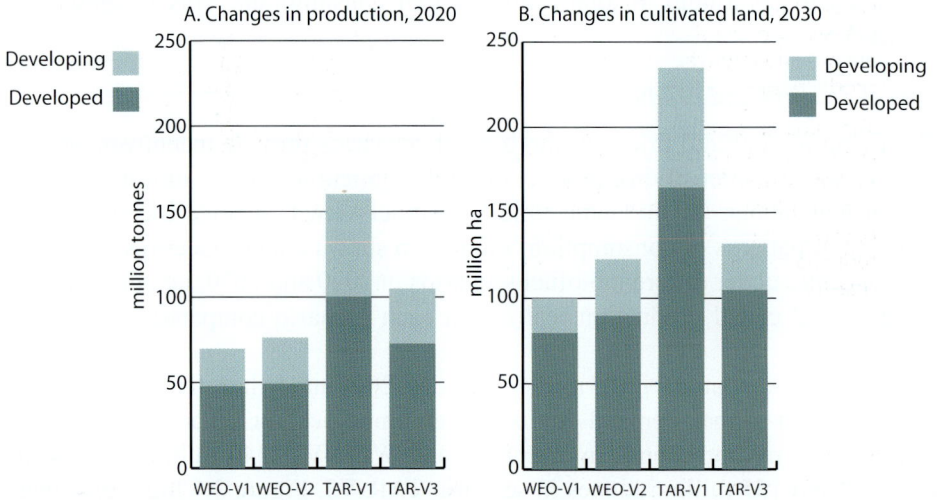

Source: IIASA WFS simulations, May 2009.

Figure 3.7
Changes in cereal use relative to baseline FAO-REF-01, 2020

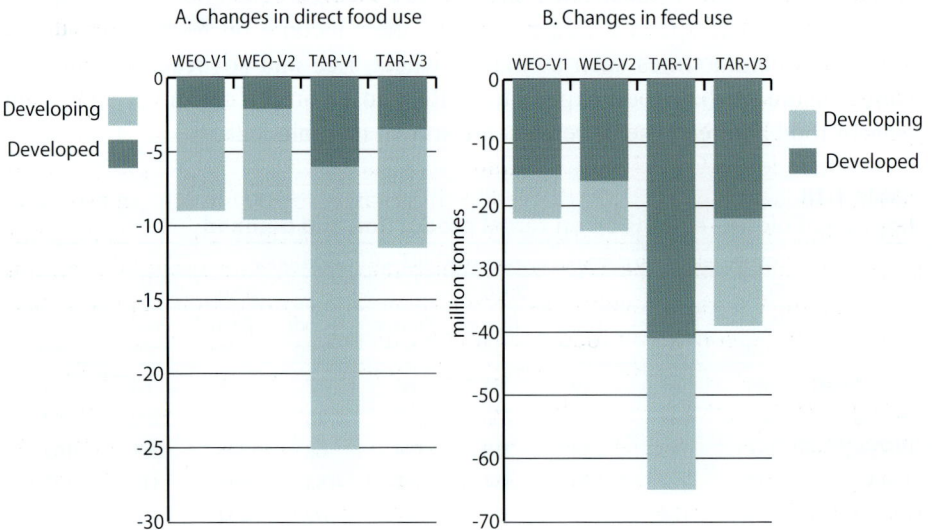

Source: IIASA world food system simulations, May 2009.

Box 3.1 - Where do the cereals needed for biofuel production come from?

On average about two-thirds of the cereals used for ethanol production are obtained from additional crop production.

The remaining one-third comes from consumption changes. The reduction in direct cereal food consumption accounts for 10 percent of the amount of cereals used for biofuel production, reduced feed use accounts for about a quarter.

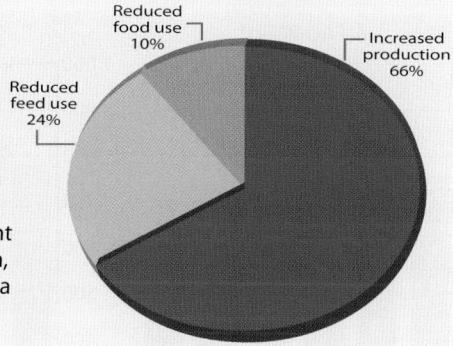

Reduced food use 10%

Reduced feed use 24%

Increased production 66%

Risk of hunger

The estimated number of people at risk of hunger used in the WFS model is based on FAO data (FAO, 2001; 2008b) and relies on a strong correlation between the share of undernourished in a country's total population and the ratio of average per capita dietary food supply to average national per capita food requirements.

The model results show that an ambitious biofuel target for 2020, as specified in the TAR scenario, causes higher prices if achieved mainly by production of first-generation biofuels. This reduces food consumption in developing countries, which results in increased numbers of people at risk of hunger. Figure 3.8 compares results until 2050 for the baseline scenario FAO-REF-01 (with no climate change and no additional biofuel demand after 2008) with the estimated numbers of people at risk of hunger in the TAR-V3 scenario (implementing an ambitious global biofuel target with swift introduction of second-generation technologies).

While in FAO-REF-01 the number of undernourished people peaks in 2009/2010 at somewhat more than 890 million and then declines to an estimated 850 million in 2020, 700 million in 2030 and 460 million in 2050, this indicator stays at a high level in the TAR-V3 scenario until 2020, with about 940 million, and only then starts to decline as second-generation production begins to take pressure off the competing food, feed and biofuel feedstock markets.

Figure 3.9 presents the simulated regional distribution of additional undernourished people under different biofuel scenarios, showing a particularly large impact in South Asia. It is worth noting that even with relatively swift deployment of second-generation technologies, as assumed in TAR-V3, the results for 2020 show an increase of 80 million undernourished people.

Figure 3.8
People at risk of hunger, developing countries

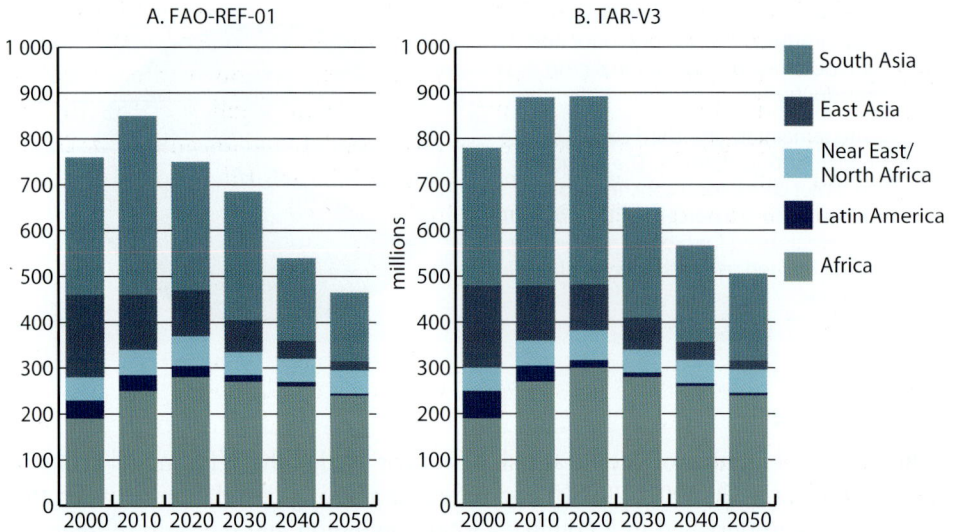

Source: IIASA WFS simulations, May 2009.

Figure 3.9
Additional people at risk of hunger relative to baseline FAO-REF-01

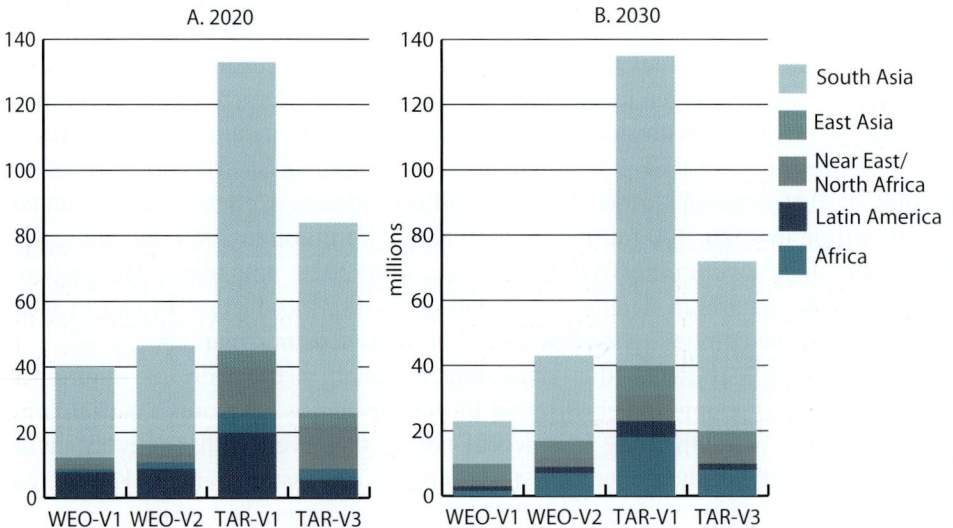

Source: IIASA WFS simulations, May 2009.

The reference scenario FAO-REF-01 (keeping biofuels consumption constant after 2008) projects that the numbers of undernourished people in developing countries will be 850 million in 2020 and 700 million in 2050. The TAR scenario estimates that an additional 132 million people will be at risk of hunger in 2020 and an additional 136 million in 2030. In the TAR scenario with accelerated second-generation biofuels deployment, the corresponding numbers of additional people at risk of hunger decrease to 85 million in 2020 and 74 million in 2030. Africa and South Asia account for more than two-thirds of these people across biofuel scenarios and in both 2020 and 2030.

Crop and livestock production value added

Biofuel development has been seen as a means to diversify agricultural production and – especially in developed economies – this has shaped agricultural support policies. This study considered the extent to which the additional production of crops as feedstocks for biofuel production will increase value added in agriculture. The percentage changes relative to the reference scenario FAO-REF-00, without any biofuels, are shown in Table 3.29.

Table 3.29
Impacts of biofuel expansion on agricultural value added

| | Change relative to reference scenario FAO-REF-00 *(%)* | | | | | | | | |
| | World | | | Developed countries | | | Developing countries | | |
Scenario	2020	2030	2050	2020	2030	2050	2020	2030	2050
FAO-REF-01	1.2	1.2	0.9	2.4	2.9	2.0	0.8	0.7	0.6
WEO-V1	2.5	3.1	3.2	4.3	6.3	5.8	1.8	1.9	2.4
WEO-V2	2.5	3.5	4.0	4.4	7.4	7.8	1.8	2.1	2.9
TAR-V1	4.4	6.6	7.1	6.9	12.1	11.4	3.4	4.4	5.7
TAR-V3	3.7	4.9	4.5	5.7	8.9	7.3	2.9	3.3	3.7

Source: IIASA WFS simulations, May 2009.

Table 3.29 highlights that for all biofuel scenarios, agricultural value added increases at the global and regional levels, as expected. For instance, under WEO-V1 (with relatively modest biofuels development), the changes in absolute terms amount to USD 41 billion in 2020, USD 57 billion in 2030 and USD 71 billion in 2050, in 1990 dollars. Developed countries initially account for about 50 percent of the global gains in agricultural value added. As the relative weight of developed countries in global agriculture decreases over time, so does their share in global gains of agricultural value added, to reach an average of 45 percent in 2050. Table 3.29 shows that agriculture sectors in developed countries also benefit more than those in developing countries in terms of percentage gains relative to the baseline. Under WEO-V1, the increase in 2020 is 4.3 percent for developed countries, compared with only about 1.8 percent for developing countries. While Africa and Latin America achieve gains of 2.4 and 3.1 percent, the gains achieved in the Near East and North Africa and in Asian regions are only 0.9 to 1.9 percent (Table 3.30).

Table 3.30
Impacts of biofuel expansion on regional agricultural value added

Region	Change in agricultural value added relative to reference scenario FAO-REF-00 *(%)*								
	WEO-V1			WEO-V2			TAR-V3		
Region	2020	2030	2050	2020	2030	2050	2020	2030	2050
North America	8.5	11.2	8.6	8.7	13.2	12.8	11.6	14.1	8.6
Europe and Russian Federation	1.8	3.5	4.6	1.7	4.1	5.3	1.9	6.1	7.3
Pacific OECD	0.8	1.6	1.7	0.8	1.4	1.6	1.7	3.0	2.8
Sub-Saharan Africa	2.4	2.4	2.9	2.4	2.6	3.4	4.2	4.8	4.5
Latin America	3.1	3.5	5.2	3.1	3.8	6.4	4.9	5.7	7.8
Near East and North Africa	1.9	2.1	2.7	2.0	2.2	2.9	3.4	3.9	3.6
East Asia	0.9	1.1	1.2	0.9	1.2	1.4	1.3	1.5	1.7
South and Southeast Asia	1.4	1.4	1.4	1.4	1.4	1.5	2.6	2.8	2.3
Rest of world	1.2	1.2	1.1	1.3	1.4	0.5	2.6	3.0	2.4
Developed	4.3	6.3	5.8	4.4	7.4	7.8	5.7	8.9	7.3
Developing	1.8	1.9	2.4	1.8	2.1	2.9	2.9	3.3	3.7
World	2.5	3.1	3.2	2.5	3.5	4.0	3.7	4.9	4.5

Source: IIASA WFS simulations, May 2009.

In the TAR-V1 scenario, with a high demand for first-generation biofuels due to high national targets and only gradual introduction of second-generation technologies, agricultural value added increases substantially, by some 6.6 percent globally by 2030. Global agricultural value added increases by USD 73 billion in 2020, USD 120 billion in 2030 and USD 155 billion in 2050, in 1990 dollars. Again, the percentage gains in TAR-V1 are higher for developed countries (averaging about 6.9 percent in 2020) than developing regions (3.4 percent), where estimated gains fall in a range of 1.7 to 5.7 percent. For developed countries, the TAR-V1 scenario estimates the increases in agricultural value added (in 1990 dollars) resulting from biofuel development at USD 33 billion in 2020 and USD 62 billion in 2030. The corresponding values for developing countries are USD 37 billion in 2020 and USD 51 billion in 2030.

Impacts on the use of cultivated land

Discussion of the extent and kind of land required for biofuel production and of the impacts on cultivated land caused by expanding biofuel production distinguishes two elements: estimating direct land-use change from the extent of land used for producing biofuel feedstocks; and estimating the indirect land-use effects that can result from bioenergy production displacing services or commodities (food, fodder, fibre products) from arable land currently in production.

This study applies a general equilibrium framework that can capture both direct and indirect land-use changes by modelling the responses of consumers and producers to price changes induced by introducing competition with biofuel feedstock production. This approach accounts for land-use changes while also considering production intensification on existing agricultural land and consumer responses to changing availability and prices of agricultural commodities.

In a baseline projection without any use of agricultural feedstocks for biofuel production, as portrayed in the FAO-REF-00 scenario, the expansion of arable land to meet growing food and feed requirements between 2000 and 2020 amounts to about 80 million ha. Africa and Latin America, with projected increases in cultivated land of 39 and 33 million ha respectively, account for more than 85 percent of total net arable land expansion.

Table 3.31
Impacts of biofuel expansion on cultivated land use

	Change in cultivated land relative to reference scenario FAO-REF-00 *(%)*								
	World			Developed countries			Developing countries		
Scenario	2020	2030	2050	2020	2030	2050	2020	2030	2050
FAO-REF-01	8	8	5	3	3	1	5	5	4
WEO-V1	19	19	21	6	6	5	12	13	16
WEO-V2	20	23	29	6	8	7	13	15	21
TAR-V1	38	46	48	12	14	11	24	30	36
TAR-V3	29	30	29	9	9	6	19	20	22

Sources: IIASA WFS simulations; FAO-REF-00 scenario, May 2009.

Table 3.31 shows the additional use of cultivated land in 2020, 2030 and 2050 in comparison with a scenario without any crop-based biofuels. For the WEO and TAR biofuel scenarios shown, this additional use in 2020 falls in the range of 19 million ha (WEO-V1) to 38 million ha (TAR-V1). For developed countries, the arable land-use increases in different biofuel scenarios during 2000 to 2020 are in the range of 6 to 12 million ha, compared with a net decrease of 3 million ha in a scenario without biofuels. In the baseline without biofuels (FAO-REF-00), the increase in arable land use between 2000 and 2020 amounts to 87 million ha; for comparison, additional crop demand with biofuel development results in a total expansion of cultivated land use of 99 to 112 million ha, and additional use of 12 to 24 million ha. The difference of 24 million ha of arable land use in developing countries in the TAR-V1 scenario (compared with results without biofuel demand) is mainly explained by additional expansions of 9 million ha in sub-Saharan Africa and 11 million ha in South America.

When looking at differences in the expansion of cultivated land for the period 2000 to 2030, the range of estimates for biofuel scenarios widens further, from an additional 19 million ha (WEO-V1) to 46 million ha (TAR-V1).

Across the full range of simulated scenarios (including SNS scenarios), the use of cultivated land in 2020 ranges from 1 643 to 1 691 million ha, a difference of 48 million ha. In 2030, it ranges from 1 676 to 1 734 million ha, representing a maximum additional use of 58 million ha.

Increases of harvested area (Table 3.32) account for both the expansion of cultivated land and increased multicropping, i.e., the intensification of cropping on existing cultivated land. For the WEO and TAR biofuel scenarios this additional harvested area falls in the range of 26 million ha (WEO-V1) to 59 million ha (TAR-V1). Under different scenarios, the harvested area in developed countries increases by 10 to 18 million ha, and in developing countries by 17 to 35 million ha. While Africa and South America account for more than 80 percent of physical land expansion (additional cultivated land), their combined share in additional harvested area is only about 45 percent, which indicates that higher agricultural prices lead to a substantial intensification of cropping also in regions with limited land resources.

Table 3.32
Impacts of biofuel expansion on harvested area

| | Change relative to reference scenario FAO-REF-00 *(%)* | | | | | | | | |
| | World | | | Developed countries | | | Developing countries | | |
Scenario	2020	2030	2050	2020	2030	2050	2020	2030	2050
FAO-REF-01	13	15	8	6	7	2	7	8	6
WEO-V1	29	33	31	10	13	6	19	20	25
WEO-V2	30	39	43	10	15	8	20	24	34
TAR-V1	57	74	71	17	23	12	38	49	57
TAR-V3	45	50	42	14	17	7	30	32	35

Sources: IIASA WFS simulations; FAO-REF-00 scenario, May 2009.

In summary, while total global arable land use is only 1 to 3 percent higher in biofuel scenarios than in a situation without biofuels, the impact becomes substantial when expressed in terms of net cultivated land expansion during 2000 to 2020, 2000 to 2030, and 2000 to 2050. From this perspective, the impact of biofuel scenarios is to increase the net expansion of cultivated land by 20 to 45 percent in 2000 to 2020, 15 to 40 percent in 2000 to 2030, and 12 to 30 percent in 2000 to 2050.

Second-generation biofuels

The previous section demonstrated that concerns about expanding the use of first-generation biofuels, especially those derived from cereals and oilseeds, are well justified in view of their possible impacts on agricultural prices, food security and land use.

In this context, second-generation biofuels, produced from woody or herbaceous non-food plant materials as feedstocks, have attracted great attention and raised hopes that the substantial technological and economic barriers that still hamper the commercial deployment of second-generation technologies can soon be resolved, to allow their use and full commercialization in the near future.

Some of the problems associated with first-generation biofuels can be avoided by the production of biofuels using agricultural and forest residues and non-food crop feedstocks. First, the energy yields per hectare achievable with second-generation feedstocks are generally higher than those of first-generation biofuels; and second, land of different quality could possibly be used for their production, thus limiting or avoiding land-use competition with food production, as lignocellulosic feedstocks are expected to be grown mainly outside cultivated land.

Following recent substantial government grants for the development of second-generation feedstocks and conversion technologies, and based on the announced plans of companies developing second-generation biofuel facilities, an optimistic view is that fully commercial-scale operations could possibly be seen as early as 2012. However, with the complexity of the technical and economic challenges involved, a more realistic expectation is that wide deployment of commercial plants is unlikely to begin before 2015 or 2020. Therefore it is still uncertain what contribution second-generation biofuels will make to meeting the global transport fuel demand by 2030 (IEA, 2008a).

Uncertainties have been included in the scenario analysis by simulating the outcomes for a range of assumptions about the expected share of biofuels that will be contributed by second-generation fuels (Table 3.33).

A recent report published by IEA states that both principal conversion processes – bio-geochemical conversion of cellulose to ethanol and thermo-chemical conversion to FT-diesel – can potentially convert 1 dry tonne of biomass (with about 20 GJ/tonne energy content) to about 6.5 GJ of energy in the form of biofuels, representing an overall biomass-to-biofuel conversion efficiency of about 35 percent (IEA, 2008a). Ranges of indicative biofuel yields per dry tonne of biomass are shown in Table 3.34.

Table 3.33
Shares and total amounts of second-generation biofuels

	Share in total transport biofuels *(%)*			Use *(million TOE)*		
Scenario	2020	2030	2050	2020	2030	2050
Global average						
WEO-V1	3	13	30	3	17	62
WEO-V2	0	0	10	0	0	21
WEO-V3	13	30	49	13	38	103
TAR-V1	2	12	26	5	37	110
TAR-V2	0	0	10	0	0	42
TAR-V3	22	38	55	41	113	234
Developed countries						
WEO-V1	4	19	40	3	15	50
WEO-V2	0	0	10	0	0	12
WEO-V3	18	36	59	11	29	73
TAR-V1	4	18	39	5	32	84
TAR-V2	0	0	10	0	0	21
TAR-V3	33	51	68	39	91	146

Source: Fischer *et al.*, 2009.

Table 3.34
Indicative biofuel yields of second-generation conversion technologies

	Biofuel yield *(litres/dry tonne)*		Energy content *(MJ/litre)*	Energy yield *(GJ/dry tonne)*		Biomass input *(dry tonne/TOE)*	
Process	Low	High	LHV	Low	High	Low	High
Biochemical enzymatic hydrolysis ethanol	110	300	21.1	2.3	6.3	18.0	6.6
Thermo-chemical FT-diesel	75	200	34.4	2.6	6.9	16.2	6.1
Syngas-to-ethanol	120	160	21.1	2.5	3.4	16.5	12.4

Source: IEA, 2008a.

Assuming that average biochemical ethanol yields of 250 litres per dry tonne of biomass will be achievable in 2020 and 300 litres in 2030, and that thermo-chemical FT-diesel conversion will produce 160 litres per dry tonne of biomass in 2020 and 200 litres in 2030, each tonne of oil equivalent of second-generation biofuels will require an average of 7.7 dry tonnes of biomass in 2020, 6.4 dry tonnes by 2030, and 6 dry tonnes in 2050. This results in the biomass demands for second-generation biofuels shown in Table 3.35.

Table 3.35
Biomass demand for second-generation biofuels

Scenario	Global demand			Demand in developed countries		
	(million dry tonnes)					
	2020	2030	2050	2020	2030	2050
WEO-V1	19	106	370	19	95	300
WEO-V2	0	0	125	0	0	74
WEO-V3	97	240	615	87	186	440
TAR-V1	35	234	660	35	207	500
TAR-V2	0	0	254	0	0	128
TAR-V3	315	725	1 402	297	583	875

Source: Fischer *et al.*, 2009.

Rapid deployment of second-generation conversion technologies after 2015 to meet the biofuel production of the TAR-V3 scenario in 2020 and 2030 would require some 315 million dry tonnes of biomass in 2020, increasing to 725 million dry tonnes in 2030. Of these, about 300 million dry tonnes in 2020 and nearly 600 million dry tonnes in 2030 would be required to meet the demand in developed countries.

Land required for second-generation biofuels

Low-cost crop and forest residues, wood process wastes and the organic fraction of municipal solid wastes can all be used as lignocellulosic feedstocks. In some regions, substantial volumes of these materials are available and may be used. In such cases, the production of biofuels requires well-designed logistics systems but no additional land. In other regions, with limited residues and suitable wastes and where large and growing amounts of feedstocks are demanded, additional land will be needed for plantations of perennial energy grasses or short-rotation forest crops. Typical yields for the most important suitable feedstocks are summarized in Table 3.36.

Taking an average typical yield of about 10 dry tonnes per hectare as possible and reasonable in 2020, the biomass requirements listed in Table 3.35, with a maximum of 315 million dry tonnes in 2020, imply that up to 32 million ha of land will be needed if all biomass is to come from plantations. In reality, the land requirement in 2020 will be much lower, owing to the availability of large amounts of cheap crop and forest residues. In this early stage of second-generation biofuel development, most of the biomass will be required in developed countries.

By 2030, assuming that research and experience increase average yields to about 15 dry tonnes per hectare (as suggested in Table 3.36), the upper limits of land required for feedstock production will be 50 million ha in the TAR-V3 scenario and less than 20 million ha in both the WEO-V3 and the TAR-V1 scenarios.

Table 3.36
Typical yields of second-generation biofuel feedstocks[a]

Feedstock	Current yields	Expected yield by 2030
	(dry tonnes/ha)	
Miscanthus	10	20
Switchgrass	12	16
Short-rotation willow	10	15
Short-rotation poplar	9	13

[a] These yields refer to generally good land; under marginal conditions, yields can be substantially lower.

Source: Worldwatch Institute, 2007.

While the conventional agricultural feedstocks currently used in first-generation biofuel production compete with food crops, second-generation lignocellulose technologies promise substantial greenhouse gas savings and may allow the use of land resources currently not or only extensively used. Acknowledging these significant advantages of second-generation lignocellulosic biofuel feedstocks over conventional agricultural feedstocks, the study employed a detailed geographical resource database (Fischer *et al.*, 2008) to estimate the land potentially available for bioenergy production under a food and environment first paradigm, i.e., excluding land currently used for food and feed production, as well as forests.

This estimation was based on a 5′ by 5′ latitude/longitude grid (of about 10 km by 10 km at the equator). It started from total land area and subtracted all land indicated as artificial and built-up surfaces, all cultivated land and current forest land. The next step was to exclude all areas indicated or designated as legally protected, and then land with very low productivity owing to cold temperatures at high latitudes or altitudes, low annual precipitation, or steep sloping conditions.

Starting with a total global land area of 13.2 billion ha (excluding Antarctica and Greenland) and subtracting all current cultivated land, forests, built-up land, water bodies and non-vegetated land (desert, rocks, etc.) resulted in 4.6 billion ha of land remaining (about 35 percent of the total). When unproductive, marginally productive (e.g., tundra, arid land) and steeply sloped land was excluded, the remaining area was estimated at 1.75 billion ha (Table 3.37), comprising grassland and woodland.

More than two-thirds of this grassland and woodland potentially suitable for biofuel feedstock production is located in developing countries, especially in Africa and South America (Table 3.37). These estimates are indicative only and are subject to the limitations and accuracy of global land cover, soil and terrain data.

Table 3.37
Regional balance of land classified as unprotected grassland and woodland potentially usable for rainfed lignocellulosic biofuel feedstock production

Region	Total grassland and woodland (million ha)	Of which (million ha)			Potential rainfed yield (dry tonnes/ha)		
		Protected areas	Unproductive or marginally productive	Balance	Average yield	Low yield	High yield
North America	659	103	391	165	9.3	6.7	21.4
Europe and Russian Federation	902	76	618	208	7.7	6.9	14.5
Pacific OECD	515	7	332	175	9.8	6.5	20.0
Africa	1 086	146	386	554	13.9	6.7	21.1
East Asia	379	66	254	60	8.9	6.4	19.0
South Asia	177	26	81	71	16.7	7.6	21.5
Latin America	765	54	211	500	15.6	7.1	21.8
Near East and North Africa	107	2	93	12	6.9	6.3	10.6
Developed	2 076	186	1 342	548	8.9	6.7	21.0
Developing	2 530	295	1 029	1 206	14.5	6.8	21.5
World	4 605	481	2 371	1 754	12.5	6.8	21.5

Source: Fischer *et al.*, 2008.

An important current use of these land resources is livestock grazing. Based on UN FAOSTAT data on feed utilization of crops and processed crop products (e.g., oilseed cakes and meals), fodder crop production, national livestock numbers and livestock production, the feed energy provided by each source was estimated for each country to determine the energy gap to be filled by grassland and pastures. The results of detailed livestock feed energy balances suggest that in 2000 about 55 to 60 percent of available grassland biomass globally was required for animal feeding. The shares are about 40 percent in developed countries and an average 65 percent for developing countries, ranging from more than 80 percent in Asian regions to about 50 percent in sub-Saharan Africa.

Hence, at current use levels, the land potentially available for bioenergy production (assuming unbiased distribution between livestock feeding and bioenergy uses) was estimated at about 700 to 800 million ha, characterized by a rather wide range of productivity levels. Of this, an estimated 330 million ha is in developed countries: about one-third each in North America; Europe, the Russian Federation and Central Asian republics; and Pacific OECD. About 450 million ha is in developing countries: 275 million ha in Africa and 160 million ha in Latin America. Regional details of the estimated land areas and potential yields of second-generation lignocellulosic feedstocks are presented in Table 3.37.

Only the demand for livestock feeding was subtracted, as it is currently the main alternative use. No allowances were included for other social or environmental land functions, such as providing a feed source for wildlife. Estimates are also subject to uncertainties regarding grass and pasture yields, which – owing to the scarcity of data – had to be estimated in model simulations with the IIASA/FAO Global AEZ Study (GAEZ) model (Fischer *et al.*, 2008).

It can be concluded that land demand for producing second-generation feedstocks as required for the most demanding TAR-V3 scenario in 2020 (about 30 million ha) and 2030 (about 50 million ha) could be met without having to compete for cultivated land. The results of the TAR scenario with accelerated second-generation biofuel deployment indicate that production of lignocellulosic feedstocks on some 100 million ha would be sufficient to achieve the target biofuel share in world transport fuels in 2050.

However, there is still need to assess and respect the current uses and functions of potentially suitable land and to regulate land use in an integrated approach across sectors, to achieve land-use efficiency, avoid conflicts and protect the rights of the weakest members of society when landownership is uncertain. Another major challenge is developing the massive infrastructure and logistics systems required for second-generation feedstock supply systems.

Combined impacts of climate change and expansion of biofuel production on world food system indicators

The previous sections reviewed the individual impacts of climate change and the expansion of biofuel production on world food system indicators. This section summarizes the results for the combined impacts of both factors, by comparing scenario outcomes with a reference simulation assuming current climate conditions and no use of crops for transport biofuel production.

Agricultural prices

Table 3.38 presents the results of scenario analysis and the deviations of price indices for cereals, all crops and agriculture (all crop and livestock sectors), for a selection of scenarios constructed by combining different climate change projections and assumptions concerning CO_2 fertilization with a range of biofuel expansion scenarios.

Table 3.38
Combined impacts of climate change and biofuel expansion on agricultural prices

Scenario	CO$_2$ fertilization	Change in price index relative to reference scenario FAO-REF-00 *(%)*			
		2020	2030	2050	2080
Cereals					
Hadley A2, FAO-REF-01	with	4	5	5	28
Hadley A2, WEO-V1	with	15	13	16	42
Hadley A2, WEO-V2	with	18	18	26	49
Hadley A2, TAR-V1	with	42	41	36	61
Hadley A2, TAR-V3	with	23	20	16	43
CSIRO A2, REF-01	with	9	10	10	28
CSIRO A2, WEO-V1	with	22	17	20	43
CSIRO A2, WEO-V2	with	24	23	30	49
CSIRO A2, TAR-V1	with	49	49	40	61
CSIRO A2, TAR-V3	with	29	26	20	45
Hadley A2, REF-01	without	10	13	16	52
Hadley A2, WEO-V1	without	20	21	30	68
Hadley A2, WEO-V2	without	24	26	42	79
Hadley A2, TAR-V1	without	49	54	53	87
Hadley A2, TAR-V3	without	25	29	31	70
Crops					
Hadley A2, REF-01	with	2	3	2	15
Hadley A2, WEO-V1	with	13	11	12	25
Hadley A2, WEO-V2	with	14	13	17	28
Hadley A2, TAR-V1	with	36	35	31	41
Hadley A2, TAR-V3	with	24	19	15	28
CSIRO A2, REF-01	with	6	6	5	14
CSIRO A2, WEO-V1	with	17	13	15	24
CSIRO A2, WEO-V2	with	18	16	20	27
CSIRO A2, TAR-V1	with	42	40	34	40
CSIRO A2, TAR-V3	with	28	23	18	27
Hadley A2, REF-01	without	7	9	12	33
Hadley A2, WEO-V1	without	17	18	24	44
Hadley A2, WEO-V2	without	19	20	30	48
Hadley A2, TAR-V1	without	44	45	45	61
Hadley A2, TAR-V3	without	28	28	27	48
Agriculture					
Hadley A2, REF-01	with	1	2	1	11
Hadley A2, WEO-V1	with	9	7	8	17
Hadley A2, WEO-V2	with	10	9	12	19
Hadley A2, TAR-V1	with	27	25	22	27
Hadley A2, TAR-V3	with	17	13	10	19
CSIRO A2, REF-01	with	4	4	4	10
CSIRO A2, WEO-V1	with	13	9	11	17
CSIRO A2, WEO-V2	with	13	12	15	19
CSIRO A2, TAR-V1	with	32	30	24	27
CSIRO A2, TAR-V3	with	21	17	12	18
Hadley A2, REF-01	without	5	7	9	23
Hadley A2, WEO-V1	without	13	13	17	31
Hadley A2, WEO-V2	without	14	15	22	34
Hadley A2, TAR-V1	without	33	33	33	42
Hadley A2, TAR-V3	without	20	20	19	33

Sources: IIASA WFS simulations; reference scenario FAO-REF-00, May 2009.

Comparing these results with outcomes in Table 3.15 (climate change impacts) and Table 3.27 (biofuel expansion impacts) indicates that the effects of both factors will combine to increase agricultural prices. For the next few decades, the most important scenario factor in determining price increases is the scale of crop use as biofuel feedstocks. In the medium and long terms, climate change becomes the overriding factor.

Taking the effects of CO_2 fertilization on crop yields into account, the simulated cereal price increases for the presented scenario combinations up to 2050 range from 15 to 40 percent when using the HadCM3 climate model outputs, and are somewhat higher when applying climate scenarios based on the CSIRO GCM. Without CO_2 fertilization effects, the cereal price increases for the decades up to 2050 range from 20 to 55 percent. Simulation results for the 2080s, when climate change impacts seriously affect crop yields, the calculated cereal price increases range from 40 to 60 percent with CO_2 fertilization and from 70 to 90 percent without CO_2 fertilization.

Cereal production and consumption

Table 3.39 lists the scenario results regarding production increases relative to the baseline scenario FAO-REF-00 (without climate change and with no crop use for biofuel production).

Table 3.39
Combined impacts of climate change and biofuel expansion on cereal production

Scenario	CO_2 fertilization	Change relative to reference scenario FAO-REF-00 *(million tonnes)*			
		2020	2030	2050	2080
Hadley A2, REF-01	with	70	65	54	-26
Hadley A2, WEO-V1	with	148	160	184	122
Hadley A2, WEO-V2	with	149	197	273	219
Hadley A2, TAR-V1	with	237	320	311	278
Hadley A2, TAR-V3	with	181	209	198	142
CSIRO A2, REF-01	with	55	48	31	-16
CSIRO A2, WEO-V1	with	126	146	161	126
CSIRO A2, WEO-V2	with	133	180	250	228
CSIRO A2, TAR-V1	with	222	299	291	291
CSIRO A2, TAR-V3	with	165	190	177	151
Hadley A2, REF-01	without	56	45	16	-98
Hadley A2, WEO-V1	without	135	138	139	41
Hadley A2, WEO-V2	without	137	176	224	144
Hadley A2, TAR-V1	without	223	294	266	193
Hadley A2, TAR-V3	without	179	183	153	66

Sources: IIASA WFS simulations; FAO-REF-00 scenario, May 2009.

Table 3.40
Combined impacts of climate change and biofuel expansion on cereal consumption (excluding biofuel use)

Scenario	CO_2 fertilization	Change (excluding biofuel feedstocks) relative to reference scenario FAO-REF-00 *(million tonnes)*			
		2020	2030	2050	2080
Hadley A2, REF-01	with	-10	-21	-25	-100
Hadley A2, WEO-V1	with	-33	-47	-60	-117
Hadley A2, WEO-V2	with	-43	-63	-99	-144
Hadley A2, TAR-V1	with	-88	-122	-128	-156
Hadley A2, TAR-V3	with	-53	-65	-61	-111
CSIRO A2, REF-01	with	-24	-38	-43	-92
CSIRO A2, WEO-V1	with	-51	-60	-78	-111
CSIRO A2, WEO-V2	with	-57	-78	-118	-133
CSIRO A2, TAR-V1	with	-102	-142	-149	-144
CSIRO A2, TAR-V3	with	-66	-83	-80	-104
Hadley A2, REF-01	without	-24	-41	-63	-170
Hadley A2, WEO-V1	without	-49	-68	-104	-191
Hadley A2, WEO-V2	without	-57	-82	-144	-221
Hadley A2, TAR-V1	without	-102	-148	-174	-232
Hadley A2, TAR-V3	without	-60	-86	-105	-183

Sources: IIASA WFS simulations; FAO-REF-00 scenario, May 2009.

Comparing these scenario results with the information in Table 3.40 indicates that up to 2050 there is relatively little climate change impact on aggregate cereal supply and consumption under the HadCM3 scenario with CO_2 fertilization; with CSIRO GCM-derived climate change impacts, the shortfall in consumption increases by about 20 million tonnes compared with biofuels only. Without CO_2 fertilization effects on crop yields, the decrease in consumption for HadCM3 in 2030 is 68 to 148 million tonnes, of which about 25 million tonnes is due to climate change. In 2050, the consumption reduction is in the range of 104 to 174 million tonnes, of which about 50 million tonnes is caused by climate change. In the long term, looking at results for 2080, climate change accounts for up to two-thirds of the reduction in cereal consumption in scenarios with CO_2 fertilization and for up to 85 percent in the HadCM3 scenario without CO_2 fertilization.

Risk of hunger

Combined scenario results regarding the number of people at risk of hunger are shown in Table 3.41. Results are consistent with the previous discussion on price changes and cereal consumption impacts. Again, the conditions portrayed by the FAO reference projections (FAO, 2006) imply a vast improvement in reducing undernourishment. Therefore, relative changes compared with the baseline FAO-REF-00 are large in 2050 and 2080 but relatively small in absolute terms.

Table 3.41
Combined impacts of climate change and biofuel expansion on risk of hunger indicator

Scenario	CO_2 fertilization	Change in number of people at risk of hunger relative to reference scenario FAO-REF-00 *(million people)*			
		2020	2030	2050	2080
Hadley A2, REF-01	with	6	9	2	29
Hadley A2, WEO-V1	with	51	41	34	39
Hadley A2, WEO-V2	with	59	54	54	43
Hadley A2, TAR-V1	with	150	148	99	55
Hadley A2, TAR-V3	with	100	82	39	40
CSIRO A2, REF-01	with	14	23	4	21
CSIRO A2, WEO-V1	with	14	23	4	32
CSIRO A2, WEO-V2	with	82	75	60	35
CSIRO A2, TAR-V1	with	178	176	104	48
CSIRO A2, TAR-V3	with	123	108	46	32
Hadley A2, REF-01	without	33	43	41	58
Hadley A2, WEO-V1	without	75	76	78	70
Hadley A2, WEO-V2	without	85	88	102	77
Hadley A2, TAR-V1	without	179	192	153	87
Hadley A2, TAR-V3	without	117	119	88	72

Sources: IIASA WFS simulations; FAO-REF-00 scenario, May 2009.

Cultivated land

Tables 3.42 and 3.43 present the combined impacts of climate change and biofuel expansion scenarios on cultivated land use. Summarizing over all the scenarios shown in Table 3.42, the additional use of cultivated land falls by 16 to 40 million ha in 2020, 17 to 49 million ha in 2030, and 20 to 58 million ha in 2050.

For harvested area, shown in Table 3.43, the additional use ranges from 24 to 59 million ha in 2020, 28 to 78 million ha in 2030, and 28 to 85 million ha in 2050.

Conclusions

This paper reports on a large number of scenario experiments conducted to improve the understanding of how climate change and expanding bioenergy use may alter the long-term outlook for food, agriculture and resource availability.

IIASA's global and spatial agro-ecological and socio-economic assessment framework provided the analytical means and science-based knowledge for the assessment. The following is a summary of the main conclusions and implications derived from the global quantitative analysis:

Table 3.42
Combined impacts of climate change and biofuel expansion on use of cultivated land

Scenario	CO$_2$ fertilization	Change relative to reference scenario FAO-REF-00 (million ha)			
		2020	2030	2050	2080
Hadley A2, REF-01	with	4	5	3	16
Hadley A2, WEO-V1	with	16	17	20	33
Hadley A2, WEO-V2	with	17	20	26	39
Hadley A2, TAR-V1	with	35	43	47	59
Hadley A2, TAR-V3	with	26	27	27	39
CSIRO A2, REF-01	with	8	11	10	20
CSIRO A2, WEO-V1	with	20	21	26	37
CSIRO A2, WEO-V2	with	21	25	33	43
CSIRO A2, TAR-V1	with	40	48	53	63
CSIRO A2, TAR-V3	with	30	33	33	44
Hadley A2, REF-01	without	8	12	14	33
Hadley A2, WEO-V1	without	19	22	31	50
Hadley A2, WEO-V2	without	20	25	37	56
Hadley A2, TAR-V1	without	39	49	58	75
Hadley A2, TAR-V3	without	29	33	38	57

Sources: IIASA WFS simulations; FAO-REF-00 scenario, May 2009.

Table 3.43
Combined impacts of climate change and biofuel expansion on harvested area

Scenario	CO$_2$ fertilization	Change relative to reference scenario FAO-REF-00 (million ha)			
		2020	2030	2050	2080
Hadley A2, REF-01	with	7	9	3	22
Hadley A2, WEO-V1	with	24	28	28	47
Hadley A2, WEO-V2	with	25	33	38	56
Hadley A2, TAR-V1	with	51	68	67	86
Hadley A2, TAR-V3	with	39	45	38	56
CSIRO A2, REF-01	with	13	17	11	24
CSIRO A2, WEO-V1	with	30	36	34	50
CSIRO A2, WEO-V2	with	31	41	45	58
CSIRO A2, TAR-V1	with	58	75	74	89
CSIRO A2, TAR-V3	with	46	52	45	60
Hadley A2, REF-01	without	14	19	20	49
Hadley A2, WEO-V1	without	30	38	46	75
Hadley A2, WEO-V2	without	32	43	56	84
Hadley A2, TAR-V1	without	59	78	85	112
Hadley A2, TAR-V3	without	45	55	56	85

Sources: IIASA WFS simulations; FAO-REF-00 scenario, May 2009.

- At the global aggregate level, climate change projected by different GCMs causes only modest changes to world food system indicators (prices, cereal production, food consumption, cultivated land use) in the period up to 2050.

- These findings assume full agronomic adaptation by farmers and do not take into account climate variability, which is expected to increase over the coming decades and may be an important destabilizing factor in the short to medium term.

- The capacity to adapt to climate change impacts is strongly linked to future development paths. The socio-economic and, even more so, the technological characteristics of different development futures strongly affect societies' capability to adapt to and mitigate climate change.

- Assumptions regarding yield increases due to increased atmospheric CO_2 concentrations (the CO_2 fertilization effect) play an important role in scenario outcomes. When disregarding these effects, negative climate change impacts on crop yields and world food system indicators become noticeable even in the short term, and are very substantial in the medium and long terms.

- Scenario results confirm that, with and without CO_2 fertilization, the impacts of climate change on crop yields and production could become severe in the second half of this century.

- If expansion of biofuel production continues to rely mainly on agricultural crops, and if expansion follows the pace projected by IEA in 2008 or achieves the levels implied by mandates and targets set in many countries, this additional non-food use of crops will have a significant impact on the world food system.

- While biofuels could have an especially large impact in the period up to 2030, the aggregate impact on the food system is likely to decrease over time. The opposite is to be expected for climate change impacts.

- For the range of scenarios analysed in this assessment, the combined impact of climate change and biofuel expansion on aggregate crop prices is in the range of a 10 to 45 percent increase. Decrease of cereal consumption typically falls initially within 35 to 100 million tonnes, increasing to a range of 60 to 150 million tonnes by 2050. Regarding cultivated land, additional use in the range of 20 to 50 million ha by 2030 and 25 to 60 million ha in 2050 can be expected.

References

Bruinsma, J., ed. 2003. *World agriculture: Towards 2015/30, an FAO perspective.* London, Earthscan and Rome, FAO. www.fao.org/es/esd/gstudies.htm.

FAO. 2001. *The State of Food Insecurity in the World 2001.* Rome.

FAO. 2006. *World agriculture: towards 2030/2050 – Interim report.* Rome. www. fao.org/es/esd/at2050web.pdf.

FAO. 2008a. *The State of Food and Agriculture. Biofuels: prospects, risks and opportunities.* Rome. 138 pp.

FAO. 2008b. *The State of Food Insecurity in the World 2008.* Rome.

FAOSTAT. Time-series and cross-sectional data relating to food and agriculture. http://faostat.fao.org/default.aspx.

Fischer, G., Shah. M. & van Velthuizen, H. 2002. *Climate change and agricultural vulnerability.* Special report for the World Summit on Sustainable Development, Johannesburg 2002. Laxenburg, Austria, IIASA. 152 pp.

Fischer, G., Frohberg, K., Keyzer, M.A. & Parikh, K.S. 1988. *Linked national models: a tool for international policy analysis.* Berlin, Kluwer Academic Publishers. 214 pp.

Fischer, G., Frohberg, K., Parry, M.L. & Rosenzweig, C. 1994. Climate change and world food supply, demand and trade: Who benefits, who loses? *Global Environmental Change,* 4(1): 7–23.

Fischer G., van Velthuizen, H., Shah, M. & Nachtergaele, F.O. 2002. *Global agro-ecological assessment for agriculture in the 21st century: Methodology and results.* IIASA RR-02-02. Laxenburg, Austria, IIASA.

Fischer, G., Shah, M., Tubiello, F.N. & van Velhuizen, H. 2005. Socio-economic and climate change impacts on agriculture: an integrated assessment, 1990–2080. *Phil. Trans. Royal Soc. B.*

Fischer, G., Nachtergaele, F., Prieler, S., Teixeira, E., van Velthuizen, H.T., Verelst, L. & Wiberg, D. 2008. *Global agro-ecological zones assessment for agriculture (GAEZ 2008).* Laxenburg, Austria, IIASA, and Rome, FAO.

Fischer, G., Teixeira, E., Tothne-Hizsnyik, E. & van Velthuizen, H. 2009. Land use dynamics and sugarcane production. *In* P. Zuurbier and J. van de Vooren, eds. *Sugarcane ethanol, Contributions to climate change mitigation and the environment.* Wageningen, Netherlands, Wageningen Academic Publishers, and Laxenburg, Austria, IIASA.

Gordon, H.B. & O'Farrell, S.P. 1997. Transient climate change in the CSIRO coupled model with dynamic sea ice. *Monthly Weather Review,* 125(5): 875–907.

Gordon, C., Cooper, C.A. Snr, Banks, H., Gregory, J.M., Johns, T.C., Mitchell, J.F.B. & Wood, R.A. 2000. The simulation of SST, sea ice extents and ocean heat transports in a version of the Hadley Centre coupled model without flux adjustments. *Climate Dynamics*, 16: 147–168.

Hirst, A.C., Gordon, H.B. & O'Farrell, S.P. 1997. Response of a coupled ocean-atmosphere model including oceanic eddy-induced advection to anthropogenic CO_2 increase. *Geophys. Res. Lett.*, 23(23): 3361–3364.

IEA. 2008a. *From 1st- to 2nd-generation biofuel technologies. An overview of current industry and RD&D activities.* Paris, OECD/IEA. www.iea.org.

IEA. 2008b. *World energy outlook 2008.* Paris, OECD/IEA. 578 pp.

Nakicenovic, N. & Swart, R., eds. 2000. *Special report on emissions scenarios.* Cambridge, UK, Cambridge University Press and IPCC. 570 pp.

New, M., Lister, D., Hulme, M. & Makin, I. 2002. A high-resolution data set of surface climate over global land areas. *Climate Research*, 21.

Pope, V.D., Gallani, M.L., Rowntree, P.R. & Stratton, R.A. 2000. The impact of new physical parametrizations in the Hadley Centre climate model – HadAM3. *Climate Dynamics*, 16: 123–146.

Rosenzweig, C. & Parry, M.L. 1994. Impacts of climate change on world food supply. *Nature*, 367: 133–138.

Tubiello, F.N. & Fischer, G. 2006. Reducing climate change impacts on agriculture: Global and regional effects of mitigation, 2000–2080. *Technological Forecasting and Social Change,* doi:10.1016/j.techfore.2006.05.027.

UN. 2009. *World Population Prospects: The 2008 Revision.* New York.

Worldwatch Institute. 2007. *Biofuels for transport. Global potential and implications for sustainable energy and agriculture.* London, Earthscan.

The study is based on a state-of-the-art ecological-economic modelling approach. The scenario-based quantified findings of the study rely on a modelling framework that includes the FAO/IIASA Agro-Ecological Zone (AEZ) model and the IIASA World Food System (WFS) model as components. The modelling framework encompasses climate scenarios, agro-ecological zoning information, demographic and socio-economic drivers, and production, consumption and world food trade dynamics.

AEZ methodology

The AEZ modelling uses detailed agronomic-based knowledge to simulate land resources availability, assess farm-level management options and estimate crop production potentials. It employs detailed spatial biophysical and socio-economic datasets to distribute its computations at fine-gridded intervals over the entire globe (Fischer *et al.*, 2002; 2005). This land resources inventory is used to assess, for specified management conditions and levels of inputs, the suitability of crops under both rainfed and irrigated conditions, and to quantify expected attainable production levels of cropping activities relevant to specific agro-ecological contexts. The characterization of land resources includes components of climate, soils, land form and current land cover. Crop modelling and environmental matching procedures are used to identify crop-specific environmental limitations under various levels of inputs and management conditions.

In summary, the AEZ framework contains the following basic elements:

- land resources database, containing geo-referenced climate, soil and terrain data;

- land utilization types (LUTs) database of agricultural production systems, describing crop-specific environmental requirements and adaptability characteristics, including input levels and management;

- mathematical procedures for matching crop LUT requirements with agro-ecological zones data and estimating potentially attainable crop yields, by land unit and grid-cell; the AEZ global assessment includes 2.2 million land grid cells at 5′ by 5′ latitude/longitude;

- assessments of crop suitability and land productivity;

- applications for agricultural development planning.

WFS model

The WFS model comprises a series of national and regional agricultural economic models. It provides a framework for analysing the world food system and for viewing national food and agricultural components as embedded in national economies and interacting with each other at the international trade level. The model consists of 34 national and regional geographical components covering the world. The individual national/regional models are linked by means of a world market, where international clearing prices are computed to equalize global demand with supply (Box A3.1).

Simulations with the WFS model generate a variety of outputs. At the global level, these include world market prices, global population, global production and global consumption. At the country level, they include producer and retail prices, levels of production, use of primary production factors (land, labour and capital), intermediate input use (feed and fertilizer), human consumption, use for biofuel production, commodity trade, value added in agriculture, investment by sector and income by group and/or sector.

Population growth and technology are key external inputs to the WFS model system. Population numbers and projected incomes are used to determine the demand for food for the period of study. Technology affects yield estimates, by modifying the efficiency of production per given units of inputs and land. For simulations of historical periods up to the present, population data are taken from official UN country-level data, while the rate of technical progress is estimated from past agricultural performance.

To assess agricultural development over the next decades to 2050, it was necessary first to make some coherent assumptions about how key socio-economic drivers of food systems might evolve over that period. For the analysis reported in this chapter, population projections were taken from the UN database for world population prospects (UN, 2009). Economic growth of countries and regional groups in the WFS model were calibrated according to information provided by the World Agriculture Towards 2030/2050 study group at FAO (J. Bruinsma, 2009, personal communication).

Another external input to the WFS model system is projected climate change, which affects region-specific crop suitability and attainable yields. The economic model uses this spatial agronomic information (derived from AEZ) in an aggregate form as an input in allocating land and agricultural inputs (Fischer *et al.*, 2005). In this study, results of the coupled atmosphere-ocean GCM developed by the United Kingdom's Hadley Centre for Climate Prediction and Research and Australia's CSIRO were used to take into account climate change impacts on land suitability and productivity (Fischer, Shah and van Velthuizen, 2002).

Box A3.1 - How does the world food system work?

The WFS model is an applied general equilibrium model system. While focusing on agriculture, this necessitates that all other economic activities are also represented in the model. Financial and commodity flows within a country and at the international level are kept consistent in that they must balance, by imposing a system of budget constraints and market clearing conditions. Whatever is produced will be demanded, for human consumption, feed, biofuel use or as an intermediate input. Alternatively, commodities can be exported or put into storage. Consistency of financial flows is imposed at the level of the economic agents in the model (individual income groups, governments, etc.), nationally and internationally. This implies that total expenditures cannot exceed total income from economic activities and from abroad, in the form of financial transfers, minus savings. On a global scale, no more can be spent than what is earned.

Each individual model component focuses primarily on the agriculture sector, but also includes a simple representation of the entire economy, which is necessary for capturing essential dynamics among capital, labour and land. For the purpose of international linkage, production, consumption and trade of goods and services are aggregated into nine main agricultural sectors: wheat; rice; coarse grains; bovine and ovine meat; dairy products; other meat and fish; oilseed cakes and protein meals; other food; and non-food agriculture. The rest of the economy is coarsely aggregated into one simplified non-agricultural sector. In the model, agricultural commodities may be used for human consumption, feed, biofuel feedstock, intermediate consumption and stock accumulation. Non-agricultural commodities also contribute as investment and as inputs for processing and transporting agricultural goods. All physical and financial accounts are balanced and mutually consistent: production, consumption and financial accounts at the national level; and trade and financial flows at the global level.

Linkage of country and country group models occurs through trade, world market prices and financial flows. The system is solved in annual increments, simultaneously for all countries in each period. Within each one-year period, demand changes with price, and commodity buffer stocks can be adjusted for short-term supply response. Production in the following marketing year is affected by changes in relative prices (owing to time lags in the agricultural production cycle). This feature makes the WFS model a recursively dynamic system.

The market clearing process results in equilibrium prices, i.e., a vector of international prices such that global imports and exports balance for all commodities. These market clearing prices are then used to determine value added in the production and income of households and governments.

Within each regional unit, the supply modules allocate land, labour and capital as functions of the relative profitability of the different crop and livestock sectors. In particular, actual cultivated area is computed from both agro-climatic land parameters (derived from AEZ) and profitability estimates. Once area, labour and capital are assigned to cropping and livestock activities, yields and livestock production are computed as a function of fertilizer applications, feed rates and available technology.

The IIASA WFS model has been calibrated and validated over past time windows, and reproduces regional consumption, production and trade of major agricultural commodities in 2000. Several applications of the model to agricultural policy and climate change impact analysis have been published (e.g., Fischer *et al.*, 1988; 1994; Rosenzweig and Parry, 1994; Fischer, Shah and van Velthuizen, 2002; Fischer *et al.*, 2005; Tubiello and Fischer, 2006).

AGGREGATION OF WORLD FOOD SYSTEM COMPONENTS TO WORLD REGIONS

Economic group	Region	WFS component
Developed	North America	Canada, United States
	Europe and Russian Federation	Austria, EU9, Eastern Europe, former Soviet Union, Turkey
	Pacific OECD	Australia, Japan, New Zealand
Developing	Sub-Saharan Africa	Kenya, Nigeria Africa oil exporters Africa medium-income/food exporters Africa low-income/food exporters Africa low-income/food importers
	Latin America	Argentina, Brazil, Mexico Latin America high-income/food exporters Latin America high-income/food importers Latin America medium-income
	Near East and North Africa	Egypt Africa medium-income/food importers Near/Middle East oil exporters Near/Middle East medium- and low-income countries.
	East Asia	China Far East Asia high- and medium-income/food importers
	South and Southeast Asia	India, Pakistan, Indonesia, Thailand Asia low-income countries Far East Asia high- and medium-income/food exporters
Rest of the world	Rest of the world	Rest of the world

Aggregate regional country group models

African oil exporters: Algeria, Angola, Congo, Gabon.

Africa medium-income/food exporters: Ghana, Côte d'Ivoire, Senegal, Cameroon, Mauritius, Zimbabwe.

Africa medium-income/food importers: Morocco, Tunisia, Liberia, Mauritania, Zambia.

Africa low-income/food exporters: Benin, the Gambia, Togo, Ethiopia, Malawi, Mozambique, Uganda, the Sudan.

Africa low-income/food importers: Guinea, Mali, the Niger, Sierra Leone, Burkina Faso, Central African Republic, Chad, Democratic Republic of the Congo, Burundi, Madagascar, Rwanda, Somalia, United Republic of Tanzania.

Latin America high-income/food exporters: Costa Rica, Panama, Cuba, Dominican Republic, Ecuador, Suriname, Uruguay.

Latin America high-income/food importers: Jamaica, Trinidad and Tobago, Chile, Peru, Venezuela.

Latin America medium-income: El Salvador, Guatemala, Honduras, Nicaragua, Colombia, Guyana, Paraguay, Haiti, Plurinational State of Bolivia.

South and Southeast Asia high- and medium-income/food exporters: Malaysia, the Philippines.

Southeast Asia high- and medium-income/food importers: Republic of Korea, Democratic People's Republic of Korea, Lao People's Democratic Republic, Viet Nam, Cambodia.

Asia, low-income: Bangladesh, Myanmar, Nepal, Sri Lanka.

Near/Middle East oil exporters: Libyan Arab Jamahiriya, Islamic Republic of Iran, Iraq, Saudi Arabia, Cyprus, Lebanon, Syrian Arab Republic.

Near/Middle East medium- and low-income: Jordan, Yemen, Afghanistan.

The rest of world aggregate includes both more and less developed countries. Although the aggregate variables are dominated by more developed countries in OECD, these countries are not included in the respective broad regional aggregates, developed and developing.

PART II

GROWTH, POVERTY AND MACROECONOMIC PROSPECTS

CHAPTER 4

POVERTY, GROWTH AND INEQUALITY OVER THE NEXT **50** YEARS

Evan Hillebrand[1]

Global poverty has fallen dramatically over the last two centuries, and the fall has intensified in recent decades, raising hopes that poverty could be eliminated within the next 50 years. After industrialization, specialization and trade increased economic growth and living standards in Western Europe and the European offshoots in the nineteenth century, much of the rest of the world also started growing rapidly after 1950.

Poverty reduction, however, has been very uneven across countries. Since 1980, China alone has accounted for most of the world's decline in extreme poverty. Even though there has been a huge rise in income inequality within China, economic growth has been so strong that hundreds of millions of people have risen out of extreme poverty and the poverty ratio has plummeted. Sub-Saharan Africa, at the other extreme, has seen its poverty headcount continue to rise; the negative impact of low economic growth has far outweighed modest improvements in within-country income inequality.

Strong economic growth is the key to future poverty reduction. If the lagging non-OECD[2] countries are able to transition to a sustainable higher growth path, the global poverty ratio will fall from about 21 percent in 2005 to less than 2.5 percent in 2050, and the number of people living in absolute poverty will decline by another billion. Although the historical record is clear that market-friendly policies and competent governance are critical to growth, few economists are bold enough to claim they know the precise combination of policies, and how to implement and sustain these policies to achieve such an economic transition.

1. This research received support through a grant from FAO for the How to Feed the World in 2050 project. Parts of this chapter represent a revision and extension of a previous paper by the author (Hillebrand, 2008).

2. For simplicity, this chapter divides countries into two groups: the OECD countries as of 1981 (Austria, Australia, Belgium, Canada, Denmark, Finland, France, Germany, Greece, Iceland, Ireland, Italy, Japan, Luxembourg, Portugal, New Zealand, the Netherlands, Norway, Spain, Sweden, Switzerland, the United Kingdom and the United States of America); and the non-OECD countries as of the same year (although some of these countries are now part of OECD).

Forecasts of future economic growth rates and poverty rates are necessarily speculative and depend on a large number of assumptions about human behaviour and policy decisions that are impossible to know in advance. This chapter reviews the poverty estimates available in the literature, analyses the changes behind the trends, and models poverty trends to 2050.

Poverty measurement

Before modern economic growth took off in a few Western Europe countries, a few European offshoots and Japan – a group of countries hereafter referred to as OECD – living standards in all countries were very low on average, by modern standards. Maddison (2003) estimated OECD gross domestic product (GDP) per capita in 1820 at about USD 1 571 in 2005 purchasing power parity (PPP) dollars versus an average of USD 730 in non-OECD countries.[3] Rising economic growth in OECD countries over the following century increased incomes and cut poverty dramatically, leaving the non-OECD countries far behind. Bourguignon and Morrisson (2002) attempted to combine measures of income distribution within countries with cross-country GDP measures, to obtain a measure of the global distribution of income and a global measure of poverty. Their paper tells a dramatic and straightforward story: global poverty rates have fallen sharply, from 85.2 percent in 1820 to 31.3 percent in 1980, as economic growth everywhere far outpaced population growth. However, these authors also show that the global distribution of income became much more unequal. Global inequality was high in 1820 (with a Gini coefficient of 0.50) and rose over the next 160 years, to reach 0.658 in 1980. In the early nineteenth century, most inequality was due to differences within countries, but most of the rise in equality since 1820 has been due to differences in growth rates among countries. Economic growth per capita in the OECD countries was twice as fast as in the non-OECD countries from 1820 to 1980. The figures shown in Table 4.1 present an introduction to the historical data on growth and poverty, based mainly on the work of Maddison (2001) and Bourguignon and Morrisson (2002), on recently updated work on poverty by Chen and Ravallion (2008), and on long-run poverty forecasts that will be discussed in this chapter.

---

3. Maddison actually estimated USD 1 109 and USD 578 in 1990 PPP prices but all his figures have been revised into 2005 prices in this chapter. To compare GDP and living standards across countries at widely different levels of development, economists usually prefer to use PPP ratios (among all currencies), which seek to estimate how much of any given currency will be required to buy an equivalent amount of the same quantity and quality of goods in any country. The International Comparison Project (ICP) undertakes a massive international survey every few years to create new estimates of these PPP ratios at a given point in time (see World Bank, 2008 for details).

Table 4.1
Long-run estimates of growth and poverty

						Alternative forecasts	
						Market first	Trend growth
Region	1820	1950	1980	1981	2005	2050	
World							
GDP *(billion 2005 PPP USD)*	913	7 006		26 825	56 593	309 569	193 318
Population *(millions of people)*	1 041	2 525		4 511	6 458	9 301	9 301
GDP per capita *(2005 PPP USD /year)*	876	2 775		5 947	8 764	33 285	20 785
average annual change[a] (%)		0.9		2.5	1.6	3.0	1.9
Absolute poverty headcount *(millions)*	887	1 376	1 390	1 896	1 377	245	1 120
Absolute poverty ratio *(%)*	85.2	54.5	31.3	42.0	21.3	2.6	12.0
Inequality index *(Gini coefficient)[b]*	0.50	0.640	0.658	0.709	0.684	0.648	0.679
Non-OECD							
GDP *(billions of 2005 PPP USD)*	628	2 702		11 324	26 008	189 980	112 177
Population *(millions of people)*	860	1 947		3 744	5 561	8 310	8 310
GDP per capita *(2005 PPP USD/year)*	730	1 388		3 024	4 677	22 861	13 498
average annual change (%)		0.5		2.5	1.8	3.6	2.4
Absolute poverty headcount				1 896	1 377	245	1 120
Absolute poverty ratio *(% of non-OECD population)*				50.6	24.8	2.9	13.5
OECD							
GDP *(billions of 2005 PPP USD)*	284	4 304		15 501	30 585	119 589	81 142
Population *(millions of people)*	181	578		767	897	990	990
GDP per capita *(2005 PPP USD/ year)*	1 571	7 446		20 222	34 089	120 756	81 933
average annual change (%)		1.2		3.3	2.2	2.9	2.0

[a] Average annual growth rates are calculated for 1821 to 1950, 1981 to 2005, and 2006 to 2050.

[b] The Gini coefficient is calculated on an individual basis: it uses information on within-country income distribution.

Sources: GDP and population for 1981 to 2005 from World Development Indicators; early years linked from Maddison, 2001; poverty headcount and ratios for 1981 to 2005 from Chen and Ravallion, 2008; for 1820 to 1980 from Bourguignon and Morrisson, 2002.

Although the poverty ratio was falling, the number of people living in absolute poverty – measured at the USD 1.25 a day standard in PPP dollars[4] –

---

4. The new standard is USD 1.25 a day, measured in 2005 PPP dollars. Previous measures of absolute poverty were USD 1 a day using 1985 price levels and USD 1.08 using 1993 price levels. Although this chapter uses USD 1.25 or USD 2.50 a day as poverty threshold figures, it should be understood that these figures are consistent with earlier literature using the USD 1 a day standard.

kept growing, from fewer than 900 million in 1820 to almost 1.4 billion in 1980 (Bourguignon and Morrisson, 2002)[5].

Subsequent work by Bhalla (2002), Sala-i-Martin (2002), Chen and Ravallion (2004) and Hillebrand (2008) extended the analysis from 1980 and found a pronounced downwards trend in poverty headcounts and poverty ratios, mainly because of very rapid economic growth in China and India. The conclusions on global inequality are more mixed. Bhalla (2002), Sala-i-Martin (2002) and Bourguignon and Morrisson (2002) show a downwards trend in global income inequality from 1980, while Milanovic (2005) and Hillebrand (2008) show little trend, at least until the late 1990s or early 2000s.

Poverty estimates made prior to late 2008 have been thrown into doubt by the release of new PPP estimates from the International Comparison Project (ICP). This new study is based on a much more complete global survey of prices (including China for the first time) and presumably gives a far more accurate measure for gauging cross-country differences in income and consumption (Heston, 2008). The major impact of this new work is that price levels for most non-OECD economies have been revised upwards, meaning that income, production and consumption levels have been revised sharply downwards, especially for China and India (Table 4.2).

Table 4.2
New and old estimates of per capita GDP in 2005

Country	2005 ICP	2005 WDI	2005 PWT63	2005 Exchange rate
	(USD in 2005 prices)			
China	4 091	6 760	6 637	1 721
India	2 126	3 452	3 536	707
Japan	30 290	30 736	27 726	35 604
United States	41 674	41 674	41 674	41 674

WDI = World Development Indicators the World Bank database; PWT63 = Penn World Tables version 63.
Source: Heston, 2008.

A new paper by Chen and Ravallion (2008) uses the 2005 ICP PPP estimates to create new estimates of global poverty for 1981 to 2005 that are hundreds of

---

5. To study incomes and poverty over time, the producers of the commonly used global economic databases – the World Bank (World Development Indicators database), Angus Maddison (2003) and the Penn World Tables (PWT) – start with PPP GDP estimates for every country at a given point in time, and then estimate past and future PPP GDP based on national income account data. This methodology has severe theoretical drawbacks, especially the implicit assumption that the PPP ratio between currencies is constant. Efforts to replace this methodology have been considered by Dowrick and Akmal (2005) and Feenstra and Rao (2008), among others, but their ideas have not yet been adopted by the global database producers.

millions of people higher than the authors' own previous calculations or other estimates appearing in the literature[6] (Table 4.3). The new Chen/Ravallion poverty numbers, while pointing in a direction consistent with the revisions of GDP per capita shown in Table 4.1, raise numerous questions of their own: Has the calculated fall in Chinese poverty really been so dramatic? Heston (2008) asserts that the implied Chinese growth going very far back is implausible. Has the fall in Indian poverty really been so small compared with Bhalla's calculations? Bhalla (2002) asserts that the household surveys underpinning the Chen/Ravallion poverty estimates badly underestimate total Indian consumption. Why are the implicit aggregate consumption figures for many countries so different from national income account figures? The aggregate consumption share figure falls dramatically in both China and India, leading to far higher estimates of poverty than consumption figures from the national accounts would suggest. Some of these questions may be answered when more details of ICP 2005 are released and when Penn World Tables (PWT) completes its analysis of the data; others will probably linger indefinitely owing to disagreements over data and methodology.

Table 4.3
New and old poverty estimates in 2005

Country/region	Chen/Ravallion 2008	WDI 2007	Hillebrand 2008
	(millions of people)		
China	208	77	131
India	456		163
Sub-Saharan Africa	391		427
World	1377	977	965

Sources: Chen and Ravallion, 2008; Hillebrand, 2008. The WDI numbers are World Bank updates of the Chen and Ravallion, 2004 calculations for 2001.

In any case, all poverty figures are estimates, based on imperfect data and on many different, challengeable assumptions about how to put the data together to come up with global inequality measures and poverty headcounts. For now, the Chen/Ravallion figures are the most up-to-date and comprehensive estimates available. The poverty numbers in the Chen and Ravallion 2008 paper, and the underlying

---

6. The data revision, and not changed economic circumstances, accounts for the huge jump in the number of people living in absolute poverty in 1981 as estimated by Chen and Ravallion compared with in 1980, as estimated by Bourguignon and Morrisson (2002). The new price data will presumably cause the 1820 to 1980 poverty estimates to be revised upwards too, but this work has not yet been done.

estimates for 119 countries made available through the World Bank's Povcal website[7] constitute the starting point for this chapter's estimates of poverty to 2050.

Explaining changes in poverty, 1981 to 2005

World poverty fell dramatically between 1981 and 2005, according to estimates by all the sources cited in the previous section, including the latest Chen and Ravallion (2008) work. All sources also agree that most, if not all of the gains were due to huge decreases in the Chinese poverty headcount. According to Chen and Ravallion (2008) the world absolute poverty headcount declined by more than 500 million people from 1981 to 2005,[8] and the world poverty headcount ratio fell from 42 to 21.3 percent (Table 4.4). The poverty headcount in China alone, however, fell by more than 600 million people. In only 24 years, China went from having 84 percent of its people living below the USD 1.25 a day absolute poverty level to having less than 17 percent of its people impoverished. Some other large countries (Brazil, India, Indonesia, Mexico, Pakistan, South Africa and Viet Nam) also showed dramatic reductions in the poverty ratio and, sometimes, the poverty headcount.[9]

Sub-Saharan Africa, on the other hand, saw a huge increase in the number of people living in absolute poverty and only a small decrease in the poverty ratio. Only four (out of 42) sub-Saharan African countries (Cape Verde, Mauritania, Senegal and South Africa) recorded falls in poverty headcounts, while a dozen countries recorded increases in poverty headcount ratios, and a few (the Democratic Republic of the Congo, Nigeria and the United Republic of Tanzania) showed tens of millions more people living in absolute poverty in 2005 than in 1981. However, faster economic growth in the last decade has led to a slight decline in the sub-Saharan Africa poverty ratios since 1996 (Figure 4.1).

Changes in the poverty headcount of any country can be ascribed to one of three factors: aggregate per capita economic growth; changes in the share of aggregate GDP going to private consumption versus the other components of GDP;[10] and distribution of consumption among individuals within the country.[11] For example,

---

7. http://web.worldbank.org/wbsite/external/extdec/extresearch/extprograms/extpovres/extpovcal net/0,,contentmdk:21867101~pagepk:64168427~pipk:64168435~thesitepk:5280443,00.html.

8. From this point onwards, all historical poverty figures (i.e., prior to and including 2005) included in this chapter are taken from Chen and Ravallion (2008) or from the World Bank's Povcal website, which contains more details than included in the 2008 paper.

9. Results for all countries, both historical and forecast, are available from the author.

10. Results for all countries, both historical and forecast, are available from the author.

11. Measured by estimated Lorenz curves and the standard accounting procedure (SAP) methodology.

---

if the share of GDP going to consumption remained the same in 2005 as in 1981, and the distribution shares across the population remained the same, all the differences in poverty levels could be explained by changes in economic growth.

Figure 4.1
Trends in GDP per capita and poverty headcount ratio, sub-Saharan Africa

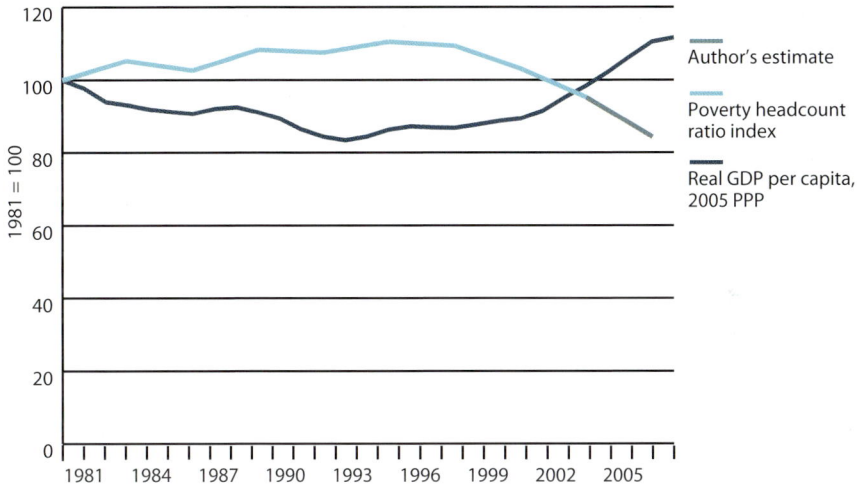

Sources: Poverty headcount ratio from Chen and Ravallion, 2008; estimate for 2008 by author; GDP per capita from World Development Indicators.

Economic growth in non-OECD countries overdetermines the estimated fall in poverty headcounts (Table 4.4). Had Lorenz curves[12] and consumption ratios remained constant, the world poverty headcount would have fallen from 1 896 million people in 1981 to 791 million in 2005, and not the 1 377 million people estimated by Chen and Ravallion. Declines in the aggregate consumption ratio and shifts in distribution combined to increase the poverty headcount by almost 600 million people from what it would have been if aggregate and by-person distribution had remained at 1981 levels.

---

12. The Lorenz curve is a widely used technique for showing inequality in income (or any other quantity distributed across a population). It shows the cumulative shares of income held by cumulative shares of the population. If income is distributed evenly, each 10 percent of the population gets 10 percent of the total income, and the curve is a straight line with a 45 percent slope. The more unequal the distribution, the greater the bow in the curve to the right of the 45 percent line. The Gini coefficient is a summary statistic that measures the area between the 45 percent line and the Lorenz curve. In principle, Gini coefficients range between 0 (perfect equality of income) and 1.0 (perfect inequality – one person in the population gets all the money). In practice, GDP per capita or consumption per capita Gini coefficients range from the mid-0.20s (in some Scandinavian countries) to the 0.60s and 0.70s (in some African countries).

Table 4.4
World poverty headcounts and poverty ratios

Region/country	Headcount (*millions*)		Ratio (%)	
	1981	2005	1981	*2005*
World	1 896	1 377	42.0	21.3
East Asia	1 072	316	77.7	16.8
China	835	208	84.0	15.9
Indonesia	108	47	71.5	21.4
Viet Nam	49	19	90.4	22.8
South Asia	548	596	59.4	40.3
India	421	456	59.8	41.7
Pakistan	62	35	72.9	22.6
Latin America	42	46	11.5	8.4
Brazil	21	14	17.1	8
Mexico	6.8	2	9.8	1.7
Sub-Saharan Africa	214	391	53.7	51.2
Congo, Dem. Rep.	9	35	31.9	59.2
Nigeria	35	88	47.2	62.4
South Africa	10	10	34.9	20.6
East Europe and Central Asia	7	17	1.7	3.7
Near East/North Africa	14	11	7.9	3.6

Source: Chen and Ravallion, 2008, with world headcount divided by world population.

Although China started with an extremely high rate of absolute poverty, its rate of real per capita economic growth was so high (8.8 percent a year)[13] that even the estimated consumption of the lowest 10 percent of the population would by 2005 have far surpassed the USD 1.25 a day per person absolute poverty standard if the overall amount of GDP going to consumption had not dropped sharply and the inequality of distribution of that total amount of consumption had not increased sharply.[14] Poverty headcounts were also down in most other East Asian countries. Indonesia and Viet Nam cut their poverty headcounts sharply by combining strong economic growth without adversely affecting consumption ratios. The Philippines was the worst performer in the East Asian region: the poverty headcount went up by 3.7 million people, mainly because of low economic growth.

---

13. 1982 to 2005, see World Development Indicators database, 2009, using GDP per capita in 2005 PPP dollars.

14. The World Income Inequality Database suggests that aggregate Chinese Gini coefficient rose about 15 points, from 0.29 to 0.44, over this period, while the Indian Gini coefficient rose about 4 points, from 0.32 to 0.36. www.wider.unu.edu/research/database/en_gb/database/.

India had high economic growth, at 3.3 percent per year, which would have been fast enough to raise 364 million people out of absolute poverty if the distribution of income and consumption had not changed so greatly. However, the ratio of aggregate consumption to GDP fell by about 20 percentage points over this period, and aggregate consumption was distributed more unevenly, with the overall Gini coefficient on household consumption rising about 4 percentage points. Pakistan performed better than India. Its poverty headcount went down and its poverty ratio dropped dramatically, from 72.9 to 22.6 percent, according to the Chen/Ravallion numbers. Its economic growth was weaker than India's, but Pakistan did not have the dramatic decline in the ratio of private consumption to GDP.

Table 4.5
Impacts of economic growth and distribution shifts on poverty headcounts

Region/country	1981	2005	Total change	Change due to GDP growth	Change due to shifts in aggregate consumption ratio	Change due to income distribution shifts (shifts in Lorenz curves)
				(millions of people)		
World	1 896	1 377	-520	-1 105	344	241
East Asia	1 072	316	-755	-957	21	181
of which China	835	208	-627	-835	38	170
South Asia	548	596	47	-389	362	75
of which India	421	456	35	-364	324	75
Sub-Saharan Africa	214	391	177	252	-63	-11
Latin America and Caribbean	41	44	3	-3	21	-15

Sources: 1981 and 2005 poverty headcounts from Chen and Ravallion, 2008; growth and distribution shifts estimated by author (sums may not total owing to rounding).

Sub-Saharan Africa had very negative results. Average real GDP growth was slower than population growth and would – without favourable distributional changes – have caused poverty headcounts to double. The worst performers were Côte d'Ivoire and the Democratic Republic of the Congo. These two conflict-torn countries had average negative GDP per capita growth of 2 and 4 percent per year, respectively. The ratio of consumption to GDP soared, but not enough to compensate for the growth effects. Nigeria also had very negative results, with the poverty headcount rising by almost 54 million people and the poverty ratio rising from 18 to 62.4 percent. Nigeria had a toxic combination of low growth in GDP per capita (0.7 percent per year), a sharp fall in the ratio of private consumption to GDP (from 42 to 28 percent) and a rise in consumption inequality (the Gini coefficient rose from 0.387 to 0.429). South Africa was one of the best performers on the continent. It had low economic growth (-0.2 percent), but a large increase in the consumption ratio (from 43 to 53 percent) and a slight decrease in inequality (the Gini coefficient fell from 0.59 to 0.58).

Latin America has higher average incomes and less absolute poverty than Asia and sub-Saharan Africa. Because it did not have much absolute poverty to begin with in 1981, it did not take much per capita GDP growth to push more people above the poverty threshold, as long as distribution did not change adversely. Per capita real GDP growth was only 0.7 percent per year from 1981 to 2005, but the regional consumption ratio average rose by two percentage points, and the population-weighted regional Gini coefficient rose only slightly. Mexico and Brazil have made dramatic progress since 1981. Brazil brought its poverty headcount down by almost 7 million people and its poverty ratio shrank from 17 percent in 1981 to 7.8 percent in 2005. Mexico reduced its poverty headcount by 4.9 million people, while reducing its poverty ratio from 9.8 percent in 1981 to 1.7 percent in 2005. Argentina, the Plurinational State of Bolivia, Peru and Venezuela all saw sharp increases in their poverty ratios between 1981 and 2002, but both the headcounts and the poverty ratios showed large decreases between 2002 and 2005, according to the World Bank Povcal database.

Trends in global inequality, 1981 to 2005

Chen and Ravallion (2008) do not report any calculations of global inequality. Hillebrand (2008) reports several different estimates of global inequality (Table 4.6). Most of these (Milanovic is the exception) use estimates of within-country income or consumption distributions and multiply these by the value of income or consumption taken from the national income accounts. For Milanovic (2005), a better measure would be to distribute the total consumption by country inferred from the household consumption surveys. Table 4.6 is based on data for the 119 countries included in the Povcal database, plus consumption figures – from various sources, but mostly using aggregate consumption data in 2005 PPP terms – for the additional 63 countries covered in the author's database, and used Bhalla's (2002) SAP to calculate world Gini coefficients, which fell slightly from 1981 to 2005, mainly because of strong economic growth in Asia.[15]

---

15. Using the 2005 ICP, Milanovic (2008) also revised upwards his estimate of global inequality. His new estimate for the global Gini coefficient in 2002 is 0.699 compared with a previous estimate of 0.653.

Table 4.6
Estimates of world Gini coefficients

	1820	1970	1980	1981	1988	1992	1993	1998	2005
Bourguignon/Morrisson	0.50	0.65	0.657			0.657			
Bhalla			0.686			0.678		0.654	
Sala-i-Martin			0.662			0.645		0.633	
Milanovic					0.619		0.652	0.642	
Hillebrand			0.653						0.634
2009 estimate using 2005 ICP data				0.709					0.684

Sources: Bourguignon and Morrisson, 2002; Bhalla, 2002; Sala-i-Martin, 2002; Milanovic, 2005; Hillebrand, 2008; and author's estimates.

Forecasting economic growth

Forecasting poverty 40 years into the future is mainly a matter of forecasting economic growth. Bourguignon and Morrisson (2002) claimed that economic growth had by far the greatest impact on global poverty inequality for 1820 to 1992. Ravallion (2001) and Dollar and Kraay (2002) show that the poor, on average, tend to share proportionately in the gains from economic growth; and the previous analysis of the Chen/Ravallion poverty data set shows that economic growth far outweighed the impact of the other two proximate causes – the distribution of national output between consumption and other uses, and changes of distribution by person, in each country.

Economists have long relied on the neoclassical growth model (Solow, 1956) to think about economic growth. In Solow's framework, economic growth depends on changes in the capital stock (machinery, buildings, roads, communication lines, etc.), changes in the labour force, and changes in technology. In this model, diminishing returns eventually set in and growth slows, unless technological change intervenes to keep productivity increasing.

According to empirical research by Abramowitz (1956) and many others, changes in technology have contributed the major part of long-run economic growth in OECD countries, and thus should be important to forecasts of the future. While changes in capital and labour are relatively simple to model and forecast, technology is not. Solow treated the technological change component as a residual or exogenous factor, not explainable by growth theory. Later researchers, especially Romer (1987; 1990), Grossman and Helpman (1991) and Barro and Sala-i-Martin (1995) have attempted to "endogenize" growth theory by trying to explain theoretically (and demonstrate empirically) the causal forces underlying technological progress, especially investment in research and development (R&D), but also institutional factors such as protection of property rights, regulation of international trade, and taxation.

169

An important corollary of the extended neoclassical growth model for poverty analysis is the convergence concept. It is implicit in the neoclassical growth model that poor countries should grow faster than rich countries and should eventually catch up – converge – with the latter's per capita output and income. According to Barro (1998: 1): "If all economies were intrinsically the same except for their starting capital intensities, ... poor places would tend to grow faster per capita than rich ones." Because rich countries are limited by diminishing returns and poor countries can grow faster by increasing capital stocks and adopting best practice technology, incomes ought eventually to converge. Lucas (2000) makes use of this convergence concept to predict rapid non-OECD growth and a convergence of incomes by 2100.

On the other hand, North (2005) believes that neoclassical economic theory by itself is not much help in explaining the process of economic change – institutions are more important. Economies are composed of institutions that provide incentives for work, trade, saving and investment, or not. Institutions that stifle competition and encourage predation might arise and persist, counter to the convergence hypothesis, because institutions poorly designed for economic growth might be well suited for maintaining the power and prosperity of those in command or be based on cultural beliefs that do not value economic growth highly. Collier (2007) warns that bad governance is only one of the four poverty traps that can keep countries down.[16] Olson (1982) suggests that even rich and prosperous countries that have achieved prosperity through good institutions are constantly at risk of economic sclerosis, as special interests accrue power over time, through lobbying and politics, to undermine the institutions that spur competition and investment.

Most long-run economic growth forecasts that appear in the literature are based on modelling exercises that use neoclassical and endogenous growth theory, the convergence concept and some reference to the institutional ideas of North, Olson and others. While there is much to criticize and debate in the theoretical literature, it is important to note that the empirical estimates of the underlying relationships are also contentious, with the magnitude of the relationships and even the direction of causality often in dispute. Any forecasting effort requires many assumptions about policy choices by future governments over long periods; long-run forecasting efforts are necessarily speculative.

---

16. The other three are: i) conflict and political violence; ii) abundance of natural resource wealth that distorts economic growth; and iii) geographical disadvantages such as being land-locked, poor in resources or harried by bad neighbours.

Forecasting poverty and inequality

What will global poverty look like in ten, 20 or 45 years? Not many explicit forecasts appear in the literature. Using the old ICP data, Chen and Ravallion (2004: 33) suggest global poverty will drop, but their estimate is based on two time series regressions (one each for East Asia and South Asia) based on past changes in the poverty headcount relative to assumptions about long-term economic growth. They assume that the poverty ratio in Africa will continue to be 45 percent. Their modelling and assumptions add up to a world poverty rate of 15 percent in 2015, thus meeting the Millennium Development Goals (MDGs).

Bhalla (2002) concludes that the world poverty rate has already dropped below 15 percent and will continue to decline. Bhalla estimates a reduced-form equation to calculate the elasticity of the poverty headcount ratio to growth in incomes or consumption, and uses this regression model to forecast future poverty levels assuming the distribution of income or consumption within countries remains the same.

The World Bank has been making forecasts of the 2015 world poverty rate in its Global Economic Prospects series since 2001. In the latest edition (World Bank, 2009), the 2015 forecast is revised upwards from 10.2 to 15.5 percent, because of the ICP revisions. These forecasts apparently use a cross-country regression that posits a constant elasticity of poverty reduction to per capita income growth adjusted by estimates of changes in within-country inequality. The constant elasticity assumption is not very reliable for extending projections far into the future, given that this is about movements below or above a fixed poverty threshold. A country with incomes just below the threshold can cross the threshold with only a low level of growth, and a country with incomes far below the threshold can have high rates of growth without moving many people out of extreme poverty. A different forecasting methodology is clearly needed.

In a major new study, Hughes *et al.* (2008) review past poverty forecasting efforts in detail and present their own set of forecasts to 2055, using the "lognormal" distribution to convert estimates of average income and the Gini coefficient into poverty headcounts. This methodology has the advantage of embedding the poverty estimates directly into a long-range macroeconomic simulation model (the International Futures Model, see Hughes and Hillebrand, 2006), so Hughes *et al.* or any user of the model can not only test directly the impact of alternative assumptions about economic growth on poverty futures, but also simulate the effects of changes in a wide variety of policy levers on economic growth, and hence on poverty. The Hughes estimates are based on the old ICP data and so are not directly comparable with the new Chen/Ravallion (2008) numbers that form the basis of this study.

This chapter uses Bhalla's SAP methodology to help forecast future poverty levels. If there are estimates of future GDP, assumptions that the within-country distribution of income and consumption remains constant, and assumptions that the ratio of consumption to income is constant, the percentiles of income and consumption can simply be read off, using the same accounting framework as used in the historical analysis. All three of these key "ifs" are problematic. There is no scientifically sound methodology for forecasting global incomes and consumption decades into the future. Most long-term projections, including this one, rely on scenarios: the researcher posits a set of assumptions about the key drivers of growth, uses a model that relates these factors to economic outcomes, and produces projections that are presumed to be part of a range of plausible outcomes. The assumption of unchanging within-country distribution is often made in long-run forecasts (Chen and Ravallion, 2004), mainly because there is little scientific basis for predicting long-range changes, and the existing empirical work on the subject shows such divergent results (see World Bank, 2007, versus Higgins and Williamson 2002). Consumption-to-GDP ratios could also change for endogenous economic reasons or because of political decisions, but in this chapter they are assumed to remain constant.

The World Bank poverty estimates give good news about global poverty from 1981 to 2005, but it is likely that the very high economic growth recorded by non-OECD countries drove poverty headcounts down even further up until 2008. Using actual GDP growth rates for between 2005 and 2008, and assuming no changes in within-country distributions, Table 4.7 shows how the global poverty headcount may have fallen by more than 200 million people, and the poverty headcount ratio declined to about 18 percent.

Table 4.7
Poverty estimates

	2005		2008		2006–2008
Region/country	Headcount *(millions)*	Ratio *(%)*	Headcount *(millions)*	Ratio *(%)*	Average annual growth in real per capita GDP *(%)*
Non-OECD	1 377	21.3	1 132	17.6	4.6
East Asia	316	16.8	247.2	15.9	5.7
China	208	15.9	148	12.3	7.5
South Asia	596	40.3	467	30.2	4.4
India	456	41.7	339	29.9	4.9
Latin America and Caribbean	46	8.4	36	6.6	3.0
Sub-Saharan Africa	391	51.2	352	44.3	2.9

Sources: Poverty estimates for 2005 from Chen and Ravallion, 2008; for 2008, author's calculations based on SAP methodology adjusted upwards based on World Bank, 2009: 117; growth rates for 2006 to 2008 from World Development Indicators and *The Economist*.

The analysis relies on the SAP methodology and a spreadsheet model that estimates average consumption by percentile of population for 182 countries. The poverty and inequality estimates from the SAP model are driven by population and economic growth numbers that are derived from scenarios produced with the International Futures (IF) model. The IF model is convenient because it contains detailed growth models for 182 states, numerous policy levers that have been calibrated based on recent empirical work at the World Bank and elsewhere, and numerous well thought-out long-range growth scenarios. It will become clear that slightly varying assumptions about a small number of key parameters can have very large effects on global poverty and inequality. The poverty forecasts presented in the following sections are based on two scenarios: the market first scenario assumes rapid technological change in OECD countries and a strong tendency towards convergence in non-OECD countries, based on globalization, pro-growth policies and institutional change; the trend growth scenario assumes less technological change, less globalization and less improvement in economic governance in slow-growth regions.

The market first scenario

The market first scenario is based on the IF default scenario as of October 2008. It was compiled by the IF team at the University of Denver, United States of America, using an optimistic set of assumptions consistent with global analysis from the United Nations (UN) and the National Intelligence Council (see, in particular, National Intelligence Council, 2008; UN, 2004; UNEP, 2007). The World Bank (2007) elaborated a similar scenario. As in the World Bank work, the numbers used here are not a forecast but a scenario based on assumptions about changes in population, capital stock and productivity gains. High growth is based on assumptions of strong technological change brought about in OECD countries by continuing R&D. Non-OECD countries advance by catch-up economic growth fostered by high investment, improved governance, efficiencies from expanded trade and financial linkages, and rising investment in human capital. There is clearly much scope for catch-up growth in non-OECD countries, but there is also no scientific way of forecasting how much convergence will be achieved or what growth enhancing or retarding policies will be followed in each country.

The assumptions used here produce another golden age of growth, with world growth and growth in most regions higher than in the last 25 years. With economic growth at this high pitch, world poverty shrinks dramatically. The number of people in extreme poverty shrinks from 1 377 million people in 2005 (the Chen/Ravallion starting point) to 964 million in 2015 and 245 million in 2050 (Table 4.8). Strong economic growth leads to the eradication of extreme poverty in India, but not China. China's far more unequal distribution of income and consumption put it at a disadvantage in eliminating poverty. Sub-Saharan Africa cuts its poverty rate substantially but, assuming continuing high population growth rates,[17] the number of people living in extreme poverty continues to grow after 2015. A few countries in East and South Asia (Afghanistan, Bangladesh, Nepal, Pakistan and the Democratic People's Republic of Korea) and Haiti account for most of the rest of the people still living in extreme poverty in 2015. By 2050, assuming per capita income growth of more than 2 percent a year, the poverty headcount in sub-Saharan Africa has started to fall but is still more than 200 million people. In this high-growth scenario, by 2050, the global poverty rate is only 2.5 percent.

Table 4.8
Poverty estimates in the market first scenario

Region/country	Average annual growth of real GDP per capita 2006–2050 (%)	Headcount at USD 1.25 a day Constant within-country distributions			Poverty ratio		
		2005 (millions)	2015 (millions)	2050 (millions)	2005 (%)	2015 (%)	2050 (%)
World	3.0	1 377	964	245	21.3	13.3	2.5
OECD	2.9						
Non-OECD	3.6	1 377	964	245	24.8	15.3	3.0
East Asia and Pacific	4.3	316	126	15.6	16.8	5.3	0.7
China	4.8	207	106	12.4	15.9	7.6	0.8
South Asia	4.3	596	249	14.1	40.3	15.4	0.6
India	3.9	456	243	0	42.0	19.8	0.0
Sub-Saharan Africa	2.5	391	395	205	50.9	41.1	11.7
Latin America	3.4	46	35	7.8	8.2	5.6	1.0
Near East and North Africa	3.3	11.0	8.7	0.7	3.6	2.2	0.1
Eastern Europe and former Soviet Union	3.4	17.3	13.5	2.1	3.7	3.8	0.4
World Gini coefficient		0.684	0.680	0.648			

Sources: Historical data from World Development Indicators, with estimates from Maddison, 2003 for missing data; scenario data from simulations with the IF model.

The world Gini coefficient falls to 0.648 in 2050, but still remains high compared with most within-country distributions, because economic growth is assumed to continue to be strong in OECD and other rich countries. Continued high global inequality and high Gini coefficients within many countries are

---

17. The population growth rates embedded in the IF forecasts closely track the UN's mid-range population forecast.

troublesome features even in this low-poverty scenario, and may prevent the poverty falls from occurring. Alesina and Perotti (1993) found that income inequality hurts growth by increasing political instability and thereby decreasing investment. Rowan (1996) believes that inequality heightens class conflict, produces capital flight and encourages redistributive policies that can be self-defeating. Chua (2004) believes that global inequalities provoke resentment of the poor towards the rich countries, at best inhibiting cooperation and trade, and at worst provoking violence.

Sub-Saharan Africa performs relatively poorly in the market first scenario, but even there the poverty headcount eventually starts to decline. Economic growth in this scenario is not low by world historical standards, and is good by Africa standards – per capita GDP is projected to rise by 2.5 percent per year for the region. The average of country growth rates is similar, but the IF projections show a wide range of country growth rates,[18] from -0.8 percent per year in Togo to 5.9 percent in the United Republic of Tanzania. These rates of growth are enough to bring the poverty rate down sharply in the region, but population growth is so high and the starting level of income so low in most countries that it takes a GDP per capita growth rate of at least approximately 2 percent per year to bring the poverty headcount down. Cameroon, the Democratic Republic of the Congo and Liberia are among the weakest performers, and eight of the 38 countries projected show higher poverty headcounts in 2050 than in 2005. High projected economic growth in Ethiopia, Mozambique, Nigeria and the United Republic of Tanzania accounts for about 70 percent of the fall in the regional poverty headcount. South Africa nearly succeeds in eliminating extreme poverty, not because of high economic growth but because its poverty headcount ratio in 2005 was so low that it did not require much positive per capita economic growth to push almost all of the population above the poverty threshold.

The IF model also produces estimates of food supply and demand, by country, which are consistent with its demographic and economic projections. World food demand in this high economic, medium population growth scenario increases by about 1.3 percent a year to 2050. World supply rises somewhat less because substantial improvements in technology and transportation infrastructure are assumed to cut crop losses sharply. Land devoted to crop production is assumed to rise only slightly, while technological advances increase world average crop yields by about 0.9 percent per year (Table 4.9). Calories available per person rise everywhere, particularly in sub-Saharan Africa. If alternative assumptions were made, by assuming a reduction in the technological advances that aid food

---

18. Mainly because of different assumptions about policy changes by country, and between-country historical differences in translating policy changes into economic growth.

production, the relative price of foodstuffs would increase, some countries would be advantaged and some disadvantaged, but overall world economic growth would slow and poverty increase.[19]

Table 4.9
World food supply and demand in the market first scenario

	World crop production (million tonnes)	Crop land (million ha)	Yield (tonnes ha)	Crop loss ratio (%)
2005	4 190	1 544	2.71	30.3
2050	6 584	1 617	4.07	22.3
Change (%)	57.1	4.7	50.0	
Average annual change (%)	1.0	0.1	0.9	

	Calories available per person			
	World	OECD	Non-OECD	Sub-Saharan Africa
2005	2 800	3 421	2 662	2 256
2050	3 207	3 635	3 135	2 588
Change (%)	14.5	6.3	17.8	14.7
Average annual change (%)	0.3	0.1	0.4	0.3

Source: IF model.

How might distribution shifts affect future poverty headcounts?

It has been seen that economic growth is not the only factor that matters for changes in poverty levels; shifts in the amount of production made available for consumption (the consumption-to-GDP ratio) and shifts in the distribution of consumption among a population (Lorenz curve shifts) can also have large impacts on poverty.

Lorenz curve shifts: Kuznets (1955) suggested that economic development itself made income distributions more unequal, by increasing returns to capital and leaving the rural poor lagging further behind workers in the modernizing sectors of the economy. More recent work by Ravallion (2001) and Dollar and Kraay (2002) rebuts the idea that growth has negative or any systematic effects on distribution. However, Barro (2000) suggests that income inequality tends to rise until a country reaches a per capita income of USD 4 815 (in 2000 PPP dollars), when it starts to fall.[20]

---

19. More interactions between growth, inequality and food supply and demand could be generated for a revised version of this chapter.

20. This idea could be explored empirically in another version of this chapter.

Some researchers have attempted to forecast changes in within-country income distributions based on demographic shifts. Using data from the 1960s to the 1990s, Higgins and Williamson (2002) find a strong relationship between trends in income equality and demographic shifts: inequality decreases as higher-earning middle-age cohorts grow in proportion to the rest of the population. The authors forecast very large decreases in within-country inequality over the next 50 years, with the weighted average African Gini coefficient falling from 0.464 in the 1990s to 0.378 in 2050, and the Latin American and Pacific Rim regions experiencing similar proportionate declines. Higgins and Williamson also report estimated changes in the ratio of income of the highest to the lowest quintiles (Q5-to-Q1) for the three regions.

Although the Higgins and Williamson regional income distribution estimates do not give a clear linkage to the country income and consumption distributions used in this chapter, their forecast of the declines in Gini coefficient and Q5-to-Q1 ratios can be used to generate forecasts of country distributions. The resulting headcounts can then be calculated to show the sensitivity of the poverty and Gini coefficient numbers to the Higgins and Williamson forecast. The new country distribution estimates used in this simulation capture the essence of the Higgins and Williamson estimates: the three regional Gini coefficients fall by the same ratio, and the Q5-to-Q1 ratios fall by the same amounts. The postulated change in within-country inequality, motivated by shifting demographics, reduces the global poverty headcount estimate for 2050 from 245 million to 127 million people.

However, researchers at the World Bank (2007) have recently used other empirical work suggesting a conclusion opposite to that of Higgins and Williamson: as the shares of older workers rise in proportion to the total workforce, inequality rises "since wage dispersion within these groups tends to be high" (World Bank, 2007: 85). The World Bank suggests an increase of about 0.04 in the African regional Gini coefficient by 2030, and an increase of 0.016 in the Asian Gini coefficient. From rough estimates of what the World Bank numbers would mean to the percentile distributions used in this chapter – with the inferred Q5-to-Q1 ratios rising in Asia and Africa, instead of falling as in the Higgins and Williamson case – the shifting within-country distribution pushes up the 2050 global poverty headcount to 328 million people.

Thus the two conflicting views of the endogenous future of Lorenz curve shifts create a band of about a 100 million people on either side of the market first scenario projected poverty headcount of 245 million in 2050. Of course, governments may also undertake policy measures that, explicitly or unintentionally, shift the Lorenz curve in either direction.

Shifting consumption-to-GDP ratios: Consumption-to-GDP ratios average about 56 percent in OECD countries, and fluctuated around a narrow range from 1981 to 2005. The average consumption-to-GDP ratio for non-OECD countries is similar, but much more variable among countries, with numbers ranging from 14 to 171 percent of GDP using PPP data from the World Bank's World Development Indicators database. Using implicit consumption figures from the household surveys reported on the Povcal website, and dividing by the GDP figures from the World Development Indicators database, the range becomes even greater, from 7 to 237 percent. Some very large ratios occur in war-torn countries, where investment is probably very low and foreign aid very high. Some very small ratios occur in countries with substantial mineral export wealth. It is also possible that some of the large and small numbers are due to data errors in the household surveys, national income accounts data or both.

The analysis presented in Table 4.5 shows that the poverty estimates were significantly affected by past shifts in consumption-to-GDP ratios, particularly the huge implicit decline in the Indian consumption figures. In a long-run scenario such as this, with very high growth rates over time, it could plausibly be assumed that the non-OECD consumption rates ought to converge and stabilize near the present OECD levels. Such an experiment was not conducted for this chapter, but its results would probably not have had a great impact on the overall numbers – because the starting point for non-OECD countries was not very dissimilar from that of OECD countries – although they could dramatically affect those countries that are now far from the OECD average.

However, this analysis also suggests that the conventional concept of pro-poor growth that looks at just the shift in income Lorenz curves and economic growth (Kakwani and Pernia, 2000; Chen and Ravallion, 2001) is inadequate – shifts in the consumption ratio must also be considered, and should not be treated as independent of either growth or the Lorenz curve. An increase in the consumption ratio, other things being equal, reduces the poverty headcount. If an increase in the ratio comes at the expense of productive investment, however, the long-term effect could be anti-poor.

This chapter's poverty measures rely on household consumption surveys that reflect changes in aggregate consumption figures with little correlation to changes in consumption and GDP figures in the national income accounts. This use of sometimes inconsistent data weakens an important analytical link between poverty and economic growth. For example, according to national income accounts data (converted into 2005 PPP data by the World Bank), India's real GDP per capita grew at an average annual rate of 3.8 percent from 1981 to 2005, and private consumption per capita grew at 2.9 percent per year. The Povcal database per capita consumption figures, based on household survey data, grew at just 1.0 percent per year over this period. If Chen and Ravallion (2001) are correct in stating that the household surveys are a better measure of consumption than the national income accounts, it should probably be concluded that the GDP growth estimates are not reliable. More detailed analysis is required, illuminating not just the forces behind shifts in the Lorenz curve but also the connection between consumption measured by the household surveys and economic growth.

Setting aside these analytical problems, the numbers in the market first scenario tell a good-news story. The extreme poverty headcount is shrinking, in most regions by 2015, and in all regions by 2050. The original MDG global poverty headcount ratio – 15 percent by 2015 – should be reached easily.[21] While this chapter focuses on the numbers at the USD 1.25 a day standard, the improvements at the more generous USD 2.50 a day standard are even more impressive: from 3 085 million people (48 percent of world population) in 2005, to 710 million (7.3 percent) in 2015. Even in the pessimistic scenario, in which demographic shifts lead to worsening within-country distributions (the World Bank scenario), the global poverty headcount still shrinks dramatically because of good economic growth.

The trouble with this good-news story, however, is that it is just a scenario; there is no way of knowing whether world economic growth rates will be anywhere near this high, or how within-country distributions will change. The growth rates assumed in the market first scenario are almost all higher than those that actually occurred in the post-Second World War "golden age" of global growth, when so many of the poverty rate reductions calculated by Bourguignon and Morrisson (2002) occurred.

Economic growth of more than 3 percent per year in real per capita terms in non-OECD countries is certainly possible over the next 40 years. Most of the countries in this group are so far behind the OECD countries in productivity levels that they have enormous growth potential through adopting modern techniques and gradually converging towards OECD-level productivity. The long-term growth rates envisioned in the market first scenario for Africa, Latin America and the Near East are actually quite close to the growth rates achieved in 2002 to 2007, coinciding with an unusually high period of world economic growth. Even assuming that war, resource constraints or climate difficulties do not intrude, maintaining such high growth rates will involve enormous changes in governance, institutions and attitudes in many countries.

---

21. Chen and Ravallion (2008) suggest that as the original goal was a "halving of the extreme poverty ratio from 1990 to 2015", the upwards revision of the historical numbers implies that the new goal should be closer to 20 than 15 percent. This goal is also easily reached in the market first scenario.

Comparison with other long-range growth and poverty projections: Nobel-Prize winning economist Robert Lucas (2000) has produced a similar scenario. He believes that non-OECD countries will converge with OECD countries over the course of this century, and cites three major reasons:

- Technology diffusion (Tamura, 1996) – the idea that knowledge produced anywhere benefits producers everywhere;

- improvements in governance (Parente and Prescott, 1994) – "governments in the unsuccessful economies can adopt the institutions and policies of the successful";

- diminishing returns and flows of resources – "high wages in the successful economies lead to capital flows to the unsuccessful economies, increasing their income levels".

Lucas' world growth model suggests that the long period of rising global income inequality that began with the industrial revolution in 1800, slowed down or ended in recent decades and will reverse itself in this century: "I think the restoration of inter-society income equality will be one of the major economic events of the century to come." [22]

Rowan (1996) predicts that within a generation most of the world's population will be rich or at least much closer to being rich than it is today. Not only will incomes converge across countries, but the world will also become more peaceful and democratic. Rowan does not deny the existence of enormous problems in every part of the non-OECD world, but he believes that better policies and growing social capabilities will spur growth:

A major reason why there are still poor countries is that their economic policies have produced unstable prices and employment, domestic prices out of line with world ones, inefficient nationalized and regulated industries, low trade shares, little foreign capital and technology, and obstacles for the creation of new industries. Such errors are now widely being corrected. Import-substitution policies are being replaced by export-oriented ones, countries hitherto hostile to foreign investment are encouraging it, regulations being reduced, firms privatized, and more. (Rowan, 2006: 93)

Maddison (2007) has also produced a bullish long-run economic forecast to 2030, although one with more diverse regional results than the market first scenario. He forecasts that between 2003 and 2030, non-OECD countries will grow almost twice as fast as OECD countries (at 3.0 compared with 1.7 percent a year) in real per capita terms. He assumes that technological advances will keep

---

22. The Lucas arguments and quotes cited are from Lucas, 2000: 164–166.

growth high in the mature economies, and expects that convergence forces will allow China and India to average about 4.5 percent per capita growth in real terms. Growth will slow over the period, as these countries approach the technological frontier and are forced to devote more resources to environmental and welfare issues. He assumes that Latin America will continue on a slow growth path, owing to outright rejection or half-hearted implementation of pro-growth policy reforms. He projects only 1 percent per capita growth in sub-Saharan Africa.

The global growth optimism in the market first scenario is replicated in the Intergovernmental Panel on Climate Change (IPCC) A1 global warming scenarios, which envision very rapid economic growth of 3.1 percent per year in real world per capita GDP for 2001 to 2050, based on increased globalization and rapid introduction of new technology (IPCC, 2009).

The bullishness of all these scenarios comes from their sharing of similar concepts about economic growth. The projections are based, implicitly or explicitly, on the extended neoclassical growth model and assumptions about the same factors that are presumably growth-promoting, such as the institutional and policy factors that promote or discourage convergence. Economists at the World Bank and elsewhere are in general agreement on the nature of the governance and institutions that work best to promote long-run economic growth:

- Free markets and private property are better at generating growth than centralized government control of production, but a strong government is essential to enforce the rules of peaceful economic behaviour and alleviate inevitable market failures.

- Trade and financial market liberalization is needed to spur competition and the flow of investment funds, including increased access to developed country goods and capital markets.

- Democratic accountability of government is helpful, to keep both corruption and predation from destroying incentives to work, save and invest, and to encourage pro-growth spending on education, health and infrastructure.[23]

Despite wide, but not universal, acceptance of these principles, however, there is little agreement on how countries can or should transition to modernity and on what outsiders can do to help. It took hundreds of years for Western Europe and North America to develop, from within, the institutions that propel the modern economy, and the Washington Consensus ideas provide only general principles, not specific policy guidance. No well-meaning expert has the ability

---

23. This list stems from the original Washington Consensus list proposed by Williamson (1989). For a more up-to-date discussion, see Commission on Growth and Development (2008) and Rodrik (2008). Olson (1996), provides a discussion on overcoming the collective action problem.

to design a fail-safe programme that guarantees economic success, even in countries with governments willing to reform. In addition, the application of policies aimed at converting these principles into practice under the guidance of the International Monetary Fund and the World Bank has lead to numerous policy failures, few successes and much bitterness (Easterly, 2001). There is also some outright political opposition to many of the tenets of this market-oriented approach to economic governance, and it is very easy for political leaders to resist or overthrow reform efforts for reasons of intellectual disagreement, ignorance, domestic politics or personal (or group) advantage (for more on this, see Acemoglu and Robinson, 2006).

The market first scenario also assumes that OECD countries continue to grow at high rates in per capita terms, compared with historical norms. This is not implausible. Despite the severe recession of 2008/2009, the OECD countries have economic and political institutions designed to generate good economic growth, and large expenditures for R&D expand the knowledge frontier in a way that could well lead to significant productivity gains for decades to come. Growth in countries at the technological frontier depends mainly on human capital development, and there is no physical limit on that. (For optimistic discussion about the future of technology, see Schwartz, 1999; and Duesterbeg and London, 2001.)

High OECD growth by itself probably hurts the global inequality numbers, but it is helpful to economic growth, and hence poverty reduction, in non-OECD countries. OECD countries face their own set of problems, however, especially in dealing with a rapidly ageing population that threatens to undermine the social contract underpinning economic success. It is easy to imagine a scenario with much lower economic growth in both OECD countries and the rest of the world.

The trend growth scenario

An alternative scenario calculates what would happen to global poverty if the benign assumptions that drove convergence of the non-OECD countries in the market first scenario did not occur. Instead, most countries are assumed to continue on the same trajectory they have been on for the last 25 years. For some countries, notably China and India, this is a very good trajectory, but for Latin America, Africa and the Near East, recent economic history has not been favourable, apart from for a few years in the early 2000s, when almost all countries participated in an unsustainable global boom.

In the trend growth scenario, the per capita growth rate in non-OECD countries as a whole is about half a percentage point per year less than in the market first scenario, but the growth assumptions are cut drastically in the countries where most of the poverty is – those in sub-Saharan and North Africa and a few in Asia and Latin America. As demonstrated, the market first scenario assumes very large increases in these countries' economic growth, compared with the past two decades.

What happens to global poverty if economic growth rates do not improve from levels recorded in 1981 to 2005? In some regions, the trend growth assumptions do not do much to raise poverty, even at the USD 2.50-a-day definition, because there is not much extreme poverty to begin with (as in Latin America, although some countries such as Haiti are badly hurt) or because the trend rates of economic growth are high (as in India and China). However, sub-Saharan Africa – which was helped in the market first scenario by some extremely favourable assumptions about policy, or even regime, changes – is seriously hurt. By 2050, the extreme poverty rate is almost five times what it was estimated to be in the market first scenario (Table 4.10).

Table 4.10
Poverty in the slow-growth regions: comparison of scenarios

	2005		2050 market first scenario		2050 trend growth scenario	
	USD 1.25/ day	USD 2.50/ day	USD 1.25/ day	USD 2.50/ day	USD 1.25/ day	USD 2.50/ day
	(millions of people below the poverty threshold)					
Latin America	46.1	122	7.8	21.1	56.9	147
Near East and North Africa	11	86.7	0.7	2.5	9.4	48.1
Sub-Saharan Africa	391	614	205	533	930	1 364
World	1 377	3 085	245	710	1 120	1 948
	(% of population)					
Latin America	8.4	22.1	1.0	2.7	7.4	19.2
Near East and North Africa	3.6	28.4	0.1	0.6	1.7	8.9
Sub-Saharan Africa	51.2	80.5	11.7	30.5	53.1	77.9
World	21.3	47.7	2.6	7.6	12.0	20.9

Sources: 2005 figures from Chen and Ravallion, 2008; 2050 figures from author's calculations.

In the trend growth scenario, the trend towards global income equality is stalled. A global Gini coefficient of 0.684 in 2005 is pushed down to 0.648 in 2050 in the market first scenario, but is barely shifted – to 0.679 in 2050 – in the trend growth scenario.

Expanding absolute income gaps in both scenarios

The absolute income gap between OECD and non-OECD countries does not shrink in either scenario. In the optimistic market first scenario, this gap rises from almost USD 30 000 per person in 2005 (in PPP dollars, 2005 price levels) to USD 98 000 in 2050, even though the non-OECD per capita GDP growth rate is almost a percentage point higher than the OECD average annual growth rate over the 45 years of the scenario. The ratio of OECD to non-OECD per capita income falls sharply, from 7.4 to 5.3, but the absolute gap more than triples.

However lamentable, a widening of the gap in absolute terms is almost inevitable, unless OECD countries stop growing. If they failed to grow at all for the next 45 years (versus 2 percent or more per year in these scenarios), it would take non-OECD countries 57 years at 3.6 percent a year growth (as in the market first scenario) to catch up with the average OECD GDP per capita income figure of USD 34 359. Even though this could be thought a desirable result, it is likely that lower growth in OECD countries would lead to lower growth in the rest of the world – it is difficult to imagine non-OECD countries growing robustly if OECD countries are stagnant.

Simulations with the IF model suggest that long-run sub-Saharan African growth would fall by between 40 and 140 percent as much as OECD growth falls, depending on assumptions about protectionism and technology. African economic growth in the IF model is also quite sensitive to the level of foreign aid. Gradually raising foreign aid contributions to 0.75 percent of OECD GDP has no discernible impact on OECD growth, but it increases sub-Saharan African growth by almost 1 percentage point a year and reduces the sub-Saharan poverty headcount by 120 million people by 2050. The model simulations implicitly assume that most of the aid (an extra USD 6.5 trillion over 45 years) is productively invested in physical and human capital.[24]

In all of these scenarios, extreme poverty becomes much more highly concentrated in sub-Saharan Africa, because higher economic growth in Asia, particularly India and China, removes hundreds of millions of people from the global poverty headcount. Assuming 2 percent per year population growth, sub-Saharan Africa needs 2 percent per year per capita GDP growth (and constant within-country distributions) just to prevent the extreme poverty headcount from rising. Faster growth – 2.5 percent per year in the market first scenario – cuts the headcount from 391 million in 2005 to 205 million in 2050, and higher growth rates are possible. However, in addition to raising GDP growth, lowering population growth or flattening within-country distributions could also help reduce the poverty headcount. If, somehow, sub-Saharan Africa could cut its population growth by half but still manage GDP per capita growth of 2.5 percent a year, the 2050 poverty headcount would fall to fewer than 100 million people. If the 2.5 percent GDP per capita growth rate is combined with the low population growth rate, and with the Higgins and Williamson (2002) favourable distribution forecast, extreme poverty in sub-Saharan Africa would almost disappear.

---

24. Of course, there is no guarantee that aid will be well spent. Sachs (2005), Easterly (2001), Collier (2007) and Moyo (2009) give differing views on the utility of foreign aid.

Comparison with other long-range growth and poverty projections

One of the most famous pessimistic scenarios in the literature was created and is periodically revised by Meadows and her associates (Meadows, 1972; Meadows, Randers and Meadows, 1992; 2004), who claim that present trends in population, industrialization, pollution and resource depletion will make current world economic growth rates unsustainable. They use a very different sort of model from the neoclassical growth model. The World3 model[25] is based on the idea that world systems, especially the agricultural system, have a finite carrying capacity that has nearly been reached. In the authors' reference scenario, global output per capita peaks around 2025, then goes into irreversible decline, mainly because of the collapse of world agriculture. This model contains no country detail or poverty estimates, but it clearly portrays a much poorer planet than that envisioned in even the trend growth scenario. The major difference between the limits to growth scenarios and the more optimistic ones discussed in this chapter is pessimism about the possibility of technological change to overcome perceived physical constraints.

Another line of thinking is represented by Wallerstein's (2004) world systems analysis. In this approach, instead of the world moving towards improved and globalized capitalism as envisioned in the market first scenario, the capitalist world economy collapses, owing mainly to underconsumption and resentment of the peripheral countries towards the core. Unfortunately for the purposes of this chapter, Wallerstein presents no scenario of future developments after the collapse.

Bremer (2009) does not predict the collapse of global capitalism, but he does worry about a retreat from the market principles reflected in the Washington Consensus and a growing embrace of "state capitalism". He discusses the rise of state-owned energy companies, the renationalization of strategic industries in many non-OECD countries, and the growth of sovereign wealth funds: "The free-market tide has now receded. In its place has come state capitalism, a system in which the state functions as the leading economic actor and uses markets primarily for political gain" (Bremer, 2009: 41).

Bremer sees this development as anti-poor. By distorting incentives and creating vast new opportunities for corruption and rent-seeking, state capitalism will inevitably slow growth and limit poverty reduction. State capitalism promotes protectionism and subsidies that will further restrict growth. Eichengreen and Irwin (2007) argue that, at best, there will be a long pause in United States trade policies geared towards liberalization and that "past gains from liberalization will get whittled away as countries backslide on previous commitments" (Eichengreen and Irwin, 2007: 25). A recent paper by Hillebrand (2010) using empirical estimates by Estavadeordal and

---

25. The computer model is available from the publishers: www.chelseagreen.com.

Taylor (2008) estimates that a global retreat into protectionism (with tariff levels returning to pre-Uruguay Round levels) might improve income inequality in a few countries, but would cut economic growth by almost one percentage point a year to 2035, and raise the global poverty headcount by at least 170 million people.

Conclusions

This chapter has taken a long view of economic growth, poverty and inequality, from 1820 to 2050. Although the data are far from perfect, and the methodology for filling the gaps requires a substantial amount of guesswork, key contributions in the literature, especially Maddison (1995; 2001; 2003) and Bourguignon and Morrisson (2002), have established that world economic growth has been, on average, very high since 1820 – high enough to cause global poverty to fall dramatically. More recent work, especially by Chen and Ravallion (2004; 2008), has shown that the downwards trend in the global poverty rate accelerated after 1980, and that even the poverty headcount has started to show a significant decline.

This chapter has projected world poverty rates, headcounts, inequality measures and absolute income gaps to 2050, based on two different scenarios for global economic growth. In the optimistic growth scenario, the global poverty rate at the USD 1.25 a day standard falls sharply, from 21.3 percent in 2005 to 2.5 percent in 2050, and the number of people living in extreme poverty falls by 1.1 billion. However, the absolute gap between per capita incomes in OECD and non-OECD countries, and the global Gini coefficient remain high.

An alternative scenario assumes that the regions that have been lagging (sub-Saharan Africa, the Near East and Latin America) do not transition to a high growth path. This results in much higher poverty levels: almost 900 million more people living in absolute poverty in 2050 than in the optimistic scenario. The chapter considers, but does not explore empirically, even more depressing scenarios. Resource constraints, if not met by technological solutions, will surely make the poverty estimates shown here worse. A breakdown of the world capitalist system, as envisioned by Wallerstein (2004), or even a gradual turning away from the system that has done so much to reduce global poverty over the last two centuries, would be disastrous.

References

Abramowitz, M. 1956. Resource and output trends in the United States since 1870. *American Economic Review*, 46: 5–23.

Acemoglu, D. & Robinson, J. 2006. *Economic origins of dictatorship and democracy.* Cambridge, UK, Cambridge University Press.

Alesina, A. & Perotti, R. 1993. *Income distribution, political instability and investment.* NBER Working Paper No. 3668. Cambridge, Massachusetts, USA, National Bureau of Economic Research (NBER).

Barro, R. 1998. *Determinants of economic growth.* Boston, Massachusetts,USA, MIT Press.

Barro, R. 2000. Inequality and growth in a panel of countries. *Journal of Economic Growth*, 5(1): 5–32.

Barro, R. & Sala-i-Martin, X. 1995. *Economic growth.* New York, USA, McGraw-Hill.

Bhalla, S.S. 2002. *Imagine there's no country: Poverty, inequality and growth in the era of globalization.* Washington, DC, Institute for International Economics.

Bourguignon, E. & Morrisson, C. 2002. Inequality among world citizens: 1820–1992. *American Economic Review*, 92(4): 727–744.

Bremer, I. 2009. State capitalism comes of age: The end of the free market? *Foreign Affairs*, May/June 2009, 40–55.

Chen, S. & Ravallion, M. 2001. How did the world's poorest fare in the 1990s? *Review of Income and Wealth*, 47(3): 283–300.

Chen, S. & Ravallion, M. 2004. *How have the world's poorest fared since the early 1980s?* Washington, DC, World Bank Development Research Group. (mimeograph)

Chen, S. & Ravallion, M. 2008. *The developing world is poorer than we thought, but no less successful in the fight against poverty.* World Bank Policy Research Paper No. 4703. Washington, DC, World Bank.

Chua, A. 2004. World on fire: How exporting free market democracy breeds ethnic hatred and global instability. New York, Random House.

Collier, P. 2007. *The bottom billion: Why the poorest countries are failing and what can be done about it.* Oxford, UK, Oxford University Press.

Commission on Growth and Development. 2008. *The growth report: strategies for sustained growth and inclusive development.* Washington, DC, World Bank, Spence Commission.

Dollar, D. & Kraay, A. 2002. Growth is good for the poor. *Journal of Economic Growth*, 7: 195–225.

Dowrick, S. & Akmal, M. 2005. Contradictory trends in global income inequality: A tale of two biases. *Review of Income and Wealth*, 51(2): 201–229.

Duesterberg, T. & London, H., eds. 2001. *Riding the next wave: Why this century will be a golden age for workers, the environment, and developing countries.* Fishers, Indiana, USA, Hudson Institute Publications.

Easterly, W. 2001. *The elusive quest for growth: Economists' adventures and misadventures in the tropics.* Cambridge, Massachusetts, USA, MIT Press.

Eichengreen, B. & Irwin, D. 2007. *The Bush legacy for America's international economic policy.* www.econ.berkeley.edu/~eichengr/bush_legacy.pdf.

187

Estavadeordal, A. & Taylor, A. 2008. *Is the Washington Consensus dead? Growth, openness, and the great liberalization*, 1970s–2000s. NBER Working Paper No. 14264. Cambridge, Massachusetts, USA, National Bureau of Economic Research (NBER).

Feenstra, R. & Rao, D.S. 2008. *Consistent comparisons of real incomes across time and space*. Draft manuscript produced for PWT Workshop 2008. http://pwt.econ.upenn.edu/workshop2008/time_space_real_income_comparisons_v3.pdf.

Grossman, G. & Helpman, E. 1991. *Innovation and growth in the global economy.* Cambridge, Massachusetts, USA, MIT Press.

Heston, A. 2008. *The 2005 ICP benchmark world implications for PWT*. PowerPoint presentation from Center for International Comparisons of Production, Income, and Prices workshop. http://pwt.econ.upenn/edu/workshop2008/PennMay_bha.pdf.

Higgins, M. & Williamson, J.G. 2002. Explaining inequality in the world round: Kuznets curves, cohort size, and openness. *Southeast Asian Studies,* 40(3): 269–288.

Hillebrand, E. 2008. The global distribution of income in 2050. *World Development*, 36(5): 727–740.

Hillebrand, E. 2010. Deglobalization scenarios: Who wins? Who loses? *Global Economy Journal*, 10(3): 45–65.

Hughes, B. & Hillebrand, E. 2006. *Exploring and shaping international futures.* Boulder, Colorado, USA, Paradigm Press.

Hughes, B., Irfan, M.T., Khan, H., Kumar, K., Rothman, D. & Solorzano, R. 2008. *Patterns of human progress, Volume 1, Reducing global poverty.* New Delhi, Oxford University Press.

IPCC Data Distribution Center. 2009. *The SRES emissions scenarios.* http://secad.ciesin.columbia.edu/ddc/sres.index.html.

Kakwani, N. & Pernia, E. 2000. What is pro-poor growth? *Asian Development Review*, 18(1): 1–16.

Kuznets, S. 1955. Economic growth and income inequality. *American Economic Review,* 45(1): 93–106.

Lucas, R. 2000. Some macroeconomics for the 21st century. *Journal of Economic Perspectives*, 14(1): 159–168.

Maddison, A. 1995. *Monitoring the world economy, 1820–1992.* Paris, OECD.

Maddison, A. 2001. *The world economy: A millennial perspective.* Paris, OECD.

Maddison, A. 2003. *The world economy: Historical statistics.* Paris, OECD.

Maddison, A. 2007. *Contours of the world economy, 1–2030.* Paris, OECD.

Meadows, D. 1972. *The limits to growth.* New York, Universe Books.

Meadows, D., Randers, J. & Meadows, D. 1992. *Beyond the limits: Confronting global collapse, envisioning a sustainable future.* White River Junction, Vermont, USA, Chelsea Green Publishing.

Meadows, D., Randers, J. & Meadows, D. 2004. *Limits to growth: The thirty year update.* White River Junction, Vermont, USA, Chelsea Green Publishing.

Milanovic, B. 2005. *Worlds apart: Measuring international and global inequalities.* Princeton, New Jersey, USA, Princeton University Press.

Milanovic, B. 2008. Even higher global inequality than previously thought: A note on global inequality calculations using the 2005 International Comparison Program results. *International Journal of Health Services,* (38)3: 421–429.

Moyo, D. 2009. *Dead aid.* New York, Farrar, Straus, Giroux.

National Intelligence Council. 2004. *Mapping the global future, NIC 2004–2013.* www.dni.gov/nic/nic_2020_project.html.

National Intelligence Council. 2008. *Global trends 2025: A world transformed, NIC 2008–2003.* www.nic.gov/nic/nic_2025_project.html.

North, D.C. 2005. *Understanding the process of economic change.* Princeton, New Jersey, USA, Princeton University Press.

Olson, M. 1982. *The rise and decline of nations: Economic growth, stagflation, and social rigidities.* New Haven, Connecticut, USA, Yale University Press.

Olson M. 1996. Distinguished Lecture on Economics in Government: Big Bills Left on the Sidewalk: Why Some Nations are Rich, and Others Poor. *The Journal of Economic Perspectives*, 10 (2): 3–24.

Pardee Center for International Futures. 2009. *Reducing global poverty.* Boulder, Colorado, USA, Paradigm Publishers.

Parente, S. & Prescott, E. 1994. Barriers to technology adoption and development. *Journal of Political Economy*, 102(2): 298–321.

Ravallion, M. 2001. Growth, inequality and poverty: Looking beyond averages. *World Development*, 29(11): 1803–1815.

Rodrik, D. 2008. *A Washington Consensus I can live with.* http://rodrik.typepad.com/dani_rodriks_weblog/2008/06/a-washington-consensus-i-can-live-with.html.

Romer, P. 1987. Growth based on increasing returns due to specialization. *American Economic Review*, 77(2): 55–62.

Romer, P. 1990. Endogenous technological change. *Journal of Political Economy,* 98(5, part II): S71–S102.

Rowan, H. 1996. World wealth expanding: Why a rich, democratic, and (perhaps) peaceful era is ahead. *In* R. Landau, T. Taylor and G. Wright, eds. *The mosaic of economic growth*, pp. 93–125. Stanford, California, USA, Stanford University Press.

Sachs, J. 2005. *The end of poverty.* New York, Penguin Press.

Sala-i-Martin, X. 2002. *The world distribution of income (estimated from individual country distributions)*. NBER Working Paper No. 8933. Cambridge, Massachusetts, USA, National Bureau of Economic Research (NBER).

Schwartz. P. 1999. *The long boom: A vision for the coming age of prosperity.* Reading, Massachusetts, USA, Perseus Books.

Solow, R.M. 1956. A contribution to the theory of economic growth. *Quarterly Journal of Economics*, 70(1): 65–94.

Tamura, R. 1996. From decay to growth: A demographic transition to economic growth. *Journal of Economic Dynamics and Control*, 20(6,7): 1237–1264.

UN. 2004. *Global Environmental Outlook (GEO4)*. www.unep.org/geo4/media.

UNEP. 2007. *Global Environmental Outlook: GEO4.* New York United Nations.

Wallerstein, I. 2004. *World systems analysis, an introduction.* Durham, North Carolina, USA, Duke University Press.

Williamson, J. 1989. What Washington means by policy reform. *In* J. Williamson, ed. *Latin American readjustment: How much has happened?* Washington, DC, Institute for International Economics.

World Bank. 2007. *Global economic prospects 2007: Managing the next wave of globalization.* Washington, DC.

World Bank. 2008. *Global purchasing power parities and real expenditures: 2005 International Comparison Program.* Washington, DC.

World Bank. 2009. *Global economic prospects 2009: Commodities at the crossroads.* Washington, DC.

CHAPTER 5

MACROECONOMIC ENVIRONMENT AND COMMODITY MARKETS: A LONGER-TERM OUTLOOK

Dominique van der Mensbrugghe
Israel Osorio-Rodarte
Andrew Burns
John Baffes

By most accounts, the recent commodity boom has been the longest and broadest (in terms of commodities involved) of the post-Second World War period (World Bank, 2009). Between 2003 and 2008, nominal energy and metal prices increased by 230 percent, food and precious metal prices doubled, and fertilizer prices increased fourfold. Although most prices have declined sharply since their mid-2008 peak, they are still considerably higher than their 2003 levels.

Apart from broad and sustained economic growth, the boom has been fuelled by a host of other factors, both macro and long-term as well as sector-specific and short-term. These include low past investment in extractive commodities, reflecting a prolonged period of declining prices due to excess capacity left after the collapse of the Soviet Union and weak demand after the 1997 East Asian (and other countries') financial crisis; a weak United States dollar (the currency of choice in most international commodity transactions); fiscal expansion and loose monetary policies in many countries; and investment fund activity by financial institutions, which chose to include commodities in their portfolios. In addition, the diversion of some food commodities to the production of biofuels (notably maize in the United States of America, and edible oils in Europe), adverse weather conditions (e.g., three droughts in Australia between 2001 and 2007, a heat-wave in central Asia during the summer of 2010), global stock declines of several agricultural commodities to historical lows, and government policies (e.g., export bans and prohibitive taxes) further contributed to the boom. Geo-political concerns played a key role as well, especially in energy markets.

In some sense, these factors created the "perfect storm", which reached its zenith in July 2008 when crude oil prices averaged USD 133 per barrel (up 94 percent from a year earlier) and rice prices doubled within just five months (from USD 375 per tonne in January to USD 757 per tonne in June 2008). The weakening and/or reversal of some of these factors, coupled with the financial crisis that erupted in September 2008 and the subsequent global economic downturn, induced sharp price declines across most commodity sectors. However, following the pick-up of growth in developed countries and the resilience of emerging economies, commodity prices began increasing again and, in February 2011, most key price indices had reached (or even exceeded) their 2008 peaks.

The recent boom has generated renewed interest in the determinants of commodity prices, including the role of commodity-specific factors, macro-economic fundamentals, and questions regarding whether a permanent shift in price trends has taken place. At the same time, food availability and food security concerns have generated calls for coordinated policy actions at the national (and perhaps international) level, reminiscent of actions taken in earlier booms. With this context in mind, this chapter identifies and analyses the dominant forces that are likely to shape long-term developments in commodity markets. Such forces include (but are not limited to) the increased interdependence between energy and non-energy markets; growth prospects, especially in developing countries, where most consumption growth is expected to take place; the effect of climate change in the production and trade of commodities; and, at the outset, what all this implies for poverty.

The following section provides a brief discussion of recent price trends, including the causes of the recent commodity price boom. This is followed by an analysis of the link between energy and non-energy prices. The next three sections deal with the issues of growth prospects, global warming and their implications for poverty. The last section concludes with a summary and a policy discussion.

The nature of the recent commodity boom

The recent commodity boom shares a number of similarities with earlier booms, but also has some differences. It involved almost all commodities (Figures 5.1 and 5.2), unlike earlier booms, which involved only agriculture (the Korean War) or agriculture and energy (the 1970s energy crisis). It was not associated with high inflation, as opposed to the 1970s boom, which was associated with inflationary pressures. On the other hand, all three booms took place against the backdrop of high and sustained economic growth. Furthermore, they all generated discussion of coordinated policy actions, owing to concerns about food security and energy availability.

The reasons behind the recent boom are numerous and, as many analysts have argued, they created a "perfect storm". On the one hand, most countries enjoyed sustained economic growth for a long period; during 2003 to 2007,

Figure 5.1
Commodity groups affected by booms, 1948 to 2008 (real prices, Manufacturing Unit Value index [MUV]-deflated)

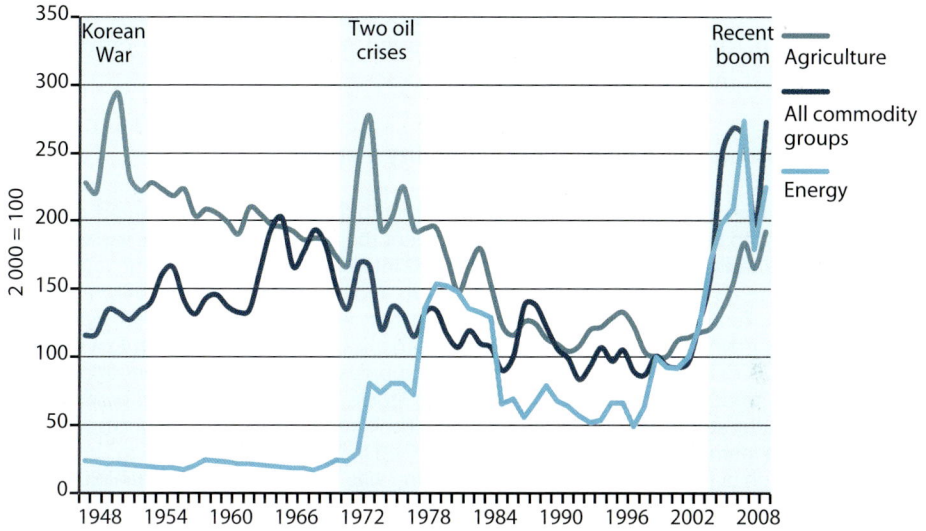

Source: World Bank.

Figure 5.2
Commodity price fluctuations, before, during and after the financial crisis (real prices, MUV-deflated)

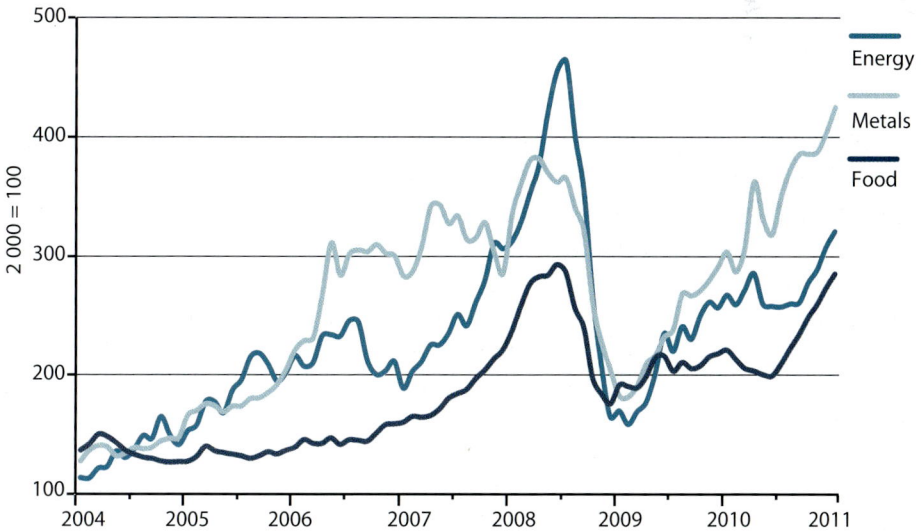

Source: World Bank.

growth in developing countries averaged 6.9 percent, the highest five-year average in recent history (the second highest five-year average, of 6.5 percent, occurred between 1969 and 1973). Fiscal expansion in many countries and low interest rates created an environment that favoured high commodity prices. The depreciation of the United States dollar played a role, as it is the currency of choice for most international transactions.

Box 5.1 - Experience with managing commodity markets

Long-term declines in and high variability of commodity prices have prompted many governments to take collective measures to prevent the decline or reduce the variability. Led by Brazil, coffee producers organized the 1962 International Coffee Agreement (and a subsequent series of agreements) to restrict exports and boost coffee prices. Similar efforts were undertaken by cocoa producers, while attempts were also made in other markets (e.g., cotton, grains). Oil producers formed the Organization of the Petroleum Exporting Countries (OPEC) in 1960, to raise prices through supply controls. Similar organizations of commodity producing countries also used buffer stocks to stabilize prices. Tin producers managed buffer stocks through the International Tin Agreement, to maintain prices within a range. The International Cocoa Agreement, formed in 1972, also attempted to stabilize prices through buffer stocks, but was suspended in 1988. The International Natural Rubber Organization was formed to stabilize rubber prices, but major producers withdrew from the organization following the East Asian financial crisis of 1997. With the exception of OPEC, all these agreements failed to achieve their stated objectives, as coordination and monitoring among many sovereign nations turned out to be a difficult task. Prior to the post-Second World War commodity agreements, another wave of agreements had been formed in response to the low prices following the Great Depression.

In the extractive sectors, especially energy commodities, underinvestment during the late 1980s and the 1990s left limited room for supply response. For example, during the early 1980s, total investment expenditures by the major United States multinational oil and gas companies averaged more than USD 130 billion per annum (in real 2006 terms). For the next 15 years, however, the annual average dropped by half (Figure 5.3). Similar reductions in investment took place in most metal sectors.

Another factor believed to have played a key role in the recent boom is the decision by managers of various investment, pension and sovereign wealth funds to include commodities in their holdings as a way of diversifying their portfolios away from traditional asset classes such as equities and bonds. Although evidence on the effect of investment fund activity on commodity prices has been mixed, many experts believe that such funds have been a key force behind the 2008 and 2010/2011 rallies (see discussion in Boxes 5.1 and 5.2, and Table 5.1, on different types of speculation, including investment fund activity).

Box 5.2 - The role of speculation during the recent commodity boom

In the early 2000s, managers of various investment, pension and sovereign wealth funds began investing in commodity markets, either as a way to diversify away from existing assets (e.g., equities and bonds) or owing to their search for higher yields. Although estimates of how much money has been invested in commodity markets are not precise, a major investment bank put the figure (as of October 2010) at about USD 350 billion; it was about USD 250 billion in 2008 according to Masters (2008). Almost two-thirds of this is invested in energy commodities. Although such transactions are not associated with real demand for commodities, they may have influenced prices, for a number of reasons. First, because investment in commodities is a relatively new phenomenon, there have been mostly inflows (not outflows) of funds, implying that some markets may have been subject to extrapolative price behaviour (i.e., high prices leading to more buying by investment funds, leading to even higher prices, and so on). Second, these funds invest on the basis of fixed weights or past performance criteria, so investment often takes place in contrast to what market fundamentals would dictate. Third, the large size of these funds compared with commodity markets may exacerbate price movements. Their influence on prices is especially likely if the rapid expansion of these markets contributes to expectations of rising prices, thereby exacerbating swings, as argued by Soros (2008: 4), who called commodity index buying "intellectually unsound, potentially destabilizing and distinctly harmful in its economic consequences". Similar views are shared by numerous authors (e.g., Eckaus, 2008; Wray, 2008).

However, the empirical evidence regarding whether or not such funds contributed to the price boom has been, at best, mixed. In the non-ferrous metals market, Gilbert (2008) found no direct evidence of the impact of investor activity on the prices of metals, but some evidence of extrapolative price behaviour resulting in price movements that were not fully justified by market fundamentals. He also found strong evidence that the futures positions of index providers over the past two years have affected soybean (but not maize) prices in the United States futures exchanges. Plastina (2008) concluded that between January 2006 and February 2008, investment fund activity might have pushed cotton prices 14 percent higher than they would otherwise have been. On the other hand, two International Monetary Fund studies (IMF, 2006; 2008) failed to find evidence that speculation has had a systematic influence on commodity prices. A similar conclusion was reached by a series of studies undertaken by the Commodities Futures Trading Commission, the agency that regulates United States futures exchanges (Büyükşahin, Haigh and Robe, 2008; CFTC, 2008).

Although the empirical evidence regarding the effect of investment fund activity is mixed and inconclusive, the consensus among experts is that the large amount of money that goes into commodities certainly has an effect on prices. On the other hand, market fundamentals will determine the long-term trends of commodity prices, which implies that investment fund activity has induced higher price variability.

Figure 5.3
Energy investment (left axis) and prices (right axis) by major multinational oil companies

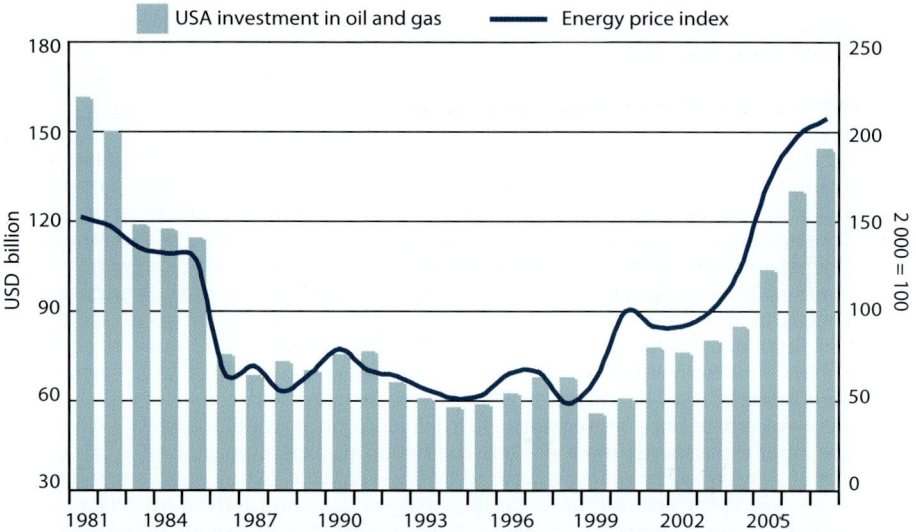

Sources: International Energy Authority (IEA); World Bank.

Table 5.1
The simplistic (and compartmentalized) view of speculation

Activity	Function	Effect
Speculation on futures exchanges	Important activity for the functioning of futures markets	Injects liquidity into the market and improves price discovery
Market manipulation	Isolated cases, such as cornering of the copper and silver markets	These are illegal activities
Building up of inventories	Accumulation of physical stocks with the expectation that price increases will generate profits	Traders buy at current prices to sell later, when the market is tight, thus balancing the market and reducing price variability
Commodity trading accounts	Professionally managed commodity investment vehicles taking into consideration market fundamentals	Enhanced price discovery through careful examination of the fundamentals and use of technical analysis
Hedge funds	Short-term profit seeking	Believed to induce short-term volatility (i.e., day-to-day)
Investment funds	Long positions in futures exchanges taken by investment, pension and sovereign wealth funds	May amplify commodity cycles owing to the size and nature of investment, but unlikely to affect long-term trends

Source: Authors.

Figure 5.4
Rice, wheat and maize consumption in China and India

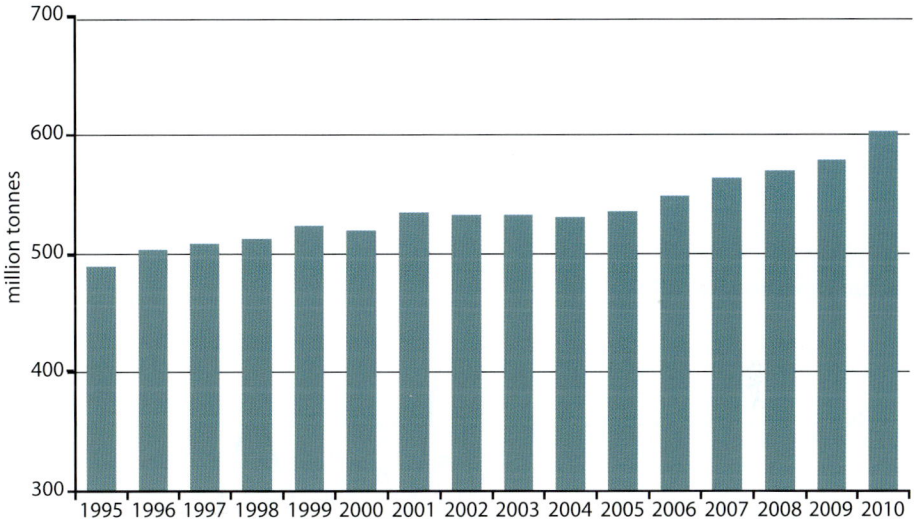

Source: World Bank calculation based on USDA data.

Figure 5.5
Rice, wheat and maize consumption in China and India, as percentage of world total

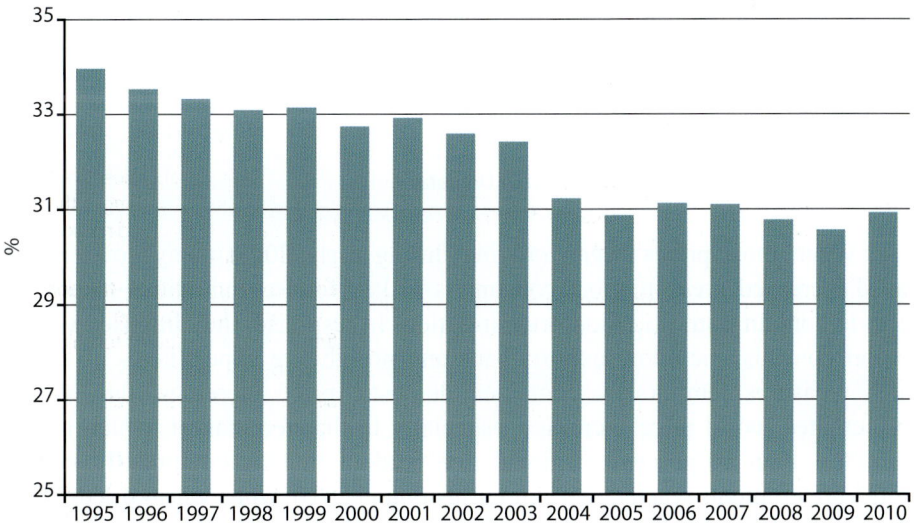

Source: World Bank calculation based on USDA data.

The diversion of considerable quantities of some food commodities for the production of biofuels has been another factor behind the boom. Almost 28 percent of United States maize area (corresponding to about 1.33 percent of global grain area) was diverted to ethanol production during 2008/2009. While the combined maize and oilseed area diverted to biofuel production corresponds to about 2 percent of global grain and oilseed area, the sharp increase in diversion of the last two to three years came at a time when global grain stocks were at historical lows, thus leaving limited room for adjustment by bringing more land into productive uses (see Figure 5.6 for historical stock-to-use ratios).

Figure 5.6
Global stock-to-use ratios, 1960 to 2010

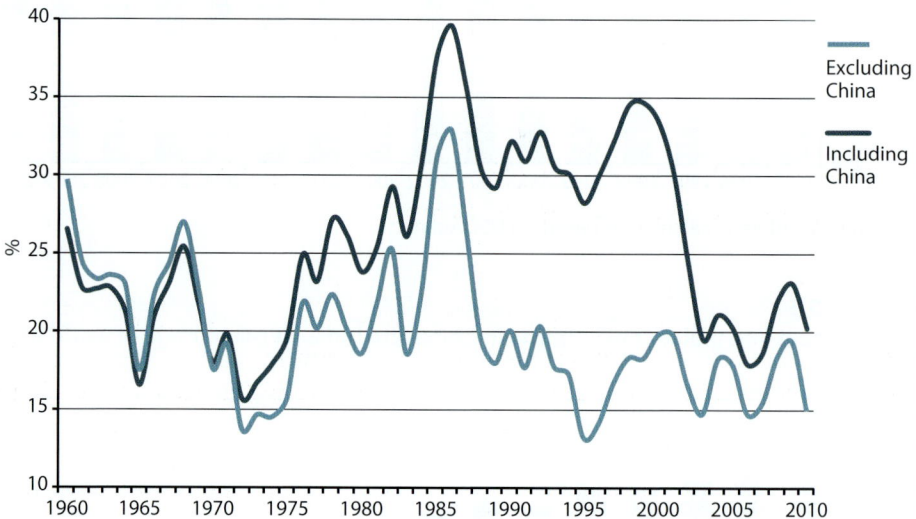

Source: World Bank calculation based on USDA data.

When most prices began rallying during early 2008, many governments faced increased pressure from consumers of key food commodities (especially rice) to contain domestic food price inflation. In response, they imposed various export controls, including exports bans and prohibitive export taxes. Although such measures temporarily contained domestic price increases, they further exacerbated world price increases, especially in the rice market, which is very thin (less than 10 percent of global rice production is internationally traded). Some governments reacted in a similar fashion in 2010, when wheat prices began spiking, but the overall trade policy response has been much more muted.

In addition to these factors, increased grain consumption by low- and middle-income countries (especially China and India), due to rising incomes and changing diets (from grain to meat consumption), has often been cited as a key driving force of the boom, including the 2008 rally. However, as Figures 5.4 and 5.5 indicate, the combined grain consumption (for both human and animal use) by China and India increased only slightly after 1995, a period in which both countries enjoyed strong economic growth. More important, when expressed as a share of global consumption, grain consumption in these two countries declined between 1995 and 2007. This should not be surprising given the low income elasticity of grains even at low per capita incomes (Table 5.2).

Table 5.2
Income elasticities

Commodity group	Low income	Lower-middle income	Upper-middle income	High income
Grains	0.15	0.10	0.05	-0.01
Vegetable oils	0.50	0.65	0.78	0.41
Meats	0.31	0.51	0.68	0.38

Source: Authors' estimates based on panel estimation.

The energy/non-energy price link

It has become increasingly clear that the energy price increases of the last few years will reshape not only energy markets but also most other markets, including agriculture. For almost 20 years, the price of crude oil averaged about USD 20 per barrel (real 2000 terms). Most analysts and researchers now believe that the "new" equilibrium price of oil will be three to four times as much, with proportional changes expected to take place in all other types of energy. High energy prices, along with the high energy intensity of most commodities (especially agriculture), imply that developments in non-energy (especially food) markets will depend on the nature and degree of the energy/non-energy price link. The rest of this section elaborates on this issue.

The channels through which energy prices affect other commodities are numerous. On the supply side, energy enters the aggregate production function of most primary commodities through the use of various energy-intensive inputs and, often, transportation over long distances, which is an equally energy-demanding process. Some commodities have to go through an energy-intensive primary processing stage. Others can be used to produce substitutes for crude oil (e.g., maize and sugar for ethanol production, or edible oils for biodiesel production). In other cases, the main input may be a close substitute for crude oil, such as nitrogen fertilizer, which is made directly from natural gas. (The various transmission channels from energy to non-energy prices are discussed in, among others, Baffes, 2007; 2010; FAO, 2002; World Bank, 2009.)

This section examines the energy/non-energy price link by estimating the following relationship:

$$\log(NON\_ENERGY_t) = \mu + \beta_1 \log(ENERGY_t) + \beta_2 \log(MUV_t) + \beta_3 TIME + \varepsilon_t. \quad (5.1)$$

where NON_ENERGY_t denotes the various non-energy United States dollar-based price indices at time t; $ENERGY_t$ denotes the energy price index; MUV_t denotes the deflator; TIME is the time trend; ε_t denotes the error term; and μ, β_1, β_2, and β_3 denote parameters to be estimated. Annual data for a number of commodity indices and prices covering the period 1960 to 2008 are used in the analysis. Although the signs and magnitudes of the coefficients are not dictated by economic theory, β_1 and β_2 are expected to be positive because energy, as well as other goods and services (reflected by the measure of inflation), constitutes a key input to the production process of all commodities. On the other hand, β_3 is expected to be negative, at least for agricultural commodities – consistent with the long-term impact of technological progress on production costs, and the low income elasticity of most food commodities, especially cereals.

The estimates presented in Table 5.3 indicate that energy prices and, to a lesser extent, inflation and technological change explain a considerable part of commodity price variability (the adjusted R2 of all regressions averaged 0.85). Specifically, the parameter estimate of the non-energy index (top row of Table 5.3) is 0.28, implying that a 10 percent increase in energy prices is associated with a 2.8 percent increase in non-energy commodity prices, in the long run. Three earlier studies (Gilbert, 1989; Borensztein and Reinhart, 1994; Baffes, 2007) reported elasticities of 0.12, 0.11 and 0.16, respectively (Table 5.4). When the sample of the current analysis is adjusted to match the samples of these studies, the pass-through coefficients become remarkably similar (0.13, 0.12 and 0.18, respectively).

However, the transmission elasticity of the non-energy index masks some variations. The highest pass-through elasticity among the sub-indices is in fertilizer, estimated at 0.55; this is not surprising as nitrogen-based fertilizers are made directly from natural gas. Note that the fertilizer and energy price increases during the recent boom were in line with the increases experienced during the first oil shock: from 1973 to 1974 phosphate rock and urea prices increased fourfold and threefold, respectively, very similar to the crude oil price increase during that period, from USD 2.81 to USD 10.97 per barrel.

Table 5.3
Parameter estimates

Index	μ	β_1	β_2	100*β_3	Adj-R^2	ADF
Non-energy	3.03[a]	0.28[a]	0.12	-0.01	0.9	-3.35[c]
	-6.54	-5.24	-0.68	-0.02		
Metals	3.77[a]	0.25[a]	-0.17	1.93[a]	0.82	-3.30[c]
	-4.8	-3.14	-0.6	-2.31		
Fertilizers	3.58[a]	0.55[a]	-0.3	0.39	0.81	-3.97[d]
	-4.12	-4.79	-0.95	-0.48		
Agriculture	2.51[a]	0.26[a]	0.33[a]	-0.99[a]	0.9	-3.81[d]
	-6.9	-5.54	-2.43	-2.73		
Beverages	1.83[a]	0.38[a]	0.55[a]	-3.12[a]	0.76	-4.95[d]
	-3.1	-4.87	-2.63	-5.22		
Raw materials	1.85[a]	0.11[a]	0.51[a]	0.08	0.91	-3.15[c]
	-4.16	-2.15	-3.15	-0.19		
Food	2.91[a]	0.27[a]	0.21	-0.71	0.85	-3.85[d]
	-7.11	-4.93	-1.39	-1.8		
Cereals	3.13[a]	0.28[a]	0.17	-0.87	0.78	-3.83[d]
	-5.94	-4.23	-0.89	-1.76		
Edible oils	3.33[d]	0.29[a]	0.12	-0.8	0.8	-2.02[b]
	-6.16	-4.51	-0.58	-1.5		
Other food	1.86[a]	0.22[a]	0.45[a]	-0.42	0.89	-3.60[d]
	-6.28	-3.81	-4.44	-1.18		
Precious metals	-1.40[a]	0.46[a]	1.05	-1.75	0.98	-3.91[d]
	-3.58	-9.4	-7.61	-3.68		

[a] Parameter estimate significant at the 5-percent level.

Rejection of the existence of one unit root at: [b] 10-percent level; [c] 5-percent level; and [d] 1-percent level of significance (the respective t-statistics are -2.60, -2.93 and -3.58). The lag length of the ADF equations was determined by minimizing the Schwarz-loss function.

Numbers in parentheses are absolute t-values (the corresponding variances have been estimated using White's method for heteroskedasticity-consistent standard errors).

ADF = the MacKinnon one-sided p-value based on the Augmented Dickey-Fuller equation (Dickey and Fuller, 1979).

Source: Authors' estimates.

The agriculture pass-through, estimated at 0.27, reflects a wide-ranging average: beverages (0.38), food (0.27) and raw materials (0.11). However, the elasticity estimates of the food price index components fall within a very narrow range: cereals (0.28), edible oils (0.29) and other food (0.22). The estimates for the key food commodities also fall within a relatively narrow range, from a low of 0.25 in rice to a high of 0.36 in soybeans (Table 5.5).

Table 5.4
Long-run transmission elasticities

Commodity	Holtham (1988) 1967:S1–1984:S2	Gilbert (1989) 1965:Q1–1986:Q2	Borensztein & Reinhart (1994) 1970:Q1–1992:Q3	Baffes (2007) 1960–2005	This study 1960–2008
Non-energy	—	0.12	0.11	0.16	0.28
Food	—	0.25	—	0.18	0.27
Raw materials	0.08	—	—	0.04	0.11
Metals	0.17	0.11	—	0.11	0.25

Holtham uses semi-annual data; Gilbert and Borensztein and Reinhart use quarterly data; and Baffes and the present study use annual data. Gilbert's elasticities denote averages based on four specifications; Holtham's raw materials elasticity is an average of two elasticities based on two sets of weights.

— = estimate not available.

Sources: Holtham, 1988; Gilbert, 1989; Borensztein and Reinhart, 1994; Baffes, 2007; authors' estimates.

Table 5.5
Parameter estimates

Index	μ	β_1	β_2	$100*\beta_3$	Adj-R^2	ADF
Wheat	3.27[a]	0.30[a]	0.12	-0.49	0.84	-4.35[c]
	(6.50)	(5.02)	(1.49)	(1.07)		
Maize	3.15[a]	0.27[a]	0.13	-0.74	0.80	-3.49[c]
	(6.23)	(4.66)	(0.70)	(1.58)		
Soybeans	3.58[a]	0.26[a]	0.25	-0.82	0.82	-3.85[d]
	(8.11)	(4.92)	(1.51)	(1.83)		
Rice	3.57[a]	0.25[a]	0.32	-1.62[a]	0.58	-4.05[d]
	(5.14)	(2.67)	(0.26)	(2.78)		
Palm oil	4.94[a]	0.35[a]	-0.01	-0.95	0.63	-3.16[c]
	(6.44)	(3.72)	(0.02)	(1.38)		
Soybean oil	5.25[a]	0.36[a]	-0.09	-0.42	0.70	-2.56
	(7.83)	(4.13)	(0.39)	(0.53)		

[a] Parameter estimate significant at the 5-percent level.

Rejection of the existence of one unit root at: [b] 10-percent level; [c] 5-percent level; and [d] 1-percent level of significance (the respective t-statistics are -2.60, -2.93 and -3.58). The lag length of the ADF equations was determined by minimizing the Schwarz-loss function.

Numbers in parentheses are absolute t-values (the corresponding variances have been estimated using White's method for heteroskedasticity-consistent standard errors).

ADF = the MacKinnon one-sided p-value based on the Augmented Dickey-Fuller equation (Dickey and Fuller, 1979).

Source: Authors' estimates.

Three key conclusions emerge from these results. First, most commodities respond strongly to energy prices, and the response appears to strengthen in periods of high prices, as confirmed by the considerable increases in the values of estimated elasticities observed when the recent boom is included in the analysis. The implication is that as long as energy prices remain elevated, not only are non-energy commodity prices expected to be high, but also analysing the respective markets requires understanding of the energy markets.

Second, while the transmission elasticities are broadly similar, this is not the case for the inflation coefficients, estimates of which vary considerably in terms of sign, magnitude and level of significance. The inflation coefficient is positive and significantly different from zero only for agriculture (and some of its sub-indices), while being effectively zero for metals and fertilizers. All this implies that the relationship between inflation and nominal commodity prices is much more complex and, perhaps, changeable over time. This may not be surprising, considering that during 1972 to 1980 (a period that includes both oil shocks) the MUV increased by 45 percent, while during 2000 to 2008, it increased by half as much. The nominal non-energy price index increases during these two eight-year periods were identical, at 170 percent.

Third, the trend parameter estimates are spread over an even wider range than the energy pass-through and inflation are. For example, the non-energy price index shows no trend at all, while the metal price index exhibits a positive annual trend of almost 2 percent, and the agriculture index shows a 1 percent negative annual trend. Furthermore, the trend parameter estimates of the agriculture sub-indices vary considerably, from 0.08 for raw materials to -3.12 for beverages, a result that confirms Deaton's (1999: 27) observation that what commodity prices lack in trend, they make up for in variability. The trend estimate of the food index, -0.71, significant at the 10 percent level, may add another dimension to the debate on the long-term decline of primary commodity prices, often discussed in the context of the Prebisch-Singer hypothesis (Spraos, 1980, and others).

The macroeconomic environment

A number of factors will shape the macroeconomic environment and agricultural supply and demand balances over the medium term (to 2030) and the longer term (to 2050). The starting point of any such analysis is demographics. Between 1950 and 2000, the world saw a huge expansion in global population, with an increase of some 3.6 billion people, or 250 percent (Figure 5.7). Over the next 50 years, the expansion will slow down considerably although, according to the United Nations (UN) medium variant, an increase of 50 percent over 2000 will be coming off a much higher base, so will still represent a rise of 3 billion people. The distributional implications of the population rise are also important. There will be nearly no increase in high-income countries, but a 150 percent increase in the least-developed countries.[1] Many of the least-developed countries have been under significant stress to feed their growing populations, owing to both natural and human-incurred reasons. On the other hand, high-income countries have both stagnating populations and food demand, and robust agriculture. This combination could lead to increased reliance on food imports among the least-developed countries, with other developing regions lying somewhere in between – some with surpluses, such as many Latin American countries, and others with potentially growing deficits, such as some in Asia. The bottom line is that agricultural production has to increase at an average of 0.8 percent per annum simply to accommodate population growth, and in the least-developed countries it will have to grow at an average of 1.8 percent per annum over the 50-year period.

Figure 5.7
Population expansion: history and projection

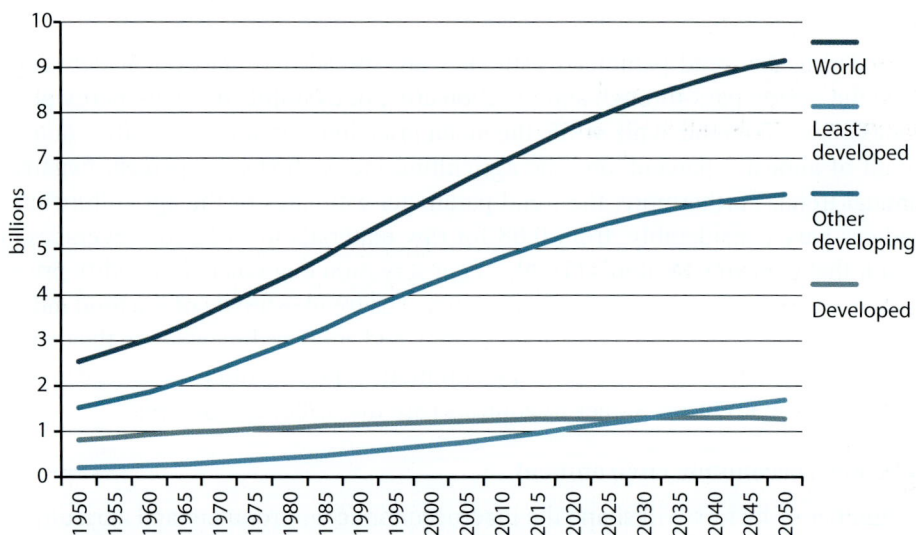

Source: UN Population Division. http://esa.un.org/unpp/index.aspl.

The economic factors that will determine food supply and balances can be divided into the two categories of demand and supply factors, which will be regionally differentiated. Historically, demand has been conditioned by two factors: income growth and shifts in tastes (often derived from income growth), examples of which include switches from diets based largely on grains to more reliance on meat- and dairy-based proteins. In most high-income and some

---

1. Using today's definition of least-developed.

developing countries, the income elasticity is nearly 0 for many food commodities as saturation points have been reached.[2] There is nonetheless a substantial portion of the global population that would potentially demand relatively more food as incomes rise. The World Bank's most recent estimate of the incidence of poverty (at the USD 2 per day level) in developing countries was about 47 percent in 2005, declining to about 33 percent by 2015. In addition, the intensification of meat and dairy consumption would raise the demand for grain-based feed in larger proportion than any relative drop in household-based grain demand.

Although income growth is regularly projected over the medium- and long-term horizons, it should be kept in mind that these are strictly scenario-based (or "what if?") projections, and not statistically based as the more standard short-term forecasts of economic growth are. The projections in this paper use a hybrid system, which in the short and medium terms relies more on estimates of potential growth using statistical techniques, but over the longer term switches to a more judgemental forecast that relies on two assumptions: i) long-term per capita growth in high-income countries will slow to 1.0 to 1.5 percent per annum; and ii) developing countries will converge towards the per capita incomes of high-income countries, but at different rates.

Figure 5.8
GDP growth scenario

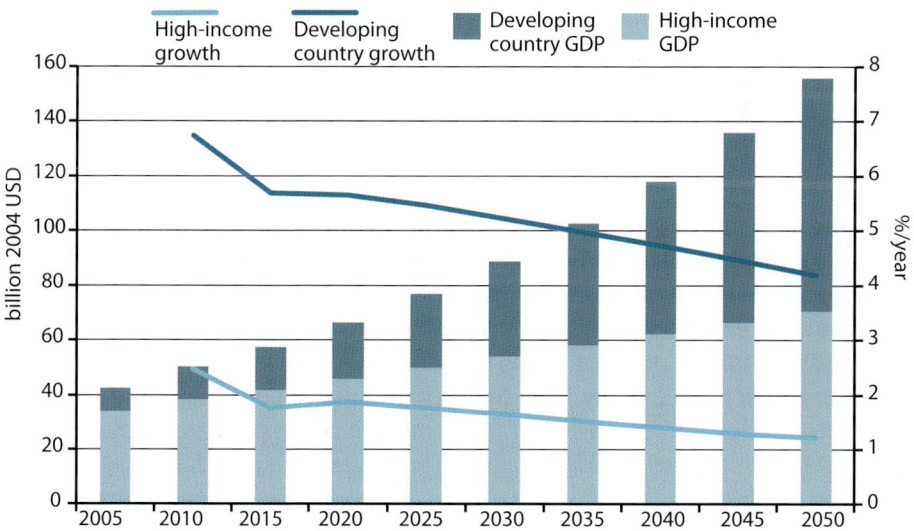

Source: Simulation results with the World Bank's ENVISAGE model.

2. It could be argued that demand may even decline as health and environmental concerns lead to changing dietary habits and lower overall food consumption.

The baseline projection has the global economy increasing at an average rate of about 2.9 percent between 2005 and 2050 (Figure 5.8). This breaks out into 1.6 percent for high-income countries and a brisk 5.2 percent for developing countries. One of the key consequences of this differential in growth rates is a very large shift in share of global output. In 2005, developing countries had a roughly 20 percent share in global output (at market exchange rates). By 2050, this jumps to about 55 percent. On a per capita basis, the growth differential narrows, as population growth is near zero in high-income countries. At market exchange rates, there is a narrowing of the income gap, but it remains substantial. In 2005, per capita incomes were some 20 times higher in high-income than in developing countries. This ratio drops to six by 2050, but varies greatly across regions, with a low of 3.5 in East Asia and the Pacific and a high of 20 in sub-Saharan Africa.

With average per capita incomes rising by 2.2 percent per annum between 2005 and 2050, an income elasticity of 0.5 would yield an increase in food demand of 1.1 percent, to be added to the 0.8 percent increase in population for a total increase of 1.9 percent per annum. This simple estimate may be an overstatement, as income elasticity for food would be expected to decline as incomes rise and is already near zero in most high-income countries. On the other hand, counterbalancing factors that would lead to a rise could include an increasing demand for meat and dairy and new competition emerging from biofuels.

The factors behind demand growth are likely to be relatively stable compared with supply-side variables. Ultimately, supply growth will be driven by the different degrees of intensification (getting more from the same amount of land) and extensification (expanding the land area under cultivation). The cost and availability of other inputs, notably water, are also important factors, but are more difficult to integrate into the current analysis.

Based on the latest available FAO data, there is significant scope for extensification in many regions of the world (Figure 5.9). Whether this potential supply is exploited or not will depend on, among other factors, the affordability of expansion in terms of infrastructure development, and the potential negative externalities of expansion (e.g., environmental degradation). Which regions expand land use will also influence changes in the patterns of food trade. For example, Latin America, which has relatively large tracts of productive non-forest land available, could see a fairly rapid expansion of its production and exportable surplus.

The huge increase in world population but stagnant or even falling agricultural prices of the last few decades have been supported by sizeable improvement in agricultural productivity growth (Coelli and Rao, 2005; World Bank, 2009), particularly in Asia, but also in North America. This rapid growth has recently tapered somewhat. For example, yield growth in wheat and rice declined from about 2 percent between 1965 and 1999, to less than 1 percent between 2000 and

2008. This is cause for concern about the future, particularly as this decline has trended well with the decline in expenditures on research and development (R&D). There are opportunities available, in part because many regions are well behind the frontier – such as Europe and Central Asia and sub-Saharan Africa – and also because the frontier can still be pushed out, notably with state-of-the-art gene-based R&D.

Figure 5.9
Land under cultivation and potentially suitable

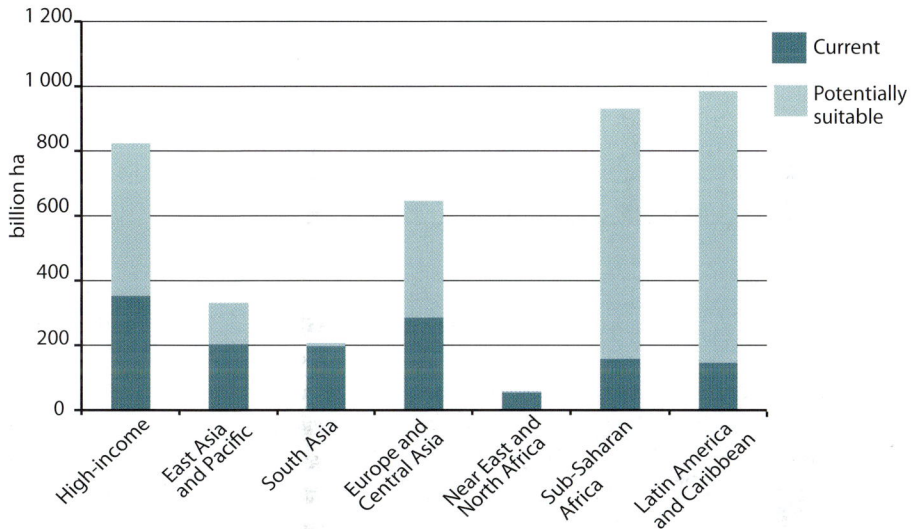

Source: FAO.

Part of the analysis of long-term trends relies on an analytical framework that allows integration of the various components – demographics, income growth, structural and taste changes, productivity and evolving factor supplies – into a consistent model of the global economy. The World Bank's ENVironmental Impact and Sustainability Applied General Equilibrium (ENVISAGE) model is a dynamic computable general equilibrium (CGE) model, described in greater detail in Annex 5.1. ENVISAGE has several advantages: first, it is global, with supply/demand balances guaranteed at the global level – differences between domestic production and demand are met through exporting surpluses or importing to meet deficits; second, it encompasses all economic activity, so if a country becomes a net importer of food, it must export more of other commodities; and third, it is based on a consistent microeconomic underpinning that facilitates what-if analysis. For example, What if productivity is higher or lower? What if demand for meat and

Figure 5.10
Changes in world agricultural prices under different productivity assumptions

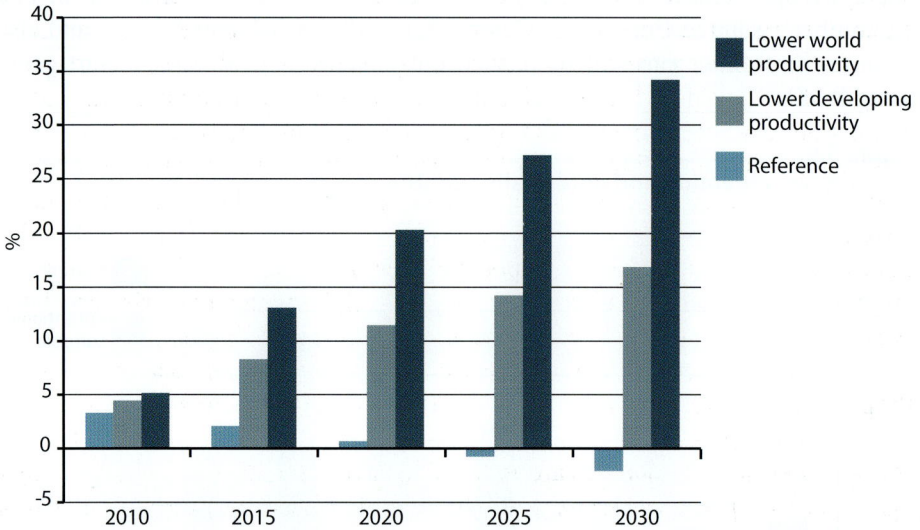

Source: Simulation results with the World Bank's ENVISAGE model.

Figure 5.11
Net agricultural trade under different productivity assumptions, 2030

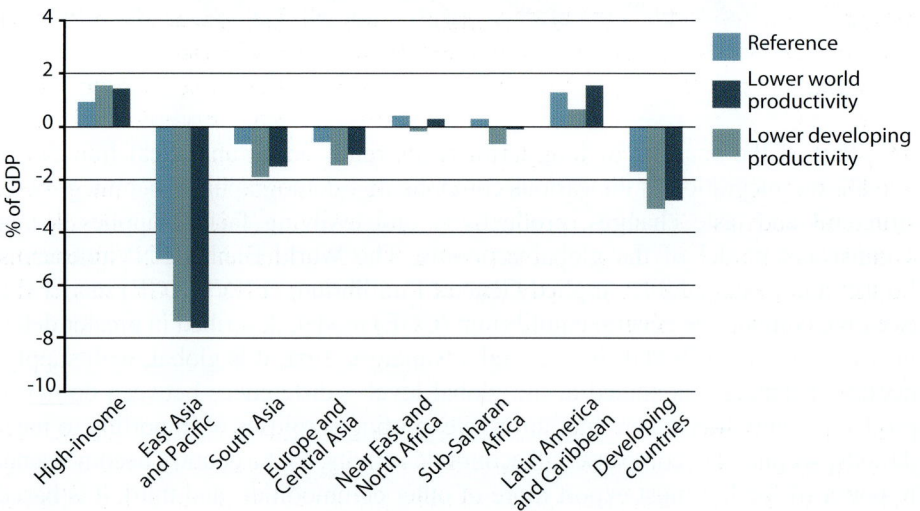

Source: Simulation results with the World Bank's ENVISAGE model.

dairy in developing countries follows a different pattern from that in high-income countries? What if energy prices rise? How does this affect the cost structure of food supply? Will it induce more demand for biofuels? The remainder of this section explores some of these fundamental questions with the assistance of the model.

The baseline scenario, with productivity growth of 2.1 percent per annum in agriculture, yields a benign price pattern for overall agriculture, i.e., there is a small negative trend over the long term, with global supply/demand balances more or less lined up (Figure 5.10). This has been the pattern for the last 30 to 40 years. Supply/demand balances at a regional level may widen, as some countries have little room for expansion and see a shift in comparative advantage for other goods. In the absence of new support policies, East Asia could see a relatively large increase in net agricultural imports, with high-income countries and Latin America and the Caribbean having exportable surpluses (Figure 5.11).

As noted earlier, assumptions regarding productivity are key to determining the potential stress on food markets. To assess the impact of the baseline assumption on agricultural productivity, two additional scenarios are undertaken. In the first scenario, developing countries are assumed to have half the productivity growth in agriculture of the baseline assumption. This could be driven by a number of factors, including failure to ramp up research and development expenditures, resistance to genetically modified organism technology, reduced effectiveness of inputs, lower land productivity (due to increasing salinity, for example) or inadequate supply of water. The model suggests that in this case, global agricultural prices would rise modestly compared with today's levels. However, developing countries' reliance on agricultural imports would also increase, again with rising dependence in Asia. Latin America and the Caribbean remains a net agricultural exporter.

If global productivity is halved, agricultural prices will rise by significantly more – nearly 35 percent above the base year in 2030, compared with about 16 percent when only developing country agriculture is subjected to the lower productivity growth. The impact on trade balances is more mixed, in most cases lying between the baseline level and the scenario in which only developing country agriculture is affected. Note that the net trade numbers are in value terms, so the change in net trade is partly the result of higher agricultural prices, and is not simply a volume phenomenon.

Climate change

One issue that might be looming large in the next few decades is the impact of climate change on global agriculture. Some estimates suggest that a rise of 2.5 °C could lower agricultural productivity by up to 40 percent, including in some very large countries such as India (Cline, 2007). The net impact of climate change on agriculture, at least at the global level, is still being debated. Some regions, notably the higher latitudes, could benefit from longer growing periods, largely offsetting the damage in regions in the lower latitudes, but also inducing further changes in trading patterns. There is also uncertainty regarding the impact of carbon fertilization. There is some evidence that higher concentrations of carbon may induce growth, at least to a certain point, and this could offset higher temperatures. Finally, although the general circulation models (GCMs) have a relatively high degree of consistency regarding temperature increases, there is much less consensus on rain patterns and the overall supply of water for agricultural purposes. In the longer run, appropriate adaptation policies may allow many regions to adapt to incremental changes in weather; however, extreme weather events may be more damaging and much more costly to cope with.

One of the features of the ENVISAGE model is that it incorporates the full cycle of greenhouse gas emissions from human activities, atmospheric concentrations and radiative forcing, and changes in temperature. This class of model is known as an "integrated assessment model", and also couples changes in global temperature to economic damage. Currently, damage to agriculture is incurred only through impacts on agricultural productivity.[3]

Figure 5.12
Impact of climate change on agricultural production, without the carbon fertilization effect

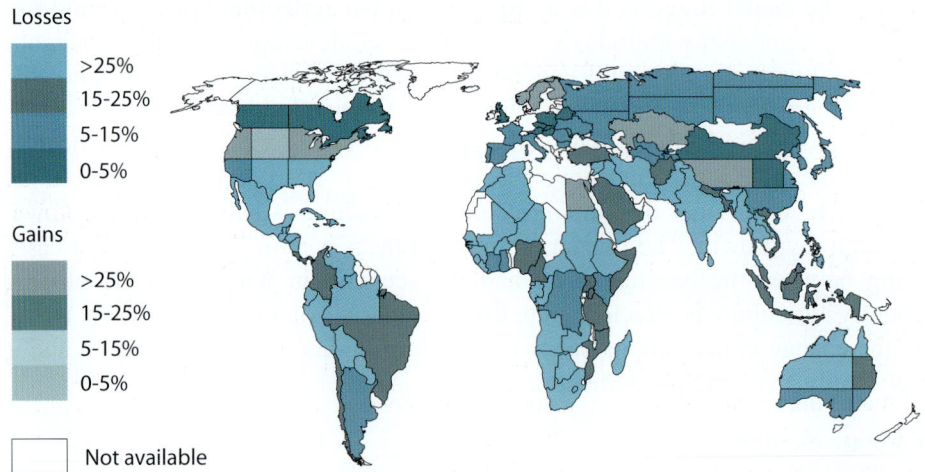

Losses
>25%
15-25%
5-15%
0-5%

Gains
>25%
15-25%
5-15%
0-5%

Not available

Source: Cline, 2007.

---

3. The ENVISAGE model is currently being modified to handle a broader set of climate change impacts, including sea-level rise, health and labour productivity effects, and water stress.

Figure 5.12 depicts how climate-induced agricultural damage is allocated across the globe, based on the estimates produced by Cline (2007). The figure clearly shows the concentration of damage in lower latitudes and largely in developing countries. In a way, it represents a "worst-case" scenario in that it estimates damage in the absence of the carbon fertilization effect. For the baseline scenario, the damage has been assumed to be the average of the situations with and without the carbon fertilization effect. Cline's estimates are based on the assumption that the increase in temperature of 2.5 °C will occur around 2080. This is based on scenarios developed at the end of the 1990s that assumed a lower profile of emissions than that observed over the last decade, in spite of the current crisis. The damage functions in ENVISAGE are calibrated to Cline's estimated impacts for a temperature change of 2.5 °C. For technical reasons this study has specified and calibrated linear damage functions. This may overstate damage in the short term, particularly in certain regions where warming could be beneficial, such as in higher latitudes, and understate damage in the long run, as many damage functions in the literature are assumed to be non-linear (e.g., Nordhaus, 2008).

For the purposes of climate analysis the model runs until 2100, but this chapter focuses on the period up to 2030. The projected atmospheric emissions profile used in this chapter is significantly higher than most of those that form the basis of the climate change analysis recently presented in the Fourth Assessment Report (AR4) of the Intergovernmental Panel on Climate Change (IPCC) (Metz *et al.*, 2007). The scenarios in AR4 were generated around 2000 and largely underestimated both output and emission growth over the last decade, notwithstanding the recent financial crisis. As a result, the baseline scenario shows much greater emission growth and, if this pattern continues, puts the world on a trajectory of much higher temperature changes than the AR4 median of about 3 °C by the end of the century (Figure 5.13). With a higher temperature profile than the AR4 median, estimated climate change impacts on agriculture occur much earlier than assumed in the Cline study, as the 2.5 °C level is reached in 2050 rather than 2080.

Climate damage is built into the standard baseline. To isolate the impact of climate change, an alternative scenario is simulated that assumes no climate change damage. All other exogenous assumptions are the same in the two scenarios. In this alternative scenario, agricultural productivity matches the exogenous assumption of 2.1 percent uniform growth with no deviation. The impacts on real income from climate damage even in 2030 could be substantial. South Asia would take the most significant hit, with a loss in real income of more than 2 percent in 2030, more than double the loss of the next hardest-hit region, sub-Saharan Africa (Figure 5.14). The relatively large losses in these two regions reflect two factors:

Figure 5.13
Baseline emission concentrations (left axis) and temperatures (right axis)

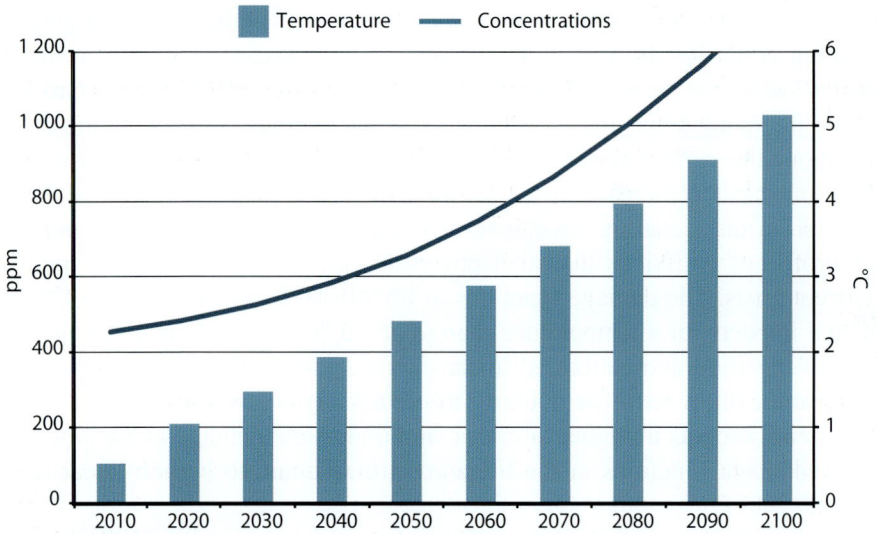

ppm = parts per million
Source: Simulation results with the World Bank's ENVISAGE model.

Figure 5.14
Potential impacts of climate change on real incomes, by region, 2030

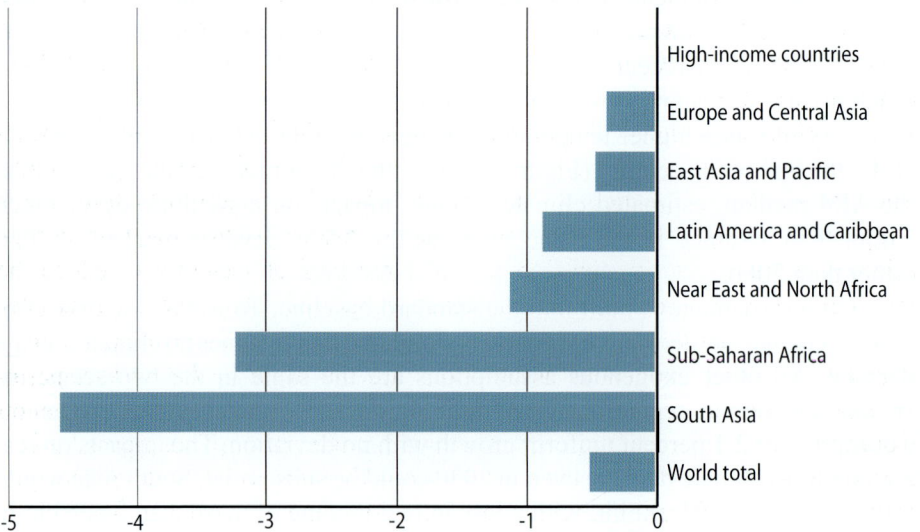

Source: Simulation results with the World Bank's ENVISAGE model.

212

first, agriculture remains important despite relatively rapid economic growth; and second, existing studies suggest that the greatest damage is occurring in these two regions, as summarized in Cline's (2007) estimates.

In this alternative scenario, the impact on high-income countries is negligible in the short term. This arises partly from gains in the terms of trade, as world prices rise in the with-damage scenario. The net trade position of all developing regions deteriorates in the with-damage scenario, albeit somewhat modestly by 2030, and improves (modestly) for high-income countries. In the long run, climate damage is bound to increase, both because the climate will deteriorate and because of non-linear effects (not currently captured in the model).

Biofuels

The expansion of ethanol based on grain feedstock is quite different from that of sugar cane-based ethanol, especially in Latin America. In the latter, the trade-off between food and fuel is somewhat limited. Moreover, sugar cane expansion will occur first in Latin America and then in other countries with low-cost sugar production. Most of this expansion will occur on land where there is limited competition among crops. In contrast, ethanol based on grains has a direct effect on several important competing crops, including oilseeds. The expansion of biodiesel has a strong and direct implication for vegetable oil prices, and feedstock and food demands are in direct competition. Large-scale biodiesel expansion will push vegetable oil prices higher. Hence, the expansion of biofuel based on grains and oilseed products is a potential exacerbating factor for higher food prices and could compromise access to food for the poorest people on the planet. The most affected food prices would be those for grains, vegetable oils, meat and dairy products, which are intensive in feedstocks.

If cellulosic/biomass ethanol can become profitable, the trade-off between food and fuel may be less important and be confined to oilseed-based biofuels. The development of biofuels is also determined by their return, which in turn is largely determined by fossil energy prices and feedstock prices. Low fossil energy prices will undermine the development of large biofuel sectors and reduce the trade-off between food and fuel. Of course, large and forced biofuel mandates could change this result. A recent study suggests that the existing biofuel mandates for 2020 would have only modest impacts on global food prices, partly because they are not particularly ambitious (Timilsina *et al.*, 2010). Sugar prices would rise the most (by 7 percent), with grain and oilseed prices rising by less than 4 percent – although potentially with greater impacts on trade patterns, as countries have different mandate targets and comparative advantage in biofuels production varies across regions. These are long-term equilibrium effects, it is likely that the short-term impacts could be more significant. Over the longer term, it is still difficult to know what policies will prevail in 2050. Biofuels, both first- and second-generation, are an area of active research.

Poverty implications

The assumptions in the baseline scenario explained in the previous section were used to "roll" the global economy to 2050. This section concentrates on the global distributional effects of the expected changes in per capita incomes and income distribution within countries.[4] Evaluation of these distributional effects is based on the World Bank's Global Income Distribution Dynamics (GIDD) model. This macro-micro simulation framework is overviewed in Box 5.3 and explained in detail in Bussolo, de Hoyos and Medvedev (2010).

Box 5.3 - The global income distribution dynamics model

The World Bank Development Economics Prospects Group developed the GIDD model, the first global CGE-micro-simulation model. The GIDD model takes into account the macro nature of growth and economic policies and adds a microeconomic – household and individual – dimension.

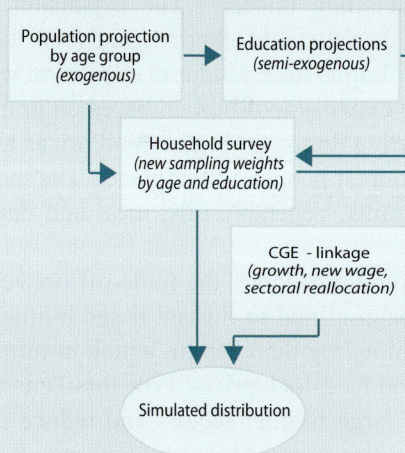

Population projection by age group *(exogenous)* → Education projections *(semi-exogenous)*

Household survey *(new sampling weights by age and education)*

CGE - linkage *(growth, new wage, sectoral reallocation)*

Simulated distribution

The GIDD model includes distributional data for 121 countries and covers 90 percent of the world population. Academics and development practitioners can use the model to assess the growth and distribution effects of global policies such as, among others, multilateral trade liberalization, and policies dealing with international migration and climate change. The GIDD model also allows analysis of the impacts on global income distribution from different global growth scenarios, and distinction between changes due to shifts in average income between countries and those attributable to widening disparities within countries.

---

4. This section relies on the methodology used in Bussolo *et al.* (2008), which projects the global economy to 2030. Nevertheless, it has some minor variations: it uses the latest version of the GIDD model (December 2010), which has 2005 instead of 2000 as the base year, and uses the latest purchasing power parity (PPP) conversion factors. As a result, slight differences may emerge between the two documents, but these will not compromise the messages and authors' conclusions in either of them.

The macro-micro modelling framework described here explicitly considers long-term time horizons during which changes in the demographic structure may become a crucial component of both growth and distribution dynamics. The GIDD model's empirical framework is schematically represented in the Figure above.

The expected changes in population structure by age (upper left of the Figure) are exogenous, meaning that fertility decisions and mortality rates are determined outside the model. The change in shares of the population by education level incorporates the expected demographic changes (the arrow linking the top left box to the top right box). New sets of population shares by age and education subgroup are then computed, and household sampling weights are rescaled, according to the demographic and educational changes (the larger box in the middle of the Figure). The impact of changes in the demographic structure on labour supply (by skill level) is incorporated into the CGE model, which then provides a set of link variables for the micro-simulation:

a) change in the allocation of workers across sectors in the economy;

b) change in returns to labour, by skill and occupation group;

c) change in the relative prices of food and non-food consumption baskets;

d) differentiation in per capita income/consumption growth rates across countries.

The final distribution is obtained by applying the changes in these link variables to the reweighted household survey (bottom link in the Figure).

Figure 5.15 plots Lorenz curves for the observed global income distribution in 2005 and the projected distribution in 2050. It appears that the largest changes in income distribution between 2005 and 2050 are expected to be around the middle of the income distribution rather than towards the upper or lower tails. In fact, because the two Lorenz curves intersect at these tails, it is not possible to say that the 2050 distribution Lorenz curve dominates that of 2005. In other words, it cannot be claimed that inequality in 2050 is lower than in 2005, regardless of the inequality measure being used. However, using standard inequality statistics such as the Gini, the Theil and the mean logarithmic deviation – i.e., indicators that do not give too much weight to the extreme parts of the distribution – a marked reduction of inequality, as shown in Tables 5.6 and 5.7, is recorded during the period considered here.

Table 5.6
Global income inequality

Index	2005	2050	Dispersion only	Convergence only
Gini	0.697	0.616	0.701	0.616
Theil	1.046	0.717	1.059	0.719
Mean log deviation	0.942	0.723	0.954	0.723

Source: Authors' estimates.

Table 5.7
Income inequality, by region

Region	Gini		Theil		Mean Log Dev	
	2005	2050	2005	2050	2005	2050
Developed countries	0.394	0.378	0.270	0.245	0.277	0.257
Developing countries	0.552	0.588	0.623	0.664	0.529	0.629
East Asia and Pacific	0.421	0.479	0.311	0.399	0.293	0.411
Eastern Europe and Central Asia	0.394	0.513	0.257	0.441	0.280	0.490
Latin America and Caribbean	0.599	0.605	0.714	0.707	0.699	0.719
Near East and North Africa	0.399	0.405	0.284	0.298	0.261	0.271
South Asia	0.297	0.326	0.156	0.183	0.141	0.176
Sub-Saharan Africa	0.495	0.488	0.499	0.481	0.425	0.410

Source: Authors' estimates.

Figure 5.15
Changes in the Lorenz curve dominance for 2005 and 2050 distributions (cumulative income share)

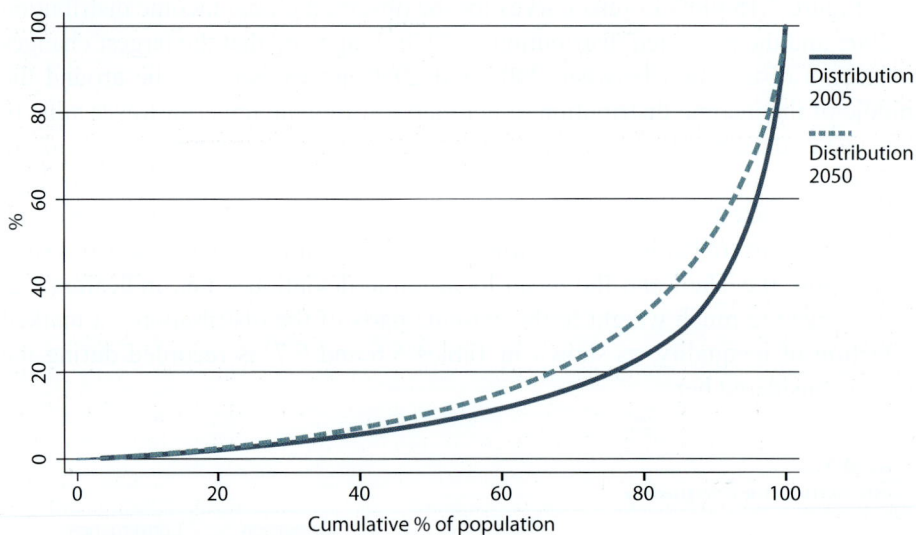

Source: Authors' calculations.

The remainder of this section analyses the drivers of these expected distributional changes by means of three complementary approaches. First, the analysis is conducted in terms of the convergence and dispersion components, i.e., the changes in income disparities among and within countries. This is taken up in the next two subsections, which show that the reduction in global income inequality between 2005 and 2050 is the outcome of two opposing forces: the inequality-reducing convergence effect and the inequality-enhancing dispersion effect. In other words, poor countries will catch up, but at a cost in terms of higher within-country and within-region income inequality. Second, the expected poverty effects of the new income distribution in 2050 are analysed using two approaches: the standard absolute poverty line of USD 1.25 a day; and a weakly relative poverty line suggested by Ravallion and Chen (2009). Third, as global poverty is expected to be substantially reduced by 2050, the emergence of a global middle class is analysed following the methodology presented in Bussolo *et al.* (2008).

The dispersion and convergence component

The dispersion component should be understood as the outcome of all the changes outlined by the baseline scenario in the previous section, but keeping constant average incomes in each country. Within countries, income distribution is expected to be altered by demographic changes, changes in skilled-to-unskilled wage remuneration, and rural-urban migration. Figure 5.16 plots non-parametric kernel densities of the global income distribution in 2005, together with the hypothetical distribution for the dispersion component, capturing only the changes in within-country inequality between 2005 and 2050. This hypothetical distribution was created by dividing household incomes in 2050 by the country-specific growth rate of average incomes between 2005 and 2050. At the global level, distributional changes within countries in this hypothetical distribution almost match the original distribution, having an almost neutral inequality effect at the global scale, with the income distribution barely increasing in Gini points (Table 5.6).

On the other hand, the convergence component takes into account each country's income variation as projected from the baseline scenario, but maintains constant global average income. Three aspects determine the existence, sign and magnitude of each country's contribution to the convergence component: i) a country will have a global distributional impact if its rate of growth differs from the global average; ii) if the country satisfies this condition, the sign of the distributional effect will depend on the country's initial position in the global distribution; and iii) the magnitude of the impact is determined by the size of the growth rate differentials (with respect to the global average) and the country's share in the global population. Hence, initially poor countries with higher-than-average growth rates will have an inequality-reducing effect whose magnitude will be determined by the size of the country's population.

Figure 5.16
Global income inequality reduction, 2005 to 2050

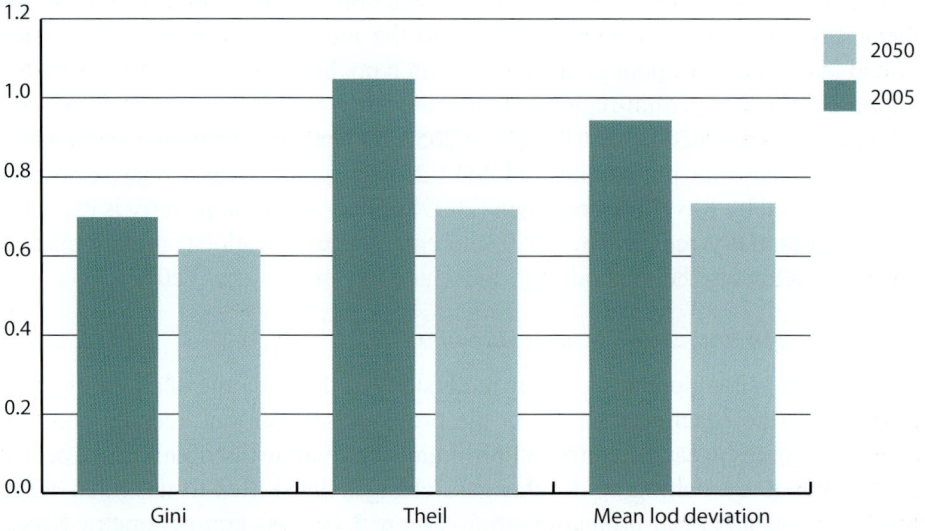

Source: Authors' calculations.

Figure 5.17
Within-region income inequality

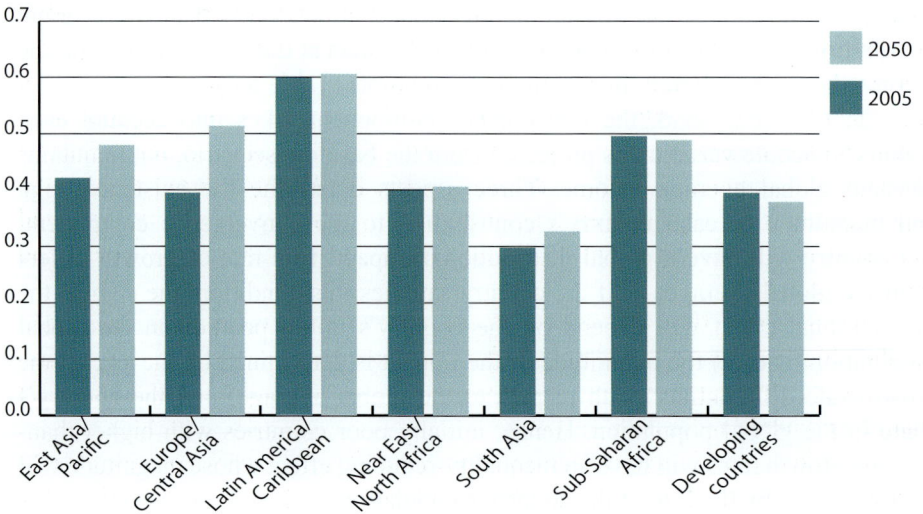

Source: Authors' calculations.

Figure 5.17 shows the change of the global income distribution due to differences in growth rates among countries when global average income is kept constant. Had the convergence effect been the only change taking place between 2005 and 2050, global inequality would have been reduced by 8.0 Gini points (Table 5.6). This means that the improvement in global income distribution reported can be explained mainly by growth rate differentials across countries, with poor countries catching up with middle- and high-income countries.

Poverty

Measurement of global poverty in developing countries has typically been based on absolute poverty measures. The typical practice for an absolute measure is to set a monetary quantity, called the poverty line, which represents the minimum income needed to acquire a set of goods that will suffice for some established basic human needs. Poverty lines are typically based on the food needed to attain a recommended daily caloric ingestion. In addition to these basic poverty lines, some countries draw complementary ones that set the minimum income needed to satisfy more complex human needs, such as health and education. At the global level, the World Bank's USD 1.25 and USD 2 a day are the best-known examples of absolute poverty lines.

Alternatively, the common practice in Organisation for Economic Co-operation and Development (OECD) countries is to use relative poverty lines. These monetary quantities are periodically adjusted, not as the minimum income needed to acquire a given basket of goods, but as a constant proportion of the countries' mean or median incomes. The first argument for using relative rather than absolute poverty measures relies on the "welfarist" assumption that people attach value to their own income relative to the average in their own society – often cited as the "theory of relative deprivation" or the "relative income hypothesis". The second argument is that relative poverty lines allow for differences in the cost of social inclusion. Following Ravallion and Chen (2009), these are defined as the expenditure needed to cover certain commodities that are deemed to have a social role in assuring that a person can participate with dignity in customary social and economic activities.

Despite these two arguments, relative poverty lines have not been used for the study of poverty in very-low-income countries because they are scale-independent; in other words, if all incomes in a society grow at the same rate, no change in poverty will occur.

Ravallion and Chen (2009) discuss all these aspects rigorously and outline an alternative measure. Using a large sample of poverty lines collected by the World Bank, they calibrate a new measure for studying global poverty called the weakly relative poverty line. The proposed weakly relative poverty line is, in general terms, a combination of the two previous approaches: i) for very low levels of income, it functions as an absolute poverty line set at the World Bank's USD 1.25 a day level (at 2005 PPP); and ii) for medium and higher incomes, it functions as a relative poverty line. Empirical implementation applied the following formula:

$$Z_i \equiv \max \left[\$1.25, a + \frac{M_i}{3}\right] \qquad (5.2)$$

where Z_i is the value of the poverty line; M_i is the mean daily income in country i; and α is estimated by Ravallion and Chen (2009) to be PPP USD 0.60. The advantage of using the weakly relative poverty line is that it will provide a better understanding about poverty and exclusion in the projected income distribution for 2050 than the absolute poverty measure will (Figures 5.18 and 5.19). Table 5.8 summarizes the regional headcount ratios of absolute and weakly relative poverty in 2005 and 2050. While absolute poverty vanishes in all regions, weakly relative poverty still accounts for a large share of the population, especially in underperforming Latin America. According to the baseline scenario, the increase in weakly relative poverty reported by Ravallion and Chen (2009) experienced during the late 1980s and until 2000 is reversed by 2050 in almost all regions. Table 5.8 shows the headcount indices for absolute and weakly relative poverty in 2005 and 2050, and changes in the number of poor in both periods.

Table 5.8
Poverty estimates

Region	Absolute poverty (USD 1.25 per day PPP)			Weakly relative poverty		
	Headcount index 2005	Headcount index 2050	-Δ poverty (millions)	Headcount index 2005	Headcount index 2050	-Δ poverty (millions)
Developing countries	21.9	0.4	1,185	31.96	12.4	843
East Asia and Pacific	15.8	0.0	-87	30.4	12.1	277
Eastern Europe and Central Asia	4.4	0.0	20	12.6	5.5	35
Latin America and Caribbean	8.1	1.0	35	33.3	31.3	(67)
Near East and North Africa	4.1	0.0	8	19.0	10.5	5
South Asia	40.5	0.0	583	40.8	4.0	499
Sub-Saharan Africa	51.7	2.8	252	55.5	20.3	104

Source: Authors' estimates.

Figure 5.18
Income distribution diversity, 2005 and 2050

USD 1.25 poverty lne

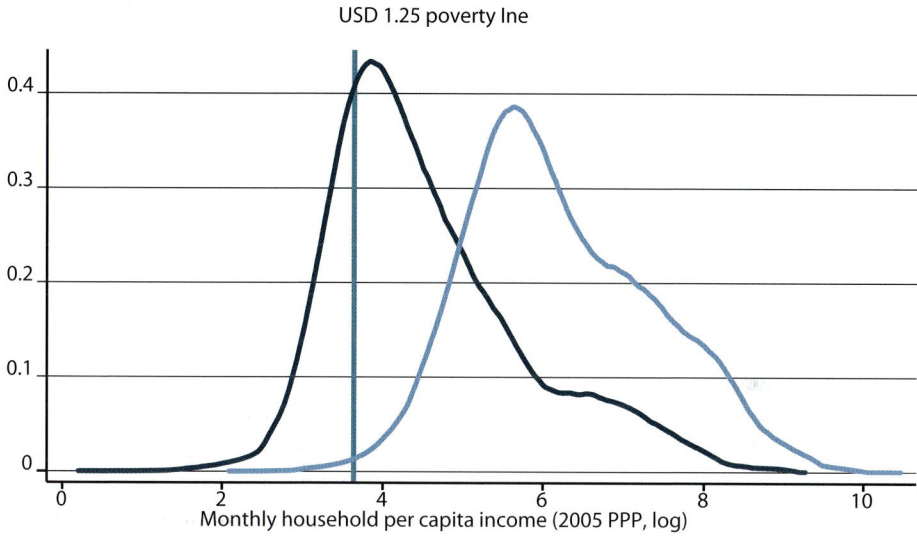

Monthly household per capita income (2005 PPP, log)

Source: Authors' calculations.

Figure 5.19
Changes in absolute and relative poverty

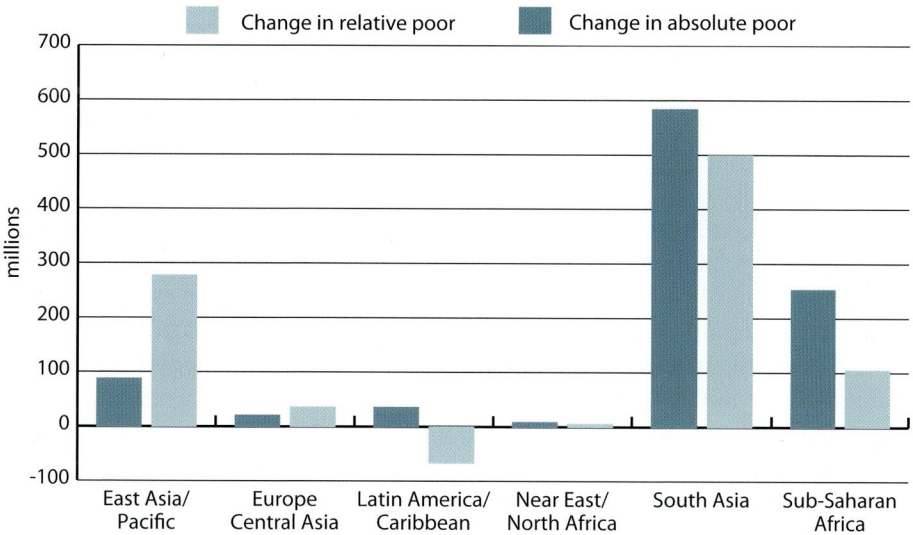

Sources: Relative poverty measured from Ravallion and Chen, 2009; authors' calculations.

The most interesting result is that while other regions perform relatively well, in Latin America the number of weakly relative poor actually increases (by 67 million), partly reflecting that this is the world's most unequal region. Within this Latin America and the Caribbean, the large majority of countries will see increases in the number of people living in relative poverty, Mexico being the most affected. Mexico alone accounts for half the increase in the number of relative poor in Latin America, followed by Brazil (11 million), Ecuador (4.8 million) and Colombia (4 million).

In sub-Saharan Africa, absolute poverty is expected to be reduced from 51.2 to 2.8 percent of the population and, remarkably, weakly poverty from 55.5 to 20.3 percent. The country that will perform the best is the United Republic of Tanzania, which will reduce its relative poverty rate by almost 70 percent, with 20 million fewer people living in absolute poverty. Nigeria and Ethiopia will reduce their net numbers of poor drastically, by 34 and 20 million respectively; however, in relative terms, the best performers are Malawi, Burundi, Guinea and Rwanda, all of which will reduce relative poverty by more than 50 percentage points.

The new middle class and beyond

In addition to the analysis of global poverty, the emergence of countries in the new middle class is of high importance because of the changes in global consumption patterns expected to accompany economic growth. Individuals in 2050 will be healthier and more educated, with higher expectations about their role in life, greater political participation and increasingly complex needs. As a result, the demand for more and better goods and services will rise as a vast number of families in developing countries emerge from poverty. This study uses the definition of absolute global middle class (GMC) used by Bussolo *et al.* (2008) to quantify the number of people who will be part of this group in the hypothetical income distribution for 2050. The GMC is defined as all the world citizens living with incomes between the current Brazilian and Italian averages.

The GMC will grow from about 450 million in 2005 to 2.1 billion in 2050, and from 8.2 to 28.4 percent of the global population (Table 5.9). Furthermore, the composition of this group of consumers is likely to change radically: while in 2005, developing country nationals accounted for 56 percent of the GMC, by 2050 they are likely to represent nearly all of this group. The biggest contributors to the increase in GMC numbers are the most populous Asian countries, led by China and India. These two countries alone are responsible for nearly two-thirds of the entire increase in the GMC, with China accounting for 30 percent and India another 35 percent. More surprisingly, as a result of sustained economic growth in China and according to the scenario described in the previous section, by 2050, 40 percent of the Chinese population will surpass GMC status.

Table 5.9
Composition of the global middle class

Region	2005 (millions)	(%)	2050 (millions)	(%)
Developed countries	190.8	33.0	27.1	4.3
Developing countries	260.2	6.4	2 117.3	29.7
East Asia and Pacific	41.1	2.3	785.7	35.0
Eastern Europe and Central Asia	85.9	19.7	117.9	30.5
Latin America and Caribbean	107.5	20.3	245.9	31.8
Near East and North Africa	18.3	8.9	151.2	47.0
South Asia	0.6	< 0.1	657.6	29.2
Sub-Saharan Africa	6.8	1.3	159.1	16.6
Total	451.0	8.15	2 144.3	28.4

Source: Authors' estimates.

There are several reasons for the projected dramatic increase in the GMC and the major shift in composition in favour of low- and middle-income countries. Faster population growth in the developing world is responsible for some of the change in composition. Thus, regions with population growth above the world average (e.g., South Asia and sub-Saharan Africa) will increase their shares in the GMC. However, the main determinant for joining the middle class is not population growth, but income growth. Although East Asia's population will grow more slowly than the world average, this region is projected to increase its share of GMC residents by more than 30 percentage points, compared with 15 percentage points in sub-Saharan Africa. This difference arises because annual per capita income growth in Asia is forecast to be more than twice that in sub-Saharan Africa, easily offsetting the decline in the former's population share.

Most developing country members of today's (2005's) GMC earn incomes far above the averages of their own countries of residence. In other words, being classified as middle class at the global level is equivalent to being at the top of the income distribution in many low-income countries. For example, in the study sample, in 2005, 180 million (out of the total 260 million) developing country citizens in the GMC are in the top 20 percent of earners within their own countries. Thus, for many nations, the correspondence between the GMC and the within-country middle class is quite low. The situation will change dramatically by 2050. A full 60 percent of developing country members of the GMC will be earning incomes in the seventh decile or lower of the national scale. For example, in China, in 2005, 27 million people belonged to the GMC, all earning more than what 90 percent of all Chinese citizens earned. By 2050, there will be 517 million

Chinese in the GMC, their earnings ranging from the fifth to the ninth deciles of the Chinese national income distribution.

Consistent with these data, by 2050 the middle class, together with the rich, will account for a larger share of the population in a greater number of countries. In 2005, the members of these two groups exceeded 40 percent of the population in only six developing countries, which were home to 3.0 percent of the population of the developing world. By 2050, the middle class and rich will exceed 40 percent of the population in 58 developing countries (as classified today), which will account for 72 percent of the world's developing country population.

Conclusions

At a minimum, the price spikes of 2007/2008 shook global complacency regarding agriculture, after a period of neglect driven in part by globally benign price changes and no major supply disruptions. Experts were aware of the falls in agricultural productivity growth and expenditures on R&D, but in a crowded field of international economic policy issues, the warning signs were largely ignored. As regards agriculture, the focus has been far more on farm support policies and trade barriers than on fundamental supply issues. Is the world now witnessing a structural shift, with higher and growing agricultural prices, or are the events of 2007/2008 and 2011 just bumps in the road? This paper suggests that the answer lies somewhere between these two extremes. There is a structural shift in agricultural markets, with greater linkages to energy markets than in the past. Higher energy prices could induce a stronger shift to biofuels, generating competing pressures on resources and higher food prices. Potentially this linkage could be strengthened if climate mitigation policies raise the end-use prices of conventional fossil fuels and induce further substitution by biofuels. At the same time, there are reasons to believe that the world can adjust to these imminent changes. Declining population growth and food saturation will temper food demand growth in the future, and health and environmental concerns could even induce a shift in tastes that would temper demand even further. There is also sufficient land to allow some expansion, if managed appropriately and sustainably. This will require investment in infrastructure, which could be onerous, particularly in the poorer parts of the world. The ability to raise productivity is also a concern, particularly in an environment with growing climate stress. This too will require resources to enhance R&D, perhaps with an emphasis on regions where productivity lags far behind best practices.

However, even if there is manageable stress at the global level, changing environments at the regional level are likely to have repercussions on distribution, both across and within countries. Managing these stresses may be more difficult, as food security at both the household and national levels is often a priority for policy-makers. And as witnessed in the most recent crisis, policy-makers naturally make the most rational decisions for their own stakeholders, even if better overall policies could be implemented with the right coordination.

References

Armington, P. 1969. A theory of demand for products distinguished by place of production. *IMF Staff Papers*, 16: 159–178.

Baffes, J. 2007. Oil spills on other commodities. *Resources Policy*, 32: 126–134.

Baffes, J. 2010. More on the energy/non-energy commodity price link. *Applied Economics Letters*, 17: 1555–1558.

Borensztein, E. & Reinhart, C.M. 1994. The macroeconomic determinants of commodity prices. *IMF Staff Papers*, 41: 236–261.

Bussolo, M., de Hoyos, R. & Medvedev, D. 2010. Economic growth and income distribution: Linking macroeconomic models with household survey data at the global level. *International Journal of Microsimulation*, 31: 92–103.

Bussolo, M., de Hoyos, R., Medveded, D. & van der Mensbrugghe, D. 2008. *Global growth and distribution: Are China and India reshaping the world?* UNU-WIDER Research Paper No. 2008/29. Washington, DC, United Nations University.

Büyükşahin, B., Haigh, M.S. & Robe, M.A. 2008. *Commodities and equities: A market of one?* Washington, DC, United States Commodity Futures Trading Commission.

CFTC. 2008. *Interagency Task Force on Commodity Markets releases interim report on crude oil.* Washington, DC, Commodities Futures Trading Commission (CFTC).

Cline, W.R. 2007. *Global warming and agriculture: Impact estimates by country.* Washington, DC, Center for Global Development and Peterson Institute for International Economics.

Coelli, T.J. & Prasada Rao, D.S. 2005. Total factor productivity growth in agriculture: a Malmquist index analysis of 93 countries, 1980–2000. *Agricultural Economics*, 32(s1): 115–134.

Deaton, A. 1999. Commodity prices and growth in Africa. *Journal of Economic Perspectives*, 13: 23–40.

Dickey, D. & Fuller, W.A. 1979. Distribution of the estimators for time series regressions with unit roots. *Journal of the American Statistical Association*, 74: 427–431.

Eckaus, R.S. 2008. *The oil price is a speculative bubarrele.* Center for Energy and Environmental Policy Research Working Paper 08-007. Washington, DC, Center for Energy and Environmental Policy Research.

FAO. 2002. *Commodity market review 2001–2002.* Rome.

Gilbert, C.L. 1989. The impact of exchange rates and developing country debt on commodity prices. *Economic Journal,* 99: 773–783.

Gilbert, C. 2008. Commodity speculation and commodity investments. Revised version of paper presented at the Globalization of Primary Commodity Markets Conference, Stockholm, 22–23 October 2007.

Hertel, T.W., ed. 1997. *Global trade analysis: Modeling and applications.* New York, Cambridge University Press.

Holtham, G.H. 1988. Modeling commodity prices in a world macroeconomic model. *In* O. Guvenen, ed. *International commodity market models and policy analysis.* Boston, Massachusetts, USA, Kluwer Academic Publishers.

Houthakker, H.S. 1975. Comments and discussion on *The 1972–75 commodity boom* by Richard N. Cooper and Robert Z. Lawrence. *Brookings Papers on Economic Activity,* 3: 718–720.

IMF. 2006. *World Economic Outlook 2006.* Washington, DC.

IMF. 2008. *World Economic Outlook 2008.* Washington, DC.

Lluch, C. 1973. The extended linear expenditure system. *European Economic Review,* 4: 21–32.

Masters, M.W. 2008. Testimony before the Committee of Homeland Security and Government Affairs, 20 May 2008. Washington, DC.

Metz, B., Davidson, O.R., Bosch, P.R., Dave, R. & Meyer, L.A, eds. 2007. Contribution of Working Group III to the Fourth Assessment Report of the Intergovernmental Panel on Climate Change. *In* IPCC. *Climate change 2007: Mitigation of climate change.* Cambridge, UK and New York, Cambridge University Press.

Nordhaus, W. 2008. *A question of balance: Weighing the options on global warming policies.* New Haven, Connecticut, USA, Yale University Press.

Plastina, A. 2008. Speculation and cotton prices. *International Cotton Advisory Committee, Cotton: Review of the World Situation,* 61: 8–12.

Radetzki, M. 2008. *A handbook of primary commodities in the global economy.* Cambridge, UK, Cambridge University Press.

Ravallion, M. & Chen, S. 2009. *Weakly relative poverty.* World Bank Policy Research Working Paper Series No. 4844. Washington, DC, World Bank.

Rimmer, M.T. & Powell, A.A. 1996. An implicitly additive demand system. *Applied Economics,* 28: 1613–1622.

Soros, G. 2008. Testimony before the US Senate Commerce Committee Oversight: Hearing on FTC Advanced Rulemaking on Oil Market Manipulation, 3 June 2008. Washington, DC. www.georgesoros.com/files/sorosfinaltestimony.pdf.

Spraos, J. 1980. The statistical debate on the net barter terms of trade between primary commodities and manufactures. *Economic Journal*, 90: 107–128.

Timilsina, G.R., Beghin, J.C., van der Mensbrugghe, D. & Mevel, S. 2010. *The impacts of biofuel targets on land-use change and food supply*. World Bank Policy Research Working Paper Series No. 5513. Washington, DC, World Bank.

van der Mensbrugghe, D. 2009. *The ENVironmental Impact and Sustainability Applied General Equilibrium (ENVISAGE) Model*. Washington, DC, World Bank. (mimeograph)

World Bank. 2009. *Global economic prospects: Commodities at the crossroads*. Washington, DC.

World Bank. 2011. *Global monitoring report*. Washington, DC.

World Bank. (various years). Commodity price data. Washington, DC, World Bank Development Prospects Group.

Wray, R.L. 2008. *The commodities market bubble: Money manager capitalism and the financialization of commodities*. Public Policy Brief No. 96. Annandale, New York, USA, Levy Economics Institute of Bard College.

THE MODEL USED FOR CLIMATE CHANGE SIMULATIONS

The quantitative analysis of the climate change section of this chapter relies extensively on the World Bank's dynamic global computable general equilibrium model, ENVironmental Impact and Sustainability Applied General Equilibrium Model (ENVISAGE) (van der Mensbrugghe, 2009). Underlying this model is the 2004-based Release 7 of the Global Trade Analysis Project (GTAP) database, which divides the world economy into 112 countries/regions (of which 94 are countries) and 57 commodities.[5] For modelling purposes, the underlying database is typically aggregated to a more manageable set of regions and sectors, which are selected according to the objectives of the particular study. In this chapter, the focus is on the agriculture and food sectors, but also energy, to capture the emergence of biofuels and the linkage between energy and agriculture. ENVISAGE has been designed for climate change studies, so the standard GTAP data are supplemented by several satellite accounts. These include energy data in volume, carbon emissions linked to the burning of fossil fuels, and emissions from the other Kyoto greenhouse gases – methane, nitrous oxides and the fluorinated gases. Both methane and nitrous oxides are linked to agricultural production. The other greenhouse gases differ from carbon emissions. First, they have a more exhaustive set of drivers, because they can be associated with all intermediate inputs, not simply fossil fuels, as well as factor inputs (e.g., land in the case of methane generated by rice production) and output. Second, technologies for their abatement are more complex than those for fossil fuel-based carbon emissions. With current technologies, the latter can only be abated by either lowering consumption of fossil fuels or substituting with lower- or zero-emission fuels. For the other greenhouse gases, abatement technologies may involve different production methods, although presumably at a higher cost.

In this chapter, the GTAP data were supplemented with a more exhaustive set of electricity activities, splitting the single GTAP electricity sector into five production activities: coal-fired, oil and gas-fired, nuclear, hydroelectric, and other (including all existing renewables). For long-term scenario analysis, several new energy technologies were introduced. These initially have low penetration,

---

5. More on the GTAP data can be found at www.gtap.org.

but under certain circumstances they could potentially replace conventional technologies. They include first- and second-generation biofuels as potential substitutes in the transport sector, and coal and gas carbon capture and storage in the power sector.

In most respects, the ENVISAGE model is a rather classical recursive dynamic global CGE with a time horizon spanning 2004 to 2100. Production is based on the capital-labour substitution with capital and energy near-complements in the short term and with substitutes in the longer term. A vintage production structure is employed that allows for partial capital mobility across sectors in the short term, or a putty-semi-putty technology. Vintage capital is associated with lower production flexibility, whereas new capital is more flexible; aggregate flexibility thus depends on the share of vintage capital in total capital, with greater flexibility associated with those economies with the highest savings rates. Factor payments accrue to a single representative household in each region, and this household allocates income among savings and expenditures on goods and services. The model allows for significant flexibility in specifying consumer demand. The top-level utility function can be specified using one of three demand systems: constant difference in elasticities (Hertel, 1997), extended linear expenditure system (Lluch, 1973) and Almost Ideal Directly Additive Demand Systems (AIDADS, Rimmer and Powell, 1996). The top-level utility function can be specified at a different commodity aggregation than production. A transition matrix, which allows for commodity substitution, converts consumer goods to produced goods. Energy demand is specified as a single bundle for each agent in the economy. Energy demand is then split into demand for specific types of energy using a nested constant elasticity of substitution (CES) structure. Trade is specified using the ubiquitous Armington assumption (Armington, 1969), although the model allows for homogeneous commodities as well. Government plays a relatively passive role, collecting taxes and spending on goods and services. The government's fiscal balance is fixed in any given year (and declines towards 0 from its initial position by 2015), and the household direct tax schedule shifts to achieve the fiscal target. The latter implies that changes in indirect taxes (e.g., import tariffs or carbon taxes) are recycled to households in lump-sum fashion. Investment is savings-driven and savings rates are influenced by the overall growth rate and by demographic factors such as dependency ratios. The current account balance for each region is fixed in any given year. The base year balances converge towards zero at some date (currently set at 2025). An *ex-ante* shift in either import demand or export supply influences the real exchange rate. Thus, for example, if a country is forced to import more food owing to climate damage to its agriculture, this would normally entail a real exchange rate depreciation that increases demand for its exports to pay for the additional food imports.

ENVISAGE has been developed as an integrated assessment model. Emissions of the greenhouse gases generated by the economic part of the model lead to changes in atmospheric concentrations. A simple reduced-form atmospheric model converts changes in the stock of atmospheric concentrations into changes in radiative forcing and global mean temperature. The resulting changes in global mean temperature feed back into the economy through damage functions that affect various economic drivers. In the current version of the model, the only feedback is through changes in agricultural productivity. The agricultural damage functions have been calibrated to the estimates from the recent study by Cline (2007).

Dynamics in ENVISAGE are driven by three key factors. The first is demographics, which describe population and labour force rates of growth. Following common practice, the baseline in this chapter uses the medium variant from the UN populations forecast, with growth of the labour force equated to growth of the working-age population (defined as those between 15 and 65 years of age). The second key driver is formed by savings and investment, which jointly determine the overall level of capital stock (along with the rate of depreciation). In ENVISAGE the savings function is partially determined by demographics. Generally speaking, savings will rise as dependency ratios (both under-15 and over-65) fall.

The third driver is productivity. ENVISAGE differentiates productivity across broad sectors: agriculture, energy, manufacturing, and services. Agriculture's productivity growth has two components to be calibrated: the exogenous component is calibrated to 2.1 percentage points per year, consistent with recent trends (World Bank, 2008); and the endogenous component comes from a linear damage function that links increases in global temperature to declines in agricultural total factor productivity (TFP) and is calibrated according to Cline's average estimates with and without carbon fertilization (Cline, 2007).

Productivity in other sectors is unaffected by climate change, and is calibrated through 2015 to match the World Bank's medium- and long-term forecast. After 2015, productivity growth in the United States of America is calibrated to achieve a long-term average (2004 to 2100) growth in real GDP per capita of 1.2 percent per year, with faster growth in the first half of the century, while productivity in other countries/regions is calibrated based on simple convergence assumptions.

PART III

LAND, WATER AND CAPITAL REQUIREMENTS

CHAPTER 6

THE RESOURCES OUTLOOK: BY HOW MUCH DO LAND, WATER AND CROP YIELDS NEED TO INCREASE BY 2050?

Jelle Bruinsma[1]

The recent food crisis was characterized by sharp food price surges and caused, in part, by new demands on agriculture, such as demand for biomass as feedstock in biofuel production (Alexandratos, 2008). It made fears that the world is running out of natural resources (foremost among them land and freshwater resources) come back with a vengeance (e.g., Brown, 2009). Concerns are voiced that agriculture might, in the not too distant future, no longer be able to produce the food needed to feed a still growing world population at levels sufficient to lead a healthy and active life.

Such fears are by no means new and continually keep coming back, prompting a series of studies and statements concerning how many people the earth can support. The continuing decline of arable land per person (Figure 6.1) is often cited as an indicator of impending problems.[2] The underlying cause for such problems is perceived to be an ever-increasing demand for agricultural products facing finite natural resources such as land, water and genetic potential. Scarcity of these resources would be compounded by competing demand for them, originating in urbanization, industrial uses and use in biofuel production, and by forces that change their availability, such as climate change and the need to preserve resources for future generations (environmentally responsible and sustainable use).

This chapter addresses some of these issues by unfolding the resource use implications of the crop production projections underlying the latest FAO

---

1. The author gratefully acknowledges substantial contributions by Gerold Boedeker, Jean-Marc Faures, Karen Frenken and Jippe Hoogeveen, as well as comments by FAO staff on an earlier draft.

2. Of course, declining land per person, combined with increasing average food consumption could also be interpreted as a sign of ever-increasing agricultural productivity.

perspective study (FAO, 2006b).[3] These projection results are also presented in Chapter 1. They can be considered as representing a baseline scenario, but do not take into account additional demand for agricultural products and for land needed by biofuel production, nor do they explicitly account for land-use changes due to climate change. This is not to say that such demands on agriculture would be additive to demand on agriculture and natural resources for food and feed purposes. There will be competition for resources and substitution among the final uses of agricultural products. These issues are discussed by Fischer, in Chapter 3 of this volume.

Figure 6.1
Arable land per capita

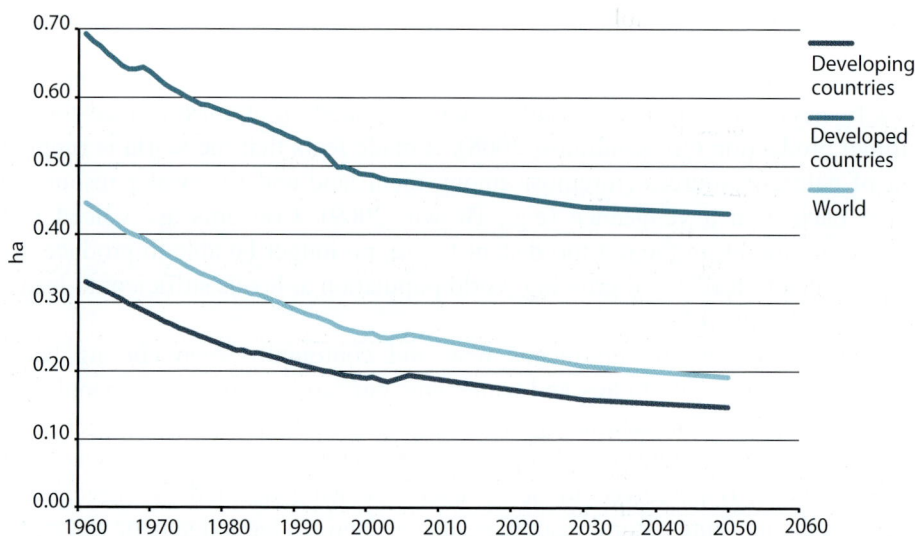

Sources: FAOSTAT and author.

In discussing the natural resource implications, this chapter focuses on the physical dimensions of natural resource use in agriculture. While acknowledging the validity and importance of environmental and sustainability concerns such as deforestation, land degradation and water pollution, the chapter does not explicitly deal with then, owing to space and time constraints.

The FAO (2006b) study has as base year the three-year average 1999/2001 based on FAOSTAT data as known in 2002 to 2004. At present, FAOSTAT offers

---

3. Unlike the preceding study (Bruinsma, 2003), for various reasons, the 2006 interim study did not deal with resource use issues, such as land and yield expansion, and water use in irrigation.

published data up to 2003 for supply utilization accounts, and up to 2007 for land use and production by crop; although time constraints and the non-availability of published food balance sheet data after 2003 meant that no new base year and projections could be derived, production and land-use data for the latest three-year average (2005/2007) are taken into account in the work underlying this chapter.

Another limitation is that while this chapter was being prepared, the results of the 2009 Global Agro-Ecological Zone (GAEZ) study were not yet available, so the results of the 2002 GAEZ study (reported in Fischer *et al.*, 2002) had to be used instead.

The chapter is based on analytical work for 146 countries: 93 developing, and 53 developed,[4] with 42 of the latter grouped into four country groups (Annex 6.1). At present, these countries account for almost 98 percent of the world's population and 100 percent of its arable land.

How much more needs to be produced?

FAO's (2006b) baseline projections show that by 2050 the world's average daily calorie availability could rise to 3 130 kcal per person, an 11 percent increase over the 2003 level. This would still leave 4 percent of developing countries' populations chronically undernourished in 2050.[5]

For these projections to materialize, annual world agricultural production would need to increase by 70 percent over the period from 2005/2007 to 2050 (Table 6.1). World population is projected to rise by about 40 percent over this period, meaning that per capita production would rise by some 22 percent. The reason this would translate into an increase of only 11 percent in per capita calorie availability is mainly[6] because of expected changes in diet, with a shift to higher-value foods of often lower calorie content (e.g., vegetables and fruits) and to livestock products, implying an inefficient conversion of calories from the crops used in livestock feeds. For example, per capita meat consumption would rise from 37 kg per year in 1999/2001 to 52 kg in 2050 (from 27 to 44 kg in developing countries), implying that much of the additional crop (cereal) production will be used as feed for livestock production.

---

4. Developed countries include the industrialized countries and countries in transition.

5. A partial update of the projections presented by Alexandratos in Chapter 1 of this volume shows a lower average calorie availability for 2050, of 3 050 kcal per person per day, and a slightly higher share of the developing countries' population chronically undernourished, at 5 percent.

6. Because total agricultural production is measured by weighing individual products with average international prices, the price-based index of the volume of production grows faster than the aggregates expressed in physical units or using a calorie-based index as diets change away from staples to higher-value commodities (FAO, 2006b; Box 3.1).

Table 6.1
Increases in agricultural production

Region	1961/1963	2005/2007	2050	1961/1963– 2005/2007	2005/2007– 2050
World (146 countries)					
Population[a] *(million people)*	3 133	6 372	8 796	103	38
Total production *(value)*				148	70
Crop production *(value)*				157	66
Cereals[b] *(million tonnes)*	843	2 012	3 009	139	49
Livestock production *(value)*				136	76
Meat production *(million tonnes)*	94	249	461	165	85
Developing (93 countries)					
Population[a] *(million people)*	2 139	5 037	7 433	135	48
Total production *(value)*				255	97
Crop production *(value)*				242	82
Cereals[b] *(million tonnes)*	353	1 113	1 797	215	61
Livestock production *(value)*				284	117
Meat production *(million tonnes)*	42	141	328	236	132
Developed (53 countries)					
Population[a] *(million people)*	994	1 335	1 362	34	2
Total production *(value)*				63	23
Crop production *(value)*				64	30
Cereals[b] *(million tonnes)*	490	900	1 212	84	35
Livestock production *(value)*				62	17
Meat production *(million tonnes)*	52	108	133	108	23

[a] Population figures for 2005/2007 are population in 2005; for 2050 from the United Nations 2002 assessment; the 2050 projection from the United Nations 2008 assessment amounts to 9 056 million for the 146 countries covered.
[b] Including rice in milled form. The latest country balance sheet (CBS) cereal data data show a world cereal production of 2 138 million tonnes for 2006/2008, implying an increment to 2050 of less than 900 million tonnes if measured from the 2006/2008 average.
Source: Author.

Table 6.1 shows the increments in production for the past and future 44-year periods. It brings out the drastic slowdown in expected production growth, compared with the past, for the country and commodity groups shown. This mirrors the projected deceleration in demand for agricultural products, which in turn reflects the decelerating growth of population and the ever-increasing share of population gradually attaining medium to high levels of food consumption (FAO, 2006b).

This slowdown is particularly pronounced for the group of developed countries, but the group of better-off developing countries (defined as having a daily calorie supply of more than 3 000 kcal per person in 2005) is expected to follow a similar pattern.

Although the annual growth of world agricultural production is projected to fall from 2.2 percent over the last decade to 1.5 percent by 2030 and to 0.9 percent from 2030 to 2050 (Table 6.2), the incremental quantities involved are still very considerable: an additional billion tonnes of cereals and 200 million tonnes of meat would need to be produced annually by 2050. The additional meat production would require ample increases in the production of concentrate feeds. For example, 80 percent of the additional 480 million tonnes of maize produced annually by 2050 would be for animal feeds, and soybean production would need to increase by a hefty 140 percent, to reach 515 million tonnes by 2050. As mentioned, these increments do not include the additional production needed as feedstock for biofuel production.

Table 6.2
Annual crop production growth (percentage per annum)

Region	1961–2007	1987–2007	1997–2007	2005/2007–2030	2030–2050	2005/2007–2050
Developing countries	3.0	3.0	2.9	1.5	0.9	1.2
excluding China and India	2.7	2.8	3.1	1.8	1.3	1.6
Sub-Saharan Africa	2.5	3.2	2.9	2.5	1.7	2.1
Near East and North Africa	2.6	2.3	2.1	1.7	1.0	1.4
Latin America and Caribbean	2.6	2.9	3.6	2.1	1.3	1.8
South Asia	2.6	2.2	2.0	1.6	0.9	1.3
East Asia	3.5	3.4	3.3	1.0	0.5	0.8
Developed countries	0.9	0.2	0.7	0.9	0.4	0.7
World	2.2	2.1	2.2	1.3	0.8	1.1
14 developing countries with > 3 000 kcal/person/day in 2005[a]	3.3	3.3	3.2	1.3	0.7	1.0

[a] These account for 40 percent of the population in developing countries.
Source: Author.

Regarding natural resource use in agricultural production, it should be borne in mind that the bulk of the foods consumed are produced locally. At present, an average of only 16 percent of world production[7] (15 percent for cereals and 12 percent for meat) enters international trade, with wide variations among individual countries and commodities.

7. Measured as ((gross imports + gross exports)/2)/production.

What are the sources of growth in crop production?

Growth in crop production comes through growth in crop yields and/or expansion in the physical area (arable land) allocated to crops, which – together with increases in cropping intensities, such as increased multiple cropping and/or shortening of fallow periods – leads to an expansion in the area harvested.

For this chapter, a detailed investigation was made of present and future land/yield combinations for 34 crops under rainfed and irrigated cultivation conditions, in 108 countries and country groups. The informal method applied took into account whatever information was available, but the investigation is based mainly on expert judgement (see Box 6.1 for a brief description of the approach followed).

The summary results shown in Table 6.3 should be taken as rough indications only. For example, yields here are weighted yields (international price weights) for 34 crops; historical data for arable land are unreliable for many countries; and data on cropping intensities for most countries are non-existent, and for this study were derived by comparing data on harvested land, aggregated over all crops, with data on arable land, and so on.

Table 6.3
Sources of growth in crop production (percentages)

Region	Arable land expansion		Increases in cropping intensity		Yield increases	
	1961–2007	1987–2007	1997–2007	2005/2007–2030	2030–2050	2005/2007–2050
All developing countries	23	21	8	8	70	71
Sub-Saharan Africa	31	25	31	6	38	69
Near East and North Africa	17	-7	22	17	62	90
Latin America and Caribbean	40	30	7	18	53	52
South Asia	6	5	12	8	82	87
East Asia	28	2	-6	12	77	86
World	14	9	9	14	77	77
Developing countries with < 40 percent of potentially arable land in use in 2005[a]		30		15		55
Developing countries with > 80 percent of potentially arable land in use in 2005[b]		2		9		89

[a] 42 countries accounting for 15 percent of the total population in developing countries.
[b] 19 countries accounting for 35 percent of the total population in developing countries.
Source: Author.

Box 6.1 - Projecting land use and yield growth

This box gives a brief account of the approach followed in making projections for land use and future yield levels. (Bruinsma, 2003: Appendix summarizes the methodology applied.)

As a starting point, the projections took the crop production projections for 2030 and 2050 from the FAO (2006b) study. These are based on demand and trade projections (including for livestock and feed commodities), which together make consistent commodity balances and clear the world market. The baseline scenario presents a view of how key food and agricultural variables may evolve over time, not how they should evolve from the normative perspective of solving nutrition and poverty problems. To the maximum extent possible, use was made of the in-house knowledge available from various disciplines within FAO. The quantitative analysis and projections were therefore considerably detailed, to provide a basis for making statements about the future concerning individual commodities, groups of commodities and groups of countries, as well as about agriculture as a whole. The analysis was carried out for as many individual commodities and countries as practicable: 108 countries/country groups covering a total of 146 countries (Annex 6.1); 34 crops (Annex 6.2); and two land classes – rainfed and irrigated agriculture.

A major part of data preparation is the unfolding of production data – the FAOSTAT data for area harvested and average yield for each crop and country for the three-year average 2005/2007, converted into the crop classification used in this study – into its constituent components of area, yield and production for rainfed and irrigated land. Such detailed data are not generally available in standard databases, so it was necessary to piece them together from the fragmentary information in both published (e.g., EUROSTAT for the European Union [EU] countries) and unpublished documents (e.g., giving areas and yields by irrigated and rainfed land at the national level or by administrative district), supplemented by a good deal of estimation. For a number of countries, such as the United States of America, China, the EU15 countries, India and Indonesia, data for irrigated agriculture are assembled at the subnational level.

No data exist on total harvested land, but a proxy can be obtained by summing the harvested areas reported for different crops. Data are available for total arable land in agricultural use (physical area, called in FAOSTAT "arable land and land under permanent crops"). It is not known whether these two sets of data are compatible with each other, but this can be evaluated indirectly by computing the cropping intensity, i.e., the ratio of harvested area to arable land. This is an important parameter that can signal defects in the land-use data. For several countries, particularly in sub-Saharan Africa, the implicit values of the cropping intensities do not seem to be realistic. In such cases, the harvested area data resulting from crop statistics were accepted as being the more robust (or the less questionable), and those for arable area were adjusted (see FAO, 1995 for discussion of these problems).

Data reported in FAOSTAT on arable irrigated land refer to "area equipped for irrigation". However, it is the "irrigated land actually in use" that is needed, and this is often between 80 and 90 percent of the area equipped. Data for the area in use were taken from FAO's AQUASTAT database.

The bulk of the projection work concerned unfolding the projected crop production for 2030 and 2050 into (harvested) area and yield combinations for rainfed and irrigated land, and making projections for total arable land and arable irrigated area in use.

Initial mechanically derived projections for rainfed and irrigated harvested areas and yield by crop (constrained to arrive at exactly the projected production) were evaluated against such information as recent growth in area and yield (total by crop) and the attainable yield levels for most crops, obtained from the GAEZ study (Fischer *et al.*, 2002) and adjusted where needed. Similar projections were made for total arable rainfed and irrigated areas, which were then evaluated against estimates for the (maximum) potential areas for rainfed (from GAEZ) and irrigated agriculture (from AQUASTAT) and adjusted where needed. In addition, irrigated area projections were checked against cropping patterns and made to obey certain cropping calendars (i.e., not all crops can be grown in all months of the year). A final step was to derive the implicit cropping intensities for rainfed and irrigated agriculture, by comparing harvested land for all crops with the arable area, and to adjust areas and yields where needed. Normally this required several iterations before an acceptable picture of the future was arrived at.

As the whole exercise depends on expert judgement and requires evaluation of each and every number, it is time-consuming. The projections presented in this chapter are not trend extrapolations, as they take into account all the knowledge currently available regarding expected developments that might make evolutions in major variables deviate from their trend paths.

About 80 percent of the projected growth in crop production in developing countries would come from intensification in the form of yield increases (71 percent) and higher cropping intensities (8 percent, Table 6.3). Intensification's share goes up to 95 percent in the land-scarce region of South Asia, and to more than 100 percent in the Near East and North Africa, where increases in yield would also have to compensate for the foreseen decline in arable land area. Arable land expansion will remain an important factor in crop production growth in many countries of sub-Saharan Africa and Latin America, although less so than in the past.

These summary results mask wide variation among countries. The actual combination of the factors used in crop production (e.g., land, labour and capital) in each country will be determined by their relative prices. Taking the physical availability of land as a proxy for its relative scarcity, and hence price, land would be expected to play a greater role in crop production the less scarce it is. For the 42 developing countries currently using less than 40 percent of their land estimated as having some rainfed crop production potential, arable land expansion is projected to account for almost one-third of crop production growth. At the other end of the spectrum, in the 19 land-scarce countries (with more than 80 percent of their suitable land already in use), the contribution of further land expansion to crop production growth is estimated to be almost nil, at 2 percent (Table 6.3).

In developed countries, the area of arable land in crop production peaked in the late 1960s, then remained stagnant for some time and has been declining since the mid-1980s. Growth in crop yields therefore accounted for all these countries' growth in crop production, and also compensated for declines in their arable land areas. This trend is foreseen to continue for the period to 2050. As a result, intensification (higher yields and more intensive use of land) is seen to contribute more than 90 percent of growth in crop production at the world level over the projection period.

It is interesting to see that growth in rice production in developing countries will increasingly have to come (at least on average) entirely from gains in yield (Table 6.4), which will also have to compensate for a slight decline in harvested land allocated to rice. This could be a sign that consumption of certain food commodities in some countries will reach saturation levels by 2050.

Table 6.4
Sources of growth in production of major cereals, developing countries (percentages)

Crop	Period	Annual growth			Contribution to growth	
		Production	Harvested land	Yield	Harvested land	Yield
Wheat	1961–2007	3.77	1.04	2.70	28	72
	2005/2007–2050	1.05	0.29	0.75	28	72
Rice, paddy	1961–2007	2.32	0.51	1.80	22	70
	2005/2007–2050	0.48	-0.11	0.59	-23	123
Maize	1961–2007	3.43	0.99	2.42	29	71
	2005/2007–2050	1.41	0.63	0.78	44	56

Source: Author.

In developing countries, the bulk of wheat and rice is produced in the land-scarce regions of Asia and the Near East and North Africa, while maize is the major cereal crop in sub-Saharan Africa and Latin America, regions where many countries still have room for area expansion. Expansion of harvested land will therefore continue to be a major contributor to the production growth of maize.

As discussed in FAO (2006b), an increasing share of the increment in cereal production, mainly coarse grains, will be used as livestock feed. As a result, maize production in developing countries is projected to grow at 1.4 percent per annum against 1.1 percent for wheat and only 0.5 percent for rice. Such contrasts are particularly marked in China, where wheat production is expected to grow only marginally and rice production to fall, while maize production grows by some 60 percent over the projection period. Hence there will be corresponding declines in the areas allocated to wheat and rice but a considerable increase in the maize area.

This study attempted to unfold crop production by rainfed and irrigated land, to analyse the contribution of irrigated to total crop production. In developing countries, irrigated agriculture is estimated to account for about a fifth of all arable land, 47 percent of all crop production and almost 60 percent of cereal production (Table 6.5). It should be emphasized that except for some major crops in some countries, there are only limited data on irrigated land and production by crop, and the results presented in Table 6.5 are in part based on expert judgement (Box 6.1). Nevertheless, they suggest a continuing importance of irrigated agriculture.

Table 6.5
Shares of irrigated land and production (percentages)

	All crops			Cereals	
Share	Arable land	Harvested land	Production	Harvested land	Production
World					
2005/2007	15	23	42	29	42
2050	16	24	43	30	43
Developing countries					
2005/2007	19	29	47	39	59
2050	20	30	47	41	60

Source: Author.

By how much does the arable land area need to increase?

At present, about 12 percent (more than 1.5 billion ha; Figure 6.2) of the globe's land surface (13.4 billion ha) is used for crop production (arable land and land under permanent crops). This area represents slightly more than a third (36 percent) of the land estimated to be to some degree suitable for crop production. The remaining 2.7 billion ha with crop production potential suggests that there is scope for further expansion of agricultural land. However, there is also a perception in some quarters that no more, or very little, additional land could be brought under cultivation. This section attempts to shed some light on these contrasting views, first by briefly discussing some estimates of land with crop production potential and some constraints to exploiting these suitable areas, and then by presenting the projected expansion of agricultural area over the period up to 2050.

How much land is there with crop production potential?[8]

Notwithstanding the predominance of yield increases in the growth of agricultural production, land expansion will continue to be a significant factor in those

---

8. This section is an adaptation of a similar section in Bruinsma (2003). It is based on the GAEZ study published in 2002 (Fischer *et al.*, 2002). This study has recently been completely revised, but the results from the revision were not yet available when this chapter was being prepared.

developing countries and regions where the potential for expansion exists and the prevailing farming systems and more general demographic and socio-economic conditions are favourable. A frequently asked question in the debate on world food futures and sustainability is: How much land is there that could be used to produce food to meet the needs of the growing population?

Figure 6.2
World land area (million ha in 2005)

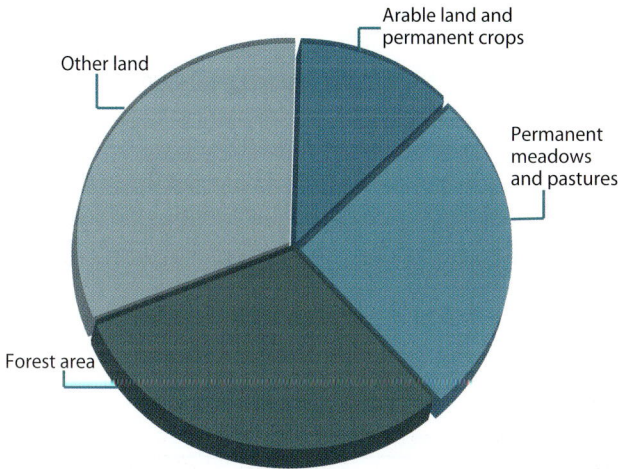

Source: FAOSTAT.

The GAEZ study published in 2002 (Fischer *et al.*, 2002), combining soil, terrain and climate characteristics with crop production requirements, estimates the suitability (in terms of land extents and attainable yield levels) for crop production of each land grid cell at the 5-arc-minute level, at three input levels – low, intermediate and high.

Summing over all the crops covered in GAEZ and the technology levels considered, an estimated 30 percent of the world's land surface, or 4.2 billion ha,[9] is to some extent suitable for rainfed agriculture (Table 6.6). Of this area, some 1.6 billion ha is already under cultivation (Table 6.7). Developing countries have 2.8 billion ha of land of varying qualities with potential for growing rainfed crops at yields above an acceptable minimum level, of which nearly 970 million ha is

---

9. Fischer *et al.* (2002: Table 5.15) report a lower 3.56 billion ha for the gross extent of land with rainfed crop production potential. This is based on a different version of the GAEZ 2002 from that used by Bruinsma (2003). OECD/FAO (2009), based on the GAEZ 2002, reports a total of 4.3 billion ha for the gross extent of land with rainfed crop production potential.

Table 6.6
Land with rainfed crop production potential

Region	Total land surface (million ha)	Share of land suitable (%)	Total land suitable (million ha)	Very suitable[a]	Suitable[b]	Moderately suitable[c]	Marginally suitable[d]	Not suitable[e]
Developing countries	7 302	38	2 782	1 109	1 001	400	273	4 520
Sub-Saharan Africa	2 287	45	1 031	421	352	156	103	1 256
Near East and North Africa	1 158	9	99	4	22	41	32	1 059
Latin America	2 035	52	1 066	421	431	133	80	969
South Asia	421	52	220	116	77	17	10	202
East Asia	1 401	26	366	146	119	53	48	1 035
Industrial countries	3 248	27	874	155	313	232	174	2 374
Transition countries	2 305	22	497	67	182	159	88	1 808
World[f]	13 400	31	4 188	1348	1 509	794	537	9 211

Attainable yields: [a] 80 to 100 percent of the maximum constraint-free yield; [b] 60 to 80 percent; [c] 40 to 60 percent; [d] 20 to 40 percent; [e] < 20 percent.
[f] Including some countries not covered in this study.
Source: Author.

Table 6.7
Total arable land: data and projections

Region	Arable land in use (million ha)						Annual growth (%)			Balance (million ha)	
	1961/ 1963	1989/ 1991	2005	2005 adjusted	2030	2050	1961- 2005	1990- 2005	2005- 2050	2005	2050
Sub-Saharan Africa	133	161	193	236	275	300	0.80	1.07	0.55	786	723
Latin America	105	150	164	203	234	255	1.01	0.64	0.52	861	809
Near East and North Africa	86	96	99	86	84	82	0.34	-0.02	-0.11	13	16
South Asia	191	204	205	206	211	212	0.15	0.07	0.07	14	7
East Asia	178	225	259	235	236	237	0.99	1.12	0.02	131	129
excluding China	73	94	102	105	109	112	0.85	0.71	0.15	78	75
Developing countries	693	837	920	966	1 040	1 086	0.67	0.65	0.27	1 805	1 684
excluding China and India	426	536	594	666	740	789	0.75	0.66	0.39	1 730	1 609
Industrial countries	388	401	388	388	375	364	-0.02	-0.21	-0.15	486	510
Transition countries	291	277	247	247	234	223	-0.32	-0.90	-0.23	250	274
World[a]	1 375	1 521	1 562	1 602	1 648	1 673	0.30	0.17	0.10	2 576	2 503

[a] Includes a few countries not included in the other country groups shown.
Source: Historical data from FAOSTAT, January 2009.

already under cultivation. The gross land balance of 2.6 billion ha (1.8 billion ha for developing countries) would therefore seem to provide significant scope for further expansion of agriculture. However, this favourable impression is qualified by a number of considerations and constraints.

First, the calculation ignores land uses other than for growing crops, so forest cover, protected areas and land used for human settlements and economic infrastructure are not taken into account. Alexandratos (1995) estimated that forests cover at least 45 percent, protected areas some 12 percent and human settlements some 3 percent of the gross land balance, so the net land balance for developing countries would be only 40 percent of the gross balance. Naturally, there are wide regional differences. For example, in the land-scarce region of South Asia, some 45 percent of the land with crop production potential that is not yet in agricultural use is estimated to be occupied by human settlements. This leaves little doubt that population growth and further urbanization will be significant factors in reducing land availability for agricultural use in this region. A more recent estimate by Nachtergaele and George (2009) shows that at the world level, urban areas take up 60 million ha of the gross land balance, protected areas 200 million ha, and forests 800 million ha, so the net land balance would be 1.5 billion ha.

Second, and probably more important than allowing for non-agricultural uses of land with crop production potential, is the method used to derive the estimates: it is enough for a piece of land to support a single crop at a minimum yield level for it to be classified as suitable land. For example, large tracts of land in North Africa that permit the cultivation of only olive trees (and a few other minor crops) are counted as suitable, even though there may be little use for them in practice. The notion of overall land suitability is therefore of limited meaning, and it is more realistic to discuss suitability for individual crops.

A third consideration is that the land balance (land with crop production potential not in agricultural use) is very unevenly distributed among regions and countries. Some 90 percent of the remaining 1.8 billion ha in developing countries is in Latin America and sub-Saharan Africa, and half is concentrated in just seven countries: Brazil, the Democratic Republic of the Congo, Angola, the Sudan, Argentina, Colombia and the Plurinational State of Bolivia (Figure 6.3). At the other extreme, there is virtually no spare land available for agricultural expansion in South Asia and the Near East and North Africa. In fact, a few countries in these two regions have negative land balances, with land classified as not suitable made productive through human intervention – such as terracing of sloping land and irrigation of arid and hyper-arid land – and put into agricultural use. Even within the relatively land-abundant regions there is great diversity of land availability, in terms of both quantity and quality, among countries and subregions.

Fourth, much of the remaining land suffers from constraints such as ecological fragility, low fertility, toxicity, high incidence of disease or lack of infrastructure. These reduce its productivity, and require high input use and management skills to permit its sustainable use, or prohibitively high investments to make it accessible or disease-free. Fischer *et al.* (2002) show that more than 70 percent of the land with rainfed crop production potential in sub-Saharan Africa and Latin America suffers from one or more soil and terrain constraints. Natural causes and human intervention can also lead to deterioration of the land's productive potential, for example through soil erosion or salinization of irrigated areas. Hence the evaluation of suitability may contain elements of overestimation (see also FAO, 2000), and much of the land balance cannot be considered as a resource that is readily usable for food production on demand.

Figure 6.3
Developing countries with the highest (gross) land balance

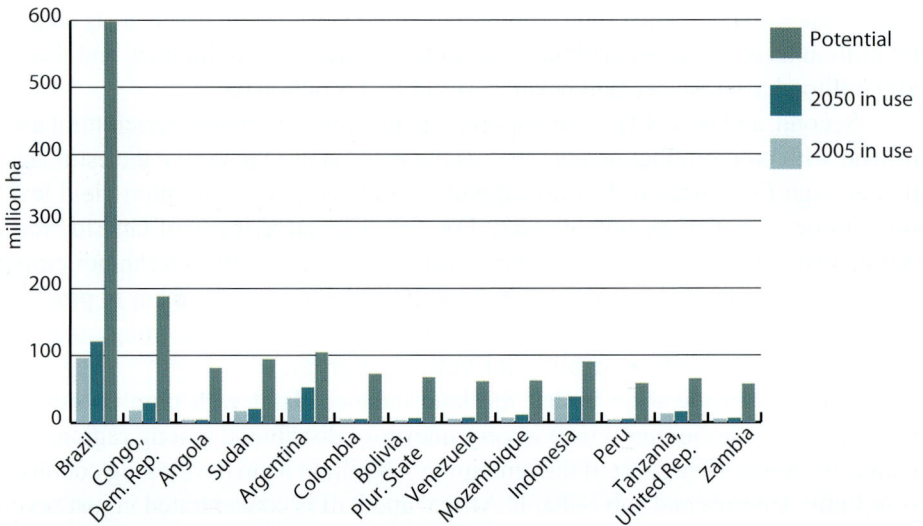

In 2005, these 13 countries with gross land balance of more than 50 million ha accounted for two-thirds of the total gross land balance in developing countries.
Source: Author.

These considerations underline the need to interpret estimates of land balances with caution when assessing land availability for agricultural use. Cohen (1995) summarizes and evaluates all the estimates of available cultivable land, together with their underlying methods, and shows their extremely wide range. Young (1999) offers a critique of the estimates of available cultivable land, including those given in Alexandratos (1995), stating that they often represent gross overestimates.

Expansion of land in crop production

The perception that there is no more or very little new land to bring under cultivation might be grounded in the specific situations of land-scarce countries and regions such as South Asia and the Near East and North Africa, but may not apply, or may apply with much less force, to other parts of the world. As discussed, there are large tracts of land with varying degrees of agricultural potential in several countries, most of them in sub-Saharan Africa and Latin America, with some in East Asia. However, this land may lack infrastructure, or be partly under forest cover or in wetlands that should be protected for environmental reasons, or the people who would exploit it for agriculture lack access to appropriate technological packages or the economic incentives to adopt them.

Figure 6.4
Arable land and land under permanent crops, past developments

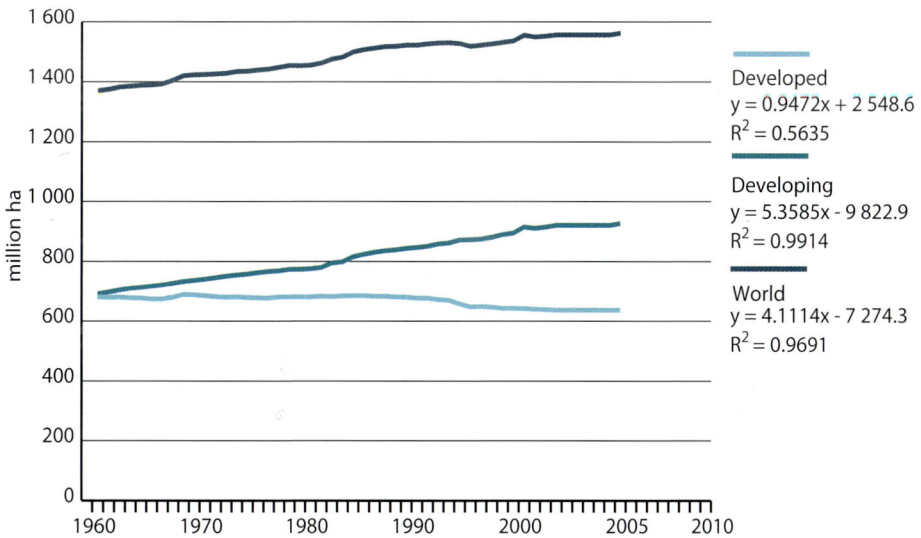

Developed
$y = 0.9472x + 2\,548.6$
$R^2 = 0.5635$

Developing
$y = 5.3585x - 9\,822.9$
$R^2 = 0.9914$

World
$y = 4.1114x - 7\,274.3$
$R^2 = 0.9691$

Source: Author.

In reality, land in agricultural use continues to expand (Figure 6.4), mainly in countries where there are growing needs for food and employment but limited access to the technology packages that could increase intensification of cultivation on land already in agricultural use. The data show that expansion of arable land continues to be an important source of agricultural growth in sub-Saharan Africa,

Latin America and East Asia (Table 6.7). This includes countries that have ample land resources with potential for crops and that face fast demand growth, particularly for exports and non-food uses, such as for soybeans in South America and oil-palm in Southeast Asia. Indeed, oil crops have been responsible for a good part of the increases in total cultivated land in developing countries and the world as a whole (FAO, 2006b), albeit often at the expense of deforestation.

The projected expansion of arable land in crop production shown in Tables 6.7, 6.8 and 6.9 has been derived for rainfed and irrigated land separately. As explained in Box 6.1, starting with the production projections for each crop, land and yield projections were derived from expert judgement and taking into account: i) base year (2005/2007) data on total harvested land and yield by crop; ii) data (or often estimates) for harvested land and yield by crop, for rainfed and irrigated land; iii) data on total arable rainfed and irrigated land, and their expected increases over time; iv) likely increases in yield, by crop and land class; v) plausible increases in cropping intensities; and vi) the land balances for rainfed and irrigated agriculture. As mentioned in Box 6.1, base year data for total arable land in several developing countries were adjusted (particularly for China)[10] partly to arrive at cropping intensities that seemed more meaningful. This is reflected in the column headed "2005 adjusted" in Table 6.7.

The overall result for developing countries is a projected net increase in the arable area of some 120 million ha (from 966 million ha in the base year to 1 086 million ha in 2050), or 12.4 percent (Table 6.7). Not surprisingly, the bulk of this projected expansion is expected to occur in sub-Saharan Africa (64 million ha) and Latin America (52 million ha), with almost no land expansion in East and South Asia, and even a small decline in the Near East and North Africa. The slowdown in expansion of arable land is mainly a consequence of the projected slowdown in the growth of crop production and is common to all regions.

The bulk of arable land in use is concentrated in a few developing countries (Figure 6.5). Towards the end of the projection period, a number of developing countries would witness a decline in arable land area (e.g., China, the Republic of Korea and others) and would embark on a pattern already seen in most developed countries, with production increasing only very slowly and increases in yield permitting a reduction in crop area.

Between 1961/1963 and 2005, the arable area in the world expanded by 187 million ha, as a result of two opposite trends: an increase of 227 million ha in developing countries, and a decline of 40 million ha in developed countries.

---

10. Data on arable land for China are unreliable. FAOSTAT data show an (unlikely) upwards trend from 1983 onwards, which distorts the historical growth rates in Table 6.7 for East Asia and for the total of developing countries.

The arable land area in the latter group peaked in the mid-1980s, at 684 million ha, and has declined ever since. This decline has been accelerating over time (Table 6.7). The longer-term forces determining such declines are sustained yield growth combined with a continuing slowdown in the growth of demand for the agricultural products grown in developed countries. The projections in this chapter foresee a further slow decline in developed countries' arable area, to 587 million ha in 2050 (although this may change under the impact of an eventual fast growth in biofuels). The net result for the world is an increase of 71 million ha in arable land area, consisting of an increase of 120 million ha in developing countries and a decline of 48 million ha in developed countries (Table 6.7 and Figure 6.6).

In the group of land-scarce countries,[11] arable land would practically remain constant (at 265 to 268 million ha), but irrigated land could expand by some 12 million ha, of which 9 million ha would be through conversion of rainfed land. Some of these countries are still highly dependent on agriculture and are experiencing above-average population growth. This, combined with their resource constraints, could make solving their food security problems extremely cumbersome, if not impossible, at least without external assistance and/or by finding non-agricultural development opportunities (Alexandratos, 2005).

The projected 2.75 million ha average annual increase in developing countries' arable area (120 million ha over 44 years) is a net increase. It is the total of gross land expansion minus land taken out of production for various reasons, such as owing to degradation, loss of economic viability or conversion to settlements. An unknown part of the new land to be brought into agriculture will come from land currently under forests. If all the additional land were to come from forested areas, it would imply an annual deforestation rate of 0.14 percent, compared with 0.42 percent (9.3 million ha per annum) for the 1990s, and 0.36 percent (7.5 million ha per annum) for the period 2000 to 2005 (FAO, 2006a). The latter estimates include deforestation from all causes, such as informal or unrecorded agriculture, grazing, logging and gathering of fuelwood.

What does the empirical evidence show concerning land expansion for agricultural use in developing countries? Micro-level analyses have generally established that under the socio-economic and institutional conditions prevailing in many developing countries, increases in output are – at least initially – obtained mainly through land expansion, where the physical potential for doing so exists. For example, in an analysis of Côte d'Ivoire, Lopez (1998) concludes that "the main response of annual crops to price incentives is to increase the area

---

11. These are the 19 countries with more than 80 percent of their land with rainfed and/or irrigation potential in use in 2005, of which six are in the Near East and North Africa, five in sub-Saharan Africa, and four in South Asia.

Figure 6.5
Developing countries with more than 10 million ha of arable land in use

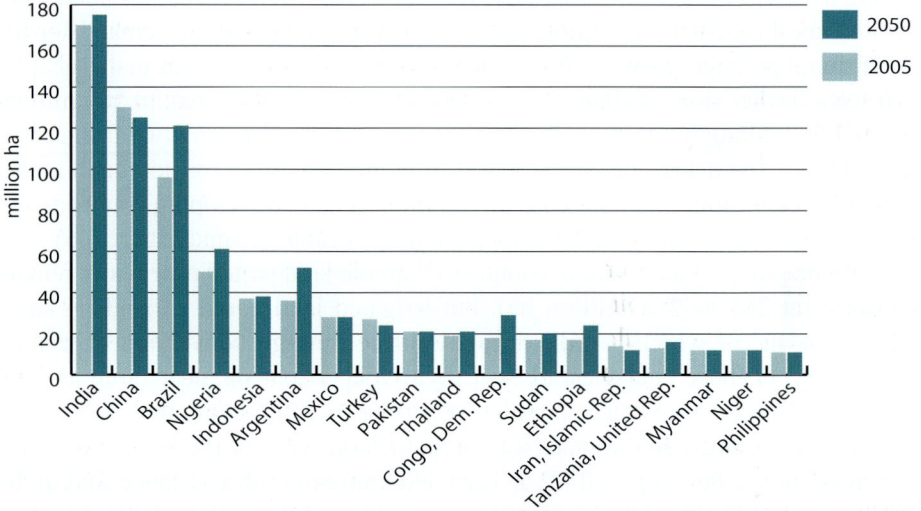

In 2005, these 18 countries accounted for 75 percent of the total arable land in use in developing countries.
Source: Author.

Figure 6.6
Arable land and land under permanent crops, past and future

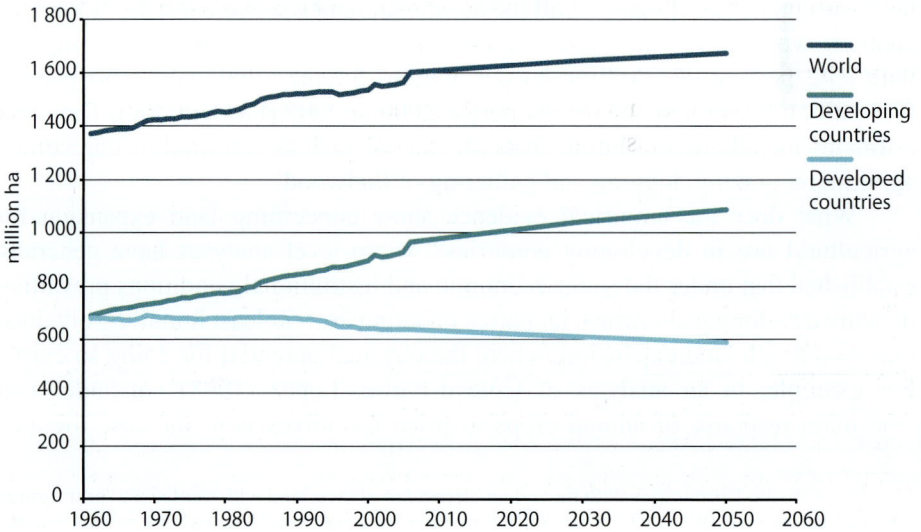

Source: Author.

cultivated". Similar findings, such as the rate of deforestation being positively related to the price of maize, are reported for Mexico by Deininger and Minten (1999). However, some land expansion takes place at the expense of longer rotation periods and shorter fallows, a practice still common in many sub-Saharan African countries, with the result that the soil's natural fertility is reduced. As fertilizer use is often uneconomic, the end-result is soil mining and stagnation or outright reduction of yields.

Although developing countries' arable area is projected to expand by 120 million ha over the projection period, the harvested area would increase by 160 million ha, or 17 percent, owing to increases in cropping intensities (Table 6.8). The overall cropping intensity for developing countries could rise by about 4 percentage points over the projection period (from 95 to 99 percent). Cropping intensities would continue to rise through shorter fallow periods and more multiple cropping. An increasing share of irrigated land in total agricultural land would also contribute to more multiple cropping. Almost one-third of the arable land in South and East Asia is irrigated, a share that is projected to rise to more than 36 percent in 2050. This high share of irrigated land is one of the reasons why average cropping intensities are considerably higher in these than in other regions. Average cropping intensities in developing countries – excluding China and India, which together account for well over half of the irrigated area in developing countries – are and will continue to be much lower.

Rising cropping intensities could be one of the factors responsible for increasing the risk of land degradation, and thus threatening sustainability, particularly when not accompanied by land conservation measures, including adequate and balanced use of fertilizers to compensate for the removal of soil nutrients by crops. This risk is expected to continue because, in many cases, socio-economic conditions do not favour implementation of the technological changes required to ensure the sustainable intensification of land use.

How much more water will be required in irrigation?

Expanding irrigated land

The area equipped for irrigation has been continuously expanding (mainly in developing countries, and only slowly in developed countries), although recently this expansion has slowed considerably (Figure 6.7). The projections of irrigation presented in this section are based on scattered information about existing irrigation expansion plans in different countries, potentials for expansion (including water availability) and the need to increase crop production. The projections include expansion in both formal and informal irrigation, the latter being particularly important in sub-Saharan Africa.

Table 6.8
Arable land in use, cropping intensities and harvested land

Region	Period	Total land in use			Rainfed use			Irrigated use[a]		
		Arable land (million ha)	Cropping intensity (%)	Harvested land (million ha)	Arable land (million ha)	Cropping intensity (%)	Harvested land (million ha)	Arable land (million ha)	Cropping intensity (%)	Harvested land (million ha)
Developing countries	2005/2007	966	95	919	777	83	649	189	143	270
	2050	1 086	99	1078	864	87	753	222	147	325
excluding China and India	2005/2007	666	82	547	582	76	442	84	124	105
	2050	785	89	697	680	83	562	106	129	136
Developed countries	2005/2007	635	74	473	584	72	422	51	100	51
	2050	587	81	478	536	80	426	51	100	51
World	2005/2007	1 602	87	1 392	1 361	79	1 070	240	134	321
	2050	1 673	93	1 556	1 400	84	1 179	273	138	377

[a] Irrigated area in use, rather than "area equipped for irrigation" (Table 6.9).
Source: Author.

Table 6.9
Area equipped for irrigation

Region	Area (million ha)					Annual growth (%)			
	1961/1963	1989/1991	2005/2007	2030	2050	1961–2005	1990–2005	1996–2005	2005–2050
Developing countries	103	178	219	242	251	1.76	1.05	0.63	0.31
excluding China and India	47	84	97	111	117	1.91	1.06	0.89	0.42
Sub-Saharan Africa	2.5	4.5	5.6	6.7	7.9	2.07	1.49	0.98	0.67
Latin America and Caribbean	8	17	18	22	24	2.05	0.62	0.27	0.72
Near East and North Africa	15	25	29	34	36	1.86	1.21	1.30	0.47
South Asia	37	67	81	84	86	1.98	1.10	0.28	0.14
East Asia	40	64	85	95	97	1.42	1.00	0.80	0.30
Developed countries	38	66	68	68	68	1.57	0.38	0.20	0.00
World	141	244	287	310	318	1.71	0.87	0.52	0.24

Source: Author.

Figure 6.7
Area equipped for irrigation, past developments

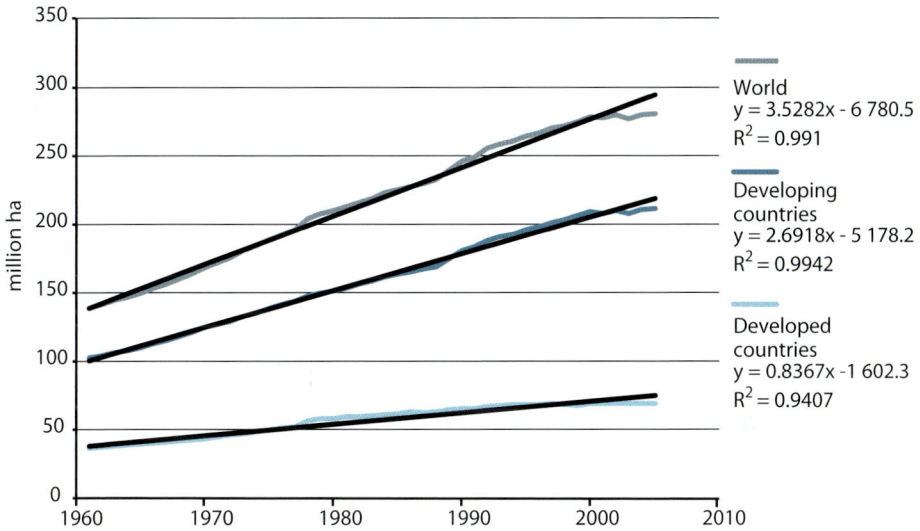

Source: Author.

The aggregate result shows that the area equipped for irrigation could expand by 32 million ha (11 percent) over the projection period (Table 6.9), all in developing countries. This means that 16 percent of the land with irrigation potential and not yet equipped in this group of countries could be brought under irrigation, and by 2050 some 60 percent of all land with irrigation potential[12] (417 million ha) would be in use.

The expansion of irrigation would be strongest (in absolute terms) in the more land-scarce regions that are hard-pressed to raise crop production through more intensive cultivation practices, such as East Asia (an expansion of 12 million ha), South Asia (8 million ha) and the Near East and North Africa (6 million ha), although further expansion in the Near East and North Africa will become increasingly difficult as water scarcity increases and competition for water from households and industry continues to reduce the share available to agriculture. China and India alone account for more than half (56 percent) of the irrigated area in developing countries. Although the overall arable area in China is expected to decrease further, the irrigated area would continue to expand through conversion of rainfed land.

---

12. Estimates of land with irrigation potential are difficult to make, and should be taken as only rough indications.

253

Developed countries account for almost a quarter of the world's irrigated area, with 68 out of 287 million ha (Table 6.9). Annual growth of these countries' irrigated area reached a peak of 3.0 percent in the 1970s, dropping to 1.1 percent in the 1980s and to only 0.2 percent in the last decade for which data are available (1996 to 2005). For the developed countries as a group, only a marginal expansion of the irrigated area (supplemented with improvements on existing areas) is foreseen over the projection period, so the world irrigation scene will remain dominated by events in developing countries.

For this study, a distinction was made between the area equipped for irrigation and the irrigated area actually in use (which is the area used in the production analysis). Areas equipped might be temporarily or even permanently out of use, for various reasons, including maintenance, degradation of irrigation infrastructure or lack of need in a particular year. The percentage of the area equipped actually in use differs from country to country, ranging from 60 to 100 percent and averaging 86 percent over all countries. (This is expected to increase very slightly to 88 percent in 2050.) Of the 219 million ha equipped for irrigation in the developing countries in 2005/2007, some 189 million ha was assumed to be in use, increasing to 222 million ha in 2050 (out of 251 million ha equipped; Figure 6.8).

Figure 6.8
Arable irrigated area, past and future

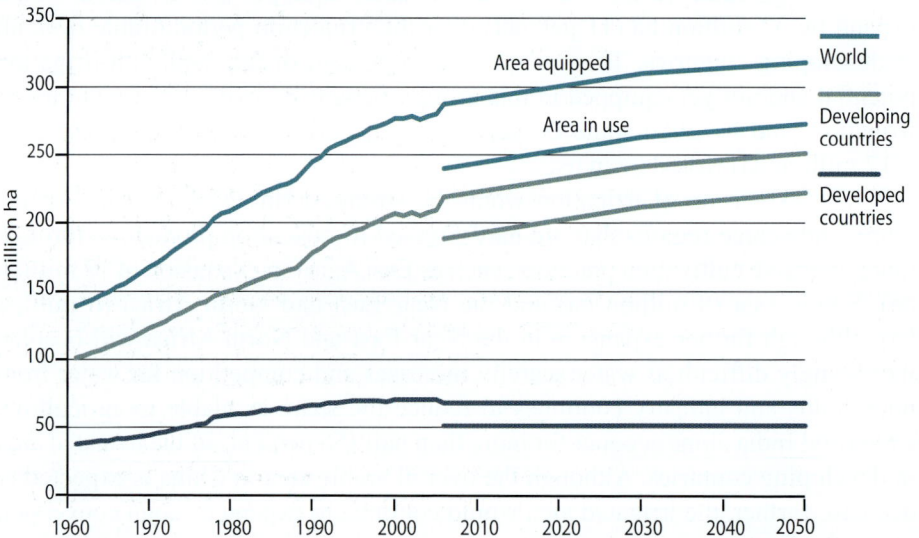

Source: Author.

The importance of irrigated agriculture was discussed in the preceding section. Owing to continuing increases in multiple cropping on both existing and newly irrigated areas, the harvested irrigated area could expand by 56 million ha (17 percent), to account for well over a third of the total increase in harvested land (Table 6.8).

The projected expansion of irrigated land, by 32 million ha, is an increase in net terms. The projection assumes that losses of existing irrigated land, such as those due to water shortages or degradation resulting from salinization and waterlogging, will be compensated for by rehabilitation or substitution of other areas. The few existing historical data on such losses are too uncertain and anecdotal to provide a reliable basis for drawing inferences about the future. Regarding investments, the rehabilitation of existing irrigation schemes will represent the bulk of future expenditure on irrigation: if it is assumed that 2.5 percent of existing irrigation must be rehabilitated or substituted by new irrigation each year – in other words, the average life of an irrigation scheme is 40 years – the total irrigation investment activity in developing countries must encompass some 173 million ha over the projection period, of which more than four-fifths (141 million ha) would be for rehabilitation or substitution, and the balance for net expansion.

The projected net increase in land equipped for irrigation (32 million ha) is less than a quarter of the increase over the preceding 44 years (145 million ha). This implies an annual growth of only 0.24 percent, well below the 1.7 percent of the historical period. The slowdown projected for most countries and regions reflects the projected lower growth rate of crop production, combined with the increasing scarcity of suitable areas for irrigation and of water resources in some countries, and the rising costs of irrigation investment.

Most of the expansion in irrigated land will be achieved by converting land in use in rainfed agriculture into irrigated land. However, irrigation also takes place on arid and hyper-arid (desert) land, which is not suitable for rainfed agriculture. Of the 219 million ha currently irrigated in developing countries, an estimated 40 million ha is on arid and hyper-arid land, which could increase to 43 million ha in 2050. In some regions and countries, irrigated arid and hyper-arid land forms an important part of the total irrigated land currently in use: 19 million out of 28 million ha in the Near East and North Africa, and 15 million out of 70 million ha in South Asia.

Water use in irrigation and pressure on water resources

A major question concerning the future is whether there will be sufficient freshwater to satisfy the growing needs of agricultural and non-agricultural users. Agriculture already accounts for about 70 percent of freshwater withdrawals in the world, and is usually seen as a major factor behind the increasing global scarcity of freshwater.

The estimates of expansion of land under irrigation presented in the preceding subsection provide a partial answer to this question, because the assessment of irrigation potential takes water limitations into account, and the projections to 2050 assume that agricultural water demand will not exceed available water resources.[13]

Renewable water resources available for irrigation and other uses are commonly defined as that part of precipitation that is not evaporated or transpired by plants, including grasses and trees, and that flows into rivers and lakes or infiltrates into aquifers. Under natural conditions, without irrigation, the annual water balance for a given area can be defined as the sum of annual precipitation and net incoming flows (transfers through rivers from one area to another) minus evapotranspiration and runoff.

shows the renewable water resources for the world and major regions. Average annual precipitation varies from 160 mm per year in the most arid region (the Near East and North Africa) to about 1 530 mm per year in Latin America. These figures give an impression of the wide range of climatic conditions facing developing countries, and the resulting differences in water scarcity: countries with low precipitation, and therefore most in need of irrigation, are also those where water resources are naturally scarce. In addition, the water balance is expressed in yearly averages and does not reflect seasonal and intra-annual variations. Unfortunately, such variations tend to be more pronounced in arid than in humid climates.

The first step in estimating the pressure of irrigation on water resources is to assess irrigation water requirements and withdrawals. Precipitation provides part of the water that crops need to satisfy their transpiration requirements. Acting as a buffer, the soil stores part of the precipitation water, and returns it to the crops in times of deficit. In humid climates, this mechanism is usually sufficient to ensure satisfactory growth in rainfed agriculture. In arid climates or during dry seasons, irrigation is required to compensate for the deficit due to insufficient or erratic precipitation. Consumptive water use in irrigation is therefore defined as the volume of water needed to compensate for the deficit between potential evapotranspiration and effective precipitation over the crop's growing period. It varies considerably with climatic conditions, season, crop and soil type. In this

---

13. The concept of irrigation potential has severe limitations, and estimates can vary over time, according to the country's economic situation, or as a result of competition for water for domestic and industrial use. Estimates of irrigation potential are based on estimates of renewable water resources, i.e., the resources replenished annually through the hydrological cycle. In arid countries where mining of fossil groundwater represents an important part of water withdrawal, the area under irrigation is usually larger than the irrigation potential.

study, consumptive water use in irrigation has been computed for each country, based on the irrigated and harvested areas, by crop, estimated for the base year (2005/2007) and projected for 2050 (see Box 6.2 for a brief explanation of the methodology applied).

Box 6.2 - Estimating irrigation water requirements

Estimation of the water balances for any year is based on five sets of data: four digital geo-referenced data sets – for precipitation (New *et al.*, 2002), reference evapotranspiration (FAO, 2004), soil moisture storage properties (FAO, 1998) and areas under irrigation (Siebert *et al.*, 2007) – and irrigated areas for all major crops for 2005/2007 and 2050. Water balances are computed by grid cell (of 5 arc minutes, 9.3 km at the equator) and in monthly time steps. The results can be presented in statistical tables or digital maps at any level of spatial aggregation (country, river basin, etc.). They consist of annual values by grid cell for actual evapotranspiration, water runoff and consumptive water use in irrigation.

For each grid cell, actual evapotranspiration is assumed to be equal to reference evapotranspiration (ET_0, in millimetres; location-specific and calculated with the Penman-Monteith method; FAO, 1998; New *et al.*, 2002) in periods of the year when precipitation exceeds reference evapotranspiration or when there is enough water stored in the soil to allow maximum evapotranspiration. In drier periods of the year, lack of water reduces actual evapotranspiration to an extent that depends on the available soil moisture. Evapotranspiration in open water areas and wetlands is considered equal to a fixed fraction of the reference evapotranspiration.

For each grid cell, runoff and groundwater recharge is calculated as that part of the precipitation that does not evaporate and that cannot be stored in the soil. In other words, the sum of the runoff and groundwater recharge is equal to the difference between precipitation and actual evaporation. Runoff is always positive, except for areas identified as open water or wetland, where actual evapotranspiration can exceed precipitation.

Consumptive use of water in irrigated agriculture is defined as the water required in addition to water from precipitation (soil moisture) for optimal plant growth during the growing season. Optimal plant growth occurs when the actual evapotranspiration of a crop is equal to its potential evapotranspiration.

Potential evapotranspiration of irrigated agriculture is calculated by converting data or projections of irrigated (sown) area by crop (at the national level) into a cropping calendar, with monthly occupation rates of the land equipped for irrigation.[1] The following table gives an example of the cropping calendar for Morocco in the base year 2005/2007.[2]

The (potential) evapotranspiration (ET_c in millimetres) of a crop under irrigation is obtained by multiplying the reference evapotranspiration with a crop-specific coefficient ($ET_c = K_c * ET_0$). This coefficient has been derived (FAO, 1998) for four different growing stages: the initial phase, just after sowing; the development phase; the mid-phase; and the late phase, when the crop is ripening to be harvested. In general, these coefficients are low during the initial phase, high during the mid-phase and lower again in the late phase. It is assumed that the initial, development and late phases each take one month for any crop, while the mid-phase lasts several months. For example, the growing season for wheat in Morocco starts in October and ends in April: initial phase, October ($K_c = 0.4$); development phase, November ($K_c = 0.8$); mid-phase, December to March ($K_c = 1.15$); and late phase, April ($K_c = 0.3$).

The surface equipped for irrigation of each grid cell is then multiplied by the sum of all crops' evapotranspiration and the cropping intensity per month, to result in the potential evapotranspiration of the irrigated area in that grid cell. The difference between the calculated evapotranspiration of the irrigated area and actual evapotranspiration under non-irrigated conditions is equal to the consumptive use of water in irrigated agriculture in the grid cell.

Crop under irrigation	*Irrigated area ('000 ha)*	Crop area as share of total area equipped for irrigation, by month (%)											
		Jan.	Feb.	Mar.	Apr.	May	Jun.	Jul.	Aug.	Sep.	Oct.	Nov.	Dec.
Wheat	618	46	46	46	46						46	46	46
Maize	119			9	9	9	9	9					
Potatoes	61					5	5	5	5	5			
Sugar beet	36			3	3	3	3	3	3				
Sugar cane	14	1	1	1	1	1	1	1	1	1	1	1	1
Vegetables	145					11	11	11	11	11			
Citrus	80	6	6	6	6	6	6	6	6	6	6	6	6
Fruits	89	7	7	7	7	7	7	7	7	7	7	7	7
Groundnut	6					1	1	1	1	1			
Other crops	124	9	9							9	9	9	9
Sum over all crops[a]	1 292	69	69	69	72	42	42	42	32	41	69	69	69

[a] Including crops not listed in the table.

The method has been calibrated by comparing calculated values for water resources per country (i.e., the difference between precipitation and actual evapotranspiration under non-irrigated conditions) with data on water resources for each country as given in FAO AQUASTAT.[3] In addition, the discharge of each major river, as given in the literature, was compared with the calculated runoff for the drainage basin of that river. If the calculated runoff value did not match the value as stated in the literature, correction factors were applied to one or more of the basic input data on soil moisture storage and open waters.

The water balance for each country and year is defined as the difference between the sum of precipitation and incoming runoff on the one hand, and the sum of actual evapotranspiration and consumptive use of water in irrigated agriculture in that year on the other hand. This water balance therefore does not account for water withdrawals for other needs (industry, household and environmental purposes).

---

[1] India, China, Indonesia, the United States of America and the EU15 have been subdivided into two to four subregions with different cropping calendars, to distinguish different climate zones in these countries.

[2] For example, wheat is grown from October to April and occupies 46 percent (618 000 ha) of the 1 292 000 ha of irrigated land in use.

[3] www.fao.org/nr/aquastat.phase, October (Kc = 0.4); development phase, November (Kc=0.8); mid-phase, December to March (Kc = 1.15); and late phase, April (Kc = 0.3).

However, *water withdrawal for irrigation* – the volume of water extracted from rivers, lakes and aquifers for irrigation purposes – should be used to measure the impact of irrigation on water resources. Irrigation water withdrawal normally far exceeds consumptive water use in irrigation because of the water lost during transport and distribution from its source to the crops. In rice irrigation, additional water is used for paddy field flooding, to facilitate land preparation, protect the plants and control weeds.

Water-use efficiency is defined as the ratio between the estimated consumptive water use in irrigation and irrigation water withdrawal. Data on country water withdrawal for irrigation were collected within the framework of the AQUASTAT programme (e.g., FAO, 2005a; 2005b). These data were compared with the consumptive use of irrigation to estimate water-use efficiency[14] at the country level. For the world, average water-use efficiency was estimated at about 44 percent in 2005/2007, varying from 22 percent in areas of abundant water resources (sub-Saharan Africa) to 54 percent in South Asia, where water scarcity calls for higher efficiencies (Table 6.10).

To estimate the irrigation water withdrawal in 2050, assumptions were made about possible developments in the water-use efficiency in each country. Unfortunately, there is little empirical evidence on which to base such assumptions. However, two factors have an impact on the development of water-use efficiency: the estimated level of water-use efficiency in the base year, and water scarcity.[15] A function was designed to capture the influence of these two parameters, bearing in mind that improving water-use efficiency is a very slow and difficult process. The overall result is that efficiency could increase by 2 percentage points, from 44 to 46 percent (Table 6.10). Such an increase in efficiency would be more pronounced in water-scarce regions (e.g., a 10 percentage point increase in the Near East and North Africa) than in regions with abundant water resources (e.g., increases of 3 percentage points or less in Latin America and sub-Saharan Africa). It is expected that under pressure from limited water resources and competition from other uses, demand management will play an important role in improving water-use efficiency in water-scarce regions. In contrast, in humid areas, the issue of water-use efficiency is much less relevant and is likely to receive little attention.

At the global level, irrigation water withdrawal is expected to grow by about 11 percent, from the current 2 620 km^3 per year to 2 906 km^3 in 2050 (Table 6.10), with an increase in developing countries of 14 percent (or 298 km^3)

---

14. It should be noted that although the term "water-use efficiency" implies losses of water between source and destination, not all of this water is actually lost as much flows back into the river basin and aquifers and can be reused for irrigation.

15. Or "stress", measured as consumptive water use in irrigation as a percentage of renewable water resources.

being offset by a decline in developed countries of more than 2 percent (12 km^3). This increase in irrigation water withdrawal should be seen against the projected 17 percent increase in harvested irrigated area (from 321 million ha in 2005/2007 to 377 million ha in 2050; Table 6.8). The difference is due in part to the expected improvement in water-use efficiency, leading to a reduction in irrigation water withdrawal per irrigated hectare, and in part to changes in cropping patterns for some countries such as China, where a substantial shift in the irrigated area from rice to maize production is expected: irrigation water requirements for rice production are usually twice those for maize.

Table 6.10
Annual renewable water resources and irrigation water withdrawal

Region	Precipitation	Renewable water resources[a]	Water-use efficiency ratio		Irrigation water withdrawal		Pressure on water resources due to irrigation	
			2005/2007	2050	2005/2007	2050	2005/2007	2050
	(mm/year)	*(km^3)*	*(%)*		*(km^3)*		*(%)*	
Developing countries	990	28 000	44	47	2 115	2 413	8	9
Sub-Saharan Africa	850	3 500	22	25	55	87	2	2
Latin America and Caribbean	1 530	13 500	35	35	181	253	1	2
Near East and North Africa	160	600	51	61	347	374	58	62
South Asia	1 050	2 300	54	57	819	906	36	39
East Asia	1 140	8 600	33	35	714	793	8	9
World	800	42 000	44	46	2 620	2 906	6	7
Developed countries	540	14 000	42	43	505	493	4	4

[a] At the regional level, includes incoming flows.
Source: Author.

Irrigation water withdrawal in 2005/2007 was estimated to account for only 6 percent of total renewable water resources in the world (Table 6.10). However, there are wide variations among countries and regions, with the Near East and North Africa using 58 percent of its water resources in irrigation, while Latin America uses barely 1 percent of its. At the country level, variations are even higher. In the base year (2005/2007), 11 countries used more than 40 percent of their water resources for irrigation, creating a situation that can be considered critical. Another eight countries consumed more than 20 percent of their water resources, a threshold sometimes used to indicate impending water scarcity. The situation is expected to worsen by 2050, with two more countries crossing the 40 percent and four the 20 percent thresholds. If the expected additional water withdrawals needed for non-agricultural use are added, the picture does not change much, as agriculture represents the bulk of water withdrawal.

Nevertheless, for several countries, relatively low national figures may give an overly optimistic impression of the level of water stress: for example, China is facing severe water shortage in the north, while the south still has abundant water resources. In 2005/2007, four countries – the Libyan Arab Jamahiriya, Saudi Arabia, Yemen and Egypt – used volumes of water for irrigation that were larger than their annual renewable water resources. Groundwater mining also occurs in parts of some other countries in the Near East, and in South and East Asia, Central America and the Caribbean, even if the water balance at the national level may still be positive.

In conclusion, for the developing countries as a whole, water use in irrigation currently represents a relatively small part of their total water resources and there remains significant potential for further irrigation development. With the relatively small increase in irrigation water withdrawal expected between 2005/2007 and 2050, this situation will not change much at the aggregate level. However, locally and in some countries, there are already very severe water shortages, particularly in the Near East and North Africa.

By how much do crop yields need to rise?

As discussed, it is expected that growth in crop yields will continue to be the mainstay of crop production growth, accounting for some 70 percent of production growth in developing countries, and for 100 percent in developed countries. Although the marked deceleration in crop production growth foreseen for the future (Table 6.2) could point to a similar deceleration in growth of crop yields, such growth will continue to be needed. Questions often asked are: Will yield increases continue to be possible? and What is the potential for continuing such growth? There is a realization that a new green revolution or one-off quantum jumps in yields are unlikely to occur, and some believe that yield ceilings for some major crops have been, or are rapidly being, reached. Empirical evidence shows that the accumulation of slower, evolutionary annual increments in yields has been far more important than quantum jumps in yields, for all major crops (e.g., Byerlee, 1996).

Harvested land and yields for major crops

As mentioned, the production projections for the 34 crops covered in this chapter are unfolded into and tested against FAO experts' perceptions of feasible land-yield combinations, based on whatever knowledge is available for each agro-ecological rainfed and irrigated environment. Major inputs into this evaluation are the GAEZ-based (Fischer *et al.*, 2002) estimates regarding the availability of land suitable for growing crops, and the yields attainable in each country and each agro-ecological environment. In practice, such estimates are introduced as constraints to land and yield expansion, but they also act as a guide to what can be grown where. The resulting land and yield projections, although partly based on past performance, are not mere extrapolations of historical trends, as they take into account current knowledge about changes expected in the future.

The overall result for yields of all the crops covered in this study (aggregated with standard price weights) is that the global average annual rate of growth over the projection period will be roughly half that of the historical period: 0.8 percent per annum from 2005/2007 to 2050, against 1.7 percent per annum from 1961 to 2007. For developing countries, the equivalent annual growth rates are 0.9 and 2.1 percent. This slowdown in yield growth is a gradual process that has been under way for some time; for example, yield growth from 1997 to 2007 was 1.3 percent per annum for the world, and 1.6 percent for developing countries. The slowdown reflects the deceleration in crop production growth explained earlier.

Table 6.11
Areas and yields for major crops in the world

	Production (million tonnes)			Harvested area (million ha)			Yield (tonnes/ha)		
Crop	1961/ 1963	2005/ 2007	2050	1961/ 1963	2005/ 2007	2050	1961/ 1963	2005/ 2007	2050
Wheat	235	611	907	206	224	242	1.14	2.72	3.75
Rice (paddy)	227	641	784	117	158	150	1.93	4.05	5.23
Maize	210	733	1 153	106	155	190	1.99	4.73	6.06
Soybeans	27	218	514	24	95	141	1.14	2.29	3.66
Pulses	41	60	88	69	71	66	0.59	0.84	1.33
Barley	84	138	189	59	57	58	1.43	2.43	3.24
Sorghum	44	61	111	48	44	47	0.93	1.39	2.36
Millet	25	32	48	43	36	34	0.58	0.86	1.43
Seed cotton	30	71	90	32	36	32	0.92	1.95	2.80
Rape seed	4	50	106	7	31	36	0.56	1.61	2.91
Groundnuts	15	36	74	17	24	39	0.86	1.49	1.91
Sunflower	7	30	55	7	23	32	1.00	1.29	1.72
Sugar cane	417	1 413	3 386	9	21	30	48.34	67.02	112.34

Crops selected and ordered according to (harvested) land use in 2005/2007.
Source: Author.

Although discussing yield growth at this level of aggregation is not very helpful, the overall slowdown reflects a pattern common to most of the crops covered in this study. Exceptions include citrus and sesame, for which strong demand growth is foreseen in the future, or which are grown in land-scarce environments. The remarkable growth in soybean area and production in developing countries (Table 6.11) has been mainly due to explosive growth in Brazil and Argentina. Soybean is expected to continue to be one of the most dynamic crops, albeit with a more moderate rate of production increase than in the past, bringing the developing countries' share in world soybean production to more than 70 percent by 2050, with four countries – Brazil, Argentina, China and India – accounting for 90 percent of total production in developing countries.

Table 6.12
Cereal yields, rainfed and irrigated

| | | World | | | | | | Developing countries | | | | | |
| | | Average yield (tonnes/ha) | | | Annual growth (%) | | | Average yield (tonnes/ha) | | | Annual growth (%) | | |
Crop		1961/ 1963	2005/ 2007	2050	1961– 2007	1987 –2007	2005/ 2007– 2050	1961/ 1963	2005/ 2007	2050	1961– 2007	1987 –2007	2005/ 2007– 2050
Wheat	total	1.14	2.72	3.75	2.1	1.0	0.7	0.87	2.69	4.00	2.9	1.5	0.9
	rainfed		2.37	3.17			0.7		1.67	2.57			1.0
	irrigated		3.50	5.08			0.8		3.41	5.06			0.9
Rice	total	1.93	4.05	5.23	1.8	1.1	0.6	1.82	3.98	5.18	1.9	1.1	0.6
(paddy)	rainfed		2.54	3.26			0.6		2.54	3.26			0.6
	irrigated		5.10	6.40			0.5		5.04	6.37			0.5
Maize	total	1.99	4.72	6.06	2.0	1.9	0.6	1.16	3.22	4.56	2.5	2.1	0.8
	rainfed		4.26	5.58			0.6		2.70	3.69			0.7
	irrigated		6.74	7.43			0.2		5.27	6.53			0.5
All	total	1.40	3.23	4.34	1.9	1.4	0.7	1.17	2.91	4.08	2.2	1.5	0.8
cereals	rainfed		2.64	3.58			0.7		1.97	2.80			0.8
	irrigated		4.67	6.10			0.6		4.39	5.90			0.7

Base year data for China adjusted.

Source: Historical data from FAOSTAT.

For cereals, which occupy half (51 percent) of the harvested area in the world and in developing countries, the slowdown in yield growth would be particularly pronounced: from 1.9 percent per annum in the historical period to 0.7 percent over the projection period for the world; and from 2.2 to 0.8 percent in developing countries (Table 6.12). This slowdown too has been under way for some time.

The differences in sources of growth among regions have been discussed. It should be noted that irrigated land is expected to play a more important role in the increase of maize production, almost entirely owing to China – which accounts for more than 40 percent of developing countries' maize production – where irrigated land allocated to maize could more than double. Part of the continued, albeit slowing, growth in yields is due to a rising share of irrigated production (normally with much higher cereal yields) in total production. This would lead to yield increases even if rainfed and irrigated cereal yields did not grow at all.

Yield increases are often credited (e.g., Borlaug, 1999) with saving land and thus diminishing pressure on the environment, such as by reducing deforestation. Using cereals as an example, the reasoning is as follows: if the average global cereal yield had not grown since 1961/1963, when it was 1 405 kg per hectare, 1 620 million ha would have been needed to grow the 2 276 million tonnes of cereals the world produced in 2005/2007; this amount was actually obtained from an area of only 705 million ha, at an average yield of 3 230 kg/ha; therefore, 915 million ha were saved because of yield increases for cereals alone. This conclusion should be qualified however, because if there had been no yield growth, the most probable outcome would have been much lower production, owing to lower demand resulting from higher cereal prices, and somewhat more land under cereals. Furthermore, in many countries, the alternative of land expansion instead of yield increases does not exist.

The scope for yield increases

Despite the increases in land under cultivation in land-abundant countries, much agricultural production growth has been based on the growth of yields, and will need increasingly to be so. What is the potential for continuing yield growth? In countries and localities where the potential of existing technology is being exploited fully, subject to the agro-ecological constraints specific to each locality, further growth – or even maintenance – of current yield levels will depend on further progress in agricultural research. In places where yields are nearing the ceilings obtained on research stations, the scope for raising them further is far more limited than in the past (Sinclair, 1998). Nevertheless, yields have continued to increase, albeit at a decelerating rate. For example, wheat yields in South Asia, which accounts for about a third of the developing countries' area under wheat, increased by 40 kg/ha per year between 1961 and 2007 (27 kg/ha over the last decade), and are projected to grow by 32 kg/ha per year over the period 2005/2007 to 2050. The equivalent increases for the developing countries overall are 50 kg/ha (past; Figure 6.9) and 30 kg/ha (future) per annum.

The variation in yields among countries remains very wide. Table 6.13 illustrates this for wheat, rice and maize in developing countries. Current yields in the 10 percent of countries with the lowest yields (the bottom decile, excluding countries with less than 50 000 ha under the crop) are generally less than one-fifth (24 percent for maize) of the yields of the best performers (top decile), and this gap has been worsening over time. If sub-national data were available, a similar pattern would probably be seen for within-country differences as well. For wheat and maize, the gap between worst and best performers is projected to persist until 2050, while for rice it may be somewhat narrowed by 2050, with yields in the bottom decile reaching 25 percent of those in the top. This may reflect the more limited scope for raising the yields of top rice performers than in the past. However, countries included in the bottom and top deciles account for only a minor share of total wheat and rice production; it is more important to examine what will happen to yield levels in the countries that account for the bulk of production. Current (unweighted) average yields of the largest producers[16] are about half those (40 percent for maize) achieved by the top performers (Table 6.13). In spite of continuing yield growth in these largest producing countries, this situation is expected to remain essentially unchanged by 2050.

Figure 6.9
Wheat yields

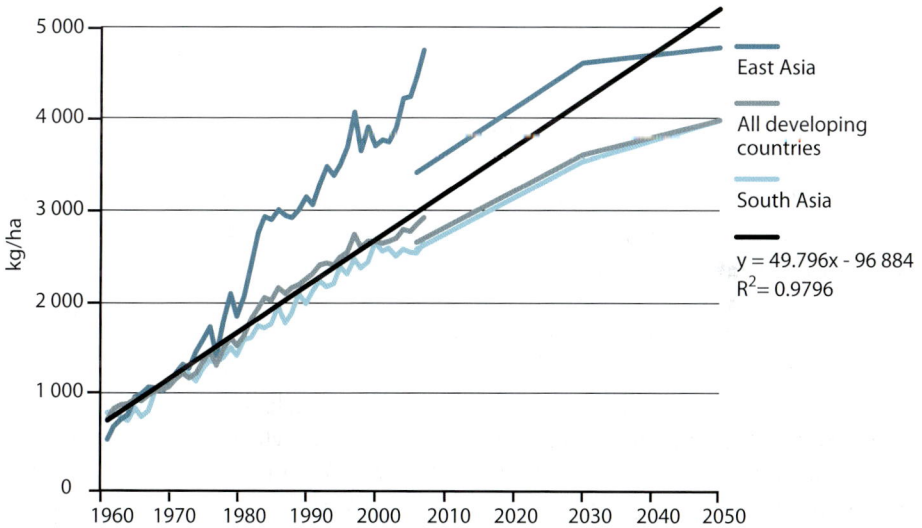

The break in the series for East Asia (and thus for all developing countries) is due to a downwards adjustment of the base year data for yields in China.
Source: Historical data from FAOSTAT.

---

16. The top 10 percent of countries ranked according to area allocated to the crop examined. For 2005/2007 these countries are China, India and Turkey for wheat; India, China, Indonesia, Bangladesh and Thailand for rice; and China, Brazil, India, Mexico, Nigeria, Indonesia and the United Republic of Tanzania for maize.

Table 6.13
Average wheat, rice and maize yields in developing countries

Crop	1961/1963 (tonnes/ ha)	(% of top decile)	2005/2007 (tonnes/ ha)	(% of top decile)	2050 (tonnes/ ha)	(% of top decile)
Wheat						
Number of developing countries included	31		32		33	
Top decile	2.15		5.65		9.02	
Bottom decile	0.40	18	0.83	15	1.50	17
Decile of largest producers (by area)	0.87	40	3.13	55	4.65	52
All countries included	0.98	46	2.35	42	3.77	42
World	1.48		2.85		3.60	
Rice (paddy)						
Number of developing countries included	44		53		56	
Top decile	4.66		7.52		9.84	
Bottom decile	0.67	14	1.06	14	2.48	25
Decile of largest producers (by area)	1.84	39	4.16	55	5.19	53
All countries included	1.90	41	3.70	49	5.15	52
World	2.19		3.74		5.33	
Maize						
Number of developing countries included	58		69		67	
Top decile	2.16		7.77		9.82	
Bottom decile	0.52	24	0.53	7	1.54	16
Decile of largest producers (by area)	1.21	56	3.15	41	4.92	50
All countries included	1.07	50	2.49	32	3.87	39
World	1.47		3.77		4.40	

Only countries with more than 50 000 ha of harvested area are included.
Countries included in each decile are not necessarily the same for all years.
Average yields are simple averages, not weighted by area.
Source: Author.

Based on this analysis, a prima facie case could be made that there has been, and still is, considerable slack in the crop yields of different countries, which could be exploited if the economic incentives are sufficient. However, the wide differences in yields among major cereal producing countries do not necessarily imply that the lagging countries have scope for yield increases equal to the inter-country yield gaps. Part of these yield differences simply reflect differing agro-ecological conditions, although not all, and perhaps not even the major part, can be ascribed to such conditions, as there are wide yield differences even among countries with fairly similar agro-ecological environments. In these cases, differences in the socio-economic and policy environments probably play a major role. The literature distinguishes two components of yield gaps: agro-environmental and other non-transferable factors, which create gaps that cannot be narrowed; and crop management practices, such as suboptimal use of inputs and other cultural practices. This second component can be narrowed – provided it makes economic sense to do so – and is therefore termed the "exploitable" or "bridgeable yield gap".

Duwayri, Tran and Nguyen (1999) state that the theoretical maximum yields for both wheat and rice are probably in the order of 20 tonnes/ha. On experimental stations, yields of 17 tonnes/ha have been reached in subtropical climates and of 10 tonnes/ha in the tropics. FAO (1999) reports that concerted efforts in Australia to reduce the exploitable yield gap increased rice yields from 6.8 tonnes/ha in 1985/1989 to 8.4 tonnes/ha in 1995/1999, with many individual farmers obtaining 10 to 12 tonnes/ha.

To draw conclusions on the scope for narrowing the yield gap, it is necessary to separate the "non-transferable" part of the gap from the "exploitable" part. One way of doing so is to compare the yields obtained from the same crop varieties grown in different locations with similar physical characteristics (climate, soil, terrain); this eliminates the non-transferable part of the comparison. This can start with an examination of data from the GAEZ analysis on the suitability of land in different countries for producing the given crop under specific technology packages. These data make it possible to derive a national maximum obtainable yield by weighting the yield obtainable in each suitability class with the estimated land area in that class. The derived national obtainable yield can then be compared with data on actual national average yields. The findings presented in Table 6.14 seem to confirm the hypothesis that a good part of the yield gap is of the exploitable type.

Countries with similar attainable averages for any given crop and technology level may be considered to be agro-ecologically similar for that crop. Naturally, any two countries can have similar attainable yields but for very different reasons; for example, in some countries the limiting factors may be temperature and radiation, in others soil and terrain characteristics or moisture availability. Nevertheless, the GAEZ average attainable yields for any crop can be taken as a rough index of agro-ecological similarity among countries for producing that crop under the specified conditions.

Table 6.14 shows the agro-ecologically attainable national average wheat yields for 16 countries,[17] and compares them with actual prevailing yields.[18] These countries span a wide range of agro-ecological endowments for wheat

---

17. Countries with more than 4 million tonnes of wheat production in 2003/2007 and rainfed agriculture accounting for more than 90 percent of total wheat production (except for Turkey, with 80 percent).

18. This comparison is somewhat distorted, as the results of the GAEZ analysis (Fischer, van Velthuizen and Nachtergaele, 2009) available at the time of writing deal with rainfed agriculture only, while the national statistics also include irrigated agriculture.

production, with some having a high proportion of their wheat land in the very suitable category (e.g., France and Poland), and others having high proportions in the suitable and moderately suitable categories (e.g., Kazakhstan and Canada). Attainable average yields in these countries range from more than 7 tonnes/ha in Hungary, Romania, France and Ukraine to less then 4 tonnes/ha in the Russian Federation, Kazakhstan and Canada.

Table 6.14
Agro-ecological suitability for rainfed wheat production, selected countries

Country	Area				Yield attainable				Actual average 2003/2007	
	Total	Very suitable	Suitable	Moderately suitable[a]	Very suitable	Suitable	Moderately suitable[a]	Average	Area	Yield
	(million ha)				(tonnes/ha)				(million ha)	(tonnes/ha)
Romania	14.4	8.3	4.2	1.9	9.0	6.9	5.2	7.9	2.0	2.6
Hungary	7.9	3.6	2.8	1.4	8.8	7.1	4.8	7.5	1.1	4.0
France	27.6	17.1	7.8	2.7	8.0	6.6	4.6	7.3	5.2	6.8
Ukraine	53.7	21.6	25.6	6.5	8.5	6.5	5.2	7.1	5.3	2.5
Poland	28.6	13.7	6.3	8.6	8.5	6.8	4.9	7.0	2.2	3.8
Germany	18.3	6.7	6.1	5.4	8.3	6.7	4.9	6.7	3.1	7.3
Italy	5.8	1.9	2.6	1.3	8.1	6.1	4.0	6.3	2.1	3.5
USA	357.8	124.9	132.2	100.7	8.4	6.0	4.1	6.3	20.3	2.8
UK	11.2	2.4	4.9	3.9	7.7	6.5	4.4	6.0	1.9	7.8
Turkey	24.8	2.5	9.4	13.0	6.6	5.8	4.7	5.3	8.9	2.2
Denmark	4.3	1.3	1.1	1.9	6.7	5.7	4.1	5.3	0.7	7.0
Argentina	87.6	8.3	36.0	43.3	6.6	5.2	3.7	4.6	5.6	2.6
Australia	47.4	3.7	15.5	28.2	6.7	5.2	3.6	4.4	12.7	1.5
Russian Federation	406.1	91.9	168.0	146.2	5.9	3.9	2.4	3.8	23.0	1.9
Kazakhstan	20.6	0.2	3.3	17.0	5.7	4.9	2.9	3.3	11.9	1.1
Canada	158.9	12.8	43.0	103.2	5.8	3.3	2.2	2.8	9.5	2.5

[a] Moderately suitable under high inputs. The data on potentials exclude marginally suitable land that in the GAEZ analysis is not considered appropriate for high-input farming.

Sources: Fischer, van Velthuizen and Nachtergaele, 2009; FAOSTAT.

The divergence between economically efficient and agro-ecologically attainable yields can be very wide. For example, the United Kingdom and the United States of America have nearly equal attainable yields (6.0 to 6.3 tonnes/ha, although the United States has much more land suitable for wheat than the United Kingdom), but actual yields are 7.8 tonnes/ha in the United Kingdom (exceeding what the GAEZ evaluation suggests as attainable on average) and 2.8 tonnes/ha in the United States. Although the United States' yields are only a fraction of those that are agro-ecologically attainable and those that prevail in the United Kingdom, the United States is not necessarily a less efficient wheat producer than the United Kingdom in terms of production costs. Other examples of economically efficient wheat producers with low yields in relation to their agronomic potentials include Argentina (2.6 tonnes/ha actual versus 4.6 tonnes/ha attainable) and Ukraine (2.5 versus 7.1 tonnes/ha).

The yield gap in relation to agronomic potential is an important element when discussing agronomic potentials for yield growth. In countries with large differences between actual and attainable yields, it seems probable that factors other than agro-ecology are responsible. Yields in these countries could grow some way towards bridging the gap if some of these factors were changed, for example, if prices rose. Once the countries with a sizeable bridgeable gap have been identified, their aggregate weight in world production of a particular crop can be assessed. If this weight is significant, the world almost certainly has significant potential for increasing production through yield growth, even on the basis of existing knowledge and technology (varieties, farming practices, etc.).

Among the major wheat producers, only some EU countries (the United Kingdom, Denmark, France and Germany) have actual yields close to, or even higher than,[19] those attainable for their agro-ecological endowments under rainfed high-input farming. In all other major producers with predominantly rainfed wheat production, the gaps between actual and attainable yields are significant (Figure 6.10). Even assuming that only half of these yield gaps (attainable minus actual) are bridgeable, the production of these countries could increase considerably without any increase in their area under wheat. As discussed, yield growth would also occur in the countries accounting for the rest of world production, including the major producers of irrigated wheat that are not included in Figure 6.10, such as China, India, Pakistan and Egypt. None of this discussion has considered the potential yield gains that could come from further improvement in varieties, as the attainable yields in GAEZ reflect the yield potential of existing varieties.

Some states in India, such as Punjab, are often quoted as examples of areas where wheat and rice yields have been slowing or are even reaching a plateau. Fortunately, India is one of the few countries for which data are available at the subnational level and distinguished by rainfed and irrigated area. Bruinsma (2003: Table 11.2) compares wheat and rice yields in major growing states with the agro-ecologically attainable yields (estimated in Fischer *et al.*, 2002), taking

---

19. That actual yield levels in the United Kingdom, Germany and Denmark exceed the average agro-ecological zone attainable yields from all suitable land can in part be explained if it is assumed that all wheat is grown only on very suitable areas (Table 6.14).

into account irrigation. This shows that, although yield growth has indeed been slowing, most actual yields are still far from the agro-ecologically attainable yield (with a few exceptions, such as wheat in Haryana). This suggests that there are still considerable bridgeable yield gaps in India.

Figure 6.10
Actual and agro-ecologically attainable wheat yields, selected countries

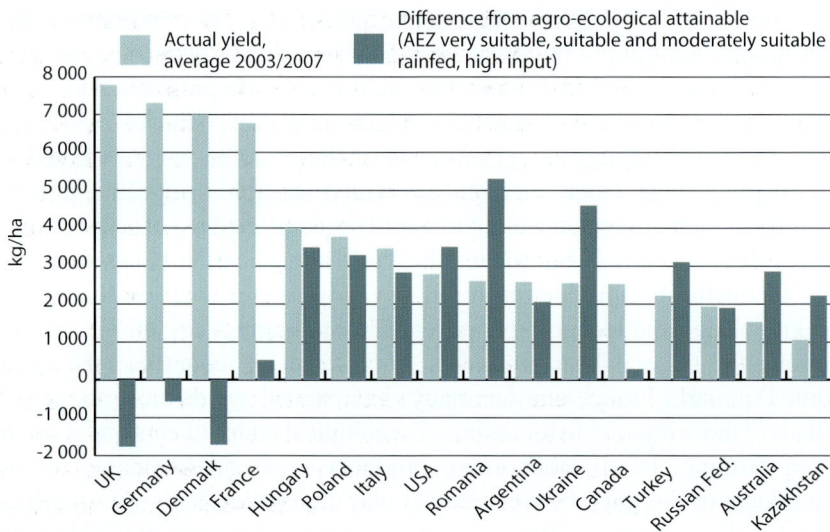

Source: Author.

This discussion gives an idea of the scope for wheat production increases through the adoption of improved technologies and practices to bridge some of the gaps between actual and obtainable yields. Wheat was used as an example, but similar analysis of other crops shows that the conclusions hold for all crops. The broad lesson from experience seems to be that if scarcities develop and prices rise, farmers quickly respond by adopting such technologies and increasing production, at least when they live in an environment with relatively easy access to improved technology, transport infrastructure and supportive policies. However, in countries with land expansion possibilities, the quickest response comes from increasing the land under cultivation, including by shifting land among crops towards the most profitable ones.

Countries use only part of the land that is suitable for any given crop. This does not mean that land lies bare or fallow, waiting to be used for increasing production of that particular crop. In most cases, such land is also suitable for other crops, and is used for them. The point being made here is that the gap between the yields actually achieved and those obtainable under high-input technology packages affords significant scope for production increases through yield growth, given conducive socio-economic conditions, incentives and policies. Although production increases may be obtained by expanding cultivation into land suitable for a particular crop, such land may not be available if it is being used for other crops.

However, even if there is sufficient slack in world agriculture to support further increases in global production, this is small consolation to food-insecure people who depend on what they themselves produce for their nutrition. Such people often live in semi-arid agricultural environments where the potential for increasing production can be very limited or non-existent. That the world as a whole may have ample potential to produce more food is of little help to them.

This discussion may create the impression that all is well regarding the potential for further production growth based on the use of existing varieties and technologies to increase yields. However, this should be heavily qualified, because: i) the exploitation of bridgeable yield gaps requires the further spread of high-external-input technologies, which might aggravate related environmental problems; and ii) perhaps more important from the standpoint of meeting future demand, the countries where there will be additional demand do not necessarily have potential for yield growth. When the potential demand is in countries with limited import capacity, as is the case of many developing countries, such potential can be expressed as effective demand only if it can be matched predominantly by local production. In such circumstances, the existence of large exploitable yield gaps elsewhere (e.g., in Argentina or Ukraine) is less important than it appears for the evaluation of potential contributions of yield growth to meeting future demand.

It follows that continued and intensified efforts are needed from the agricultural research community, to raise yields (including through maintenance and adaptive research) in the often unfavourable agro-ecological and socio-economic environments of the countries where the additional demand will be.

Summary and conclusions

This chapter has discussed the natural resource implications of the latest FAO food and agriculture baseline projections to 2050 (FAO, 2006b). These projections offer a comprehensive and consistent picture of the food and agricultural situation in 2030 and 2050, covering food and feed demand, including all foreseeable diet changes, trade and production. The main purpose of the chapter is to provide an indication of the additional demands on natural resources that will be derived from the crop production levels for 2030 and 2050 as foreseen in the FAO 2006 projections. It does not deal with additional demand for agricultural products used as feedstock in biofuel production, or the impacts of climate change (these are dealt with in Chapter 3), nor does it deal with the additional production needed to eliminate (or accelerate the elimination of) the remaining undernourishment by 2050.

Growth in agricultural production will continue to slow as a consequence of the slowdown in population growth and because an ever-increasing share of world population is reaching medium to high levels of food consumption. Nevertheless, agricultural production would still need to increase by 70 percent (and nearly 100 percent in developing countries) by 2050 to cope with a 40 percent increase in world population and to raise average food consumption to 3 130 kcal per person per day by 2050. This translates into additional production of 1 billion tonnes of cereals and 200 million tonnes of meat a year by 2050 (compared with production in 2005/2007).

Some 90 percent (80 percent in developing countries) of the growth in crop production would be a result of higher yields and increased cropping intensity, with the remainder coming from land expansion. Arable land would expand by 70 million ha (less than 5 percent), an expansion of about 120 million ha (12 percent) in developing countries being offset by a decline of 50 million ha (8 percent) in developed countries. Almost all of the land expansion in developing countries would occur in sub-Saharan Africa and Latin America.

Land equipped for irrigation would expand by 32 million ha (11 percent), while harvested irrigated land would expand by 17 percent. All of these increases would be in the developing countries. Mainly (but not only) as a result of slowly improving water-use efficiency, water withdrawals for irrigation would grow more slowly, but would still increase by almost 11 percent (or 286 km^3) by 2050.

Crop yields would continue to grow, but at a slower rate than in the past. This process of decelerating growth has already been under way for some time. On average, annual growth over the projection period would be about half (0.8 percent; 0.9 in developing countries) of its historical growth rate (1.7 percent; 2.1 percent in developing countries). Cereal yield growth would slow to 0.7 percent per annum (0.8 percent in developing countries), and average cereal yield would reach 4.3 tonnes/ha in 2050, up from 3.2 tonnes/ha at present.

Are the projected increases in land, water use and yields feasible? The GAEZ study shows that ample land resources with some potential for crop production remain, but this needs to be heavily qualified. Much of the suitable land not yet in use is concentrated in a few countries in Latin America and sub-Saharan Africa, not necessarily where it is most needed, and much is suitable for growing only a few crops, not necessarily those for which the demand is highest. In addition, much of the land not yet in use suffers from constraints (chemical, physical, disease, lack of infrastructure, etc.) that cannot be overcome easily (or economically). Part of the land is under forests, protected or under urban settlements, and so on. Overall, however, although a number of countries – particularly in the Near East and North Africa and in South Asia – have reached or are about to reach the limits of their available land, at the global scale there are still sufficient land resources left to feed the world population for the foreseeable future.

The availability of freshwater resources shows a very similar picture to that of land availability, with sufficient resources at the global level being unevenly distributed and an increasing number of countries or parts of countries reaching alarming levels of water scarcity. Many of the water-scarce countries in the Near East and North Africa and in South Asia also lack land resources. A mitigating factor could be that there are still ample opportunities for increasing water-use efficiency, such as through providing the right incentives to use less water.

The potential to increase crop yields (even with existing technology) seems considerable. Provided the appropriate socio-economic incentives are in place, there are still ample bridgeable yield gaps – the differences between agro-ecologically attainable and actual yields – to be exploited. Fears that yields, such as for rice, are reaching a plateau do not seem warranted, except for in a few very special instances.

Towards the end of the projection period there are signs that an increasing number of countries (not all of them among today's "developed countries") will reach saturation levels, when agricultural production ceases to increase and arable land is taken out of production. Likewise, although land allocated to crops such as maize and soybeans could still increase considerably, land allocated to crops such as rice, potatoes and pulses would decline. Naturally, apart from rising yields, this reflects slowing (or even declining) population growth, medium to high food consumption levels, and the shift in diets to livestock products resulting in more land being allocated to crops for animal feed.

Does this mean that all is well? Certainly not. The conclusion that the world as a whole produces or could produce enough food for all is small consolation to the people and countries (or regions within countries) that continue to suffer from undernourishment. The projected increases in yield, land and irrigation expansion will not come about entirely spontaneously (driven by market forces), but will require huge public interventions and investments, particularly in agricultural research and in preventing and mitigating environmental damage. In the problem countries, public intervention will continue to be required, to develop agriculture and adapt it to local circumstances, and to establish social safety nets.

References

Alexandratos, N., ed. 1995. *World agriculture: Towards 2010, an FAO study.* Chichester, UK, J. Wiley and Sons, and Rome, FAO.

Alexandratos, N. 2005. Countries with rapid population growth and resource constraints: Issues of food, agriculture, and development. *Population and Development Review*, 31(2): 237–258.

Alexandratos, N. 2008. Food price surges: Possible causes, past experiences, relevance for exploring long-term prospects. *Population and Development Review*, 34(4): 663–697.

Borlaug, N. 1999. Feeding a world of 10 billion people: the miracle ahead. Lecture presented at De Montfort University, Leicester, UK.

Brown, L. 2009. Could food shortages bring down civilization? *Scientific American*, 22 April 2009.

Bruinsma, J., ed. 2003. *World agriculture: Towards 2015/2030, an FAO perspective.* London, Earthscan and Rome, FAO. www.fao.org/es/esd/gstudies.htm.

Byerlee, D. 1996. Modern varieties, productivity and sustainability: Recent experience and emerging challenges. *World Development*, 24(4): 697–718.

Cohen, J. 1995. *How many people can the earth support?* New York, W. Norton.

Deininger, K. & Minten, B. 1999. Poverty, policies and deforestation: the case of Mexico. *Economic Development and Cultural Change,* January 1999.

Duwayri, M., Tran, D. & Nguyen, V. 1999. Reflections on yield gaps in rice production. *International Rice Commission Newsletter*, 48: 13–26.

FAO. 1998. *Crop evapotranspiration: Guidelines for computing crop water requirements,* by R. Allen, L. Pereira, D. Raes and M. Smith. FAO Irrigation and Drainage Paper No. 56. Rome.

FAO. 1999. *Bridging the rice yield gap in the Asia-Pacific region.* FAO Expert Consultation, 5–7 October 1999, Bangkok.

FAO. 2000. *Land resource potential and constraints at regional and country levels,* by A. Bot, F. Nachtergaele and A. Young. World Soil Resources Report No. 90. Rome.

FAO. 2004. *Global map of monthly reference evapotranspiration – 10 arc minutes.* FAO-GeoNetwork. www.fao.org/geonetwork/srv/en/main.home.

FAO. 2005a. *Irrigation in Africa in figures: Aquastat survey – 2005.* FAO Water Report No. 29. Rome.

FAO. 1995. *World agriculture: Towards 2010, an FAO study,* edited by N. Alexandratos. Rome, and Chichester, UK, J. Wiley and Sons.

FAO. 2005b. *Key water resources statistics in Aquastat.* Rome.

FAO. 2006a. *Global Forest Resources Assessment 2005.* FAO Forestry Paper No. 147. Rome.

FAO. 2006b. *World agriculture: towards 2030/2050 – interim report.* Rome.

Fischer, G., van Velthuizen, H., Shah, M. & Nachtergaele, F. 2002. *Global Agro-ecological Assessment for Agriculture in the 21st Century: Methodology and results.* RR-02-002. Laxenburg, Austria, IIASA.

Fischer, G., van Velthuizen, H. & Nachtergaele, F. 2009. *Global Agro-ecological Assessment – the 2009 revision.* Laxenburg, Austria, IIASA.

Lopez, R. 1998. The tragedy of the commons in Côte d'Ivoire agriculture: Empirical evidence and implications for evaluating trade policies. *World Bank Economic Review,* 12(1): 105–131.

Nachtergaele, F. & George, H. 2009. How much land is available for agriculture. Unpublished paper for FAO. Rome.

New, M., Lister, D., Hulme, M. & Makin, I. 2002. A high-resolution data set of surface climate over global land areas. *Climate Research,* 21: 1–25.

OECD/FAO. 2009. *Agricultural Outlook 2009–2018.* Paris and Rome.

Siebert, S., Döll, P., Feick, S., Hoogeveen, J. & Frenken, K. 2007. *Global map of irrigation areas version 4.0.1.* Frankfurt-am-Main, Germany, Johann Wolfgang Goethe University, and Rome, FAO.

Sinclair, T. 1998. Options for sustaining and increasing the limiting yield-plateaus of grain crops. Paper presented at the NAS Colloquium Plants and Population: Is There Time? Irvine, California, USA, 5–6 December 1998.

Young, A. 1999. Is there really spare land? A critique of estimates of available cultivable land in developing countries. *Environment, Development and Sustainability,* 1: 3–18.

Developing countries

Sub-Saharan Africa

Angola	Democratic Republic	Madagascar	Sierra Leone
Benin	of the Congo	Malawi	Somalia
Botswana	Eritrea	Mali	Sudan
Burkina Faso	Ethiopia	Mauritania	Swaziland
Burundi	Gabon	Mauritius	Togo
Cameroon	Gambia	Mozambique	Uganda
Central African	Ghana	Niger	United Republic of
Republic	Guinea	Nigeria	Tanzania
Chad	Kenya	Rwanda	Zambia
Congo	Lesotho	Senegal	Zimbabwe
Côte d'Ivoire	Liberia		

Latin America and the Caribbean

Argentina	Ecuador	Mexico	Suriname
Brazil	El Salvador	Nicaragua	Trinidad and Tobago
Chile	Guatemala	Panama	Uruguay
Colombia	Guyana	Paraguay	Venezuela
Costa Rica	Haiti	Peru	
Cuba	Honduras	Plurinational State	
Dominican Republic	Jamaica	of Bolivia	

Near East and North Africa

Afghanistan	Islamic Republic	Libyan Arab	Syrian Arab Republic
Algeria	of Iran	Jamahiriya	Tunisia
Egypt	Jordan	Morocco	Turkey
Iraq	Lebanon	Saudi Arabia	Yemen

South Asia

Bangladesh	India	Nepal	Pakistan	Sri Lanka

East Asia

Cambodia	Indonesia	Myanmar	Thailand
China	Lao People's	Philippines	Viet Nam
Democratic People's	Democratic Republic	Republic of Korea	
Republic of Korea	Malaysia		

Industrial countries

European Union 15*

Austria	France	Italy	Spain
Belgium	Germany	Luxembourg	Sweden
Denmark	Greece	Netherlands	United Kingdom
Finland	Ireland	Portugal	

Other industrial countries

Australia	Israel	Norway	United States of America
Canada	Japan	South Africa	
Iceland	New Zealand	Switzerland	

Transition countries

Russian Federation

Countries in the European Union *

Czech Republic	Latvia	Poland
Estonia	Lithuania	Slovakia
Hungary	Malta	Slovenia

Central Asia*

Armenia	Georgia	Kyrgyzstan	Turkmenistan
Azerbaijan	Kazakhstan	Tajikistan	Uzbekistan

Other Eastern Europe*

Albania	Bulgaria	Macedonia	Romania
Belarus	Croatia	Montenegro	Serbia
Bosnia and Herzegovina	Former Yugoslav Republic of	Republic of Moldova	Ukraine

* Country group treated as an aggregate in the analysis.

Wheat	Plantains	Sunflower seed
Rice, paddy	Sugar beet	Palm oil/palm-kernel oil
Maize	Sugar cane	Rapeseed
Barley	Pulses	Other oilseeds
Millet	Vegetables	Cocoa beans
Sorghum	Bananas	Coffee
Other cereals	Other fruits	Tea
Potatoes	Citrus fruits	Tobacco
Sweet potatoes and yams	Soybeans	Seed cotton
Cassava	Groundnuts	Jute and hard fibres
Other roots	Sesame seed	Rubber
	Coconuts	

INVESTMENT IN DEVELOPING COUNTRIES' FOOD AND AGRICULTURE: ASSESSING AGRICULTURAL CAPITAL STOCKS AND THEIR IMPACT ON PRODUCTIVITY

Stephan von Cramon-Taubadel
Gustavo Anriquez
Hartwig de Haen
Oleg Nivyevskyi

Between 1975 and 2005, global dietary energy supplies grew faster than the world population, which itself more than doubled. At the global average, food availability per person increased from 2 400 to nearly 2 800 kcal per day over those 30 years. In the developing countries the increase was even steeper, from 2 200 to 2 600 kcal per person per day. This was a remarkable achievement for the global food and agriculture system, and resulted from significant investment and technical progress. As a result, the share of the world's population with adequate access to food grew markedly. Most of the increases in consumption in developing countries were met by their domestic production, but food imports also expanded strongly.

Unfortunately, the growth in global per capita food supplies was not accompanied by a reduction in the number of undernourished people. Although the prevalence of undernourishment in developing countries declined from 20 to 16 percent between 1990/1992 and 2003/2005, the absolute number of undernourished individuals increased from 840 million to nearly 850 million. According to preliminary estimates by FAO, the high food prices in 2007 and 2008 may have driven this number up by a further 100 million. This rising trend could continue, as a result of the global financial crisis.

As has been confirmed by research results and numerous high-level intergovernmental bodies, there is no lack of knowledge about how to increase progress towards the reduction of hunger and poverty (World Bank, 2008). Rapid progress in cutting the incidence of chronic hunger in developing countries is possible if political will is mobilized. Nearly three-quarters of the poor in developing countries live in rural areas. They depend on agriculture for their earnings, either directly or indirectly. According to FAO (2003), a twin-track approach is required, combining the promotion of quick-response agricultural growth led by small farmers, with targeted programmes to ensure that the neediest people, who have neither the capacity to produce their own food nor the means to buy it, have access to adequate supplies. The two tracks are mutually reinforcing, as programmes to enhance direct and immediate access to food offer new outlets for expanded production.

Countries that have followed this approach have been comparatively successful in reducing the prevalence of undernourishment and achieving rapid and sustainable economic growth. A common feature of countries that have been successful in reducing hunger and poverty is that they have had higher overall rates of economic growth than the less successful countries, which they have achieved through a relatively higher growth in agriculture. These successful countries have also typically shared some other common features: absence of conflict, good governance, functioning markets, public investment in rural infrastructure, and greater integration into world markets. Such success stories can be found in all regions (FAO, 2008).

The vital role of income-earning opportunities in the rural areas of developing countries for success in improving the living conditions of the majority of the poor and hungry highlights the importance of investments in agriculture and rural development. In the World Food Summit Plan of Action of 1996, the members of FAO expressed their commitment: "to promote optimal allocation and use of public and private investments to foster human resources, sustainable food, agriculture, fisheries and forestry systems, and rural development, in high and low potential areas" (FAO, 1996: Preamble). According to the plan, many developing countries needed:

...to reverse the recent neglect of investment in agriculture and rural development and mobilize sufficient investment resources to support sustainable food security and diversified rural development. A sound policy environment, in which such food-related investment can fulfil its potential, is essential. Most of the resources required for investment will be generated from domestic private and public sources. Governments should provide an economic and legal framework which promotes efficient markets that encourage private sector mobilization of savings, investment and capital formation. They should also devote an appropriate proportion of their expenditure to investments which enhance sustainable food security. (FAO, 1996: Commitment Six)

Five years after the World Food Summit, FAO (2002) presented estimates of agricultural investments and capital stocks in developing countries since 1975, and concluded that additional resources for promoting agricultural growth are especially needed in countries where undernourishment is most prevalent. Many of the problems recognized in 2002 are still not resolved today. In addition, the pressures on world agriculture resulting from population growth, urbanization and growing demand for diversity, food quality and safety have grown, while new challenges have been added: climate change and variability, financial crisis, reduced availability of national and international finance, reduced public stockholding, fluctuating energy prices, and uncertain prospects for trade policy reforms.

Using various analytical tools, this chapter presents an update of earlier capital stock and investment estimates. It seeks to contribute information for assessing the extent to which developing countries have followed up on the commitments made more than a decade ago, and whether they are on track to achieve food security in the future. The following section provides an overview of possible approaches to measuring investment and agricultural capital stock (ACS). This is followed by a section that presents and discusses the results of the estimates made. In the next section, the capital stock estimates are used to produce new estimates of total factor productivity (TFP) changes in agriculture in different regions of the world, contrasting these with earlier TFP estimates. This section also explores the role that public expenditure on agriculture plays in encouraging both ACS and TFP growth. The final section closes the chapter with a summary and outlook.

Approaches to measuring investment and agricultural capital stock

Comprehensive analysis of the ACS and investment needs in agriculture requires data on fixed and human capital on farm, data on fixed capital in infrastructure, research and technology dissemination, and data on the industries up- and downstream from agriculture (input supply and agricultural processing) that have significant impacts on agricultural production. In addition to changes in physical and human capital, changes in natural capital can also have major effects on agricultural performance. Sustainable land-use practices such as conservation farming and integrated plant nutrition systems have contributed to considerable success in soil fertility management in many countries.

These are demanding requirements, which no existing source or compilation of data comes close to satisfying. Even if comprehensive data on all the components of ACS were readily available for all countries, or at least for a representative sample, difficult issues of allocation/attribution would remain. For example, machinery might be used for farm and non-farm purposes (e.g., transportation); apparently unrelated upstream investments in flood and erosion control can have far-reaching impacts on farming downstream; and investments in telecommunications infrastructure can have an important influence on market efficiency, production and welfare (Jensen, 2007). Measuring ACS necessarily involves finding a compromise between comprehensive coverage of countries/ geographic entities over time – which is possible only for a relatively narrow definition of ACS – and comprehensive coverage of the relevant components of ACS, which involves exhaustive work on a country-by-country basis.

To date, two main approaches to measuring ACS and investments in agriculture have been employed: one based on national accounts, which captures a relatively broad set of ACS components, but only for a relatively narrow set of countries; and the other based on physical inventories contained in the FAOSTAT database, which are available for essentially all countries over several decades, but which cover only a relatively narrow set of fixed assets in farming. Both approaches are employed in this chapter. The following subsections first review earlier estimates of investment in developing country agriculture, before describing these two approaches and their strengths and weaknesses.

Earlier estimates of investment in developing country agriculture

Various attempts have been made to take stock of ACS formation in developing countries. FAO's last estimates of fixed capital in primary agriculture (FAO, 2002) covered the period 1975 to 1999 and revealed significant differences among countries. Specifically, the regions with the lowest prevalence of chronic undernourishment, particularly Latin America and the Near East and North Africa, were found to have a much higher ACS per agricultural worker ratio than the other developing regions. Not only was the level of capital intensity highest in regions with low prevalence of hunger, but these same regions had also realized a significant increase in the ACS-to-labour ratio, whereas other developing regions displayed stagnating or, in the case of sub-Saharan Africa, even declining capital intensities.

The same FAO publication also presented calculations of average labour productivity, measured as agricultural value added per agricultural worker. Not surprisingly, countries with low capital intensity in agriculture showed low productivity per agricultural worker. The divergence of gross domestic product (GDP) per agricultural worker in country groups of different capital-to-labour ratios, and hence of different rates of hunger prevalence, seemed to be very large and widening over time. Throughout the 1990s, the value added per worker in the group of countries with less than 2.5 percent of their populations' undernourished was about 20 times higher than that in the group with more than 35 percent undernourished.

Although equally relevant for the performance of the food and agriculture sector, capital formation in upstream and downstream sectors and in rural infrastructure has been far less frequently and completely documented. According to a tentative estimate published by FAO at the time of the World Food Summit in 1996, annual gross investments in primary agriculture of developing countries amounted to approximately USD 77 billion during the preceding ten to 15 years (with net investments of USD 26 billion). Over the same period, annual gross investments in post-harvest activities amounted to USD 34 billion, and public gross investments in rural infrastructure, agricultural research and extension to USD 29 billion. According to these estimates, therefore, capital formation in up- and downstream sectors and in rural infrastructure added up to almost the same total as investments in primary agriculture. By far the largest share of this off-farm investment (60 percent) took place in Asia during this period, while Latin America and the Caribbean accounted for 20 percent, and the Near East and North Africa and sub-Saharan Africa for 10 percent each. Unfortunately, the estimates published at that time did not allow a breakdown by country, nor have they been regularly updated. However, available evidence from various research projects shows that rural infrastructure is inadequate in many low-income countries, particularly in much of sub-Saharan Africa.

While changes in natural agricultural capital cannot be inferred at the global level, progress has been made towards including the cost of natural capital depletion in national accounts. Based on these efforts, the World Bank (2005a) estimated the value of natural resources, and concluded that in low-income countries, excluding the oil states, the share of natural capital in total wealth is greater than the share of produced capital. To account for the value of depletion of this natural capital, so-called adjusted net national savings were calculated as indicators of the real growth potential of a country. The results show that "net savings per person are negative in the world's most impoverished countries, particularly in sub-Saharan Africa" (World Bank, 2005a). Depletion of soil quality is found to be a major loss in this context. It is alarming that this trend is identified in precisely those countries where agricultural development matters most for poverty and hunger reduction.

Estimating the agricultural capital stock from national accounts data

Crego *et al.* (1998) first used information on gross fixed capital formation in national accounts to generate ACS estimates for 57 countries between 1967 and 1992. This chapter draws on an expanded version of their database, produced by Anriquez and Daidone (2008), which contains more than 100 countries, of which only 76 have agricultural gross fixed capital formation series long enough to allow reasonable estimates of physical capital stocks. This expanded database has been updated to cover the period up to 2002. As in Crego *et al.* (1998), data are not available in all years for all countries, but inter- and extrapolation enable a balanced panel from 1967 to 2003 to be generated for all 76 countries. Exceptions are the transition countries of Central and Eastern Europe, for which the series begin in 1990.

The data set was generated based on the assumption that agricultural capital is composed of three components: i) physical capital; ii) livestock; and iii) tree stocks, which represent the value of planted permanent crops. The physical capital series was constructed using time series of gross fixed capital formation in agriculture as published in national accounts statistics and, in a few instances, using case studies attempting to calculate these same series. The method used to estimate physical capital stocks is a variation of the perpetual inventory method, which estimates current capital stocks by adding suitably depreciated investments from previous periods. Because capital stocks depreciate, only a finite history of investments in previous periods has to be considered to determine current capital stocks.

This study assumes a hyperbolic depreciation function (details in Crego *et al.*, 1998) and that the lifetime of each investment is normally distributed with a mean of 20 years and a standard error of eight years. This means that with 95 percent probability, each agricultural investment has a service lifetime of between four and 36 years. When applying this methodology, a long time series on gross investment is required. Where such a series was not available, previous gross investment levels were predicted (back-casted), using both agricultural value added (either available or predicted using simple log trend) and the observed gross investment-to-agricultural value added ratio. All national capital stock series were estimated in constant national currency, and converted to current United States dollars using national deflators (to convert to series in current local currency) and current exchange rates. The final comparable series in 1995 dollars were created by deflating the current dollars series using the United States agricultural value added deflator.

The value of livestock was calculated using the stock numbers reported by FAOSTAT for different types of animal. Heads of livestock were valued using United States dollar prices, which were estimated as regional weighted (by quantity) averages of implicit unit export prices, also obtained from FAOSTAT. Current dollar series were converted to constant 1995 dollars by deflating with the United States agricultural GDP deflator.

Tree stocks were valued as the present value of discounted future net revenues. First, net revenues were assumed to equal 80 percent of gross revenues, which were themselves calculated per permanent crop as the product of yield and price. Yields were calculated using area and total output data from FAOSTAT, while prices were five-year moving averages of actual producer prices reported by FAOSTAT for each country. Two simplifying assumptions were made: first, that all permanent crops were at half of their productive life spans; and second, that the life span of all permanent crops was 26 years. Future revenues were discounted using a "real" rate of return defined as the difference between the yields of ten ten-year United States bonds and the inflation of the United States GDP deflator for each period. The value of tree stocks was converted to real 1995 dollars first by converting the series to current dollars using the period's exchange rate, and then by deflating this series with the United States agricultural GDP deflator.

Estimating the agricultural capital stock using physical inventories in FAOSTAT

For many countries, national accounts data on gross fixed capital formation in agriculture are not available. As an alternative, in 1995, the FAO Statistics Division first compiled estimates of ACS based on the physical stocks of various types of agricultural asset. For each asset, physical stocks were multiplied by a constant base year unit price to produce a series of asset values over time. These values were subsequently aggregated over all assets to produce an estimate of total ACS at constant prices.

Estimates of ACS based on this method were first prepared at the regional level in 1995, as part of the World Agriculture Towards 2010 exercise (Alexandratos, 1995). These estimates were for the period 1975 to 1995 and used 1990 United States dollar prices. They were subsequently updated in 2001 for the period 1975 to 1999, using a broader set of assets and 1995 dollar prices, and covering individual developing countries rather than only regional aggregates (FAO, 2002). These are the estimates of ACS referred to in the subsection on earlier estimates. A further update to include the years to 2002 was prepared by Barre in 2006 (FAO, 2006).

For this chapter, the 1975 to 1999 estimates produced in 2001 were updated and extended to 2007.[1] The assets covered fall into four categories, as outlined in Table 7.1, and are available for 223 countries and geographic entities.[2] To convert physical inventories into asset values, the 1995 unit asset prices compiled by FAO (2002) were used. These were drawn from a number of sources, including country investment project reports prepared by and for FAO, FAOSTAT data on purchase prices of means of production such as tractors, and unit trade values. Details on

---

1. For some assets, FAOSTAT data were available only until 2005 or 2006. In these cases, the remaining years to 2007 were extrapolated.

2. The number of countries changes over time, for example, owing to the break-up of the Soviet Union. FAOSTAT data include entities such as Gaza and Greenland, which are not independent countries.

these unit prices and other aspects of the estimation are given in FAO, 2001b. Key issues include the following:

- No data on physical stocks of hand tools were available, so the stock of these tools was estimated by multiplying the number of individuals active in agriculture in each country and year by a uniform estimate of USD 25-worth of hand tools.

- Unit land prices were estimated as the incremental values of development to make land suitable for crop production, plant it to permanent crops, or provide it with irrigation services.

- No data on physical stocks of structures were available, so these were estimated as a function of the number of animals/poultry in each country and year.

Table 7.1
Agricultural assets covered in the FAOSTAT measure of ACS

Land development	Livestock	Machinery	Structures
Arable land	Cattle	Tractors	For animals
Permanent crops	Buffaloes	Harvesters	For poultry
Irrigation	Sheep	Milking machines	
	Goats	Hand tools	
	Pigs		
	Horses		
	Camels		
	Mules and donkeys		
	Poultry		

Source: FAOSTAT.

Strengths and weaknesses of the national accounts and FAOSTAT approaches

National accounts-based estimates of ACS have the important advantage of providing a considerably broader coverage of fixed capital in agriculture than the estimates based on FAOSTAT physical inventories do. The use of the permanent inventory method coupled with consistent national accounts data on investments also provides theoretically much sounder estimates of the value of ACS in each year than the FAOSTAT approach does. The use of constant prices in the FAOSTAT approach means that it essentially produces a volume index that does not account for the age of assets or for quality improvements in assets over time (e.g., the average tractor made in 2005 can do more than the average tractor made in 1975, and there have been genetic improvements in livestock over the same period).

The main disadvantage of the national accounts-based estimates is that they are only available for some countries. As might be expected, Organisation for Economic Co-operation and Development (OECD) and other industrialized countries are well represented in the national accounts database, but this is not the case for developing countries (Table 7.2). For example, China is not included in the national accounts estimates, and only ten countries in sub-Saharan Africa are, compared with 51 in the FAOSTAT physical inventory estimates.

Table 7.2
Numbers of countries/geographic entities in the national accounts and FAOSTAT physical inventory databases

	Number of countries covered	
Region	National accounts estimates	FAOSTAT estimates
East Asia and Pacific	4	42
Europe and Central Asia	6	25
Latin America and Caribbean	15	45
Near East and North Africa	7	22
South Asia	3	7
Sub-Saharan Africa	10	51
High-income OECD	24	24
High-income non-OECD	6	7
Total	75	223

Source: FAOSTAT.

This would not be of major concern if the national accounts database included a representative sample of all developing countries. However, there are indications that this is not the case. As demonstrated in the following section, there appears to be some selection bias in the sample of countries covered by the national accounts approach; countries that are able to provide the required national accounts data appear to perform better on average in terms of investment in ACS. Hence, analysis based exclusively on the national accounts method might paint an overly positive picture of ACS levels and investments over time.

Results: development of the agricultural capital stock since 1975

In the following subsections, estimates of ACS and its growth are presented for various groups and sub-groups of countries. In most cases, the FAOSTAT physical inventories estimates are presented, because the generation of consistent aggregates over time is possible only with these estimates.

Development of the agricultural capital stock by region

Figure 7.1 displays the development of total ACS between 1975 and 2007, worldwide and broken down into developed and developing countries. The worldwide rate of ACS growth (net investment in ACS) slowed around 1990; calculations confirm that the average annual rate of worldwide ACS growth fell from 1.1 percent between 1975 and 1990 to 0.5 percent between 1991 and 2007 (Table 7.3). This slowdown was caused primarily by stagnating and then falling levels of ACS in developed countries, although rates of ACS growth also fell in developing countries over time. However, rates of ACS growth did not become negative in developing countries, as they did for developed countries after 1990.

Table 7.3
Average annual rates of ACS growth before and after 1990 (percentages)

	1975-1990	1991-2007
World	1.11	0.50
Developed countries	0.60	-0.34
Developing countries	1.66	1.23

Source: Authors' calculations.

Further disaggregating these average annual growth rates reveals several interesting patterns. First, the reduction in the rate of ACS accumulation was sharpest in the second half of the 1990s, with ACS growth becoming strongly negative in developed countries and falling notably in developing countries (Figure 7.2). Since the beginning of the 2000s, the worldwide rate of ACS growth has increased again (from 0.32 to 0.52 percent per year), as the rate of ACS shrinkage in developed countries has slowed. At the same time, rates of ACS growth in developing countries have remained positive, but continued to fall. Hence, the gap between rates of ACS growth in developing and developed countries has closed from a high of slightly more than 2 percent (1.27 versus -0.76 percent) in 1995/1999 to slightly more than 1 percent (1.01 versus -0.11 percent) in 2005/2007.

The rapid reduction in rates of ACS growth in developed countries over the 1980s and 1990s was driven by episodes of significant disinvestment in different regions. In the 1980s, North America saw negative rates of ACS growth, and in the 1990s rates of ACS growth in Western Europe became negative, presumably in part owing to the effect of the 1993 so-called MacSharry reforms of the European Union's (EU's) Common Agricultural Policy. In the second half of the 1990s and into the 2000s, ACS in the transition economies of Central and Eastern Europe

Figure 7.1
Development of ACS

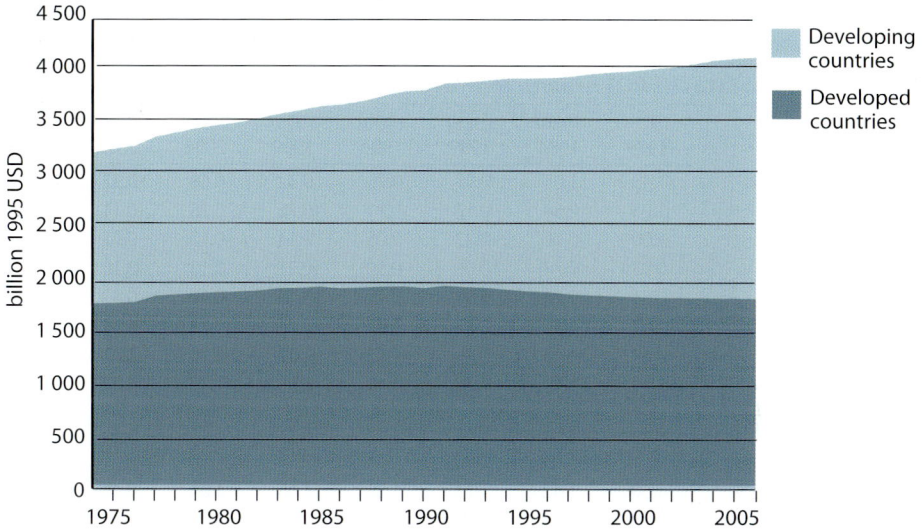

Sources: Authors' calculations with FAOSTAT physical inventories estimates.

Figure 7.2
Average annual ACS growth rates in developed and developing countries

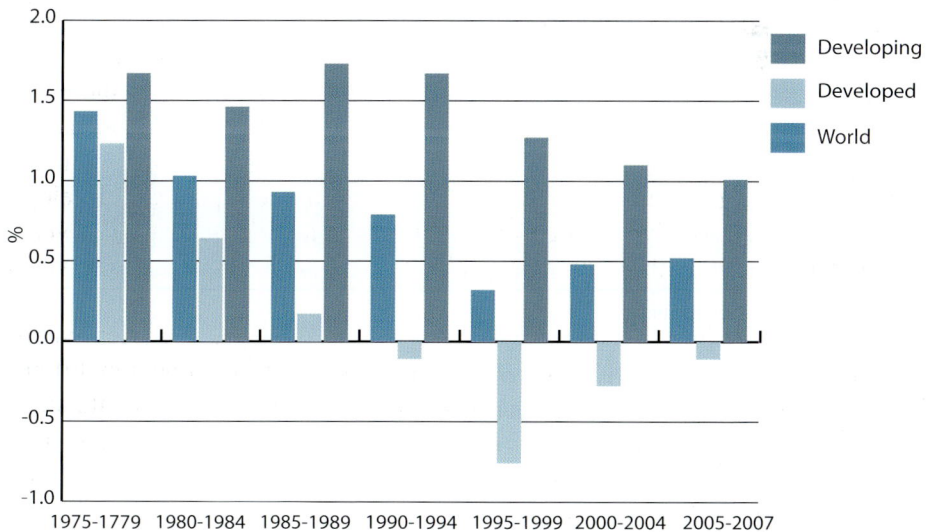

Sources: Authors' calculations with FAOSTAT physical inventories estimates.

shrank especially dramatically.[3] In developing countries, rates of ACS growth have been consistently positive across regions and sub-periods, with South Asia recording a sustained reduction in growth rate since the early 1990s.

Table 7.4
Average annual rates of ACS growth (percentages)

Region	1975/ 1979	1980/ 1984	1985/ 1989	1990/ 1994	1995/ 1999	2000/ 2004	2005/ 2007	1975/ 2007
World	1.43	1.03	0.93	0.79	0.32	0.48	0.52	0.78
Developed	1.23	0.64	0.17	-0.11	-0.76	-0.28	-0.11	0.09
North America	1.00	-0.16	-0.23	0.05	0.14	-0.12	0.02	0.08
Western Europe	0.93	0.74	0.06	-0.50	-0.27	-0.14	-0.10	0.09
Oceania	-0.84	0.24	0.51	-0.17	-0.54	0.49	0.42	0.02
Transition	2.03	1.55	0.62	0.07	-2.77	-0.71	-0.31	0.02
Developing	1.67	1.46	1.73	1.67	1.27	1.10	1.01	1.43
Latin America and Caribbean	2.15	1.40	1.76	1.40	0.39	1.16	0.22	1.24
Near East and North Africa	0.93	1.76	1.99	1.87	0.71	0.93	0.99	1.34
Sub-Saharan Africa	1.68	1.42	1.23	1.86	1.65	1.64	0.96	1.52
East and Southeast Asia	1.75	1.37	2.04	1.80	1.86	1.35	1.73	1.70
South Asia	1.61	1.49	1.19	1.42	1.22	0.34	0.32	1.11

Source: Authors' calculations.

Table 7.5
Average annual rates of ACS growth per worker in agriculture, developing country regions

Region	1975-2007 annual rate of growth (%)		
	ACS	Population active in agriculture	ACS/person active in agriculture
Latin America and Caribbean	1.24	-0.08	1.33
Near East and North Africa	1.34	0.83	0.51
Sub-Saharan Africa	1.52	1.97	-0.44
East and Southeast Asia	1.70	0.97	0.72
South Asia	1.11	1.38	-0.26

Source: Authors' calculations.

The consistently positive rates of ACS growth in developing country regions in Table 7.4 mask important changes in the availability of ACS per worker. In sub-Saharan Africa and South Asia, the growth of the population active in agriculture

---

3. The drop in the rate of ACS accumulation in developed countries is also at least partly due to improvements in input quality, which the FAOSTAT-based estimates do not pick up. See subsection on Comparing the national accounts and FAOSTAT approaches.

has outstripped the rate of ACS growth, leading to average annual reductions in ACS per agricultural worker of 0.44 percent per year in sub-Saharan Africa and 0.26 percent per year in South Asia, from 1975 to 2007 (Table 7.5). In the Near East and North Africa and in East and Southeast Asia, population growth has eroded but not completely outweighed growth in ACS. In Latin America and the Caribbean, the population active in agriculture has fallen at an average of almost 0.1 percent per year since 1975, contributing to an overall increase in ACS per agricultural worker over this period.

Figure 7.3 presents information on gross and net investments in ACS for developing and developed countries. Net investment is calculated as the simple difference between ACS in year t + 1 and year t. Gross investment is calculated assuming that in addition to net investment, 5 percent of the ACS in year t depreciated and was replaced.

Figure 7.3
Gross and net investments in ACS, developing and developed countries

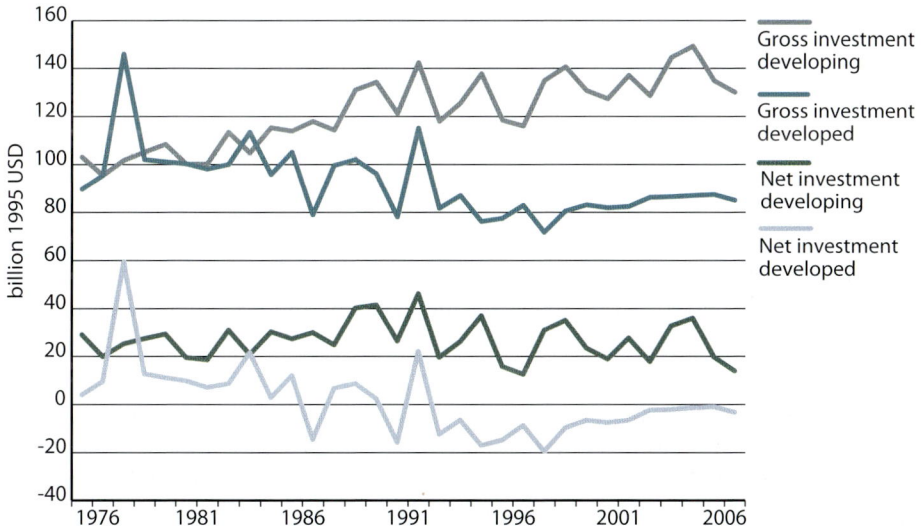

Source: Authors' calculations with FAOSTAT physical inventories estimates.

Components of agricultural capital stock

Of the four categories of agricultural capital in the FAOSTAT physical inventories estimates, land is clearly the most important, with a value share that is consistently between 52 and 55 percent of total ACS (Figure 7.4). Livestock comes next, with a share of 24 to 26 percent, followed by machinery, with 16 to 17 percent, and structures, with 5 percent.

Figure 7.4
Composition of global ACS

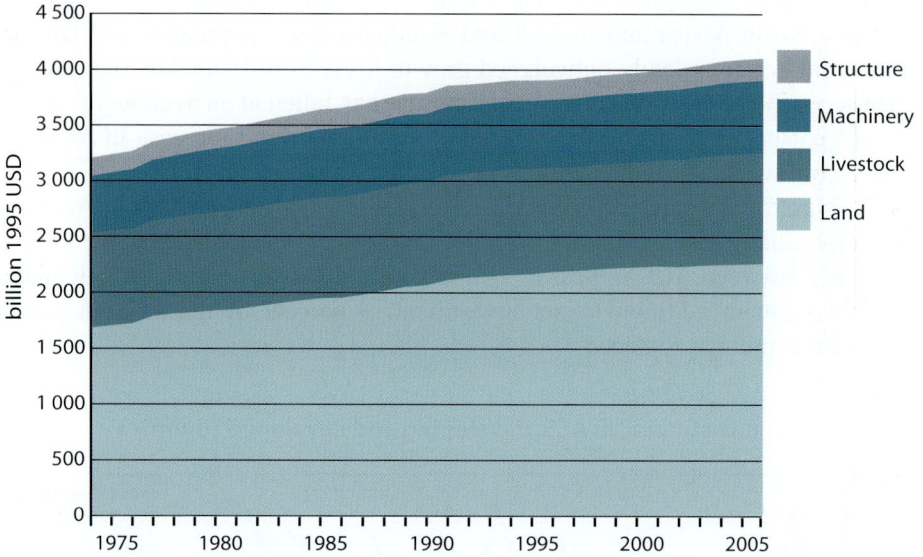

Source: Authors' calculations with FAOSTAT physical inventories estimates.

All four components of ACS increased between 1975 and 2007. Expressed in average annual rates of growth, the land component increased by 0.93 percent per annum over this period, while livestock increased by 0.50 percent, machinery by 0.75 percent, and structures by 0.66 percent. However, the individual components evolved differently over time. The rates of growth in the livestock, machinery and structures components of ACS fell during the 1980s, reaching low negative rates in the first (machinery) or second (livestock, structures) half of the 1990s (Figure 7.5). Since 2000, these components have again displayed positive rates of growth. Land has followed a different pattern. Average annual rates of growth in land stock were never negative between 1975 and 2007 (Figure 7.5). However, they declined steadily over the entire period, and have not recovered since 2000. As the land component of ACS measures the value of land improvements (investments in permanent crops, irrigation and arable land) this sustained slowdown in land growth is not necessarily due to increased scarcity of agricultural land alone. It can also reflect a reduction in the willingness to invest in improving the productivity of land.

Figure 7.5
Average annual rates of growth in global ACS components

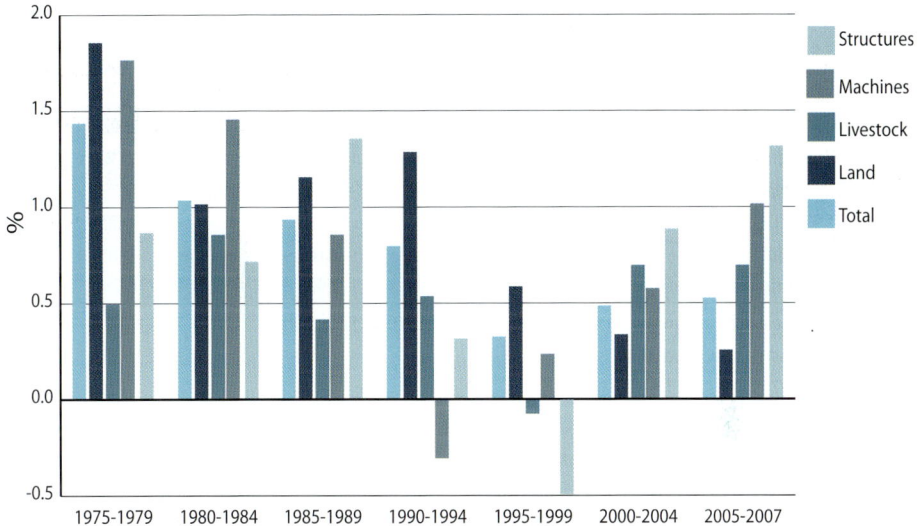

Source: Own calculations with FAOSTAT physical inventories estimates.

Agricultural capital stock and the prevalence and depth of hunger

The fact that population growth has outstripped ACS growth in sub-Saharan Africa and South Asia is cause for concern, because many countries with severe hunger problems are located in these regions. To cast more light on this issue, the study explored the relationship between ACS growth and the prevalence and depth of hunger as defined by FAO (2008). Hunger prevalence is defined according to the proportion of the population that is undernourished, and the depth of hunger according to the gap between actual calorie consumption by the undernourished and minimum dietary energy requirements (MDERs).

The estimates in Tables 7.6 and 7.7 indicate that the ACS per person active in agriculture has grown least in those countries with the highest prevalence and depth of hunger. In the countries in hunger prevalence categories 4 and 5 (more than 20 percent undernourished), growth in ACS has been outstripped by population growth, resulting in a reduction in the ACS per person active in agriculture. The same is true of countries in depth of hunger category 4 (where the average undernourished individual consumes less than 88 percent of his/her MDER). In both tables, China is listed separately because it would obscure the other countries in its hunger prevalence and depth of hunger categories.

Table 7.6
Average annual rates of growth in ACS per worker in agriculture, developing countries, by hunger prevalence category

Hunger prevalence category	1975–2007 annual rate of growth *(%)*		
	ACS	Population active in agriculture	ACS/person active in agriculture
1 < 5% undernourished	1.21	0.27	0.93
2 5–9% undernourished	1.88	-0.11	2.00
3 10–19% undernourished	1.83	1.55	0.28
4 20–35% undernourished	1.22	1.48	-0.25
5 > 35% undernourished	1.29	2.16	-0.85
China 9% undernourished	1.71	0.96	0.74

Hunger prevalence categories based on 2003/2005 data from FAO, 2008.
Source: Authors' calculations.

Table 7.7
Average annual rates of growth in ACS per worker in agriculture, developing countries, by depth of hunger category

Depth of hunger category	1975–2007 annual rate of growth *(%)*		
	ACS	Population active in agriculture	ACS/person active in agriculture
1 gap < 7% of MDER	0.73	-1.98	2.76
2 gap 7–9% of MDER	1.53	0.83	0.69
3 gap 10–12% of MDER	1.53	0.94	0.59
4 gap >12% of MDER	1.47	1.77	-0.30
China gap 12.6% of MDER	1.71	0.96	0.74

Depth of hunger categories based on 2003/2005 data from FAO, 2008.
Source: Authors' calculations.

Agricultural capital stock in countries with success in hunger reduction

If countries with high prevalence and depth of hunger are characterized by lower levels of investment in ACS, is there also evidence that countries that have been successful in reducing hunger are characterized by higher investments? Figure 7.6 presents information on annual rates of ACS growth between 1990 and 2005 for the developing countries that FAO (2008: 16) identifies as having made the most progress or having experienced the largest setbacks in achieving the 1996 World Food Summit target of halving the number of hungry people by 2015. With the exception of the Democratic People's Republic of Korea, all the developing countries that suffered notable setbacks had negative rates of ACS growth between

1990 and 2005, and with the exception of Peru (and slight exception of Ghana), all countries that made notable progress had positive rates of ACS growth.

Figure 7.6
Annual rates of ACS growth in countries making the most progress or suffering the largest setbacks towards the 1996 World Food Summit targets

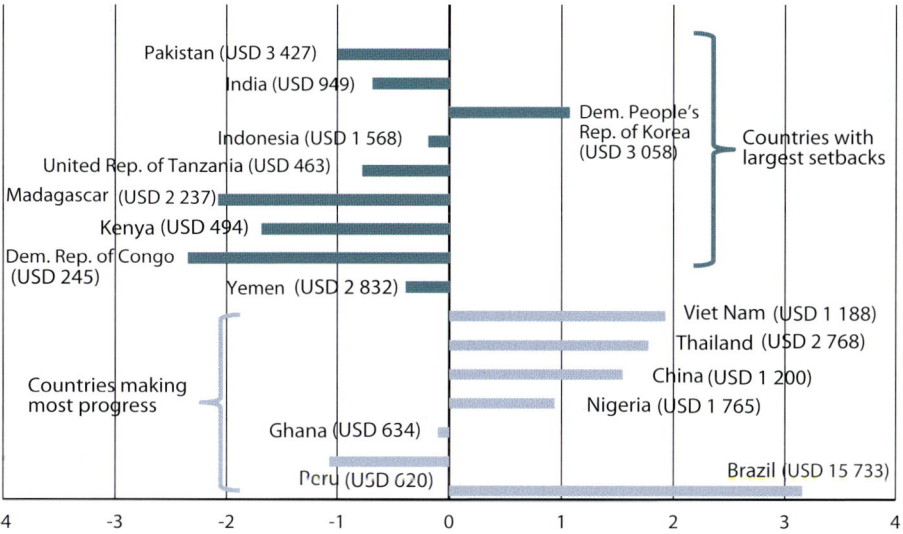

The 2005 ACS per worker is given in parentheses.
Sources: Authors' calculations with FAOSTAT physical inventories estimates; FAO, 2008.

The relatively high rate of ACS growth recorded for the Democratic People's Republic of Korea must be interpreted with caution because it is difficult to confirm official statistics in this country; Peru's progress towards the World Food Summit target, despite negative ACS growth, may be due to the resolution of internal conflicts and unrest during the 1990s. Note that according to the national accounts-based ACS estimates, Peru is one of the better-performing countries, with positive accumulation of ACS per worker for the same period. Ghana has made good progress towards the World Food Summit target and is on track for Millennium Development Goal (MDG) 1, but has witnessed slightly negative ACS growth. This suggests that the determinants of success have been outside the farm sector. A recent OECD study (Dewbre and De Battisti, 2008) concludes that Ghana's success in poverty reduction may have less to do with on-farm investments than with public investments in research, technology and infrastructure, leading to strong growth and income diversification in the rural non-farm economy.

Comparing the national accounts and FAOSTAT approaches

The results discussed so far have been drawn from the FAOSTAT physical inventories estimates of ACS. As these estimates cover essentially all countries, they lend themselves to calculation of the regional and global aggregates presented in previous subsections. However, these estimates suffer from methodological weaknesses. A comparison with estimates of ACS based on the national accounts method can cast some light on the robustness of the FAOSTAT estimates and the advantages and disadvantages of the two approaches.

Both the national accounts and FAOSTAT estimates cover almost the same set of developed countries (30 in the national accounts estimates, 31 in the FAOSTAT estimates; Table 7.2). Hence, an almost direct comparison of levels and changes is possible for these countries. Although the focus of this chapter is developing countries, the study began by comparing the national accounts and FAOSTAT estimates for developed countries. This revealed important discrepancies (Figure 7.7). First, the national accounts method produces a higher overall estimate of ACS. This is presumably owing to its more comprehensive coverage of ACS; the gross capital formation data on which the national accounts estimates are based capture investments that are not included in the limited set of assets covered by FAOSTAT.

Second, the national accounts estimates are more volatile than the FAOSTAT ones. This reflects a fundamental difference between the two. FAOSTAT estimates are calculated using a constant set of 1995 prices, while the use of changing prices, deflators and exchange rates in calculation of the national accounts-based estimates means that they capture not only changes in the volume of ACS, but also changes in the valuation. For example, the drop in the national accounts-based ACS estimates in developed countries during the first half of the 1980s (Figure 7.7) is presumably due to the strength of the United States dollar over this period, which reduced the dollar value of the ACS in other developed countries, such as in Europe. Examination of the national accounts-based data set for other regions (not shown) reveals that this drop was even more marked in Latin America, where currencies depreciated heavily against the dollar as a consequence of the debt crisis in the early 1980s; similar evidence of a fall in ACS at this time is also revealed for Asia and, to a lesser extent, Africa. There are no up-to-date numbers on the evolution of ACS in 2008/2009, but the experience of past global debt crises suggests that they can provoke large and protracted dents in the evolution of ACS.

A final discrepancy is that the national accounts estimates for developed countries trend strongly upwards over the entire period since 1975, while the FAOSTAT estimates show only a very slight increase overall, and a sustained downwards trend in the 1990s. This difference is at least partly due to the use of constant prices in the FAOSTAT approach, which means that the FAOSTAT estimates fail to capture increases in the quality of many components of ACS over time. However, Table 7.8 shows that the national accounts estimates do not trend upwards for all regions. Seven of the nine sub-Saharan African countries in the national accounts database, and all three of the South Asian countries display negative ACS trends between 1975 and 2003, while eight of the nine and the same three, respectively, display positive trends in the FAOSTAT database. It appears that the two approaches for estimating ACS produce substantially different results.

Table 7.8
Trends in the development of ACS by region, comparison of the national accounts estimates and FAOSTAT physical inventories estimates

	National accounts estimates			FAOSTAT estimates		
Region	Positive trend	Negative trend	No significant trend	Positive trend	Negative trend	No significant trend
North America (2)	0	2	0	1	1	0
Western Europe (16)	3	8	5	5	8	3
Oceania (2)	2	0	0	0	0	2
Other developed (3)	1	0	2	2	0	1
Transition economies (10)	4	4	2	4	0	6
Latin America and Caribbean (17)	3	9	5	13	2	2
Near East and North Africa (9)	6	2	1	9	0	0
Sub-Saharan Africa (9)	0	7	2	8	0	1
East and Southeast Asia (4)	4	0	0	4	0	0
South Asia (3)	0	3	0	3	0	0
Other developing (1)	1	0	0	1	0	0
Total (76)	39	24	13	50	11	15

The number of countries in the region is given in parentheses.

Sources: Authors' calculations with FAOSTAT physical inventories estimates; Anriquez and Daidone, 2008.

The comparisons in Table 7.8 are based on the 76 countries in the national accounts database. Comparisons for the many, especially developing, countries that are not included in this database are clearly not possible. Given the theoretical advantages of the national accounts approach to measuring ACS, the entire discussion in this chapter could be based on results produced from this method, if it was clear that these results were based on a representative sample of countries. However, this does not appear to be the case. Table 7.9 presents estimates of the ACS per person employed in agriculture for groups of developing countries in different hunger prevalence categories. One set of estimates is based on the national accounts approach and two on the FAOSTAT approach, using first the full sample of countries and second only the countries in the national accounts database. The estimates based on the national accounts approach are uniformly higher than those based on the FAOSTAT physical inventories approach, mirroring the result found for the aggregate of developed countries (Figure 7.7). In addition, the result that the ACS per person employed in agriculture is declining in the countries with the highest prevalence of hunger (Table 7.6) is confirmed with the national accounts-based estimates.

Table 7.9
ACS per person active in agriculture, by hunger prevalence category, comparisons across different approaches and samples of countries

Hunger prevalence category	1975– 1979	1980– 1984	1985– 1990	1991– 1994	1995– 1999	2000– 2003	Countries in sample
	(USD/capita)						(No.)
National accounts approach							
2	10 404	8 445	9 053	12 719	15 671	15 404	3
3	16 128	12 897	9 341	8 857	9 492	9 660	11
4	6 833	5 139	3 780	3 476	3 848	3 796	9
5	3 027	2 086	1 613	1 368	1 026	940	3
FAOSTAT physical inventories approach (full sample)							
2	3 660	4 122	4 535	5 104	5 315	5 820	20
3	1 636	1 668	1 675	1 906	2 070	2 076	28
4	1 391	1 389	1 371	1 398	1 397	1 353	24
5	891	880	854	820	773	724	18
FAOSTAT physical inventories approach (same sample as national accounts approach)							
2	3 524	3 860	4 430	5 283	5 863	6 569	3
3	4 192	4 409	4 343	5 084	5 644	5 862	11
4	2 338	2 322	2 246	3 001	3 692	3 888	9
5	1 470	1 493	1 480	1 434	1 355	1 266	3

Sources: Authors' calculations with FAOSTAT physical inventories estimates; Anriquez and Daidone, 2008.

However, the national accounts estimates are based on fewer developing countries than the FAOSTAT results. For example, there are nine countries in hunger prevalence category 4 in the national accounts database, compared with 24 in the FAOSTAT database (Table 9.9). If the FAOSTAT results are recalculated for only those countries that are included in the national accounts database, evidence of selection bias becomes apparent. Specifically, with the exception of hunger prevalence category 2 in the 1970s and 1980s, the FAOSTAT estimates increase, often considerably, when only the restricted sample of countries in the national accounts database is considered. In hunger prevalence category 4, for example, the estimate of the ACS per worker in agriculture is USD 1 353 in the full FAOSTAT sample of 24 countries, but increases to USD 3 888 if only the nine countries included in the national accounts data are considered.

Figure 7.7
Total ACS in developed countries, comparison of national accounts estimates and FAO physical inventories estimates

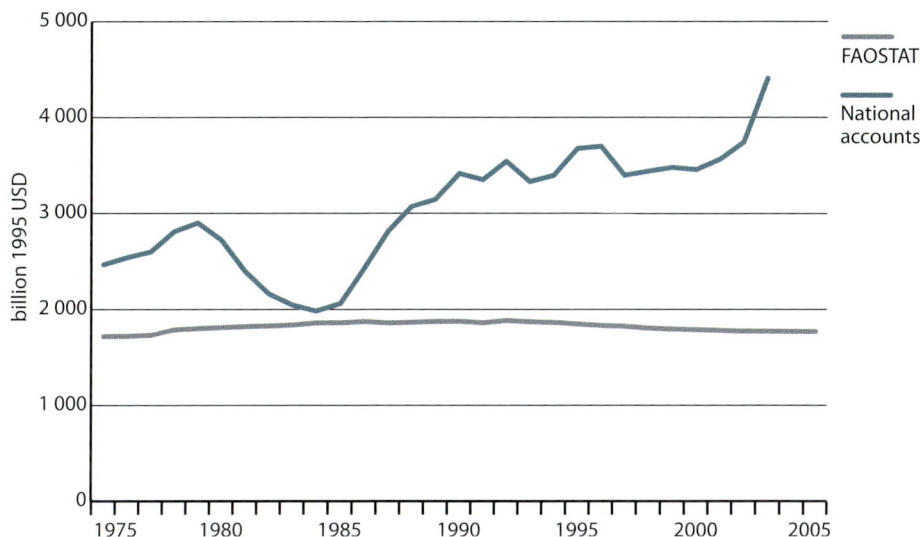

Sources: Authors' calculations with FAOSTAT physical inventories estimates; Anriquez and Daidone, 2008.

A comparison of the lists of countries in hunger prevalence category 4 included in the two databases confirms that selection bias may be playing a role. In addition to the nine countries in the national accounts database, the FAOSTAT database includes a number of countries (Bangladesh, Cambodia, Cameroon, the Democratic Republic of the Congo and Senegal) that have considerably lower levels of ACS per worker in agriculture (Table 7.10). It could be conjectured that a developing country's level of ACS per worker in agriculture is correlated with its ability to provide the detailed national accounts information required for the calculation of national accounts-based ACS estimates. If this is true, the national accounts method will tend to overestimate ACS levels at the aggregate level of groups of developing countries.

Altogether, the results of the comparisons presented in this subsection are sobering. They reveal important differences between the two sets of estimates. While both approaches point to a reduction in ACS per person employed in agriculture in the countries with the greatest prevalence of hunger, in other respects (e.g., the development of ACS in developed countries) there are large discrepancies. The inescapable conclusion is that too little is known about ACS, despite its obvious importance for efforts to combat hunger.

Table 7.10
Countries in hunger prevalence category 4 in the national accounts and FAOSTAT physical inventories databases

Countries in both databases	Countries in only the FAOSTAT database	
Botswana	Bangladesh	Grenada
Dominican Republic	Cambodia	Guinea-Bissau
India	Cameroon	Mongolia
Kenya	Democratic People's Republic of Korea	Nicaragua
Malawi	Democratic Republic of the Congo	Senegal
Niger	Djibouti	Sudan
Pakistan	Gambia	Timor-Leste
Plurinational State of Bolivia		Yemen
Sri Lanka		

Each of the approaches to estimating ACS suffers from important weaknesses that limit its usefulness as a basis for deriving robust conclusions and policy implications. The FAOSTAT-based estimates cover only certain components of ACS. The national accounts-based estimates cover only a (probably non-representative) sample of developing countries. However, each approach also has important advantages. The FAOSTAT approach provides global coverage over a long period; if the assets that it includes are representative of overall ACS, it provides a robust basis for comparisons across countries/regions and time. The national accounts estimates provide additional information on the value, as opposed to only the volume, of ACS.

For the moment, the only option is to work with both sets of estimates, interpreting them carefully. In the future, priority must be given to improving these estimates and resolving the differences between them. A first step would be to update the constant 1995 prices used to generate the FAOSTAT estimates, for example to 2000 and 2005. Efforts should also be made to expand the set of countries included in the national accounts database, with a special emphasis on developing countries characterized by high prevalence and depth of hunger. Producing robust ACS estimates that combine the coverage of the FAOSTAT approach with the greater conceptual consistency of the national accounts approach will require significant commitment of resources, but the effort must be made.

Agricultural capital stock and public expenditure on agriculture

To what extent does public expenditure in agriculture support investment in ACS? Two types of public expenditure are relevant in this regard: national government expenditure, and international expenditure in the form of official development assistance (ODA). Although these types of public expenditure are discussed separately in the following paragraphs, the available statistics do not always distinguish clearly between them, so double counting may occur. It should also be noted that not all the public expenditure that supports the production capacity of the food and agriculture sectors or, more generally, benefits the rural population is included in official agricultural budgets.

National public spending: Several studies have shown that the level of public national spending on agriculture and rural areas has fallen during the 1990s and early 2000s. In its 2001 report, FAO noted "...that in countries with a very high incidence of undernourishment, public expenditure on agriculture does not reflect the importance of the sector in overall income or its potential contribution to the alleviation of undernourishment" (FAO, 2001a). While some of the earlier decline was the result of structural adjustment programmes and has even led to more efficient resource allocation, the main effect of low public expenditure has been inadequate provision of public services, lacking infrastructure, and hence missing incentives for investment in rural areas by farmers and other private investors.

Country panel data on national government expenditure and ODA that match the ACS data presented here over time and in cross-section are not available. Fan and Rao (2003) describe the compilation of a panel on national government expenditure for a set of 43 developing countries from Asia, Africa and Latin America for 1980 to 1998. This data set has since been expanded to 44 countries, and was updated to 2005 (data from Shenggen Fan, personal communication). It points to increasing real levels of government expenditure for the aggregate of all 44 countries over time. Figure 7.8 provides evidence of a robust positive relationship between government expenditure on agriculture and growth in ACS in these countries over the period 1980 to 2005.

International assistance: External assistance to agriculture in developing countries has declined since the late 1980s. At the country level, the relationship between agricultural ODA and ACS growth (Figure 7.9) is not as clear-cut as is that between government agricultural expenditure and ACS growth. Between 1995 and 2005, the correlation coefficient between agricultural ODA receipts and growth in ACS for 118 developing countries in the ODA database is 0.48. However, the relationship between ACS and ODA is weakened by the fact that several countries (e.g., Brazil, the Sudan, Myanmar, Turkey and the Syrian Arab

Figure 7.8
Government expenditure on agriculture per worker, and ACS growth in 44 developing countries, 1980 to 2005

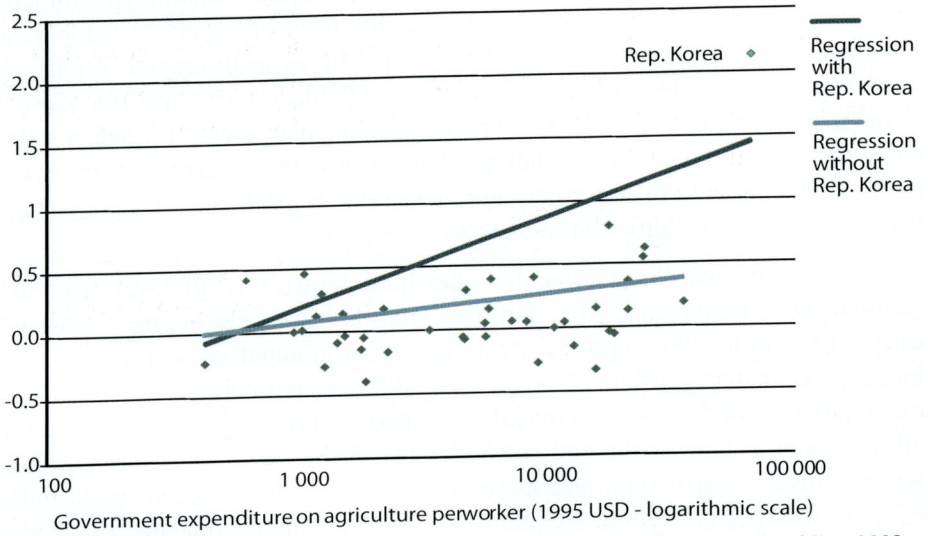

Government expenditure on agriculture perworker (1995 USD - logarithmic scale)

Sources: Authors' calculations with FAOSTAT physical inventories estimates; Fan and Rao, 2003.

Figure 7.9
Agricultural ODA and ACS growth in 118 developing countries, 1995 to 2005

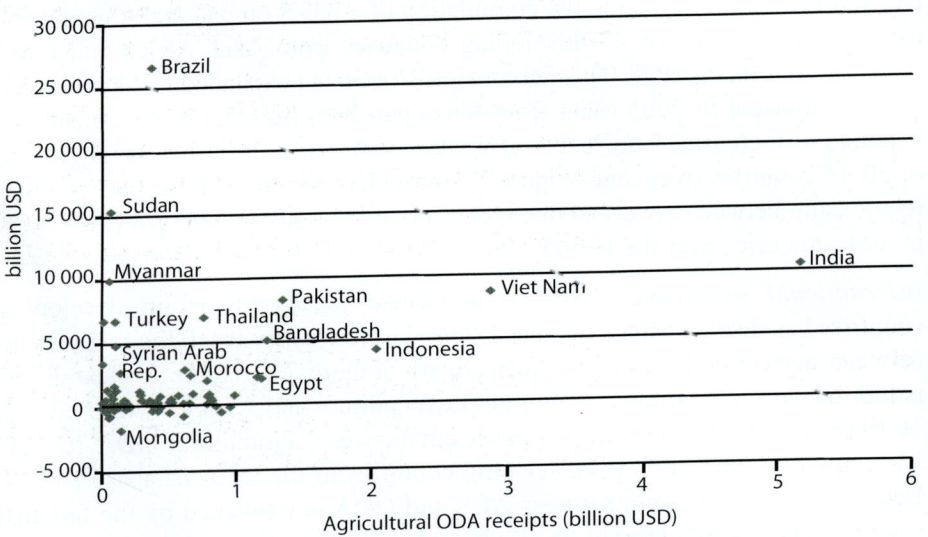

Agricultural ODA receipts (billion USD)

Source: Authors' calculations with FAOSTAT physical inventories estimates.

Republic) recorded large increases in their ACS despite receiving comparatively low amounts of agricultural ODA (Figure 7.9). (Of course, it is not surprising that a country such as Brazil does not depend on agricultural ODA.) Furthermore, if a few countries that received large amounts of ODA (e.g., India, Viet Nam, Indonesia, Pakistan, Bangladesh and Thailand) are omitted, there appears to be no significant relationship between agricultural ODA receipts and growth in national ACS for the remaining countries.

Besides direct public investment, favourable market prospects and other components of the overall investment climate – such as stability and security, regulation and taxation, finance and infrastructure, and a functioning labour market – play a decisive role in determining the rate of ACS growth. In its *World Development Report 2005*, the World Bank (2005b: xiii) observed that "more governments are recognizing that their policies and behaviours play a critical role in shaping the investment climate of their societies and they are making changes". However, the report also underlines the need for far more progress, especially in rural areas.

Agricultural capital stock and productivity

The updated estimates of ACS presented in the previous section provide a basis for generating estimates of changes in TFP in individual countries and regions. TFP analysis can cast light on the extent to which countries have translated investment in agriculture into productivity gains. The starting point for this was a study in which Rao, Coelli and Alauddin (2004) estimate TFP changes in agriculture using panel data on agricultural inputs and output in 111 countries between 1970 and 2001. Data made available by these authors were then analysed to determine whether TFP estimates change when updated and expanded estimates of ACS are used. Rao, Coelli and Alauddin (2004) used land, tractors and an aggregate of five types of livestock as capital inputs; the analysis was also able to consider four types of machinery, nine types of livestock and structures from the FAOSTAT physical inventories data.

Methods

Rao, Coelli and Alauddin (2004) employed data envelopment analysis (DEA) to estimate the technical efficiency of agriculture for each country in their data set, and to derive shadow prices for agricultural inputs and outputs. They then used the Malmquist productivity index to measure growth in TFP and to decompose this TFP growth into its two main components: technical change (shifts in the production frontier over time); and efficiency change (a production unit's ability to move closer to the production frontier). Both methods are well-established in the literature (e.g., Coelli *et al.*, 2005). The application of these methods to panel data of countries over time treats each country in its entirety as an individual production unit and assumes that all countries have access to the same technology that underlies the frontier.

These methods were applied using the FEAR package in the programming language R (Wilson, 2008). The first series of estimates presented are those reported in Rao, Coelli and Alauddin (2004), which are replicated to confirm that there are no computational discrepancies (Model I). These are followed by a series of estimates in which different aspects of the data, model and/or estimation technique are modified to produce alternative results. Modifications account for the following factors:

- The data in Rao, Coelli and Alauddin (2004) contain a minor miscoding that leads livestock output to be listed as crop output, and vice versa, for North and Central America. This is rectified in Model II.

- Zelenyuk (2006) introduces a weighted TFP estimation technique that produces consistently aggregated regional averages. He demonstrates that this technique is superior to the standard approach of calculating output-weighted aggregates of individual country TFP estimates. Model III employs Zelenyuk's method.

- The updated FAOSTAT ACS data presented earlier in this chapter begin in 1975 and extend to 2007, while Rao, Coelli and Alauddin (2004) use data from 1970 to 2001. To make subsequent comparisons possible, this chapter first presents average TFP growth rates for 1970 to 2001 that account for both the output data miscoding and the aggregation method changes (Model IVa = Model II + Model III). It then presents the results of this model for only 1975 to 2001 (Model IVb).

- Maintaining the same two-output, five-input model estimated by Rao, Coelli and Alauddin (2004), the land, tractor and livestock input data are replaced by the more comprehensive land, machinery and livestock estimates in FAOSTAT ACS data for the period 1975 to 2001 (Model V).

- In DEA estimation, the so-called "curse of dimensionality" can influence results (Daraio and Simar, 2007). Essentially, as more inputs and outputs are included in the estimation procedure, the best practice frontier becomes increasingly flexible in higher-dimensional space. This permits it to accommodate individual observations better, creating the impression that they are all close to the frontier, and distorting subsequent TFP estimates. To reduce this problem, Model V was re-estimated with the data aggregated to one output and four inputs (as opposed to two and five). The reduced 1-by-4 dimension was chosen based on recommendations from Park, Simar and Weiner (2000) and Daraio and Simar (2007: 153–154). Results are presented for 1975 to 2001 (Model VIa) and for 1975 to 2007 (Model VIb), to take advantage of the longer time period covered by the FAOSTAT ACS data.

- As an alternative means of dealing with dimensionality, partial or so-called robust frontiers can be estimated based on the m-order expected maximum output frontier proposed by Cazals, Florens and Simar (2002). The basic idea of this method is to estimate a more "taut" frontier, which does not envelop all the data points by repeated local re-sampling from the set of available observations. The advantages of the m-order method are summarized by Daraio and Simar (2007), and include robustness to outliers in the data, and less susceptibility to the curse of dimensionality. This method is employed in Model VII, which uses two outputs and all six available inputs (land and labour as in Rao, Coelli and Alauddin, 2004, and the four capital inputs in FAOSTAT ACS data).

In all estimations, labour, land and fertilizer (as a proxy for working capital) are included as inputs, along with the various measures and aggregations of capital input as outlined.

Results

A comparison of the different estimates of TFP change for the regions and the world reveals a number of results that are robust to the data, model and estimation technique alternatives outlined in Table 7.11. At the global level, estimates of annual TFP growth are quite consistent across models, ranging from 1.2 to 1.8 percent per year. Comparison of Models IVa and IVb reveals that omitting estimates for the first half of the 1970s leads to increased estimates of TFP growth in all regions except Latin America. The increase is especially large for China. This suggests that the first years of the 1970s were characterized by below-average TFP growth in most of the world. Comparison of Models VIa and VIb reveals that increasing the coverage to include 2002 to 2007 has no major impact on results.

North America and Oceania, Europe, the transition countries and China exhibit above-average rates of TFP growth that are relatively robust to the estimation method used. The Near East and North Africa, sub-Saharan Africa, Latin America and Asia (without China) exhibit below-average levels that are less robust, with estimates ranging from 0.3 to 1.8 percent for sub-Saharan Africa, and from 0.2 to 1.5 percent for Asia (without China). This is probably because the quality of the underlying data for the countries in these regions is comparatively low; DEA estimation is known to be highly sensitive to aggregation and data quality (Fuglie, 2008: 433; Daraio and Simar, 2007). This might also explain why, when Models I and II are compared, correcting the miscoded North and Central American output data has little impact on TFP growth rates for most regions, but a large impact on sub-Saharan Africa, where estimated TFP growth rates increase from 0.3 to 1.5 percent.

With this evidence of sensitivity in mind, the m-order estimates presented in Table 7.11 might be considered the most reliable, as the m-order method is far less sensitive to outliers. Because of this, the decomposition of TFP growth rates into efficiency change and technical change components in Table 7.12 is based on the m-order estimates. The results of this decomposition reveal that efficiency improvements have made relatively large contributions to TFP growth in the Near East and North Africa, China and the transition countries. In all other regions, TFP change has been largely determined by technical change. High rates of efficiency improvement in the transition countries and China are expected, as the restrictions and inefficiencies of central planning were removed over the study period.

Comparing TFP change and ACS accumulation between 1975 and 2007 reveals some interesting patterns. The industrialized countries (Europe, North America, Oceania and the transition countries) are characterized by low rates of ACS growth and relatively high rates of TFP growth (Figure 7.10). Their TFP growth over the study period is largely due to technical change (except in the transition countries, where efficiency improvements have also played a role). This is presumably a reflection of increases in input quality that are not captured by the FAOSTAT ACS estimates; TFP estimates based on the national accounts estimates of capital inputs would likely point to lower rates of TFP growth. The developing countries are characterized by much higher rates of ACS growth, but lower rates of TFP growth in Latin America and, especially, sub-Saharan Africa. China stands out as having the highest rates of both ACS and TFP growth. These results underline that high ACS growth does not necessarily lead to higher TFP growth.

All the TFP estimates in Tables 7.11 and 7.12 are based on non-parametric techniques and data on capital inputs derived from FAOSTAT. Ongoing work is looking into parametric estimation using stochastic frontier methods on the same data, and non-parametric and parametric estimates using national accounts-based capital input data. Of course, any use of national accounts-based data will have to deal with the issue of selection bias identified in the subsection on Comparing the national accounts and FAOSTAT approaches.

Table 7.11
TFP growth according to different specifications and estimation techniques

Model	Time period[a]	Specification (n = outputs; m inputs)	Near East and North Africa	Sub-Saharan Africa	North America and Pacific	Latin America	Asia without China	China	Europe	Transition	World
			Average annual rate of TFP growth (%)								
I = Rao, Coelli and Alauddin, 2004: 2 outputs (crop and livestock); 5 inputs (labour, fertilizer, land, tractors and livestock)	1970–2001	DEA n = 2, m = 5	0.6	0.3	2.2	0.7	0.3	3.0	1.9	1.4	1.5
II = I + corrected output data[b]	1970–2001	DEA n = 2, m = 5	0.8	1.5	2.3	0.7	0.5	2.8	2.1	1.9	1.4
III = I + corrected aggregation[c]	1970–2001	DEA n = 2, m = 5	0.7	0.6	1.9	0.9	0.2	2.5	1.9	2.2	1.2
IVa = II + III	1970–2001	DEA n = 2, m = 5	0.6	0.4	2.3	0.4	0.2	2.5	2.1	1.9	1.4
IVb = IVa beginning in 1975	1975–2001	DEA n = 2, m = 5	0.8	1.2	2.5	0.3	0.4	4.6	2.3	2.4	1.8
V = IVb + new FAOSTAT estimates for land, machinery (in lieu of tractors) and livestock inputs	1975–2001	DEA n = 2, m = 5	1.6	1.5	2.5	1.2	0.8	2.7	1.9	2.1	1.6
VIa = V with 1 aggregate output[d] and 4 inputs (labour, fertilizer, land and capital = structures, machinery and livestock)	1975–2001	DEA n = 1, m = 4	0.9	1.6	2.3	0.2	1.3	2.2	1.8	1.8	1.5
VIb = VIa extended to 2007	1975–2007	DEA n = 1, m = 4 m-order	0.9	1.8	2.2	0.6	1.5	2.1	1.5	1.8	1.5
VII = VI with 2 outputs (crop and livestock) and 6 inputs (labour, fertilizer, land, structures, machinery and livestock)	1975–2007	n = 2, m = 6	1.5	0.9	2.0	1.0	1.4	2.1	1.4	1.7	1.7

[a] All results exclude 1992 and 1993, as these produce highly variable estimates owing to the impact that the break-up of the Soviet Union, the Socialist Federal Republic of Yugoslavia and Czechoslovakia had on input and output statistics.

[b] In the original Rao, Coelli and Alauddin (2004) data set (made available by the authors), crop output values are listed as livestock output values, and vice versa, for North America and Central American countries. This is corrected here.

[c] Zelenyuk (2006) introduces a weighted TFP estimation technique that produces consistently aggregated regional averages, and demonstrates that this is superior to the standard approach of calculating output-weighted aggregates of individual country TFP estimates.

[d] Output series are aggregated using 1999/2001 average international prices.

Source: Authors' calculations.

Table 7.12
Estimated changes in TFP and its components by region, 1975 to 2007 (percentages)

Region	Efficiency change	Technical change	TFP change
Near East and North Africa	1.4	0.1	1.5
Sub-Saharan Africa	0.3	0.6	0.9
North America and Pacific	-0.7	2.7	2.0
Latin America	0.3	0.7	1.0
Asia, without China	-0.9	2.3	1.4
China	0.9	1.3	2.1
Europe	0.3	1.1	1.4
Transition countries	0.7	1.0	1.7
World	0.7	1.0	1.7

Results based on m-order estimates in Table 7.11. Results exclude 1992 and 1993, as these produce highly variable estimates owing to the impact that the break-up of the Soviet Union, the Socialist Federal Republic of Yugoslavia and Czechoslovakia had on input and output statistics.

Source: Authors' calculations.

Figure 7.10
Annual rates of agricultural capital stock growth and TFP growth by region, 1975 to 2001

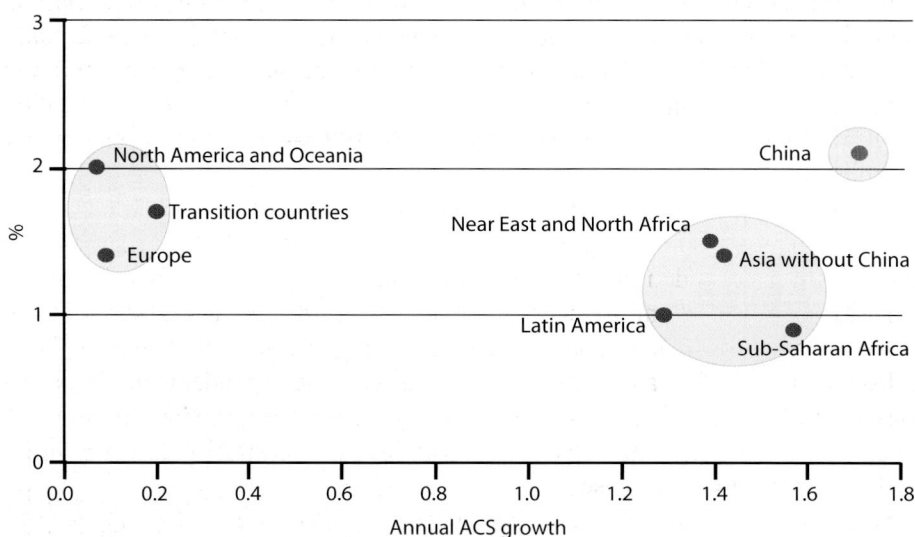

Sources: Authors' calculations with FAOSTAT physical inventories estimates; authors' TFP estimates based on data in Rao, Coelli and Alauddin, 2004, and FAOSTAT physical inventories estimates.

Factors that influence TFP growth

In a series of regressions, Rao, Coelli and Alauddin (2004) study factors that explain TFP levels across countries, such as land quality, irrigation, government expenditure, literacy rate and trade openness. They find that results are sensitive to the period that is studied and to whether or not the transition economies are included in the analysis, the latter probably being due to questionable data for these countries and to their unique circumstances. Two robust results are that both reducing illiteracy and reducing the incidence of malaria have positive impacts on TFP in agriculture. Foreign direct investment (FDI) as a share of GDP is found to have a uniformly positive impact on agricultural TFP, satisfying the expectation that FDI will be associated with improved technologies and expertise in implementing them. A surprising result of the analysis presented by Rao, Coelli and Alauddin (2004) is that both gross domestic investment as a share of GDP, and government consumption as a share of GDP have negative effects on agricultural TFP. The authors suggest that the latter result may be due to urban biases in government expenditure, so that much government consumption may actually be discriminating against agriculture in many countries. For this chapter, the issue of government expenditure and TFP was revisited using the Fan and Rao (2003) data on government expenditure on agriculture in developing countries. Although these data on government expenditure are available for only 44 countries, they have the advantage of measuring government expenditure specifically on agriculture. Hence, unlike the general government consumption data employed by Rao, Coelli and Alauddin (2004), they should be free of any urban bias.

The regression analysis is based on panel data for 37 of the 44 developing countries in the Fan and Rao (2003) database from 1980 to 2005. A lack of data for some of the independent variables described in the following leads to complete omission of seven countries in the Fan and Rao (2003) data, and omission of individual observations for some of the remaining 37 countries. The result is an unbalanced panel with a total of 761 observations. The dependent variable is the logarithm of TFP levels for country i in year j, calculated using base period TFP levels (relative to the United States of America) extrapolated with the m-order TFP estimates (for details, see Rao, Coelli and Alauddin, 2004: 29). Two base periods are used to account for the transition countries' entrance into the TFP estimation.

Drawing on Rao, Coelli and Alauddin (2004) and theoretical considerations, the following series of possible covariates is identified (descriptive statistics are given in Table 7.13):

- A dummy variable that equals 0 prior to 1994 and 1 thereafter is added to account for the transition countries' entry into the TFP estimation in 1994. Although there are no transition countries among the 37 countries included in this regression analysis, these countries affect TFP levels for all countries when they enter the TFP estimation.

- The rural population as a share of total population is included as a measure of labour abundance in agriculture, where a high share might point to surplus labour with a very low marginal product (World Bank World Development Indicators).

- The ratio of imports plus exports to GDP is a measure of economic openness that can capture access to foreign technology as well as the overall policy climate in a country (World Bank World Development Indicators).

- The share of irrigated land in total agricultural land is a proxy for land quality (FAOSTAT).

- An indicator of institutional quality that combines the quality of bureaucracy, the rule of law and the lack of corruption is added to measure the quality of government. This variable is standardized to ease interpretation (PRS Group's IRIS dataset).[4]

- A political regime index, defined as a country's degree of democracy less its degree of autocracy, is included to capture governance. This index is also standardized (POLITY IV Project).[5]

- Net FDI flows as a share of GDP are included to capture inflows of technology and expertise that might be expected to boost TFP (World Bank World Development Indicators).

- Gross fixed capital formation as a share of GDP might capture technology that is embodied in fixed capital (World Bank World Development Indicators).

- The logarithm of government expenditure on agriculture is included, in the expectation that higher levels of expenditure will be associated with improved availability of productivity-enhancing infrastructure, research and education (Fan and Rao, 2003).

---

4. www.prsgroup.com.
5. www.systemicpeace.org/inscr/inscr.htm.

Table 7.13

Results of panel regression analysis to explain differences in national TFP growth (dependent variable is *log*[TFP level] in country *i* and year *j*)

	Regression results		Descriptive statistics			
Covariate	Coefficient	Standard error	Mean	Minimum	Maximum	Standard error
Dummy1994–2005	-0.308[c]	0.023	0.47	0.00	1.00	0.50
Rural population as % of total	-0.031[c]	0.003	54.7	6.60	93.6	23.3
Exports + imports as % of GDP	0.002[a]	0.001	55.2	1.53	228.9	30.2
Share of irrigated land in total agricultural land	0.111	0.292	0.10	0.00	1.00	0.19
Institutional quality index	-0.109[c]	0.031	0.00	-2.61	2.31	1.00
Political regime index	0.032	0.040	0.00	-1.55	1.30	1.00
Institutional quality[a] political regime	-0.020[a]	0.011	0.16	-2.26	2.87	0.94
FDI inflows as % of GDP	-0.016[c]	0.006	1.58	-2.76	12.2	1.83
FDI[a] institutional quality	0.032[c]	0.006	0.53	-6.91	20.9	2.06
FDI[a] political regime	0.020[c]	0.005	0.49	10,3	14.1	2.31
Fixed capital formation as % of GDP	-0.007[c]	0.002	20.3	3.53	43.6	6.23
Fixed capital[a] institutional quality	0.003[a]	0.002	2.23	58.8	86.4	21.8
Fixed capital[a] political regime	-0.004[h]	0.002	0.21	52.2	47.3	20.8
Log [government expenditure on agriculture]	0.034[b]	0.014	1.01	-2.20	3.77	1.09
Period	1981-2005					
Number of observations	761					
R²	0.32					

Significance levels: [a] 10 percent; [b] 5 percent; [c] 1 percent.

Several interaction terms involving the institutional quality and political regime indicators are included in the final specification, which also includes country fixed effects that are found to be jointly significant at the 1 percent level. The regression model was estimated using the *plm* package in R (Croissant and Millo, 2008) and results are presented in Table 7.13.

The overall fit of the regression (R^2 = 32 percent) is good for a panel estimate with annual country data. Most of the estimated coefficients are significant and have the expected signs. A 10 percent increase in the share of the rural population reduces agricultural TFP by 0.31 percent, all other things being equal, while a 10 percent increase in the share of trade in GDP increases agricultural TFP by 0.02 percent. The share of irrigated land has the expected positive impact on TFP, but this effect is not significant. This analysis finds a surprising negative relationship between institutional quality and TFP levels, as Rao, Coelli and Alauddin (2004) do. The impact of the political regime index (increasing democracy) on agricultural TFP is positive but not significant, while the interaction between institutional quality and the political regime has a negative impact.

FDI alone has a significant negative impact on agricultural TFP, but the interactions between FDI and the institutional quality and political regime variables are significantly positive and larger in magnitude. Hence, the impact of a 10 percent increase in net FDI inflows as a share of GDP is a 0.16 percent reduction in TFP, at mean values of the institutional quality and political regime covariates. However, for a country that is 1 standard deviation above the mean in terms of its institutional quality and political regime this turns into a net 0.36 percent increase. This result suggests that the institutional and governance environment within which FDI takes place is of crucial importance. An appropriate environment can ensure that the potential productivity-enhancing impacts of FDI are realized, while under conditions of poor governance and institutional quality, FDI will be more short-term and perhaps more focused on rent extraction than on establishing capacities for adding value. Fixed capital formation has a weak negative impact on agricultural TFP, which is only partially compensated for by the interaction between fixed capital formation and institutional quality, and somewhat strengthened by its interaction with the political regime variable. The overall negative effect of capital formation on agricultural TFP may be partly due to an urban/rural bias in the economy-wide measure of capital formation employed. However, it should be recalled that no positive relationship between ACS growth and TFP growth for regional aggregates was found (Figure 7.10).

The coefficient on the logarithm of government expenditure on agriculture is positive and significant. This coefficient indicates that a 10 percent increase in government expenditure on agriculture, all other things being equal, will lead to a 0.34 percent increase in the country's agricultural TFP. This underscores the importance of national government expenditure on agriculture as a means of not only increasing rates of ACS growth as identified in Figure 7.8, but also of contributing to increased productivity through technical change and more efficient use of inputs. Of course, even a specific measure of government expenditure on agriculture such as that employed here does not take the composition of this expenditure into account. In a study of ten Latin American countries, López (2005: 18) presents econometric evidence that "while government expenditures have a positive and highly significant effect on agriculture per capita income, the structure or composition of such expenditures is quantitatively much more important and also of great statistical significance. [...] According to the estimates, a reallocation of just 10 percent of the subsidy expenditures to supplying public goods instead may cause an increase in per capita agriculture income of about

2.3 percent." Hence, a variable that isolates the public good aspect of government expenditures on agriculture, if available, could be expected to have an even higher estimated impact on agricultural TFP growth than that measured here.

Conclusions

The fixed capital stock (ACS) in primary agriculture has been growing steadily at the global level over the last three decades, although at declining rates for most of this period. Using a volume approach with country-specific constant values per asset to measure ACS, the average annual rate of worldwide ACS growth fell from 1.1 percent for the 1975 to 1990 period, to 0.5 percent for 1991 to 2007. Reductions were recorded in both developed and developing countries, although the rates of ACS growth were considerably stronger in developing than developed countries, in both sub-periods. In the latter group, some growth rates have been negative since the mid-1990s. Recently, this trend seems to have been reversed. Since reaching a point close to stagnation in the mid-1990s, global ACS growth rates have started to increase gradually, reaching 0.5 percent per year in 2005 to 2007. The reasons for this slight acceleration of capital growth need to be examined further, including whether the new demand for bioenergy has played a role. If ACS growth rates continue to improve, this may signal improving prospects for the world's aggregate capacity to meet future demand. As the data for 2007 are based on projections, and data for 2008/2009 are not yet available, it is not possible to determine the impact that the food price crisis of 2007/2008 had on rates of ACS accumulation worldwide or in developing as opposed to developed countries.

A shift in the relative shares of capital formation among regions and country groups appears to be taking place. The gap between higher rates of ACS growth in developing and lower rates in developed countries is closing. Whereas ACS shrinkage in developed countries has slowed, rates of ACS growth in developing countries have remained positive but continued to fall. As future demand growth is expected mainly in the developing countries, this shift could lead to increasing supply bottlenecks in import-dependent developing countries, unless action is taken to increase investments in these countries.

Annual rates of growth in the stock of improved agricultural land have been declining at the global level. As a consequence, the share of land, including land equipped for irrigation, in total ACS at the global level (currently about 50 percent) is gradually declining. This may in part reflect reduced willingness to invest in improving the productivity of the existing stock of land, which would be cause for concern, especially in many marginal areas where the ongoing depletion of natural capital through declining soil fertility is not accounted for.

ACS has grown the least in countries with the highest prevalence and depth of hunger. The majority of poor and hungry people live in rural areas and depend directly or indirectly on agriculture for their livelihood. Therefore, increasing the ACS per person active in agriculture has been an important factor in reducing undernourishment. However, in several of the least developed countries, particularly in sub-Saharan Africa and South Asia, the growth of the population active in agriculture has outstripped the rate of ACS growth. This development is particularly worrying because it severely limits these countries' ability to increase labour productivity in rural areas, and hence to reduce poverty and undernourishment. This result is obtained irrespective of the method used to estimate capital stock. By contrast, with few exceptions, the countries making the most progress towards the World Food Summit target of halving the number of undernourished by 2015 have realized relatively high rates of growth in ACS per worker in agriculture.

Government expenditure on agriculture is correlated with capital formation in a sample of developing countries. This correlation between national government expenditure on agriculture and growth in national ACS confirms the decisive role of public expenditure in creating an enabling environment in terms of the infrastructure and sustainable access to natural resources that provide adequate incentives for the private sector, particularly farmers, to invest in productive assets. This should be a strong signal for governments in developing countries to change priorities in budget allocations, to avoid or at least reduce any existing discrimination against agriculture. Public expenditure on agriculture can be an important ingredient in an investment climate conducive to agricultural development and the reduction of hunger.

Between 1975 and 2007, annual TFP growth in world agriculture was roughly 1.7 percent. This average masks important differences among regions, ranging from 2.1 percent in China and 1.7 percent in transition countries, to 1.4 percent in the rest of Asia, 1 percent in Latin America, and 0.9 percent in sub-Saharan Africa. These differences among regions point to substantial scope for further productivity growth. The breakdown of TFP growth into efficiency gains and technical change also varies widely among regions. Efficiency gains have contributed relatively little to overall total TFP in developing countries, but have played a significant role in transition countries. This has implications for the entry points of productivity-enhancing policies in developing countries.

Government expenditure on agriculture has a significant positive impact on TFP in a sample of developing countries. All other things being equal, increasing government expenditure on agriculture by 10 percent leads to a 0.34 percent increase in the country's agricultural TFP. FDI is also found to have a positive impact on productivity growth, but only when combined with an institutional environment characterized by efficient bureaucracy, a lack of corruption, and democratic political structures. This suggests that the investment climate in a country – including its institutional and governance structures – has an important influence on the type of FDI that it can attract and on the impact of this investment on the agricultural economy.

The estimates of ACS levels and growth presented in this chapter are based on two different approaches, which differ in many respects. Each of these approaches is characterized by important strengths and weaknesses, and the approaches do not always produce the same results. International organizations such as FAO should engage in a concerted and sustained effort to refine and reconcile estimates of fixed capital formation, including upstream and downstream sectors and rural infrastructure in developing countries. Efforts should also be made to combine the advantages of existing methodologies, and to improve the collection and processing of consistent data.

References

Alexandratos, N., ed. 1995. *World agriculture: Towards 2010, an FAO study*. Chichester, UK, J. Wiley and Sons, and Rome, FAO.

Anriquez, G. & Daidone, S. 2008. An expanded cross-country database for agricultural investment and capital. Presentation to FAO-ESA, Rome.

Cazals, C., Florens, J.P. & Simar, L. 2002. Nonparametric frontier estimation: A robust approach. *Journal of Econometrics*, 106: 1–25.

Coelli, T.J., Batteses, G.E., O'Donnell, C.J. & Rao, D.S.P. 2005. *Introduction to efficiency and productivity analysis*. Berlin, Springer.

Crego, A., Larson, D., Butzer, R. & Mundlak, Y. 1998. *A new database on investment and capital for agriculture and manufacturing*. Policy Research Working Paper Series No. 2013. Washington, DC, World Bank.

Croissant Y. & Millo, G. 2008. Panel data econometrics in R. *Journal of Statistical Software*, 27(2).

Daraio, C. & Simar, L. 2007. *Advanced robust and nonparametric methods in efficiency analysis: Methodology and applications*. New York, Springer.

Dewbre, J. & De Battisti, A.D. 2008. *Agricultural progress in Cameroon, Mali and Ghana, why it happened and how to sustain it*. Paris, OECD.

Fan, S. & Rao, N. 2003. *Public spending in developing countries: Trends, determination and impact*. EPTD Discussion Paper No. 99. Washington, DC, IFPRI.

FAO. 1996. *World Food Summit – Rome Declaration on Food Security*. Rome.

FAO. 2001a. *Mobilizing resources to fight hunger.* 27th Session of the Committee on World Food Security. Rome.

FAO. 2001b. *Note on the revision of the estimates of capital stock,* prepared by the FAO Statistics Division. Rome.

FAO. 2002. *The World Food Summit: Five Years Later. Mobilizing the political will and resources to banish world hunger.* Technical Background Documents. Rome.

FAO. 2003. *Anti-Hunger Programme. A twin-track approach to hunger reduction: priorities for national and international action.* Rome.

FAO. 2006. *Investment as a means to agricultural and rural development in Africa: A study on the investment needs for agriculture and rural development,* prepared by M. Barre. Rome.

FAO. 2008. *The State of Food Insecurity in the World. 2008: High food prices and food security – threats and opportunities.* Rome. www.fao.org/docrep/011/i0291e/i0291e00.htm.

Fuglie, K. 2008. Is a slowdown in agricultural productivity growth contributing to the rise in commodity prices? *Agricultural Economics,* 39(3, suppl.): 431–441.

Jensen, R. 2007. The digital provide: Information (technology), market performance and welfare in the south Indian fisheries sector. *Quarterly Journal of Economics,* 132(3): 879–924.

López, R. 2005. *Why governments should stop non-social subsidies: Measuring their consequences for rural Latin America.* College Park, Maryland, USA, University of Maryland, Department of Agricultural and Resource Economics.

Park, B.U., Simar, L. & Weiner, C. 2000. The FDH estimator for productivity efficiency scores: Asymptotic properties. *Econometric Theory,* 16: 855–877.

Rao, D.S.P., Coelli, T.J. & Alauddin, M. 2004. *Agricultural productivity growth, employment and poverty in developing countries, 1970–2000.* ILO Employment Strategy Paper No. 2004/9. Geneva, International Labour Organization (ILO).

Wilson, P.W. 2008. FEAR 1.0: A package for frontier efficiency analysis with R. *Socio-Economic Planning Sciences,* 42: 247–254.

World Bank. 2005a. *Beyond GDP, World Bank releases new measure of wealth.* http://web.worldbank.org/wbsite/external/news/0,,contentmdk:20648103~page pk:34370~pipk:34424~thesitepk:4607,00.html.

World Bank. 2005b. *World Development Report 2005: A better investment climate for everyone.* Washington, DC.

World Bank. 2008. *World Development Report 2008: Agriculture for development.* Washington, DC.

Zelenyuk, V. 2006. Aggregation of Malmquist productivity indexes. *European Journal of Operational Research,* 174(2): 1076–1086.

CHAPTER 8

CAPITAL REQUIREMENTS FOR AGRICULTURE IN DEVELOPING COUNTRIES TO 2050

Josef Schmidhuber
Jelle Bruinsma
Gerold Boedeker

This chapter reports on ongoing work at FAO to estimate investment requirements in developing countries' agriculture. Estimates cover most capital items, but do not single out areas for public involvement of either domestic or foreign funding sources. Neither has any attempt been made to gauge the incremental investment needs required to attain certain development goals, such as Millennium Development Goal (MDG) 1 or the target set by the World Food Summit. This means that important investment areas such as agricultural research or rural infrastructure are excluded, but they are covered in other work (Schmidhuber and Bruinsma, 2011). An item of major concern to public investment – ensuring access to food for the most needy, such as through social safety nets[1] – is also not dealt with here.

The estimates presented in this chapter embody a broad range of capital items needed to achieve the 2030 and 2050 crop and livestock production levels in developing countries that are foreseen in the baseline outlook of the latest FAO perspective study (FAO, 2006b). The majority of these capital items relate to primary agriculture. In addition, a number of the activities covered relate to downstream industries of primary agriculture, notably various forms of processing, storage and marketing.

Total investment requirements are made up of net additions to and replacement of obsolete capital stocks. Traditionally, the lion's share of capital needs has been covered by private farmers and by entrepreneurs in the related upstream and downstream industries (including capital outlays in non-monetized forms). Some capital items, such as irrigation development, rural infrastructure and agricultural research, require public intervention, but this chapter makes no effort to measure

---

1. Accounting for more than a fifth of the incremental annual public investment estimated in the FAO (2003) Anti-Hunger Programme.

317

the needed or desired level of public sector engagement. This can vary widely across capital items and countries, and any quantitative assessment would need to start from a detailed and disaggregated basis.

Methodology and measurement

Imputed versus actual

The basic goal of this assessment was to gauge the amount of capital that will be required to produce the total amounts of crops and livestock products projected in FAO's long-term outlook to 2030 and 2050, such as the hectares of land to be developed, irrigated and put under permanent crops; the numbers of tractors, combines, implements or hand tools to be acquired; and the increases in livestock herds, sheds, etc. All investments are imputed estimates, and not necessarily actual investments. Capital stocks too are imputed and not necessarily actual, as are capital stocks, i.e., net investments and depreciation.

Box 8.1 - Past FAO estimates of investment requirements

The 1981 publication *Agriculture: towards 2000* (FAO, 1981) estimated the average annual gross investment for the 20-year period 1980 to 2000 for 90 developing countries (excluding China) at USD 69 billion in 1975 dollars: USD 47 billion for investment in primary agriculture, about a third of which is for investment in replacement; and USD 22 billion for investment in supporting capital stock. Separate estimates are given for (net) investment in forestry and fisheries. These investment estimates refer to total investment required – the sum of private and public investment.

The 1988 study *World agriculture: towards 2000* (Alexandratos, 1988) is an update of the 1981 study and follows the same methodology. For 93 developing countries (excluding China) the estimate of average annual gross investment for the 17-year period 1982/1984 to 2000 amounts to USD 88 billion in 1980 dollars. Investments in primary agriculture are estimated at USD 50 billion (nearly 60 percent for investment in replacement), and investment in supporting capital stock at USD 38 billion. No estimates are given for investment in forestry and fisheries.

The investment estimates in the technical background document for the 1996 World Food Summit (FAO, 1996) are based on the study by Alexandratos (1995). These estimates refer to the group of 93 developing countries and the investment needed to achieve the production projections in this latter publication (i.e., the World Food Summit target is not considered and 637 million people are left undernourished in 2010). The estimate for average annual gross investment for 1993 to 2013, in 1993 dollars, is USD 129 billion, of which USD 86 billion is in primary agriculture (USD 61 billion for replacement) and USD 43 billion in support (or post-production) investment. To this are added USD 37 billion of investments in public support services (mainly technology generation and transfer) and rural infrastructure, two categories that were not covered in earlier studies. The total then amounts to USD 166 billion, of which about three-quarters (USD 125 billion) is private and one-quarter (USD 41 billion) public investment.

The next FAO exercise, giving investment estimates of a slightly different nature, was FAO, 1999 (which is also reported in FAO, 2001). These estimates are an update of the 1995 estimates (for developing countries only), but refer to what is needed to reach the World Food Summit target of halving the number of undernourished people by 2015. They estimate an average annual gross investment for 2000 to 2015, in 1995 dollars, of USD 140 billion, of which USD 93 billion is in primary agriculture (USD 66 billion for replacement) and USD 47 billion in support (or post-production) investment. To this are added USD 40 billion of investments in public support services and rural infrastructure. The total amounts to USD 180 billion.

The latest FAO publication giving investment estimates (FAO, 2003) refers to what is needed to reach the World Food Summit target in 2015. These estimates cover only investment incremental to expected future public investment. Average annual investment for 2003 to 2015, in 2002 dollars, is USD 23.8 billion, of which USD 2.3 billion is for productivity improvements, USD 7.4 billion for natural resource development, USD 7.8 billion for rural infrastructure, USD 1.1 billion for knowledge generation, and USD 5.2 billion for ensuring access to food.

These imputed investments and capital stocks can differ from actual investments and capital stocks for a number of reasons. For instance, if farmers work with excessively depreciated capital stocks (old tractors, tillers, threshers, sheds, etc.), actual capital stocks would be lower than the imputed ones, and vice versa. Conversely, some investments may not entirely or always translate into monetary expenditures. For instance, when a farmer builds a storage facility for cereal crops or a shed for grazing animals, these activities may not be fully reflected in the actual value of the capital stocks; they are, however, part of the imputed capital as they absorb resources with positive opportunity costs and reflect a shift away from consumption into investment.

As a consequence, the estimated investment numbers and capital stocks may not always correspond to those from other sources, such as national accounts. Although this means that deviations from actual capital stocks are unavoidable in the short run, imputed and actual capital stocks and investment requirements should converge in the longer run, at the latest after one full depreciation period of the item with the longest life span. The outlook to 2050 should thus be sufficiently long to ensure convergence. At any rate, the advantage of the calculation of imputed capital stocks is that the results are comparable across countries and over time.

Investment areas and unit costs

To derive capital needs from production projections, changes in agricultural outputs are linked to 26 different capital items. For each capital item, specific unit costs and a specific lifetime, and thus depreciation period, are chosen. The imputed values are obtained by multiplying the physical quantities (hectares, numbers, etc. in the base year and in 2030 and 2050) with an average unit cost expressed in constant 2009 United States dollars. Although the calculations are undertaken on the basis of 93 individual developing countries, specificity for unit costs and depreciation periods is limited to regional averages. Of the 26 capital items, 14 relate to primary agriculture (including some non-conventional ones, such as establishment of permanent crops, herd increases and working capital) and 12 to the agricultural downstream sector (see Box 8.2 for a list of the capital items).

Investment in agricultural downstream activities covers storage, processing and marketing of agricultural products. These are included for the sake of completeness, although they may not always be entirely attributable to agriculture and agricultural development. Investments related to manufacturing and distribution of agricultural inputs such as fertilizer are not included, and expenditures on agricultural research could not be estimated as part of the investment requirements. For all investment items – in both the primary and the downstream sectors – unit costs are identified. Obviously, the absolute levels of investment requirements are contingent on factors such as the assumed unit costs, the capital (input) absorbed per unit of agricultural activity, or the assumed life span of a capital item.[2]

Depreciation and gross investment

Additions to capital stocks between the base year (2005/2007) and 2030 and 2050 amount to the cumulative net investment requirements over the projection period. Requirements for replacement investment are then derived for the capital goods that must be replaced periodically. For each capital item, a specific lifetime is identified. For example, permanent crops are assumed to have a life span of 25 years, and tractors one of 15 years. For many capital items, replacement investments exceed net investments. Estimates for replacement investment are added to the net requirements, to obtain estimates of gross investment (see Box 8.3 for a summary explanation).

Country coverage

Capital stock and investment calculations are performed for the 93 developing countries covered in the FAO 2006 study (see list of countries in Annex 8.1; note that Central Asian countries in transition are not included).

---

2. Investments in physical units are generally more robust than those in monetary terms, as it is difficult to assemble appropriate unit value costs for the various investment items.

Box 8.2 - Capital items included

Crop production:
Development of arable land under crops
Soil and water conservation
Flood control
Expansion and improvement of irrigation
Establishment of permanent crops: citrus, other fruits, oil-palm, coconuts, cocoa, coffee, tea and rubber
Mechanization: tractors and equipment
Other power sources and equipment: increase in number of draft animals, equipment for draft animals, hand tools
Working capital: 50 percent of the increase in the cost of fertilizer and seed

Livestock production:
Increase in livestock numbers: cattle and buffaloes, sheep and goats, pigs, poultry
Housing and equipment for commercial production of pigs and poultry
Development of grazing land

Downstream support services:
Investment in milk production and processing
Investment in meat production and processing
Dry storage: cereals, pulses, oilseeds, cocoa, coffee, tea, tobacco and sugar
Cold storage: bananas, fruits and vegetables, livestock products
Rural marketing facilities
Assembly and wholesale markets for fruits and vegetables
Milling of cereals
Processing of oilseeds, sugar crops, fruits and vegetables
Ginning of seed cotton
Other processing

Endogeneity and technology shifts

The projections of future investment needs are linked to and derived from the projections of 40 individual agricultural production activities, assuming certain technologies and/or complete technology packages (frontiers). Over an outlook horizon of more than 40 years, investment requirements will not be defined by only a given, current state of technology, but will encompass shifts to new frontiers. Depending on factors such as the farm size or opportunity costs of farm labour, farmers will shift to new technology levels. Although important, these shifts have not been explicitly taken into account. Instead, links have been established indirectly by associating output levels (e.g., crop yields) with a certain package of input requirements; in many cases, this is done in a step-wise linear manner that is meant to emulate the shifts in technology (for a description of this approach, see Bruinsma *et al.*, 1983). To make assumptions more transparent, and these technology shifts more explicit, future revisions will therefore attempt to

321

include such frontier shifts directly, with links to changes in the overall level of development and/or farm size.

Box 8.3 - Derivation of investment requirement estimates

The projections to 2050 cover 40 agricultural production activities (34 relating to crop production and six to livestock production) in 93 developing countries. Each activity draws on certain amounts of current inputs and capital stock services.

For each of the 26 capital items distinguished, the value of capital stock CS is calculated for each year covered in the model (t = 2005/2007, 2010, 2015, 2030 and 2050), multiplying the physical quantity Q (hectares, numbers, etc.) with an average unit cost P, expressed in 2009 United States dollars.

For each capital item, the net investment in any year is defined as the net increase in the value of capital stock over that year, or as the growth of capital stock g times capital stock CS at the beginning of the year. The growth rate is estimated as the annual growth of capital stock over the period preceding the year in question (except for the base year):

$$I_t^n = g_t \cdot CS_t \qquad (1)$$

Replacement investment in any year t is equal to the gross annual investment of L years earlier, where L is the economic life of the capital good in question. Gross annual investment is defined as the sum of net annual investment and replacement investment in the same year:

$$I_t^g = I_t^n + I_{t-L}^g \qquad (2)$$

Equation (2) can be approximated as:

$$I_t^g = \frac{I_t^n}{1 \ (1+g_t)^{-L}} \qquad (3)$$

Cumulative net investment over any of the periods distinguished in the model (2005/2007 to 2010, 2010 to 2015, 2015 to 2030, and 2030 to 2050) is defined (and calculated) as the net increase in capital stock over that period:

$$CI_t^n = CS_t - CS_{t-1} \qquad (4)$$

Cumulative gross investment is defined (and calculated) in a manner similar to annual gross investment:

$$CI_t^g = \frac{CI_t^n}{1-(1+g_t)^{-L}} \qquad (5)$$

Total annual and cumulative net and gross investments are simply derived by adding up the 26 capital items.

Public and/or private

No distinction is made regarding the potential source of the required capital. The amounts therefore include all potential sources: private and public, and of both foreign and domestic provenance. The way capital stocks are currently financed suggests that the largest part of total investments comes from private domestic sources, and the selection of capital items in this assessment suggests that private sources (domestic and foreign) would be the prime source of capital, at least if it is assumed that public investments should be limited to activities where public goods are produced (hunger and poverty reduction, environmental sustainability, social cohesion, etc.). The public sector can play a role either in funding these investments directly or by helping link, pool and promote private flows. Typically, such investments include the creation and maintenance of infrastructure, large-scale irrigation schemes, or research and development (R&D) of new crop varieties and animal breeds. Depending on the level of public engagement, these investments can help attract further private flows (crowding in) or, if too massive, replace private engagement (crowding out). Private-public partnerships would aim to maximize the former and minimize the latter.

The results

Projected capital stocks and investment needs

Provisional results regarding investment requirements for primary agriculture and its downstream industries in developing countries show that the total over the 44-year period 2005/2007 to 2050 could amount to almost USD 9.2 trillion (2009 dollars), 46 percent of which will be for primary agriculture and the remainder for support services (Table 8.1). Within primary agriculture, almost a third (31 percent) of all capital needs will stem from projected mechanization needs, and almost a quarter (23 percent) from further expansion and improvement of irrigation.

Broken down by type of investment, 60 percent, or USD 5.5 trillion, will be needed to replace existing capital stocks, the other 40 percent, or USD 3.6 trillion, will be growth investments, and thus net additions to the existing capital stock. A detailed account of sector-specific investment projections is available in Annex 8.2.

The share of investments in primary agriculture is expected to fall in all regions, again at considerably different rates. Investments in downstream activities are expected to rise in all regions. Perhaps surprisingly at first sight, the fastest growth in downstream activities is expected for sub-Saharan Africa, albeit from a relatively low absolute level. This region's food system is the least mature, and growth reflects a gradual move away from a heavy reliance on primary production only. East Asia, by contrast, already has the most mature system, with higher levels of grain, sugar, meat and milk processing, so exhibits the smallest non-primary growth, but at much higher absolute levels (Figure 8.1).

Table 8.1
Cumulative investment from 2005/2007 to 2050 (billion 2009 USD)

	Net	Depreciation	Gross
Total for 93 developing countries	3 636	5 538	9 174
Total in primary production	1 427	2 809	4 236
Crop production	864	2 641	3 505
Land development, soil conservation and flood control	139	22	161
Expansion and improvement of irrigation	158	803	960
Establishment of permanent crops	84	411	495
Mechanization	356	956	1 312
Other power sources and equipment	33	449	482
Working capital	94	0	94
Livestock production	562	168	731
Herd increases	413	0	413
Meat and milk production	149	168	317
Total in downstream support services	2 209	2 729	4 938
Cold and dry storage	277	520	797
Rural and wholesale market facilities	410	548	959
First-stage processing	1 522	1 661	3 182

Source: Authors' calculations.

Table 8.2
Growth rates of agricultural production (percentages per annum)

Region	1961–2007	1981–2007	1991–2007	2005/2007–2030	2030–2050	2005/2007–2050
Developing countries	3.5	3.6	3.5	1.8	1.1	1.5
excluding China and India	3.0	3.0	3.1	2.1	1.4	1.8
Sub-Saharan Africa	2.6	3.3	3.1	2.7	1.9	2.3
Near East and North Africa	3.0	2.7	2.5	2.1	1.3	1.7
Latin America and Caribbean	3.0	3.0	3.4	2.1	1.2	1.7
South Asia	2.8	2.8	2.4	2.0	1.3	1.6
East Asia	4.3	4.5	4.3	1.3	0.6	1.0

Source: Authors' calculations.

A striking feature of the outlook is that the annual net additions to capital stock (growth investments) exhibit a noticeable decline over time and result in a slowdown in the annual net capital requirement. Growth investments account for an average of 40 percent of total investments, 55 percent at the beginning of the projection period and merely 30 percent towards 2050 (Figure 8.2). For the aggregate of developing countries as a whole, this reflects a number of factors.

Figure 8.1
Capital stocks in primary agriculture and downstream industries, sub-Saharan Africa and East Asia

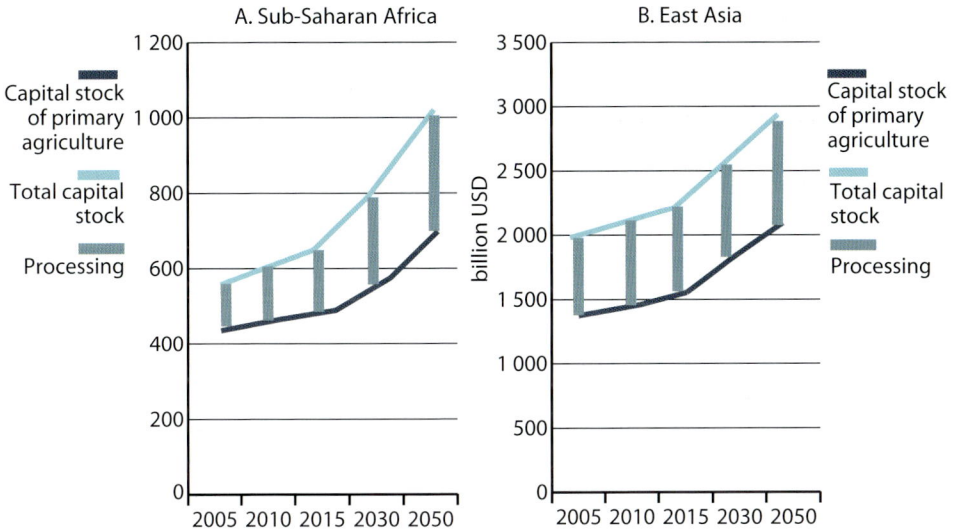

A. Sub-Saharan Africa

B. East Asia

Capital stock of primary agriculture

Total capital stock

Processing

Source: Authors' calculations.

Figure 8.2
Total annual (public and private) investment requirements in developing countries

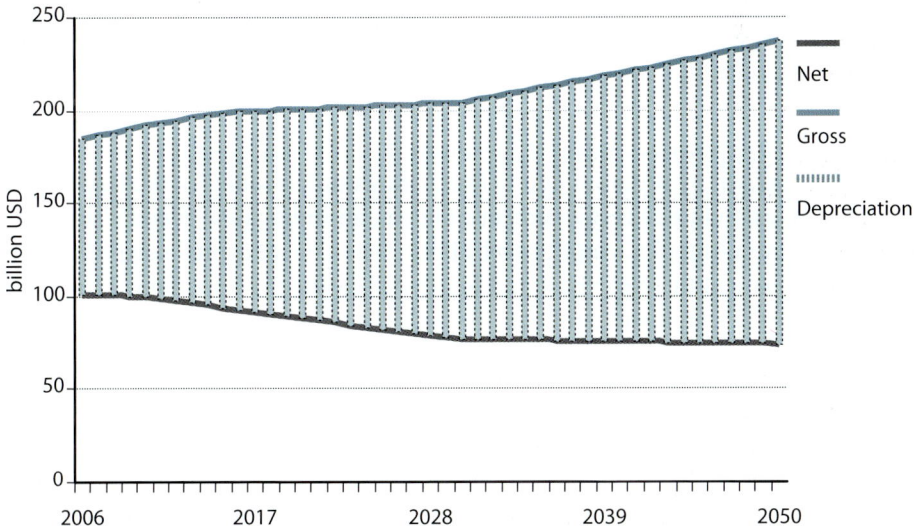

Net

Gross

Depreciation

Source: Authors' calculations.

325

First, a declining incremental production need (Table 8.2), driven by declining population growth and growing satiation levels of per capita consumption of food and fibre, also drives down incremental investment needs. For developing countries as a whole, overall agricultural production grew at a rate of 3.5 percent per year over the last 46 years, and is expected to grow at less than half that rate over the next 44 years. Second, while the decline in production dynamics supports the projected slowdown in capital needs, there will be a countervailing shift towards more capital-intensive forms of production and a growing replacement of labour by capital. This explains the more moderate decline in incremental capital needs than is suggested by the expected levelling of output growth. And third is the impact of a change in the overall efficiency of input use, or total factor productivity (TFP). This is derived as the residual element of output growth that cannot be explained by growing input use, i.e., by either changes in labour or changes in capital and land. Although no TFP accounting is available for the past, future TFP growth is expected to be moderately positive for developing countries as a whole, albeit at rates that vary considerably across regions.

For the aggregate of all developing countries, the relative importance of these factors (from 2005 to 2050) renders the following shares of (net) change: capital, + 71 percent; agricultural labour, - 16 percent; land use, + 25 percent; and TFP, + 20 percent.[3] This suggests a moderate decline in the role of labour inputs and an equally moderate replacement of labour with capital. Obviously, the aggregate hides vastly divergent developments in the various regions; for instance, there is a much larger substitution of capital for labour in Latin America (capital, + 62 percent; labour, - 73 percent; land, + 49 percent; and TFP, + 62 percent) and no such shift at all in sub-Saharan Africa (capital, + 48 percent; labour, + 59 percent; land, + 28 percent; and TFP, - 35 percent). Put colloquially, sub-Saharan Africa would continue to grow by "transpiration", while Latin America could grow further by efficiency gains or "inspiration".

A breakdown by region suggests that Asia would account for the largest part of global investment needs (57 percent); China and India alone account for some 40 percent. Latin America would absorb about 20 percent of capital needs, and sub-Saharan Africa and the Near East and North Africa for the remaining 23 percent (Table 8.3). Asia's high share reflects the region's large agricultural base, its high overall output and its relatively capital-intensive forms of agricultural production (irrigation, mechanization, terracing, etc.). However, growth rates for Asia would be more modest. This is in stark contrast to sub-Saharan Africa, where the overall level of investment requirements is expected to be relatively modest – reflecting

---

3. The underlying growth accounting approach applied here assumes a uniform, constant real wage across all income strata.

the region's generally labour-intensive, capital-saving forms of production – while growth rates are projected to be higher, reflecting a very gradual shift to a more capital-intensive form of agriculture and moderately rising per capita production levels driven by a doubling of the region's population and consumer base.

Table 8.3
Cumulative investment from 2005/2007 to 2050

Region	Net	Depreciation	Gross Crop production	Livestock production	Support services	Total	Share
			(billion 2009 USD)				*(%)*
93 developing countries	3 636	5 538	3 505	731	4 938	9 174	100
excluding China and India	2 427	3 169	2 184	384	3 029	5 596	61
Sub-Saharan Africa	479	462	319	83	539	940	10
Latin America and Caribbean	842	962	528	127	1 149	1 804	20
Near East and North Africa	451	742	619	45	529	1 193	13
South Asia	843	1 444	1 024	123	1 139	2 286	25
East Asia	1 022	1 928	1 015	353	1 582	2 950	32

Source: Authors' calculations.

Broken down into annual instalments over the 44-year outlook period, the total gross needs of USD 9.2 trillion amount to annual capital requirements of nearly USD 210 billion. A larger share of the net investment requirements will occur in the early years and decades of the outlook, reflecting (among other factors) higher incremental investment needs in these years. Thereafter the slowdown in production growth will be reflected in a levelling-off of incremental investment needs. This "front-loading" effect could have important policy implications and lend itself to important policy messages.

As indicated, this chapter does not provide an assessment of public versus private financing from either domestic or foreign sources. If current private and public shares were to be applied to the projections, 70 percent, or USD 150 billion, of the USD 210 billion would come from private sources, and the remaining 30 percent, or USD 60 billion, would have to be provided by public sources, both foreign (official development assistance [ODA]) and domestic.

Performance indicators for agricultural production, capital stocks, labour and land

How much will be produced, and by whom?

In 2005, East Asia alone accounted for nearly half the developing world's agricultural output. Measured in international commodity prices,[4] USD 554 billion dollars came from East Asia, followed by Latin America and South Asia, each producing an annual agricultural output of USD 210 to 215 billion, with the Near East and North Africa producing only USD 95 billion, and sub-Saharan Africa only USD 98 billion (Table 8.4).

Table 8.4
Gross value of agricultural production

Region	2005	2030	2050	2050/2005
	(billion 2004/2006 international dollars)			(ratio)
Developing countries	1 172	1 784	2 207	1.88
Sub-Saharan Africa	98	182	263	2.69
Latin America and Caribbean	210	343	436	2.08
Near East and North Africa	95	155	200	2.11
South Asia	216	356	459	2.12
East Asia	554	748	848	1.53

Source: Authors' calculations.

A look at the long-term growth path to 2050 suggests a dynamic that is quite different from current rates and levels. Sub-Saharan Africa, currently the region with about the lowest agricultural output, is expected to show the fastest growth, and could nearly triple its production to USD 263 billion by 2050. In contrast, East Asia, currently the largest producer, may see an increase of only 53 percent (Table 8.4). This reflects the fact that sub-Saharan Africa has to meet the food needs for the largest population increase of all regions, and may do so from its own agricultural production base. East Asia is expected to see only a very modest overall growth in its population to 2050, falling to zero growth between 2030 and 2050. Moreover, the region has already attained relatively high per capita consumption levels (2 870 kcal per day in 2000), which are expected to rise only moderately to levels somewhat above 3 200 kcal per day. Like sub-Saharan Africa, it may feed its population from its own agricultural resources, with self-sufficiency declining only very moderately. The only region that is expected to step up production significantly beyond its own needs is Latin America, with self-sufficiency rates projected to rise from 118 to 130 percent; Latin America will thus cover the moderately growing deficits of all other regions.

---

4. International commodity prices are used to avoid the use of exchange rates to obtain country aggregates, and to facilitate international comparative analysis of productivity. These international prices, expressed in "international dollars", are derived using a Geary-Khamis formula for the agriculture sector. This method assigns a single price to each commodity. For example, 1 tonne of wheat has the same price, regardless of the country where it was produced.

To meet these production increases, the various regions will have to put more money into agriculture and mobilize more capital, land and labour. The amounts of additional resources the various regions will commit and the roles that incremental capital, land and labour will play are discussed in the next subsection. The starting point for this analysis is the expected output per person, which serves as the basis for discussion of how efficiently land, labour and capital will be used. This discussion is based on an outlook for labour and capital intensity of production, and explores the scope and limits of agriculture in creating incomes and helping reduce poverty.

Output per person

From a developmental perspective, the most important indicator[5] is probably the evolution of agricultural output per person employed in agriculture – the agricultural gross value of production per capita (AGVP/PC). It is a first proxy for how much revenue people employed in agriculture generate and how revenues will evolve over the long run to 2050. It also provides hints as to how large a contribution agriculture will make to overall poverty reduction, and how rapidly the agricultural transformation is likely to evolve.

Table 8.5
Gross value of production per agricultural labourer

Region	2005	2030	2050	2050/2005
	(billion 2004/2006 international dollars)			(ratio)
Developing countries	882	1 319	1 844	2.09
Sub-Saharan Africa	475	587	700	1.47
Latin America and Caribbean	4 993	10 405	18 173	3.64
Near East and North Africa	1 827	3 157	4 888	2.68
South Asia	575	836	1 230	2.14
East Asia	845	1 398	2 221	2.63

Source: Authors' calculations.

A first inspection of the levels and trends of output per labourer across regions reveals vast divergences (Table 8.5). In 2005, by far the highest level of agricultural output per person was attained in Latin America, and despite these high initial levels no slowdown in growth per agricultural labourer is expected for the region. On the contrary, agricultural output per person is projected to rise faster than in any other region, nearly quadrupling to USD 18 173 per person by 2050. At the other end of the spectrum, in sub-Saharan Africa, output per agricultural labourer

---

5. Ideally, performance should be measured as gross margins (returns on variable costs) or net margins (returns on total costs) of production; however this would require a complete accounting for the variable and fixed costs of production.

---

is the lowest today and will remain the lowest by far over the next decades. The gap between sub-Saharan Africa and all other regions is even expected to widen, as AGVP/PC is expected to grow by less then 50 percent in 45 years.

This raises questions as to what drives these divergent regional trends and what the different paths mean for poverty reduction through agriculture. The first question can only be answered by analysing the trends in the underlying variables. The two factors involved are trends in the overall value of output and trends in the evolution of the agricultural labour force.

Table 8.6
Aggregate self-sufficiency rates (percentages)

Region	2005	2050
Developing countries	99	99
Sub-Saharan Africa	97	95
Latin America and Caribbean	118	130
Near East and North Africa	79	78
South Asia	99	98
East Asia	94	91

Source: Authors' calculations.

Growth in overall agricultural output will be highest in sub-Saharan Africa. As discussed, this reflects high growth in consumption and the fact that much of the added need is expected to be met by domestic production. Self-sufficiency is expected to decline only moderately, from 97 percent in 2005 to 95 percent in 2050 (Table 8.6). Output may also rise in Latin America, albeit less rapidly and predominantly for export markets, to make up for the slightly rising deficits of other regions. This means that the difference in the growth of output per worker is almost entirely due to changes in the agricultural labour force. The agricultural labour force of sub-Saharan Africa is projected to nearly double by 2050, while it will fall by nearly half in Latin America, to 24 million (Table 8.7).

The mere numerical description of these trends does not allow any inferences to be drawn on the desirability of the associated development paths. However, it can be concluded that even the near tripling of agricultural output in sub-Saharan Africa will not make a significant difference in revenues per person working in agriculture. When combined with the outlook for capital stocks (Table 8.8) and land available per labourer (Table 8.9.), it can also be concluded that too many people will remain dependent on a labour-intensive, capital-saving form of small-scale agriculture.[6] The poverty reduction potential of this form of agriculture remains limited because too many farmers will have too few revenues to share.

6. The capital stock available per worker will not increase in sub-Saharan Africa, while it will triple in Latin America (Table 8.8).

This is not to suggest that poverty reduction efforts and strategies should ignore small-scale agriculture. On the contrary, as more than 70 percent of the poor reside in rural areas and most depend on small-scale agriculture, poverty reduction strategies should start from and fully embrace small-scale farmers (UNDP, 2005). However, while a smallholder structure is the starting point for poverty reduction, it cannot be an objective in its own right, particularly in sub-Saharan Africa. For one thing, the expected growth in this region's domestic food markets is too limited to engender improved incomes for a growing number of farmers; for another, agricultural export markets would remain elusive for an undercapitalized form of small-scale agriculture. If market potential is limited to food needs, new markets (e.g., energy markets), new non-market income possibilities (payments for carbon offsets, climate change mitigation programmes, payment schemes for environmental services), or strategies for a complete exit from agriculture need to be found to generate income possibilities for the region's young and rapidly growing labour force.

Table 8.7
Agricultural labour force

Region	2005	2030	2050	2050/2005
	(million people)			(ratio)
Developing countries	1 330	1 353	1 197	0.90
Sub-Saharan Africa	206	310	376	1.83
Latin America and Caribbean	42	33	24	0.58
Near East and North Africa	52	49	41	0.79
South Asia	376	426	373	0.99
East Asia	655	535	382	0.58

Source: FAO Statistics Division.

The poverty reduction potential will also not be significant in Latin America's large-scale agriculture, at least not in absolute terms. Too few people in the agriculture sector today are in need of being brought out of poverty in the future. Those remaining in agriculture will produce enough agricultural output to make a living from it. In tandem with this rising output per person, Latin America will continue to pursue its current export orientation. The overall rate of self-sufficiency is expected to rise from 118 to 130 percent by 2050 (Table 8.6). The region will continue and even expand its role as the world's agricultural powerhouse, making up for the less dynamic growth in other regions.

An alternative way of attaining higher incomes and ensuring livelihoods, although not explored in this chapter, would be to raise revenues not covered by agricultural production. Options would include revenues raised from the provision of environmental services, particularly contributions to greenhouse gas abatement and the entry into the carbon market. It is important to note that agriculture, which accounts for more than 30 percent of greenhouse gas emissions (including through deforestation), is not only one of the main sources of emissions, but also has significant potential for climate change mitigation. Funds raised from these alternative sources could help farmers adopt carbon-saving production technologies, reducing the carbon footprints of traditional technologies while increasing the productivity and profitability of agricultural production. Promising options include a shift to no-tillage and conservation agriculture, more efficient milk and ruminant meat production systems (FAO, 2006a), or a transition from paddy to upland rice production.

Another income source could be increased production of agricultural feedstocks for the energy market. The energy market is so large that such production would not be subject to demand constraints and would allow more farmers to draw revenues from otherwise increasingly saturated markets. For small-scale farmers, bioenergy could help overcome the on-farm power constraint, the factor that often limits agricultural productivity growth the most. For larger-scale farmers, bioenergy offers new potential to produce for a market that is, in essence, characterized by perfectly elastic demand and that will absorb any incremental production, as long as agricultural feedstocks are competitive as inputs into the energy market – i.e., as long as energy prices are above parity prices in the energy market. This necessitates high energy prices. The perfectly elastic demand also means that food prices would be determined by energy prices and that poor food consumers could be priced out of food markets by less elastic energy consumers.

The success of such diversification into new agricultural activities will be contingent on whether or not smallholder agriculture has a comparative advantage for these new markets. Typically, smallholder agriculture is labour-intensive, capital-saving and, particularly, deficient in expertise. In contrast, many of the emerging income options require expertise and capital, and seldom require unskilled labour. Tapping into carbon offset schemes under the Clean Development Mechanism (CDM), for instance, is mostly limited to large projects and large farmers, and a large share of these CDM projects have been granted to large holdings or agricultural industries in Latin America. The administrative hurdles of such schemes are too onerous for smallholders to meet. Commercial bioenergy production is also highly expertise- and capital-intensive; for instance, Brazilian ethanol production has become more profitable as it becomes more labour-saving. The discrepancy between the factor needs and factor endowments of smallholders means that they are unlikely to have a comparative advantage for these alternative income sources; in fact, their factor endowment is precisely the opposite of the factor requirements needed for such activities.

Options for overcoming a lack of capital and expertise exist. One would be to improve or establish the institutional setting that allows a pooling of smallholder resources, to create enough human and financial capital to overcome the resource limits. Cooperatives can play an important role in pooling resources; public investments can support and foster these efforts. There are numerous examples of successful resource pooling, particularly for new bioenergy projects. In Thailand, for instance, 4 000 farmers pooled their resources in a cooperative for setting up a cassava-based bioethanol project in the Chok Chai district of Nakhon Ratchasima; through the country's Agricultural Cooperative Federation they even established a joint venture with a USA-based energy company, to overcome remaining capital constraints and attract the necessary expertise to operate a large-scale ethanol plant.

These examples suggest that the comparative disadvantages of small-scale farming in new market opportunities could be overcome, and that the new markets could be tapped by small-scale operators if their resources are pooled. In turn, this would require a strengthening of rural institutions, and thus public investments. The greatest needs, but also the greatest potentials, for institutional improvements lie in sub-Saharan Africa.

Why will outcomes be so different?

An important factor that helps explain differences in the output per worker is the capital stock available per labourer. Taking the two extreme cases of Latin America and sub-Saharan Africa, the estimates summarized in Table 8.8 suggest that a farm worker in the former region has, on average, ten times as much capital available than a farm worker in the latter. Behind the abstract aggregate of capital per farmer are a large range of tools and equipment that make agriculture in Latin America so much more productive than in Africa. These tools include more and better mechanization, tractors, tillers and combines, irrigation, storage and processing plants, and other elements of an efficient downstream sector. Although not included in the estimates, Latin American farmers also have far more support capital in better infrastructure, research institutions, roads and electricity. Equally important is the reliability of these supplies, rendering fewer off-hours because of interruptions in electricity supplies or irrigation water availability. For instance, rural roads per hectare amount to 0.017 km in Latin America compared with 0.007 km – less than half – in sub-Saharan Africa. Rural electricity supplies per worker are 50 times higher in Latin America than sub-Saharan Africa.

The outlook to 2050 suggests that the interregional differences in capital stocks per worker are likely to become more pronounced. Capital stocks per worker will roughly double in East Asia, South Asia and the Near East and North

Africa, while they will triple in Latin America and completely stagnate in sub-Saharan Africa. This means that by 2050 a worker in Latin America will have 28 times as much capital available as a worker in sub-Saharan Africa. These huge differences in capital intensity are at the heart of differences in the current output per worker and the divergent growth paths the two regions are expected to take.

Table 8.8
Capital stock per worker

Region	2005	2030	2050	2050/2005
	('000 2009 USD/person)			(ratio)
Developing countries	4.28	5.72	7.68	1.79
Sub-Saharan Africa	2.78	2.62	2.77	1.00
Latin America and Caribbean	25.24	45.70	77.77	3.08
Near East and North Africa	11.61	17.33	25.41	2.19
South Asia	3.88	4.59	6.10	1.57
East Asia	3.06	4.87	7.67	2.51

Source: Authors' calculations.

As discussed, critical elements in the divergent developments of labour productivity across regions include the diversity of developments in the agricultural labour force. Latin America, for instance, will almost halve its labour force, while sub-Saharan Africa will nearly double its. The importance of this effect can be seen when agricultural output is related to land rather than labour (Table 8.9).

Table 8.9.
Harvested land per worker

Region	2005	2030	2050	2050/2005
	(ha/person)			(ratio)
Developing countries	0.69	0.75	0.90	1.30
Sub-Saharan Africa	0.86	0.68	0.63	0.73
Latin America and Caribbean	3.47	5.53	8.62	2.49
Near East and North Africa	1.41	1.50	1.87	1.33
South Asia	0.60	0.56	0.65	1.08
East Asia	0.45	0.57	0.81	1.80

Source: Authors' calculations.

Output per hectare in Latin America is only 2.5 times higher than it is in sub-Saharan Africa, and somewhat lower than it is in East Asia. However, by 2050, a worker in Latin America will be cropping twice as much land, while arable land available per labourer will shrink in sub-Saharan Africa. This again raises the question of how sustainable the outlook is for sub-Saharan Africa, if agriculture continues to be based on a farming system in which a limited resource base has to be shared among a rising number of resource users. Even if the basis of the argument is largely arithmetical, small-scale agriculture is unlikely to provide much revenue generation and poverty reduction. Another question that arises is whether agricultural development in sub-Saharan Africa needs to be combined with exit strategies, to ensure that fewer people are left in the sector and that they have enough resources to generate sufficient income.

What bang for the buck? Incremental capital output ratios and investment rates in primary agriculture

In an increasingly globalized world, private investors, development planners and policy-makers are interested in identifying investment opportunities in agriculture at home and abroad. A broad and easy-to-calculate indicator that helps address this issue is the incremental capital output ratio (ICOR). High ICORs suggest that increases in agricultural output require high investments, and vice versa.

Table 8.10
Average ICORs and investment rates in primary agriculture, 2005/2007 to 2050 (percentages)

Region	Investment as share of AGVP	Inputs as share of AGVP[a]	Investment as share of agricultural GDP	ICOR
Developing countries	6.7	27	9.2	6.3
excluding China and India	7.5	27	10.3	5.8
Sub-Saharan Africa	6.2	11	6.9	3.1
Latin America and Caribbean	5.7	29	8.0	4.8
Near East and North Africa	11.4	40	19.0	11.1
South Asia	9.0	28	12.5	7.2
East Asia	5.2	28	7.2	7.4

[a] From Alexandratos, 1988.

Comparison of the ICORs across regions (Table 8.10) suggests that changes in agricultural capital stocks are expected to render fairly different levels of agricultural output across the main developing regions. By far the highest ICORs (averaging more than 11) are projected for the Near East and North Africa, while by far the lowest (averaging just over 3) are expected for sub-Saharan Africa. In both regions, the expected ICORs are consistent with current factor endowments and expected factor returns. High ICORs for the Near East and North Africa reflect the high level of capital intensity that this region has already attained, leaving it with few options for stepping up production through an easy expansion of cropland or irrigation water use. In fact, the Near East and North Africa has virtually exhausted its agricultural land base and is also approaching the limits of its renewable water resources. This makes further increases in production a capital-intensive endeavour and ultimately implies low returns on future additions to the existing capital stock. Extreme examples include agricultural production systems that use groundwater mining or water supplies from energy-intensive desalinization plants; ICORs are particularly high where investments have been geared towards low-value outputs such as cereals and other food staples. The unit production costs of such farming systems often exceed international commodity prices by multiples, and can only be sustained with exorbitantly high subsidies.

From a planning and policy perspective, this suggests that further expansion of production in the Near East and North Africa has to be weighed against alternatives such as increased imports of agricultural goods or investments in foreign capital stocks and cropland. While the region has focused on imports for a long time, it has recently also pursued the option of securing domestic supplies through foreign direct investment in other regions.

Inspection of the ICORs in other regions (Table 8.10) helps explain why many of these new investments are currently directed to sub-Saharan Africa. The low ICORs of just over 3 suggest that incremental capital invested in sub-Saharan African agriculture will render nearly four times as much as investments in the Near East and North Africa. This is consistent with African agriculture's abundant land and labour combined with a shortage of the capital (both working and fixed) needed to make the existing land and labour base more productive.

How will farm revenues perform compared with non-agricultural incomes?

As outlined in the previous subsection, the trends in future farm revenues exhibit vast differences across regions, and people dependent on agriculture in the various regions will see vastly different growth potentials for their agricultural incomes. A crucial question regards whether or not the projected revenue paths for agriculture are more or less favourable than those outside agriculture, or – more precisely – whether they are more or less favourable than those of the average income earner (agricultural and non-agricultural combined).

The agricultural and non-agricultural income trajectories are compared in Figure 8.3, which depicts three important features of the projected income trajectories for the various regions. First, the horizontal extension of the paths captures the projected income growth for each region. It suggests that East Asia's income growth per person is expected to be much higher than that in any other region; for example, it is expected to be three times that of sub-Saharan Africa, and the overall picture suggests a continuation of the growth patterns seen over the last three decades. Income growth is also projected to be high in South Asia, followed by Latin America and the Near East and North Africa. The second feature is captured by the slope of the trajectories. The steeper the slope, the higher the agricultural growth prospects relative to overall growth. A slope steeper than the 45° diagonal denotes that agriculture outperforms the average for the region. Clearly, this is not expected in any of the regions; instead, trajectories are flat for all regions, and move further away from the 45° diagonal as 2050 approaches. This unequal growth is particularly pronounced for all regional aggregates of sub-Saharan Africa and Asia. The third feature stems from the location of a trajectory above or below the diagonal; this denotes whether agricultural incomes are above or below average incomes, for both the starting and the end years. As can be seen immediately from Figure 8.3, the only region where agricultural incomes are above average incomes is Latin America, while the reverse is the case for all other regions. Even for Latin America, it should be noted that the vertical axis depicts AGVP rather than agricultural GDP, i.e., agricultural incomes are overstated by the amount of working capital employed. Given the relatively advanced stage of agriculture in Latin America, the effect of income overestimation could be considerable; taking this into account, it is probable that agricultural incomes are not above average incomes in any region, in either the base year or 2050.

Figure 8.3

Regional income trajectories: agricultural versus non-agricultural, 2005 to 2050

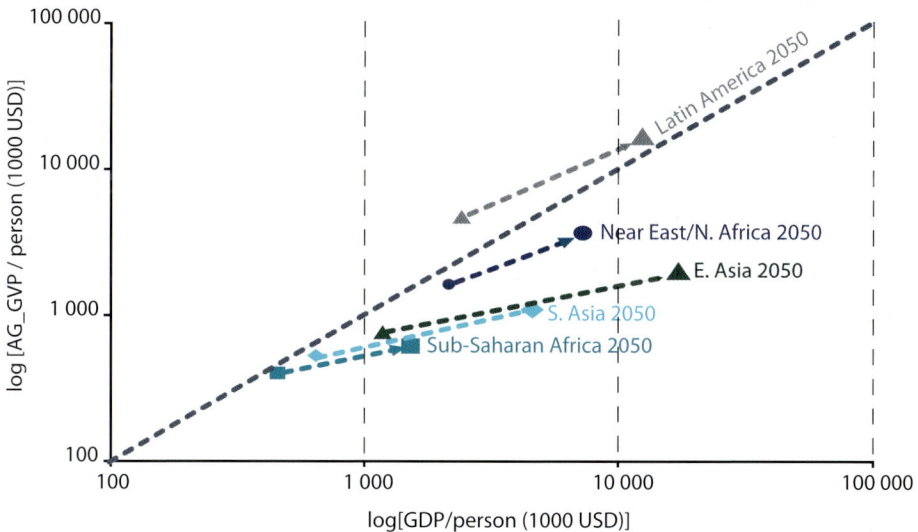

Source: Authors' calculations.

In summary, this means that the projected income trajectories suggest a largely negative outlook for agriculture. In no region will agricultural labourers be able to accomplish the same income growth as their peers outside agriculture. The only exception is Latin America, where farm revenues are slightly higher than average incomes, and growth rates in farm revenues, on average, just match those of the region's economy. The outlook also suggests a growing divergence between agricultural and non-agricultural incomes, and thus probably an even stronger concentration of poverty in rural areas. The results are likely to understate the true agricultural versus non-agricultural income gap for two reasons. First, agricultural income growth is compared with average income growth; where agriculture accounts for a large share of the total economy, the difference between agricultural and non-agricultural incomes is likely to be larger than it appears in the results. Second, the population projections for agriculture refer to the agricultural labour force, which is a subset of the overall agricultural population; if agricultural incomes were divided over the larger agricultural population, this would widen further the gap with non-agricultural incomes.

It must be emphasized that these results are only preliminary; they need to be vetted and confirmed with projections for agricultural GDP, rather than just those for AGVP. The growing divergence may also bring to the fore a possible shortcoming of the underlying partial equilibrium approach. Past developments show that considerable, and even growing, rural/urban income differences can persist over extended periods, but a growing income divergence over more than four decades may become untenable, and suggests that hitherto exogenous assumptions, such as the projections for agricultural labour force or even general population projections, may need to be endogenized. Rising income gaps would ultimately raise the pressure to leave rural areas (push), and attract cheap labour to more remunerative urban areas and non-farm environments (pull).

The prospect of a widening income gap between farm and non-farm incomes has also given rise to new initiatives for providing support to developing country farmers. FAO is currently examining various possibilities of such support measures; the decisive criterion for these measures is that they help farmers to catch up with the average incomes in an economy or region, without introducing new or augmenting existing measures that distort international competition, resource allocation and trade. The scope, options and limits of such measures were discussed at the Summit on World Food Security in November 2009.

Summary and conclusions

Cumulative gross investment requirements for developing countries' agriculture add up to a total of nearly USD 9.2 trillion over the 44 years from 2005/2007 to 2050. This amount would be necessary to remain consistent with FAO's long-term outlook for global agriculture (FAO, 2006b).

Broken down by type of investment, more than USD 5.5 trillion, or 60 percent of the total, will be required to replace the existing capital stock (or the new capital items that are added and subsequently depreciate over the 44 years to 2050); the rest, about USD 3.6 trillion, will be added to the existing capital stock to increase (nearly double) output and raise productivity. Broken down by activity, primary agriculture will account for about USD 4.2 trillion of the total, while the remaining USD 4.9 trillion will be absorbed by downstream needs (processing, transportation, storage, etc.). Within primary agriculture, mechanization will account for the single biggest investment item (31 percent), followed by expansion and improvement of irrigation (23 percent). The cumulative investments result in yearly averages of about USD 210 billion gross and USD 83 billion net. All estimates, gross and net, cumulative and annual, are in constant 2009 dollars.

A striking feature of the outlook is that annual net additions to the capital stock (growth investments) exhibit a noticeable decline over time, resulting in a slowdown in growth of the annual net capital requirement. These net investments account for 55 percent of the total at the beginning of the projection period, and for merely 30 percent towards 2050. The change in net investments reflects a number of factors. First, incremental production will need to decline alongside declining incremental needs. Partly offsetting this decline is a shift towards more capital-intensive forms of production, with a growing replacement of labour by capital. A third factor, again supporting the decline in net capital needs, is the somewhat higher overall efficiency of input use in the future.

Growth accounting results suggest that overall growth will be characterized by increasing substitution of labour with capital, and moderate TFP growth. However, there are marked regional differences; for instance, in Latin America growth will be capital- and productivity-based, with negative labour contributions, while in sub-Saharan Africa it will be heavily labour- and moderately capital-based, with limited efficiency gains.

The analysis of performance indicators suggests that there are marked regional differences in agriculture's capacity to generate incomes and reduce poverty. For instance, projections for the gross value of production suggest that revenues generated by an agricultural labourer in sub-Saharan Africa will rise by only 50 percent over the next four decades. The expected growth in food markets will not suffice to lift revenues significantly.

The analysis of expected revenues, capital stocks and land available per labourer suggests that too many people in sub-Saharan Africa will remain dependent on a labour-intensive, capital-saving form of small-scale agriculture, in which too many farmers will have to share too few resources and revenues. The poverty reduction potential in the projected revenue/capital stock trajectory in sub-Saharan Africa will thus be limited.

This raises questions regarding the alternative income sources that could be tapped. Emerging options include new opportunities arising from higher energy prices and the production of bioenergy feedstocks; income opportunities from the provision of environmental services; or a greater export orientation of production. All three growth options call for an expertise- and capital-intensive form of agriculture, and thus run counter to the factor endowment that characterizes Africa's smallholder structure. One option for overcoming these constraints would be to increase investments in resource-pooling institutions.

The available capital stock per worker was identified as an important explanatory variable for interregional differences in performance. A farmer in Latin America has on average ten times as much capital available as a farmer in sub-Saharan Africa. Behind the abstract aggregate of capital per farmer are a large range of tools and equipment that make agriculture in Latin America far more productive than in Africa. These include more and better mechanization, tractors, tillers and combines, irrigation, storage and processing plants, and other elements of an efficient downstream sector.

References

Alexandratos, N., ed. 1988. *World agriculture: towards 2000, an FAO study.* London, Belhaven Press, and New York, New York University Press.

Alexandratos, N., ed. 1995. *World agriculture: Towards 2010, an FAO study.* Chichester, UK, J. Wiley and Sons, and Rome, FAO.

Bruinsma, J., Hrabovszky, J., Alexandratos, N. & Petri, P. 1983. Crop production technology and input requirements in the agriculture of developing countries. *European Review of Agricultural Economics*, 10(3): 197–222.

FAO. 1980. *Agriculture toward 2000: Estimation of investment requirements.* Rome, Global Perspective Studies Unit. (mimeograph)

FAO. 1981. *Agriculture: towards 2000.* Rome.

FAO. 1996. *Investment in agriculture: evolution and prospects.* Technical Background Document No. 10 for the World Food Summit. Rome.

FAO. 1999. *Investment in agriculture for food security: Situation and resource requirements to reach the World Food Summit objectives.* CFS:99/Inf 7. Rome.

FAO. 2001. *Mobilizing resources to fight hunger.* CFS:2001/Inf. 7. Rome.

FAO. 2003. *Anti-Hunger Programme.* Rome. ftp://ftp.fao.org/docrep/fao/006/j0563e/j0563e00.pdf.

FAO. 2006a. *Livestock's long shadow: Environmental issues and options.* Rome. www.fao.org/docrep/010/a0701e/a0701e00.htm.

FAO. 2006b. *World agriculture: towards 2030/2050 – interim report.* Rome. www.fao.org/es/esd/at2050web.pdf.

Schmidhuber, J. & Bruinsma, J. 2011. Investing towards a world free of hunger: lowering vulnerability and enhancing resilience. *In* FAO. *Safeguarding food security in volatile global markets*, edited by A. Prakash. Rome, FAO.

UNDP. 2005. *Halving hunger: It can be done*. London, UN Millennium Project, Task Force on Hunger.

Countries Included in the Analysis

Sub-Saharan Afirca

Angola	Democratic Republic	Madagascar	Sierra Leone
Benin	of the Congo	Malawi	Somalia
Botswana	Eritrea	Mali	Sudan
Burkina Faso	Ethiopia	Mauritania	Swaziland
Burundi	Gabon	Mauritius	Togo
Cameroon	Gambia	Mozambique	Uganda
Central African	Ghana	Niger	United Republic of
Republic	Guinea	Nigeria	Tanzania
Chad	Kenya	Rwanda	Zambia
Congo	Lesotho	Senegal	Zimbabwe
Côte d'Ivoire	Liberia		

Latin America and the Caribbean

Argentina	Ecuador	Mexico	Suriname
Brazil	El Salvador	Nicaragua	Trinidad and
Chile	Guatemala	Panama	Tobago
Colombia	Guyana	Paraguay	Uruguay
Costa Rica	Haiti	Peru	Venezuela
Cuba	Honduras	Plurinational	
Dominican Republic	Jamaica	State of Bolivia	

Near East and North Africa

Afghanistan	Islamic Republic	Libyan Arab	Syrian Arab
Algeria	of Iran	Jamahiriya	Republic
Egypt	Jordan	Morocco	Tunisia
Iraq	Lebanon	Saudi Arabia	Turkey
			Yemen

South Asia

Bangladesh	Nepal	Sri Lanka
India	Pakistan	

East Asia

Cambodia	Indonesia	Myanmar	Viet Nam
China	Lao People's	Philippines	
Democratic	Democratic	Republic of Korea	
People's Republic	Republic	Thailand	
of Korea	Malaysia		

Cumulative investment requirements for 2005/2007 to 2050, by region (billion 2009 USD)

	Sub-Saharan Africa			Latin America/Caribbean			Near East/North Africa			South Asia			East Asia		
	Net	Depre-ciation	Gross	Net	Depre-ciation	Gross	Net	Depre-ciation	Gross	Net	Depre-ciation	Gross	Net	Depre-ciation	Gross
Total	479	462	940	842	962	1 804	451	742	1 193	843	1 444	2 286	1 022	1 928	2 950
Total investment in primary production	177	225	401	290	365	655	194	470	664	340	808	1 147	427	941	1 368
Crop production	101	218	319	183	345	528	151	468	619	223	801	1 024	206	809	1 015
Land development, soil conservation and flood control	45	3	48	44	4	48	5	1	7	21	5	25	25	9	33
Expansion and improvement of irrigation	14	31	45	28	69	96	52	215	267	28	236	265	36	251	288
Establishment of permanent crops	4	41	45	4	47	51	2	15	17	16	52	68	58	256	314
Mechanization	22	37	59	85	207	292	77	224	300	115	304	420	57	184	241
Other power sources and equipment	10	105	115	0	19	19	1	13	14	17	204	220	6	108	114
Working capital	6	0	6	22	0	22	15	0	15	26	0	26	25	0	25
Livestock production	76	7	83	107	20	127	42	3	45	117	6	123	220	133	353
Herd increases	67	0	67	85	0	85	37	0	37	96	0	96	128	0	128
Meat and milk production	9	7	15	22	20	42	5	3	8	21	6	27	93	133	225
Total investment in downstream support services	302	237	539	552	597	1 149	257	272	529	503	636	1 139	595	987	1 582
Cold and dry storage	41	37	78	96	88	184	20	46	66	55	109	164	65	240	305
Rural and wholesale market facilities	87	72	159	60	61	121	68	68	136	100	163	263	96	184	280
First-stage processing	174	129	302	396	447	844	169	158	326	348	364	712	435	563	998

Source: Authors.

PART IV

RAISING PRODUCTIVITY: RESEARCH AND TECHNOLOGY

CHAPTER 9

SETTING MEANINGFUL INVESTMENT TARGETS
IN AGRICULTURAL RESEARCH AND DEVELOPMENT:
CHALLENGES, OPPORTUNITIES AND FISCAL REALITIES

Nienke Beintema
Howard Elliott

This chapter is aimed at the policy-makers who ask "Is there enough investment in agricultural research and development (R&D)?" They are constantly being reminded of declarations, commitments and targets asserting that they must do more or better. To provide some analytical structure and limits to the discussion, this chapter examines underinvestment separately from the demand and supply sides, before moving on to the investments, policy actions and institutional arrangements that are needed to bring supply and demand into balance.

The chapter has four sections in addition to this introduction. The next section sets the scene by providing historical trends in human and financial investments in agricultural R&D. This is followed by a section looking at underinvestment in three ways (two technical and one political): first, evidence of a continuing high rate of return relative to the social rate of discount is a formal definition of underinvestment, as additional investment would add more to social gains than to social costs; second, failure to maintain on-farm productivity growth at its historical trend and potential level is a sign of underinvestment; and third, if there are large gaps between the resources required to attain political commitments, such as the Millennium Development Goals (MDGs) with respect to poverty and hunger, there is underinvestment with respect to political commitments. At this point, nothing is said about how quickly the gaps must be eliminated to avoid waste.

Turning to the supply side, the chapter's fourth section questions whether countries' national efforts are commensurate with their financial and human resource capacities for doing more to deliver on commitments and investment targets set in various international fora. This section examines several public finance issues on the taxation and expenditure sides (which are not independent of each other), identifies the agricultural research intensity ratio, analyses the four components that determine the ratio's value, and comments on what could be done to increase investment in R&D. On the human resources side, it identifies gaps in research and higher education that affect research institutions' ability to ramp up their efforts in response to emerging challenges. Financial resource needs cut across all scales, from the global to the local.

The final section deals with new challenges that imply the need for not only reinvestment in agricultural R&D, but also investment in other parts of the knowledge system, to ensure balanced growth. The need for more highly trained researchers to deal with climate change, price volatility in global markets or water scarcity puts demands on the university system to expand M.Sc. and Ph.D. training. Such an expanded cadre provides valuable research support to existing scientists, while learning the advanced skills needed to become senior researchers.

New challenges demand new approaches, new skills and new institutional arrangements for collaborative research. The time and process by which these new arrangements come about are necessary investments.

Trends in agricultural R&D investments[1]

Public agricultural R&D spending

Global spending on public agricultural R&D (including the government, non-profit and higher education sectors) totalled USD 24 billion in 2005 purchasing power parity (PPP) dollars in 2000,[2] the latest year for which comparable global data are available (and excluding Eastern Europe and the former Soviet Union, for which no time series data were available).[3] Total public investment had increased

---

1. This section draws on Beintema and Stads (2008; 2010), Stads and Beintema (2009) and underlying data sets of the Agricultural Science and Technology Indicators (ASTI) initiative (www.asti.cgiar).

2. Financial data in this chapter are reported in real values using gross domestic product (GDP) deflators with the benchmark year 2005 and PPP indices from the World Bank (2008a). PPPs are synthetic exchange rates used to reflect the purchasing power of currencies, typically by comparing prices among a broader range of goods and services than conventional exchange rates (see also Beintema and Stads, 2011b). These global trends differ from those reported in Pardey *et al.* (2006). These revisions were in response to the World Bank's adjustments to its comparative pricing of goods and services across countries (using PPP indices), reclassification of non-Organisation for Economic Co-operation and Development (OECD) high-income countries, and new estimates for Latin American and a number of other countries (Beintema and Stads, 2008).

3. If Eastern European and former Soviet Union countries are included, total public spending on agricultural R&D totalled USD 25.1 billion (in 2005 international prices) in 2000 (Beintema and Stads, 2010).

---

considerably from the USD 16 billion reported in 1981 (Table 9.1). However, this increase did not take place equally across all regions. Spending in Asia and the Pacific more than doubled during the period, increasing at an average of 4.2 percent per year (Figure 9.1).[4] This was largely a result of high growth in agricultural R&D spending in the two largest countries, China and India (with annual growth of 4.4 and 5.8 percent, respectively). In contrast, spending in sub-Saharan Africa grew by an average of only 0.6 percent per year from 1981 to 2000. More worrisome is that the spending for the region as a whole contracted slightly during the 1990s, with more than half of the sub-Saharan African countries for which time series data were available spending less in 2000 than they did in 1991.

Table 9.1
Total public agricultural R&D expenditures, by income class and region

Country category	Public spending	Public agricultural R&D spending		Regional share of global total		
	1981	1991	2000	1981	1991	2000
Country category	(million 2005 PPP dollars)			(%)		
Country grouping by income class						
Low-income (46)	1 410	2 010	2 566	9	10	11
Middle-income (62)	4 670	6 453	7 953	29	30	33
High-income (32)	9 951	12 806	13 456	62	60	56
Total (140)	16 032	21 268	23 975	100	100	100
Low- and middle-income countries by region						
Sub-Saharan Africa (45)	1 084	1 253	1 239	7	6	5
China	773	1 350	2 244	5	6	9
India	400	748	1 301	3	4	5
Asia and Pacific (26)	2 032	3 460	5 114	13	16	21
Brazil	1 005	1 414	1 247	6	7	5
Latin America and Caribbean (25)	2 245	2 676	2 755	14	13	12
Near East and North Africa (12)	720	1 074	1 0412	5	5	6
Subtotal (108)	6 081	8 463	10 519	36	40	43

Estimates exclude Eastern Europe and the former Soviet Union. Estimation procedures and methodology are described in Beintema and Stads, 2011b and various ASTI regional reports available at www.asti.cgiar.org. Number of countries indicated in parentheses.

Sources: ASTI datasets and other secondary sources prepared for Beintema and Stads, 2010.

As a result of these different regional growth patterns, the distribution of agricultural R&D spending changed over the two decades. Due to the high

---

4. The regional totals refer to developing countries (defined as low- and middle-income countries) only, and exclude high-income countries such as the Republic of Korea in the Asia and the Pacific region and Israel and Kuwait in the Near East and North Africa region.

increase in total spending in Asia and the Pacific, its share in the global total increased from 13 percent in 1981 to 21 percent in 2000. As a result, the shares of sub-Saharan Africa and Latin America declined to 5 and 11 percent of the total, respectively. Total agricultural R&D spending in sub-Saharan Africa was about the same as total spending in Brazil, the largest public investor in Latin America, and considerably lower than the spending levels in India and China. Although spending in high-income countries as a whole continued to grow in absolute terms, their share of total global spending declined from 62 to 56 percent. The shares of spending by low- and middle-income countries increased from 9 to 11 percent and from 23 to 32 percent, respectively.

Figure 9.1
Annual growth rates in agricultural R&D spending

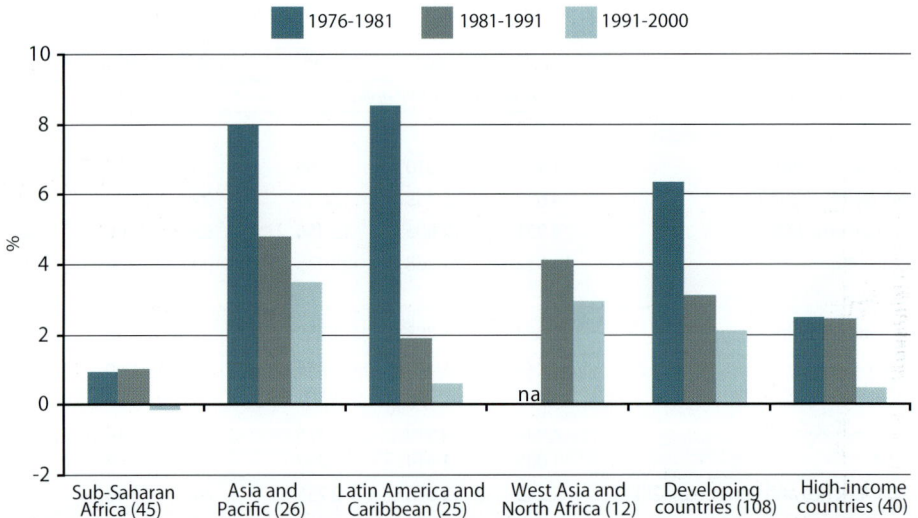

Estimates exclude Eastern Europe and the former Soviet Union. Estimation procedures and methodology are described in Beintema and Stads, 2011b and various ASTI regional reports available at www.asti.cgiar.org. Number of countries indicated in parentheses.
Sources: ASTI datasets and other secondary sources prepared for Beintema and Stads, 2010.

Although data on global public investment patterns since 2000 are still unavailable,[5] more recent data collected by the ASTI initiative show that investments continued to grow in China and India (Figure 9.2). Agricultural R&D

---

5. Data collection efforts by the ASTI initiative are under way in sub-Saharan Africa and a few other countries to ensure a new global update for 2008.

expenditures in Latin America and the Caribbean rebounded in recent years, following a period of contraction during the late 1990s, which was mostly due to financial crisis in a number of Southern Cone countries. Public agricultural R&D investment also increased in sub-Saharan Africa, by about one-fifth during 2001 to 2008. This was the result of increased government commitments in some of the larger countries, such as Ghana, Nigeria, the Sudan, the United Republic of Tanzania and Uganda. However, many other countries experienced declining agricultural spending, and even in the countries that experienced growth, the additional funds were mostly used for salary increases or augmentation of (often poorly maintained) infrastructure, and not for research. In addition, increased funding, such as in Nigeria and the Sudan, often followed many years of underinvestment, and levels are still below those needed to sustain these countries' agricultural R&D needs (Beintema and Stads, 2011a).

Figure 9.2
Public agricultural R&D investment trends in developing countries

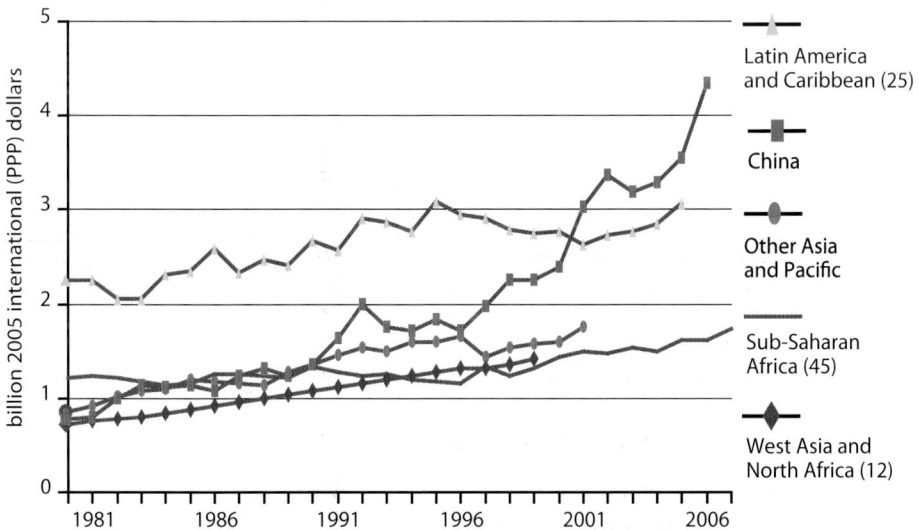

Number of countries indicated in parentheses.
Sources: Beintema and Stads 2008; 2011a; Stads and Beintema, 2009, based on ASTI data sets.

However, public agricultural R&D has become increasingly concentrated in a handful of countries (Pardey *et al.*, 2006). The top five countries in terms of agricultural R&D spending – the United States of America, Japan, China, India and Brazil – accounted for 48 percent of total global public agricultural R&D in 2000, up from 41 percent in 1991. Meanwhile, only 6 percent of agricultural R&D

investments worldwide were made in 80 (mostly low-income) countries, which had a combined population of more than 600 million people and accounted for 14 percent of the world's agricultural land area. In Latin America, about three-quarters of total public investments in agricultural R&D were spent by only three countries: Brazil, Mexico and Argentina. Since the mid-1990s, the investment gap between the region's low- and middle-income countries has widened, in part as a result of sharp cuts in research expenditures in some of the poorer, more agriculture-dependent countries, such as Guatemala and El Salvador. Asia has a similar, but less pronounced, knowledge divide between its rich and poor countries, and the gap between the scientific "haves" and "have-nots" is becoming increasingly visible. From 1981 to 2002, especially in the latter decade, both China and India intensified their agricultural research spending, while some smaller countries, such as Malaysia and Viet Nam, also realized impressive agricultural R&D spending growth. However, other countries, such as Pakistan, Indonesia and Lao People's Democratic Republic, proved sluggish and at times negative, largely owing to the Asian financial crisis, the completion of large donor-financed projects, or high rates of inflation. In Africa, agricultural research has historically been better funded in some countries, such as Kenya and South Africa, than in a large number of the very poorest countries, especially in Western Africa. There is no evidence that this divide has increased over the past few decades, partly because of the donor dependency of many countries and the erratic nature of government and donor support to agricultural research over the years.

The government sector is still the main player in public agricultural R&D, in terms of execution as well as funding. The government sector accounted for 60 percent of total full-time equivalent (FTE) R&D staff in Latin America in 2006, and for 74 percent in sub-Saharan Africa in 2008 (Figure 9.3). Despite this leading role of the government sector, the higher education sector has gained prominence in several countries, and accounted for 36 percent of total public agricultural R&D in Latin America in 2006, compared with 29 percent in 1981. The higher education share in sub-Saharan Africa increased from 12 percent in 1981 to 24 percent in 2000. In absolute terms, the total number of FTE researchers employed in the higher education sector almost doubled in Latin America and tripled in sub-Saharan Africa. In some countries, such as Argentina and Mexico, higher education's capacity approaches that of the government sector. In India, higher education has surpassed the government sector in terms of FTE agricultural research staff; this results from the integration of research, extension and education in the Indian system. Despite the increasing share of the higher education sector as a whole, the individual capacity of each faculty/school is often very small, and the agricultural higher education system is often fragmented, such as in the Sudan, the Philippines and Nigeria.

Figure 9.3
Shares of institutions in agricultural R&D, Latin America and the Caribbean and sub-Saharan Africa

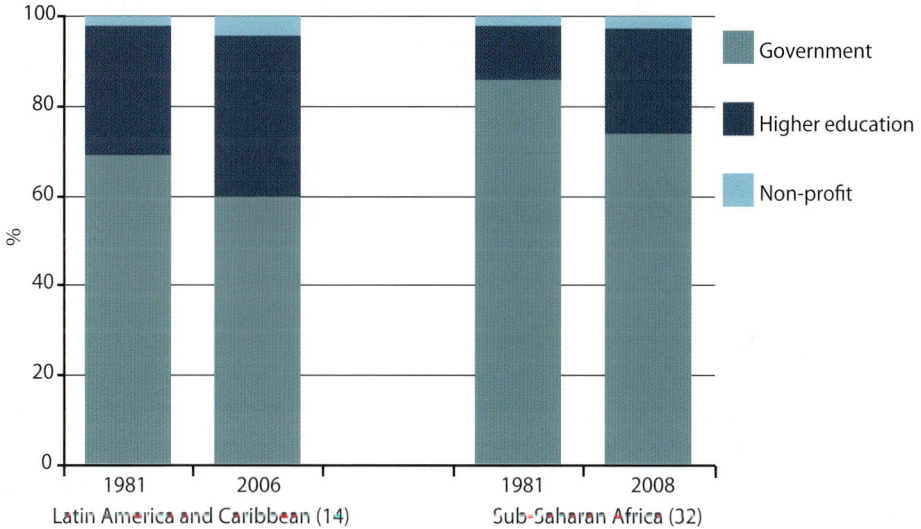

Shares measured in terms of FTE researchers. Number of countries indicated in parentheses.
Sources: Stads and Beintema, 2009; Beintema and Stads, 2011a, based on ASTI data sets.

The government sector is also still the largest contributor to public agricultural research (Figure 9.4). Government allocations accounted for an average of 81 percent of total funding received by a sample of more than 400 government agencies and non-profit institutions in 53 developing countries. Only 7 percent of total funding was received from donor contributions, in the form of loans or grants. This share was driven mostly by the high donor dependency of government agencies in sub-Saharan Africa. In 2000/2001, among the main government agencies in 23 countries for which data were available, 35 percent of funding came from donor loans and grants. Internally generated funds, including contractual arrangements with private and public enterprises, accounted for an average of 7 percent of total funding. The 36 non-profit organizations in the sample received close to two-thirds of their funding contributions from producer organizations and marketing boards. These contributions were collected mainly through taxes on the export or production of commercial crops. The non-profit organizations were also more active than the government agencies in raising income from internally generated resources, which accounted for 26 percent of their total funding and included contracts with private and public enterprises.

Figure 9.4
Composition of funding sources since 2000

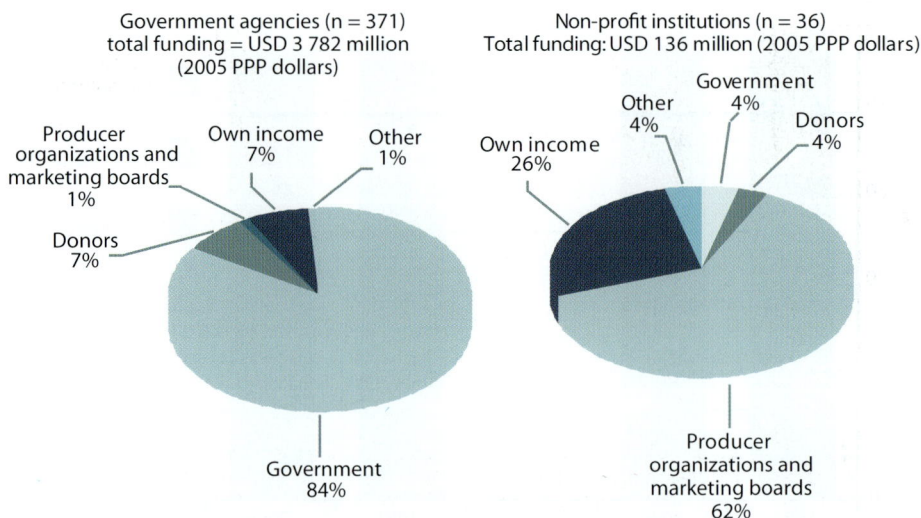

Government agencies (n = 371)
total funding = USD 3 782 million
(2005 PPP dollars)

Non-profit institutions (n = 36)
Total funding: USD 136 million (2005 PPP dollars)

Producer
organizations and
marketing boards
1%

Own income
7%

Other
1%

Donors
7%

Government
84%

Own income
26%

Other
4%

Government
4%

Donors
4%

Producer
organizations and
marketing boards
62%

Own income includes contracts with private and public enterprises. Data are for 53 developing countries; they exclude China, Nigeria and South Africa, which have large agricultural R&D investments.
Source: Echeverria and Beintema, 2009, based on ASTI data sets.

Although government allocations still represent the main source of funding, there are again considerable differences among countries. A number of developing countries depend on non-governmental sources of funding. This is mostly the result of high donor dependency, but some countries have increased the diversity of their funding sources and include considerable income from the sale of products or services, contractual arrangements with public and private enterprises, or contributions from producer organizations through taxation of exports or production.

In a sample of 62 developing countries, more than half of total FTE researchers in agricultural R&D were involved in crop research, while 16 percent focused on livestock (Figure 9.5). The remaining one-third of the researchers focused on forestry (6 percent), fisheries (5 percent), natural resources (9 percent), post-harvest (4 percent) and other agricultural disciplines. Researchers in sub-Saharan Africa and Latin America and the Caribbean spent relatively more time on livestock research than staff in Asia and the Pacific and the Near East and North Africa.

Figure 9.5
Research focus of agricultural research staff, by main subsector (shares of FTE staff)

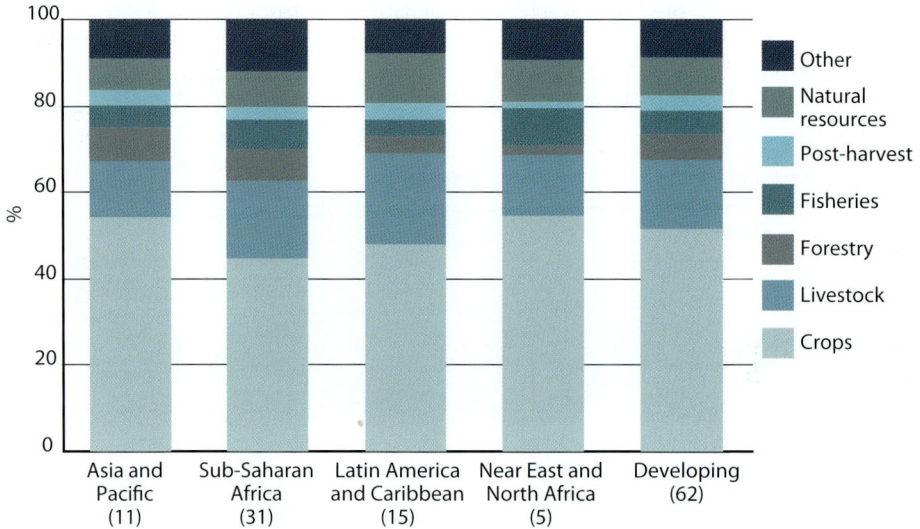

Sub-Saharan Africa data are for 2008; Asia and the Pacific for 2002; Latin America and the Caribbean for 2006; and the Near East and North Africa for 2002/2003. The "other" category includes subsectors such as forestry, fisheries, natural resources, post-harvest and socio-economics. Number of countries indicated in parentheses.
Source: Authors' calculations based on ASTI data sets.

Private agricultural R&D spending

Data on private sector investments in agricultural R&D remain very limited. In 2000 (the only year for which global estimates are available), the private sector spent an estimated USD 16 billion in 2005 PPP dollars (Figure 9.6), 41 percent of the global total. Almost all of these private sector investments were made by private companies performing agricultural R&D in high-income countries. Private sector investments in the developing world accounted for only 2 percent of total public and private agricultural R&D investments in 2000, with most being made by Asian private companies (Beintema and Stads, 2008). The private sector plays a stronger role in funding agricultural research, as many private companies contract research to government and higher education agencies, but its role in most developing countries is and will remain small, given the limited funding opportunities and incentives for private research. Most private sector research in developing countries focuses on the provision of input technologies or technical services for agricultural production, and most of these technologies are produced in high-income countries (Pardey *et al.*, 2006).

Figure 9.6
Shares of public and private agricultural research investments, circa 2000

39.6 billion in 2005 international (PPP) dollars

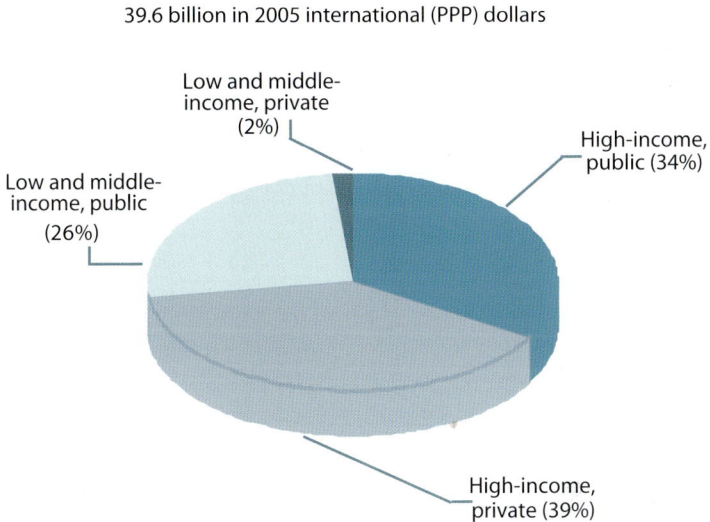

Low and middle-income, private (2%)

High-income, public (34%)

Low and middle-income, public (26%)

High-income, private (39%)

Source: Beintema and Stads, 2008, based on ASTI data sets.

There is only limited information on the level of private sector involvement over time or on the type of research that private companies are conducting. Alston *et al.* (1999) found that only 12 percent of private research in Australia, the Netherlands, New Zealand, the United Kingdom and the United States of America focused on farm-oriented technologies in 1992; the corresponding share for the public sector in these countries was 80 percent. Food and other post-harvest activities accounted for 30 to 90 percent in Australia, the Netherlands and New Zealand; and chemical research for between 40 and 50 percent in the United Kingdom and the United States of America. Pray and Fuglie (2001) found that the share of private sector investments in total agricultural R&D investments grew between the mid-1980s and the mid-1990s in China, India and Indonesia (from a sample of seven Asian countries), and that private investments grew more than public ones. However, growth in private sector investments was uneven across subsectors, with investments in the agricultural chemical and, to a lesser extent, the livestock subsectors increasing substantially, while growth in other subsectors, such as plantation crops and machinery, was slower.

International agricultural R&D investment

The majority of research is carried out by the Consultative Group on International Agricultural Research (CGIAR), which currently consists of 15 centres. The first four centres were established in the late 1950s and the 1960s, with considerable financial support from the Rockefeller and Ford Foundations. During the 1970s, the number of centres increased to 12, and the funding per centre increased over the decade. This led to a tenfold increase (in nominal terms) in total CGIAR investments. Total funding continued to increase during the 1980s, but at a slower pace. During the 1990s, however, total funding grew less than the number of centres, and per-centre spending levels could not be maintained. Since 2000, overall funding to CGIAR has increased, but a larger proportion is directed to specific research projects and programmes that also involve non-CGIAR research organizations (Beintema *et al.*, 2008; Pardey *et al.*, 2006).

There are a number of other international research providers, mostly with a regional or subregional focus. For example, the two largest non-CGIAR agencies conducting research in Africa are the International Cooperation Centre of Agricultural Research for Development (CIRAD, based in France) and the Institute of Research for Development (IRD). In Asia, the Australian Centre for International Agricultural Research (ACIAR) develops international agricultural research partnerships, but does not conduct research in the region's developing countries itself. The mandate of the Japan International Research Center for Agricultural Sciences (JIRCAS) covers all developing countries, and most JIRCAS agricultural research is in Asia. Two important regional agencies conducting agricultural research in Latin America and the Caribbean are the Tropical Agricultural Research and Higher Education Center (CATIE) and the Caribbean Agricultural Research and Development Institute (CARDI). Other agricultural research agencies are also active in these three regions (Beintema and Stads, 2006; 2008; Stads and Beintema, 2009).

Three definitions of underinvestment in agricultural R&D

As argued in the introduction, underinvestment in research can occur when: i) the rate of return on research is consistently higher than the social rate of return on alternative investments; ii) the nature of investment has changed so that the country is failing to maintain historical growth in on-farm productivity; and iii) there are gaps between current investments and the resources needed to attain goals.

Evidence from rates of return analysis

The underinvestment hypothesis is a straightforward application of marginalist economic theory: if, through policy decision or budget constraints, the social

value of the last unit of product consumed (or input employed) is greater than the social cost, there is underconsumption or underuse of the product (or input) because it would pay to borrow until the social gain and social cost are equal. If projects are ranked in descending order by their expected rates of return (the marginal efficiency of investment), and the return of the last project undertaken is higher than the social (opportunity) cost (of capital), this is prima facie evidence of underinvestment.

Hundreds of studies of the social rate of return on research consistently show that the rate of return on public investment in agricultural research is higher than either the social rate of return on capital or the other opportunities for public investment. In general, the return on public investment is higher than the private rate of return, even when allowing for the marginal excess tax burden of the tax collection system and the returns accruing to farmers. This is because it is impossible to appropriate many of the benefits associated with research by private firms (Widmer, Fox and Brinkman, 1988; Evenson and Westphal, 1995).

There is no tendency for the rate of return to decline over time, and it appears that rates of return may be higher when the research is conducted in more developed countries (Alston *et al.*, 2000).

Roseboom (2002) defines the "underinvestment gap" as the difference between the economic rate of return of a marginal R&D project and the social rate of return. Based on the distribution of projects studied, he concluded that:
Under the assumption of full information and rationality, developed countries could have invested about 40 percent more in public agricultural R&D and developing countries about 137 percent more. In terms of agricultural R&D intensity (i.e. expenditures as a percentage of agricultural GDP), developed countries could have invested 2.8 rather than 2.0 percent and developing countries 1.0 rather than 0.4 percent in the period 1980 to 1985.

Fuglie and Heisey (2007) analysed the economic returns of public agricultural R&D in the United States of America and summarized their findings as follows:

- *There appear to be significant social returns to private agricultural research. The private sector is able to capture only a share of the productivity benefits from its technology.*

- *Agricultural research generates long-term benefits. Public research undertaken today will begin to noticeably influence agricultural research productivity in as little as two years and its impact could be felt for as long as 30 years.*

- *Agricultural knowledge or research "spill-overs" across state and national boundaries are significant.*

It is important to note that the rate of return concept measures the economic benefits of agricultural investments, but not the non-economic impacts, such as environmental, social, health and cultural benefits and costs. These are important when investment decisions are being made, but are not included as they are difficult to quantify and validate (Beintema *et al.*, 2008). Furthermore, spill-overs of agricultural technologies among countries and regions account for a large share of the total social benefits of public agricultural research. When spill-overs occur, rate of return studies will overestimate the total benefits of the research investment in the receiving country, while underestimating the total benefits in the country where the technology originated (Alston, 2002; Beintema *et al.*, 2008). Pardey *at al.* (2006) state that the supply and demand for spill-over technologies are changing. Agricultural research in high-income countries is focusing increasingly on technology types that are less relevant for the agriculture sector in developing countries (especially the poorest ones). In addition, technologies have become less mobile because of stricter intellectual property rights and other regulatory policies.

Failure to maintain historical levels of productivity growth

It is sometimes necessary to reiterate the importance of productivity. Nobel Prize winner, Sir W. Arthur Lewis (1996) stated unequivocally that "an increase in agricultural productivity is fundamental to the solution of the problem of distribution since it makes possible simultaneous increases in mass consumption, saving and taxation". Although agricultural research has proved good for increasing on-farm productivity and providing spill-overs to other social goals, it is a blunt instrument for addressing these other goals directly. Other authors have underlined the importance of productivity growth. Cereal output in developing countries has grown by 2.8 percent per year for three decades, and yields, not area, were responsible for this growth. Total factor productivity has grown along with yields (Pingali, 2009). Today's investment in research drives tomorrow's growth of productivity (Fuglie and Heisey, 2007).

Recent studies point out that underinvestment in research that enhances productivity at the farm level explains the significant decline in the rate of agricultural productivity growth in developed countries. The greater share of agricultural innovations can be traced to organized, scientific and industrial R&D efforts funded by government and the private sector, but such investment has not only slowed but has also changed its focus.

Pardey (2009) notes that productivity growth in the United States of America was slower in the 1990 to 2005 period than between 1961 and 1989, and suggests several possible causes of this: bad weather, a changing regulatory environment, degradation of the natural resource base, slower growth of investment, the changing composition of agricultural research, changing private sector roles, and reduced spill-over from other countries and CGIAR. He argues that this decrease in productivity growth is partly the result of a slowdown in spending and the redirection of agricultural R&D away from maintaining or enhancing productivity.

Alston, Pardey and James (2009) point out that public investments in California's agriculture have shown benefit-to-cost ratios of 10:1, indicating substantial underinvestment in agricultural research according to the first definition. In addition to the slowing and increased variability of funding, these authors point to recent trends indicating that the extent of underinvestment in productivity-enhancing agricultural science may be worsening:

Public-sector research has drifted away from on-farm productivity enhancements toward investments emphasizing food safety and quality, human health and nutrition, and natural resources and the environment. Much of this research could have social payoffs comparable to those from farm-productivity enhancing research; but a slower rate of growth in total spending and the drift of research emphasis will result in slower rates of farm productivity growth and a decline in global competitiveness.

For developing countries, the decline in productivity-enhancing research in developed countries means that the spill-over benefits will be reduced, at the same time as climate change and economic conditions become worse. Alston, Pardey and James (2009) note that the situation will become even worse for developing countries, given the long lag before spill-over benefits occur.

This chapter identifies a "productivity growth failure", which is the difference between the historical growth rate of on-farm productivity (approximately 2 percent) and the current rate (approximately 1 percent). This situation is characterized as underinvestment when the level and composition of investment keep on-farm productivity growth below its historical trend and presumed potential.

The incremental investment needed to achieve goals and commitments

There are many prescriptive targets for investment in agricultural R&D. While they all perform useful functions in defining what should be done, their original purpose and value are often forgotten. Table 9.2 summarizes some of the most common targets and the investment needed to achieve them.

For countries with adequate policies and institutions, what are the additional needs for reducing income poverty to the desired level, and what do these imply in terms of research and other support to the agriculture sector? For countries without adequate policies, what studies and activities are needed to improve policies and institutions? If the focus is exclusively on MDG 1, the additional costs of attaining health, education and environmental goals that do not come as spill-overs from meeting MDG 1 would have to be estimated.

Table 9.2
Common prescriptive targets

Target	Underlying concept	Qualifications	Formulation[a]
Agricultural research intensity ratio (ARI)	There should be a norm for relating reinvestment in the agriculture sector to the size of the sector	Its components are more instructive than its level; there are different norms for different classes of country	ARE/AGDP from 0.2 to 2.5
Maputo Declaration: Commitment to Agriculture	Public expenditure in agriculture needs to double to achieve MDG 1	Determinants of investment needs and growth possibilities are country-specific	AE/BUD = 10%
Fiscal effort	Even low-income countries can raise the government share in the economy to 20%	Fiscal will/drag is country-specific	BUD/GDP ≈ 20%
Growth rates to achieve MDG 1: e.g., ASARECA (Omamo *et al.*, 2006)	Overall growth must be accelerated to achieve reduction in poverty and hunger	Needs to identify and prioritize sectors that can produce this growth or economy	ΔGDP/GDP = 6%
	GDP growth of 6% produces GDP per capita growth of 3% (except for in the Democratic Republic of the Congo, which starts from negative growth)	Implies threefold increase in agriculture sector and subsector growth rates. Differential growth may lead to geographic concentration	ΔAGDP/AGDP from 4.3% to 6.6%
Climate change adaptation: e.g., Oxfam, World Bank	Urgent adaptation and mitigation; net addition to current aid	Research includes more robust estimates of the economics of adaptation, study of best practices, and an intensive action-learning phase	Annual requirements of USD 10–40 billion (World Bank) and USD 50 billion (Oxfam International 2007)

[a] The formulations are discussed in detail in the next section.
ARE = agricultural research expenditure.
AGDP = agricultural GDP.
BUD = government budget (public expenditure).
AE = public expenditure on agriculture.
Δ = change in variable since the previous period.

An important target that has been elevated to a political commitment is the Commitment to Agriculture expressed in the Maputo Declaration (2003), which has been adopted by the New Partnership for Africa's Development (NEPAD) Comprehensive Africa Agriculture Development Programme (CAADP).[6] CAADP's strategy restates MDG 1 – to reduce poverty and hunger by half by 2015 – and postulates that it would require the economy to grow at 6 percent annum. As one of the largest sectors, agriculture must strive for a growth rate approaching this level (with the possibilities for growth varying widely by region and commodity subsector). The Maputo Declaration called on governments to raise their expenditures on agriculture to 10 percent of their national budgets. The simplicity of the stated target of raising the GDP growth rate to 6 percent belies the complexity of the task, and raising expenditure on agriculture to 10 percent of national budgets is not sufficient. The critical question is: If the necessary changes in policies and institutions are forthcoming, how much in additional financial resources will be needed to achieve the 2015 goals?

CAADP calls for increasing investment in four pillars:

- extended area under sustainable land management and reliable water control (USD 37 billion);

- rural infrastructure and trade-related capacity for market access (USD 37 billion);

- increased food supply through policy, technology and farm services (USD 7.5 billion) and disaster and emergency relief and safety nets (USD 42 billion);

- agricultural research, technology dissemination and adoption (USD 4.6 billion).

Under CAADP, African countries would commit themselves to:

- increasing their domestic contributions progressively, from 35 to 55 percent by 2015;

- increasing private sector contributions;

- doubling current annual spending on agricultural research within ten years; Beintema and Stads (2006) calculated that this means an average increase of 10 percent per year, which is substantially higher than the average annual growth rate of 1 percent that occurred during the 1990s;

- investing 10 percent of government budgets in agriculture.

A pioneering example of the use of CAADP targets is the strategic priorities study carried out by the Association for Strengthening Agricultural Research in

---

6. NEPAD/CAADP has become an operational programme of the African Union.

Eastern and Central Africa (ASARECA) in association with the International Food Policy Research Institute (IFPRI). This study assessed the possibility of creating a regional strategy for ASARECA's ten member countries to meet MDG 1 with respect to hunger (Omamo *et al.*, 2006).

The ASARECEA study had the beneficial effect of focusing attention on the supply side, and highlighted the information gaps for planning. In the absence of field data on the various agro-ecological zones (and lacking the time to generate them), IFPRI used crop models that predicted the expected performance of different commodities according to soil, topography and rainfall. Looking at the drivers of demand in the region, IFPRI's multi-market model helped demonstrate that regional staples, livestock products, fruit and vegetables would have the greatest impact on poverty reduction. Milk and cassava were seen as having the greatest potential for GDP growth, but such growth would be concentrated in a small number of countries. The study underlined that agricultural productivity growth alone would not be sufficient to meet poverty reduction targets; the region would also require growth in non-agricultural sectors and improvement in market conditions. This would follow identification of the best development domains for strategic investment.[7] The exercise prioritized areas of high potential, low population density and low market access, i.e., areas that require significant investment in infrastructure, markets, adaptive research and scaling up of technology.

The study concluded that under the default "business as usual" scenario, none of the ten ASARECA countries would achieve the 6 percent growth in GDP that is needed to achieve MDG 1. It was estimated that most countries would produce less than 3 percent growth in agriculture (based on historical trends and allowing for rapid growth in some countries recovering from civil war). Other development goals, such as food and nutrition security, would remain out of reach. Meeting the goals would demand a trebling of current growth rates. Not all commodities and all regions have the potential to contribute equally.

Analysing the agricultural research intensity ratio

Placing a country's agricultural R&D efforts in an internationally comparable context requires measures other than absolute levels of expenditures. The most common research intensity indicator is the agricultural research intensity ratio (ARI). This is the ratio formed by the sum of agricultural R&D investments over agricultural GDP. For two decades, the ARI was held up as an instrument of

---

7. A development domain is an area with homogenous production potential, access to markets and population density. Investment requirements differ among different development domains.

coercive comparison: if a country's neighbour with similar characteristics had a higher ARI, the presumption was that the country was not trying hard enough to support agricultural research.

The ARI first appeared in a World Bank sector paper on agricultural research in 1981. The authors were looking for a target figure for establishing a norm to which National Agricultural Research Systems (NARS) could aspire.[8] Lacking an empirical basis from the developing world, they borrowed the estimated investment in science and technology in developed countries (about 2 percent of GDP) and made this the target figure. However, this target proved to be unrealistic for low-income developing countries, largely owing to competing claims on low fiscal capacity and the large weight of the agriculture sector in the economy. Moreover, the target did not account for the more limited opportunities for innovation in developing countries (Roseboom, 2004). In addition, the expectation that agricultural R&D investments would continue to grow at the high rates of the 1980s was not met. A more realistic research intensity target of 1 percent has been recommended[9] in more recent literature (e.g., Pardey and Alston, 1995; Roseboom, 2004; Casas, Solh and Hafez, 1999).

Trends in the ratio of agricultural research intensity to GDP

The average intensity ratios for developing countries fluctuated slightly around 0.56 percent during 1981 to 2000 (Figure 9.7). This is often attributed to the fact that the denominator, agricultural output, grew at the same pace as total public agricultural research spending. In contrast, the average intensity for high-income countries increased considerably during this period. For every USD 100 of agricultural output, high-income countries spent USD 2.35 on public agricultural R&D in 2000, compared with USD 1.51 in 1981. More than half of the industrialized countries for which data are available had higher research intensity ratios in 2000 than in 1991. Most countries in the samples for Asia and the Pacific and Latin America and the Caribbean also increased their intensity ratios (Beintema and Stads, 2008; Stads and Beintema, 2009). In sub-Saharan Africa however, only six of 26 countries reported higher intensity ratios in 2000 than in 1991 (Beintema and Stads, 2006).

The use of intensity ratios is not always appropriate because they do not take into account the policy and institutional environment in which agricultural

---

8. For many observers, the 1980s was the "decade of NARS", which saw the creation of new national institutes and consolidated national systems in Africa, experiments with *fundaciones* in Latin America, and second-generation council models in Asia.

9. "Recommended" in the sense that a 1 percent target could be attained by even poor countries, if all the priority and institutional factors are functioning as desired.

---

research occurs, or the broader size and structure of a country's agriculture sector and economy. Human and capital investments have a basic fixed component, regardless of the size of a country's population, especially when facilities and services are dispersed across broad areas. Furthermore, a number of countries conduct research in areas related to the agribusiness sector, whose production value is counted as manufacturing not agriculture (and hence is not included in agricultural GDP). More important in this context, an increase in the research intensity could mean a decrease in agricultural output, rather than a higher level of investment, as was the case for a number of OECD countries during the 1990s (OECD, various years).

Figure 9.7
Intensity ratios of agricultural R&D spending (shares of agricultural GDP)

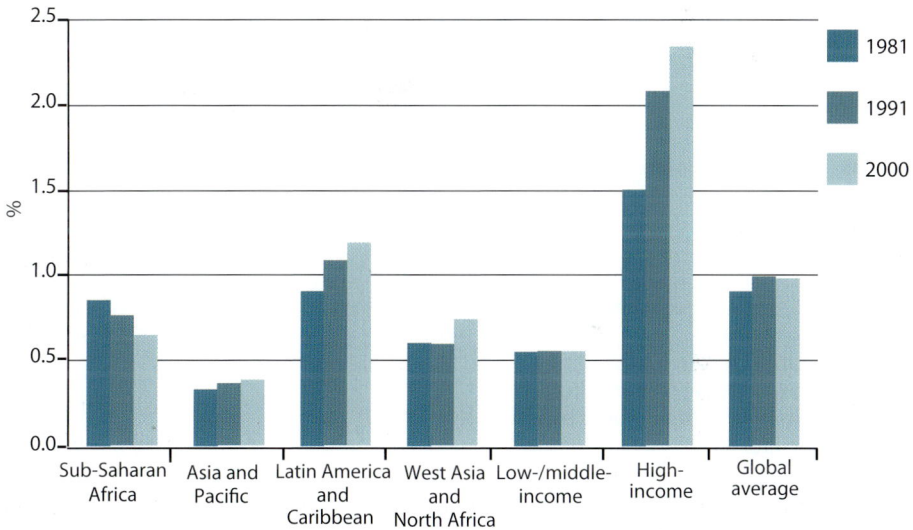

Estimates exclude Eastern Europe and the former Soviet Union. Estimation procedures and methodology are described in Beintema and Stads, 2011b and various ASTI regional reports available at www.asti.cgiar.org.
Source: Beintema and Stads, 2010, based on ASTI datasets and other secondary sources.

A number of countries, such as China and India, continue to have relatively low intensity ratios (Beintema and Stads, 2008). Nevertheless, both of these countries have increased their agricultural R&D investments significantly over the past decade or so, such that their agricultural research systems are well equipped in terms of both infrastructure and human resources. However, specific areas may require further investment. Research intensity ratios should be interpreted within the appropriate context of investment growth, human resource capacity and infrastructure.

While it is clear in cross-section that rich countries have higher ARIs than poor countries, it is necessary to examine budget details country by country to understand what drives this difference and what it implies for the contribution of research to growth and poverty reduction (Elliott, 1995).

Trends in agricultural research intensity and the research effort

The ARI alone can only be the starting point for discussion: it is necessary to create a new identity for the ARI, by decomposing it into four components, as shown in Figure 9.8.

Figure 9.8
Components of the agricultural research intensity ratio

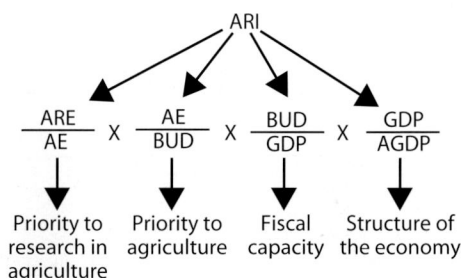

$$\text{ARI}$$

$$\frac{\text{ARE}}{\text{AE}} \quad \text{X} \quad \frac{\text{AE}}{\text{BUD}} \quad \text{X} \quad \frac{\text{BUD}}{\text{GDP}} \quad \text{X} \quad \frac{\text{GDP}}{\text{AGDP}}$$

Priority to research in agriculture   Priority to agriculture   Fiscal capacity   Structure of the economy

ARE = agricultural research expenditure.
AE = public expenditure on agriculture.
BUD = government budget (public expenditure).
AGDP = agricultural GDP.
Source: Adapted from Elliott, 1995.

The four components are:

- *priority for agricultural research:* the share of agricultural research in total agricultural expenditure;

- *priority for public agricultural expenditure:* the share of public expenditure on agriculture in total public expenditure;

- *fiscal effort (or fiscal capacity):* the share of public (government) revenue and expenditure in GDP;

- *structure of the economy:* the inverse of agriculture's share in GDP.

Each of the elements in the identity is a ratio, so the ARI itself is independent of the unit of measurement for each element. Each element has its own drivers, which are analysed as determinants of a country's efforts in agricultural research (Pardey, Kang and Elliott, 1989).

Figure 9.9
Determinants of ARI, by country income group

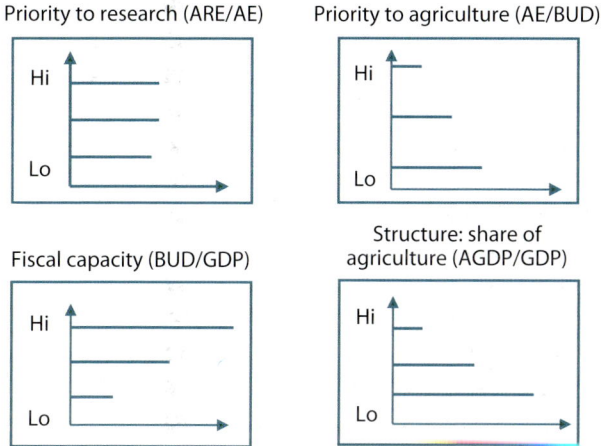

Priority to research (ARE/AE)   Priority to agriculture (AE/BUD)

Fiscal capacity (BUD/GDP)   Structure: share of agriculture (AGDP/GDP)

Y axis = level of country income per capita; x axis = ratio as defined.
Source: ASTI.

Figure 9.9 represents each of the elements in the ARI identity by country income group (low, middle and high).

The following stylized facts are illustrated in Figure 9.9:

- The share of expenditure on agricultural research in total agricultural expenditure is fairly similar across income levels (upper-left quadrant).

- Although total expenditures on agriculture are low in absolute terms, in low-income agriculture-based economies they represent a higher share of total public expenditure than in wealthier countries (upper-right quadrant).

- The fiscal capacity (tax collections, public budgets) accounts for a far smaller share of GDP in low-income than in higher-income countries (lower-left quadrant).

- The agriculture sector's share in the economy declines as the country's income rises (lower-right quadrant).

Changes in the ARI at the country level require very country-specific analysis of the drivers of each ARI element. Policy-makers' commitments to invest more in agricultural R&D can be measured against the realism of their targets, the coherence of their strategies and priorities, their political and fiscal capacity, and the weight of the sector they are trying to move. In the most developed countries, agricultural research intensity is rising. As the growth of higher education expenditures also rises as income rises, it could be asked whether this investment is all productive or includes an element of income-elastic consumption of research made possible by rising fiscal resources and a declining share of agriculture in the economy. Countries in the middle-income group, where non-agriculture is growing, have an opportunity to shift tax burdens away from agriculture, invest in infrastructure and other public goods, and improve incentives that reinforce agricultural development. These become easier as the share of agriculture in the economy falls. In low-income, agriculture-based economies, it is difficult to raise the ARI where the fiscal base is small, the agriculture sector is large and the relative cost of a researcher is high.

The following subsections highlight issues concerning underinvestment in research that originate in each of the four components of the ARI. As yet, there is no structured, cross-country information that can "unpack" each of the drivers of the ARI. This has to be done at the country level, by policy-makers seeking to understand the points of intervention that will improve investments in agriculture.

Priority for research – the share of agricultural research in public agricultural expenditure: The first determinant of the ARI is the priority given to agricultural research within the overall effort to develop agriculture.

Studies by IFPRI have suggested that agricultural research in low-income economies continues to be the most productive investment in support of the agriculture sector, followed by education, infrastructure and input credits. "Disaggregating total agricultural expenditures into research and non-research spending reveals that research had a much larger impact on productivity than non-research spending" (Fan and Rao, 2003).

Donor programmes, especially in Africa, can have an important impact on the allocation of resources. Programmes for highly indebted poor countries aim at social goals, and public expenditure reviews point out that this has affected the selection of projects within sectors, including agriculture (Bevan, 2001).

The domestic political economy of budget allocations needs to be better understood. For example, in India, overall public expenditure on agriculture has remained at approximately 11 percent of the budget, while the share of subsidies for fertilizer and electricity, and support prices for cereals, water and credit have steadily risen at the expense of investment in R&D, irrigation and rural roads (World Bank, 2008b; Beintema Stads, 2008).

In some more scientifically advanced middle-income countries, the higher education sector has become a major player in agricultural research. In Argentina, Costa Rica, Honduras, Mexico and Uruguay, for example, higher education accounted for more than 40 percent of agricultural research, with government funding coming mostly from ministries of education. In South Africa, funding for the Agricultural Research Council comes through the Council of Science and Technology (with input from the National Department of Agriculture).

In North America, the changing composition of agricultural research expenditure is a recent concern: the share of research directed to farm-level productivity enhancement has fallen to as low as 60 percent (Pardey, 2009; Alston, Pardey and James, 2009; Fuglie and Heisey, 2007).

While research investment undoubtedly has value beyond the farm-gate, the long-term slowdown in research into productivity growth at the farm level is cause for concern, for three reasons: i) cumulative loss in productivity growth translates into a significant loss of future income; ii) there is an accompanying loss of potential spill-over to neighbouring states (which may have accounted for as much as 50 percent of measured research benefits); and iii) there is the risk of missing out on research discoveries that will be needed ten to 20 years from now, as the world confronts the impact of climate change:

Given research lags that may be as long as 10-20 years, the effect of this slow-down in developed countries will become apparent in the future when scarcity of land and water, the impact of climate change, and population pressure will become major problems for developing countries. The stream of research outputs which have travelled fairly freely will be reduced significantly. (Alston, Pardey and James, 2009)

Recent studies in Canada have also documented a slowdown in productivity growth, linked to declining public research investment and structural changes in the sector, which have led to calls for more public sector research expenditure:

- Veeman, Unterschultz and Stewart (2007) found that R&D expenditure for Canadian agricultural research has shown no growth since 1990 and that total factor productivity (TFP) growth in the prairie crop sector fell to an average of 0.51 percent per year for the 1990 to 2004 period, which is much lower than historic rates of close to 2 percent per year.

- Gray and Weseen (2008) argue that the slowdown in productivity growth highlights a need for more effective research expenditure.

- While noting that the private sector has filled the applied research gap in key crops, the Canadian Grains Council (CGC, 2008) argues that the private sector has concentrated on recombinant DNA technologies, which

369

reduces the possibilities for sharing and coordination. CGC emphasizes the importance of: i) sharing discoveries; ii) developing polices that protect plant breeders, small-scale seed producers and niche developers; and ii) facilitating greater collaboration among public and private sector research partners.

The composition of agricultural research expenditure is also changing in some middle-income countries, such as Argentina, where research into food safety, food technology and processing are budgeted to the national agricultural research institute. However, as the increase in GDP occurring further down the value chain is counted in the manufacturing sector, the apparent rise in ARI is partly an accounting phenomenon.

Priority for agriculture – the share of agricultural expenditure in total public expenditure: The second component in the ARI is the share of agriculture in total public expenditure. This ratio is subject to many different drivers, including the following:

- *The influence of the domestic political economy:* In their review of medium-term expenditure frameworks, Akroyd and Smith (2007) point out the difficulties of budgeting in a "neo-patrimonial political model" and cite Birner and Palaniswamy (2006) on political challenges to increased spending on agriculture. These challenges include farmers' low political voice, lack of knowledge about agriculture's potential for pro-poor growth, and, possibly, the negative experiences that donors and governments have had of prior agricultural programmes.

- *The impact of donor programmes:* Fan and Rao (2003) point out that structural adjustment programmes have increased the size of government spending, but not all sectors have received equal treatment. In Africa, expenditures on agriculture, education and infrastructure all declined as a result of structural adjustment programmes.

In sub-Saharan Africa, CAADP reports that seven countries have reached or exceeded the Maputo target of spending 10 percent of their budgets on agriculture. For agriculture-based economies, the difficulty lies in the next two components: fiscal capacity (the share of tax collections and expenditures); and the agriculture sector's importance in the economy. To transform an economy, one opportunity is to shift tax collection to a growth base in non-agriculture and to begin net reverse flows of public funds to agriculture. In transforming economies, fiscal policies can make or break a pro-agriculture strategy. This factor leads into consideration of the fiscal effort or fiscal capacity of a country.

Fiscal effort – the share of government revenue and expenditure in the economy:
A government that can raise and spend 20 percent of GDP through tax collection
can do more than a government that can raise and spend only 12 percent. This
includes being able to spend more on agriculture and agricultural research. How
a country raises its revenues and how it spends its budget are specialized fields.
This subsection is concerned with policy decisions that should involve some input
from agricultural policy advisers.

This starts with a look at the revenue side. The question of whether a country's
fiscal effort and taxation of agriculture are appropriate can only be answered in
the light of the specific constraints facing that country. Constraints could include
the nature of the country's taxable bases, incentive structures, fiscal structure and
fiscal culture. The following are common issues in designing fiscal policies with
agricultural development in mind:

- *Taxable bases:* Countries with agriculture as their principal resource have
  historically overtaxed the sector through biased macroeconomic policies
  and export taxes and marketing board surpluses. Oil- or mineral-rich
  countries with large agricultural populations have an opportunity to free
  agriculture from poor terms of trade and local taxation that discourage
  production. Failure to do so is often the cause of countries suffering from
  the "curse of wealth".

- *Fiscal structure:* Decentralization of fiscal responsibility to the state/
  province and district levels may be a positive factor in raising revenue
  by bringing services and taxation together in the minds of the population.
  However, districts may also introduce levies on local agriculture and trade
  for revenue purposes that are unnecessary disincentives for development
  if they could be substituted by federal grants.

- *Fiscal culture:* Low revenue collection and low government services
  may result from a variety of circular problems and pathologies: low tax
  rates, excessive exemptions, lax tax administration, widespread non-
  compliance and corruption; or problems of central versus decentralized
  accountability. Turning the culture around may be a long-term effort.

- *Fiscal returns on public investment:* Easterly (2007) argues that planners
  have to be aware of the fiscal effects of public investment. Benefit-cost
  analysis focuses on social costs and benefits, but the fiscal returns and
  benefits of early pay-back from increased production and exports should
  also be kept in mind.

- *Impact of taxes on key sectors:* In the post-conflict Ugandan economy, the World Bank decided that raising Uganda's fiscal effort above its low 12 percent would have been counterproductive at a time when attracting private sector reinvestment in key agricultural activities was crucial for post-civil war recovery. Future tax collections would come from expanding the base, rather than raising the average rate of taxation. (Kreimer *et al.*, 2000).

The other aspect of the government's role is the efficiency of its expenditure. Do projects meet all the priority criteria? Does the budget process allocate funds in that direction, and is this the way the funds will be spent? The remainder of this subsection looks at the effectiveness of public expenditure in agriculture and its link back to underinvestment and proposals for dealing with it.

This discussion starts with a few general observations:

- It is easier to make progress on revenue reform than on expenditure. It only takes a handful of people to design a regulation or tax reform, but it is impossible to subject all activities to a benefit-cost analysis at the project level. Such detail is necessary because it is a mistake to lump all, for example, roads (or all agricultural projects) into one bundle and say "we do roads (or agriculture)" (Harberger, 2009).

- Agricultural research organizations have assimilated the tools of planning and priority setting, but they are largely absent from the budget discussions where trade-offs are made. Decision-makers rarely have the time or information to make informed choices among projects that have different fiscal and social profiles.

- Donor programmes, especially in sub-Saharan Africa, have had an impact on broad priorities, but have not necessarily been able to control expenditures.

The World Bank introduced medium-term expenditure frameworks as part of Poverty Reduction Strategy Programmes. They were supposed to ensure that expenditures were driven by policy priorities. Various reviews have highlighted their successes and failures:

- The medium-term expenditure framework in Uganda has been successful in shifting expenditure composition, most notably in favour of education, and protecting priority sectors against cuts. It has been less successful in ensuring that budget allocations translate reliably into actual expenditures (Bevan, 2001).

- The Nigeria Agriculture Public Expenditure Review pointed out seven areas of concern, including discrepancies between policies and expenditures, off-budget funds, lack of information about the functional areas of public spending in agriculture, and poor data quality for planning and impact analysis (Mogues *et al.*, 2008).

- The frameworks failed to link budgets with strategies and policies; spending patterns were not pro-growth or pro-poor; there was a high degree of centralization in spite of decentralization plans; there was low execution capacity; donor funding was not integrated; and there were poor tracking and monitoring (Fan, 2009).

As with any budgetary and control mechanism, there were loopholes in the process: ring-fencing of certain types of expenditure (e.g., drought relief), supplementary budgets, and donor support that bypassed the mechanism. In the final analysis, it was concluded that the reform of budgetary processes requires major cultural changes for some countries and development of the capacity for implementation.

Towards more effective financing of research – the interaction of revenue collection and allocation mechanisms: Before leaving this somewhat structuralist view of the ability to finance research, it is worth noting that the source of funding affects the nature of the research carried out. Partly in response to the problems mentioned in the previous subsection, donors have been searching for effective and innovative funding mechanisms that will result in more efficient and successful research agencies and systems. The "new" school of public administration argues that not all public goods need to be produced by the public sector itself, and that research deals with many cases of "impure" public goods. This opens up both investment in and delivery of quasi-public R&D results through many forms of partnership with interested producers and beneficiaries.

Echeverria and Beintema (2009) define effective financing as "one that increases the average returns of current levels of investment in agricultural research and that also attracts complementary investment from additional sources. An effective funding mechanism will then be the one that allows optimum use of research infrastructure to execute the research." Because of the underinvestment in agricultural R&D, policy-makers and research managers will find a right mix of various financing mechanisms in addition to the direct allocations from central and/or regional public budgets. As mentioned earlier, government support to agricultural R&D has stagnated or declined in many countries, especially when measured in inflation-adjusted terms. Governments have also hampered the performance of agricultural R&D agencies because actual disbursements have fallen behind earlier budget allocations.

Echeverria and Beintema (2009) list a number of alternative funding mechanisms that bring the sources of funding closer to the prospective beneficiaries of research, or that permit project-level control of expenditure:

- Competitive grants often complement direct government budget allocations, and have played an important role in mobilizing research actors around specific outputs and improving the efficiency and accountability of research outputs and actors. On the other hand, they may not be as effective as core funds in ensuring long-term capacity. Furthermore, competitive funding schemes mostly fund specific projects, and often cover only their operational costs and not salaries or maintenance of the institutional infrastructure.

- Producer check-offs and export levies are mostly collected through taxes raised on the export or production of commercial crops. One benefit is that farmers are more involved in setting the research priorities. They finance "club goods and services", which is a form of benefit restriction. Such para-fiscal levies come from the industry itself, but may not be available at the times of major crisis when they are most needed.

- A number of agencies and countries have been successful in commercializing their research outputs, often through partnerships with the private sector. One important downside is that in many countries the revenue from commercialization goes directly into the government treasury, so there is limited incentive for research agencies to sell research outputs and services.

- The debate about programme versus project funding continues (see also the paragraph on CGIAR investments in the subsection on International agricultural R&D investment). When donors talk about shifting financing from the supply side (institutional commitment) to financing results it is often a prelude to reducing the overall level of funding.

The structure of the economy – the inverse of the share of agriculture in GDP:
The final element underlying movements in the ARI is the inverse of the share of agriculture in GDP.

In a successful transformation, the share of agriculture in GDP, the share of population in agriculture in the total population, and the dependence on agriculture as the source of development finance all drop. This is made possible by rising productivity in agriculture and the transfer of labour to other sectors.

In transforming economies, agricultural and fiscal policy should be able to make the breakthrough to more sustainable support for agricultural research:

i) better macro policies usually improve the opportunities for agriculture; ii) new tax bases outside the agriculture sector remove some of the fiscal drag caused by agricultural taxation; and iii) a large population in non-agriculture can make significant contributions to a declining population in agriculture.

The policy lesson for governments would be to maintain a macroeconomic balance and the positive environment for agriculture that it creates. Productivity increases will free both land and labour, and policies should facilitate the movement of people out of agriculture as they are no longer needed on farms to feed the country. The point is not to maintain millions of small farmers but to eliminate poverty, with recourse to safety nets where agricultural and overall growth are not enough (Valdés and Foster, 2005).

New analytical challenges coming from ReSAKSS

Recently, the Regional Strategic Analysis and Knowledge Support System (ReSAKSS) has synthesized much of the available information on investment in agricultural R&D, to link expenditure with performance in Africa. The complete report is available in Omilola *et al.* (2010), and the findings are consistent with issues raised in this chapter. Observations in the report that merit further analysis and explanation include the following:

- While some African countries have increased their budgetary allocations to the agriculture sector, the majority have generally stayed in the same budgetary allocation grouping, especially those countries with low spending rates.

- In West Africa, actual public expenditure on agriculture increased in most countries (with the exception of those in crisis or post-conflict situations).

- Regardless of the distribution of agricultural expenditures, there are two trends emerging in West Africa:

    - First, in the Sahelian countries, the majority of expenditures were directed towards investment, while in the coastal countries, a large share of agricultural expenditure was devoted to recurrent expenses.

    - Second, in the Sahelian countries where expenditures were largely spent on investments, agricultural spending was predominantly financed by official development assistance (ODA). Internal sources accounted for less than 25 percent of agricultural funding. The share of internal resources in agricultural investment was much higher in the coastal countries, reaching 67 percent in Ghana, 57 percent in Côte d'Ivoire, and 48 percent in Benin.

- There is a negative correlation between the share of agricultural expenditure devoted to investments and the share of agricultural investments financed from internal sources. This dependence on external sources for investments is attributed to the dismantling of programmes of public support to agriculture under structural adjustments.

- The capacity to disaggregate expenditure by subsector and function varies greatly. The available data point to a lack of congruence between expenditure on crops versus livestock and their relative shares in agricultural GDP. It was even more difficult to obtain data disaggregated by function: R&D, irrigation, inputs and equipment, extension and administration.

The focus then turns to the quality of expenditures. Looking in greater detail at Malawi, Nigeria and Zambia, the authors find that agricultural spending has increased in all three, largely from government sources. Adding Ghana to their observations, they note that countries all focus on one major programme: cocoa in Ghana, and input supply in the other three (raising questions about the sustainability and balance of agricultural spending).

With the resurgence in development assistance of recent years, the need for clear links between plans, budgets and outcomes calls for improved data and analysis on the issues raised in this chapter.

Challenges and essential investments

Investment options targeting special non-productivity objectives

This chapter has argued that research oriented towards enhancing farm-level productivity has being shown to have high rates of return and make generally positive contributions to environmental and social objectives. Other policy instruments can be designed (e.g., safety nets, facilitation of migration, and payments for the true value of resources and ecosystem services) to ensure that society gains.

The International Assessment of Agricultural Science and Technology for Development (IAASTD) (Gurib-Fakim *et al.*, 2008) identifies some of the directions in which new research can make a direct impact on sustainability and social goals (Table 9.3)

Investment in basic capacity for R&D

This section highlights three basic areas of underinvestment:

- the need for basic studies and methodologies – even a country considered too small to have a full-fledged NARS needs to invest in knowing: i) its own potential in terms of water resources, soils and climate; ii) where it can obtain knowledge, science and technology to realize this potential; and iii) sufficient advanced science to be a good negotiator of partnerships and purchaser of technology;

- the need to address capacity requirements in a systemic way that includes balanced growth of research institutes, universities and upstream and downstream partners;

- the need to integrate networks at the global, regional and subregional levels, while escaping high transactions costs and dispersion of effort.

Table 9.3
Investment options outlined in the IAASTD Global Report

Goal	Investment required to	Examples and comments
Environmental sustainability	1. Reduce the ecological impact of farming systems	Management practices; reduced use of fossil fuel, pesticides, fertilizer; biological substitutes for fossil fuels and chemicals
	2. Enhance systems that are known to be sustainable	Social science research on policies and institutions.
	3. Support traditional knowledge	Non-conventional crops and breeds; traditional management systems
Hunger and poverty reduction	1. Target institutional change in organizations	Planning with a pro poor perspective
	2. Include equity in planning and pro-poor policies	Access to resources; sharing of benefits from environmental services
Improving nutrition and human health	1. Improve nutritional quality and safety of food	Coexistence of obesity and micronutrient deficiency; pesticide residues; sanitary and phytosanitary measures
	2. Control environmental externalities	Pollution; overuse of antibiotics, pesticides; on-farm diversification
	3. Ensure better diagnostic data and response to epidemic disease	Increasing zoonotic diseases and dangers of pandemics; prediction of disease and pest migration with climate change
Economically sustainable development	1. Enhance research on water use and control of pests and diseases	Both affected by population growth and climate change
	2. Carry out productivity-enhancing research to save land and water as limiting factors	TFP benefits from higher yields per hectare and more crop per drop; need to address the most limiting factors
	3. Establish prices and incentives that promote proper social use of resources	Pricing policies and payment for ecosystem services make land and water use more efficient
	4. Advance basic research in genomics, proteomics and nanotechnology	Historically high rates of return on basic research; applications may spill over to developing countries in the future

Source: Adapted from Gurib-Fakim *et al.*, 2008: 381-384, Table 6.2, and Chapter 8, Section 4.

Basic studies and methodologies: Decisions about investment are ultimately based on two judgements: What are the possibilities of advancing knowledge and technology? and What is the value to society of the new technology? (Ruttan, 1982). Processes for making decisions increasingly involve a combination of supply-led analysis of expected gains prepared by scientists, and a participatory (bottom-up) evaluation of the usefulness of the knowledge to clients and beneficiaries. Both the governance of the process and the nature of the evidence have to be appropriate to the level and nature of the decision to be made.

The need for basic studies is apparent in the three approaches being adopted to address priorities and strategies for global agricultural R&D (CGIAR, 2009):

- The trust in models approach includes definition and characterization of the systems that will form the building blocks for assessing agricultural, environmental and institutional/policy research challenges and opportunities; and evaluation of the nature and scale of potential R&D-induced impacts (by system), according to scenarios and parameter estimates established during the elicitation process.

- The trust in front-line researchers approach designs and implements a science-focused identification of appropriate technical, institutional and social variables to be used in assessing the potential impact of research-induced change.

- The trust in wisdom approach draws on consultation with highly recognized research and policy leaders as reviewers and on stakeholder and partner dialogues.

Modelling and spatial analysis tools can be used to identify homogeneous development domains. However, basic information needs to be collected and processed. In all areas, there is an urgent need for better information and tools on hydrology, meteorology and soil potential that goes beyond research strictly devoted to agriculture.

In some areas, the challenges are likely to grow with climate change, population growth and increasing resource scarcity. The current level of research can therefore be considered underinvestment even if it is congruent with the current importance of the issue. Given that agricultural and land-use practices contribute 32 percent of global greenhouse gas emissions (Stern, 2007), the need for better understanding is clear. In this respect, IAASTD highlights the need for strategic cross-disciplinary methodological research on environmental sustainability and poverty reduction:

The first important need for [agricultural knowledge, science and technology] *AKST investment is for social and ecological scientists working with other* scientists to develop methodologies and to quantify the externalities of high and low external input systems from a monetary perspective as well as from other perspectives such as the concept of energy flows used in energy evaluations. Evidence on these externalities' potential implications on food security also needs to be analysed. (Gurib-Fakim et al., 2008)

Both neoclassical economists and agro-ecologists agree that proper pricing of resources and the value of ecosystem services has been underexamined. While they may disagree on the need for markets versus the need for taxes and subsidies, their shared recognition of this continuing need argues for openness to all solutions.

Institutional capacity for agricultural research and higher education: More investment is needed to reverse the general underinvestment of the last decade, meet the various political targets, and prepare for the emerging challenges outlined in the previous two subsections. However, this presumes that there is either sufficient research capacity to address these targets or the commitment to invest in creating it. Moreover, the rate at which research capacity can grow is linked to the strength of the higher education system. In many countries, this system itself requires re-tooling. Plans that project annual growth of 10 percent or more in current research expenditures should therefore be reviewed carefully so that good intentions do not result in wasteful expenditures that put pressure on scarce human and institutional resources.

Several organizations and publications have expressed concern in this regard:

- An assessment of the national agricultural research and extension systems in Africa found many agencies with professional staff shortages, vacant staff positions and an ageing pool of professional research staff (FARA, 2006).

- A recent study by the ASTI initiative covering 14 countries showed that although the numbers engaged in agricultural research and higher education had increased by 20 percent from 2000/2001 to 2007/2008, two-thirds of this increased capacity was trained only to B.Sc. level (Beintema and Di Marcantonio, 2009).

This is a worrisome trend, especially in light of the increasing costs of postgraduate training abroad, and the diminishing relevance of these programmes to Africa. This calls for an expansion of postgraduate-level training in agricultural sciences (World Bank, 2007).

Although the number of agricultural science universities and faculties has grown substantially during the past three decades, many suffer from staff shortages, insufficient funding, declining student enrolments, outdated curricula, and a continuing focus on undergraduate studies (Beintema, Pardey and Roseboom, 1998; IAC, 2004; World Bank, 2007). Donor support for training programmes waned in the 1990s, and African governments have largely been unable to fund training themselves (Beintema, Pardey and Roseboom, 1998). Eicher (2006) has highlighted the sequential rather than balanced way in which agricultural research, extension and higher education have been addressed in sub-Saharan Africa. A more balanced development of the agricultural knowledge system is needed.

Some initiatives are already addressing this problem. The Rockefeller Foundation established a programme for training future teachers of biometrics so that African universities are able to meet the demand for this basic research skill, which has been neglected in recent years. A number of countries have recently established postgraduate training programmes, but these are generally small in terms of student enrolments. There is increasing recognition of the need to expand Africa's postgraduate training in agricultural sciences, at both the national and regional levels (World Bank, 2007).

Current discussions of capacity go beyond the usual discussion of scientific and technical skills. Reviews of agricultural education and training by the World Bank (2007) and research institutions mention three needs: i) scientific capacity; ii) "soft skills" for innovative work across institutional boundaries; and iii) institutional capacity to learn and change. For the first of these needs, M.Sc. and Ph.D. students are an essential part of the research infrastructure.

Among the soft skills needed are postgraduate education and building of the personal skills that facilitate work across ministerial and sectoral boundaries. Institutional policies that facilitate cross-institute and cross-sectoral collaboration are being put in place by research institutes and universities. Policies aimed at increasing women's participation should also be established; given the growing concern over declining agricultural research capacity, women's increased participation in agricultural R&D is important for not only gender balance, but also for tapping substantial additional human resources for agricultural R&D.

These processes are important because the training lags and transaction costs involved in adopting new agendas affect the rate at which the research system can grow without wasting resources.

Policy and the institutional architecture for research: Policies, institutional arrangements and governance of research all require investments for which they must compete with the actual performance of scientific and technical research. Attempts have been made to measure the productivity of social science and policy research, establish the value of institutional changes or management improvements, and examine the cost of governance. This is where the concept of the process being as important as the product comes up against the burden of high transaction costs.

As noted in the subsection on Basic studies and methodologies, the planning of investment requires a process in which the issues are properly framed, relevant information is collected and used, different perspectives are integrated, and some form of governance mechanism is used to oversee implementation. There is need for better basic information, methodologies and models to support decision-making.

The structure of global agricultural research is undergoing a period of important changes. The alliance of research centres supported by CGIAR is forming a consortium that will negotiate the core functions and mega-programmes to be supported by a consolidated donor fund. It is essential that this fund provide a guaranteed core on which a sustainable system can be built. In sub-Saharan Africa, bilateral and multilateral donors are promoting the creation of regional research programmes and subregional centres of excellence in areas that go beyond the previous research networks. The development of a more effective global system that meets stakeholder needs is receiving investment of time and resources.

Many emerging problems, such as climate change, migratory pests and pandemics, are transboundary in nature, and new mechanisms are needed for addressing them effectively.

Legal frameworks, particularly relating to intellectual property and biosafety are affecting both the concentration of research activity and access to strategic genes. New institutions such as the African Agricultural Technology Foundation have been set up to facilitate developing countries' access to proprietary technology. Recent attention to biosafety has resulted in regulations that may have the unintended consequences of either keeping certain potentially valuable technologies out of developing countries, or concentrating ownership further in the hands of large corporations able to bear the costs of the required processes and in countries with large enough markets to justify these costs. Research into proper frameworks for ensuring that developing countries benefit from new science and technology is a priority for investment.

Concluding remarks

The rate of growth in agricultural research investment has been declining globally, and a large number of developing countries experienced negative growth rates during the 1990s, especially in sub-Saharan Africa. Since the turn of the millennium, a number of mostly larger African countries have increased their commitments to agriculture and agricultural research, but many countries are still experiencing declines in agricultural spending, and even for those experiencing growth, levels are often still below those necessary to sustain national agricultural R&D needs. General underinvestment is evidenced by: i) the continuing high rates of return on research demonstrated in studies at the commodity level; and ii) macroeconomic studies showing that the relevant MDGs cannot be reached without a doubling or even tripling of research investment, given the estimated growth-poverty reduction elasticities, Also of concern is new evidence that a change in the composition of research away from farm-level productivity enhancement is statistically related to a decline in the growth of agricultural productivity to below historical levels in advanced economies. This trend may be considered another form of underinvestment that reduces potential spill-overs for the future. Policy-makers are reminded that growth in agricultural productivity provides the consumption, savings and taxes needed for development and the attainment of social goals.

Capacity in agricultural research is increasingly concentrated in a few leading countries in each region. While efforts are under way to create new structures and mechanisms for collaboration across the global, regional and national levels, policy-makers are reminded that no country is too poor or too small to support a national effort that is sufficient for it to gain from global knowledge. Various investment targets have been adopted over the years, such as CAADP's for making public expenditure on agriculture equal to 10 percent of the national budget. Seen from the results side, investment should be sufficient to produce 6 percent growth in agricultural production (or to meet MDG 1). However, such targets do not provide guidance on their own feasibility or on how rapidly the institutional and human capacity can be built up to achieve them.

One of the main indicators for comparing relative R&D investment levels is the ratio of agricultural research investment to agricultural output – the ARI. Many have suggested an ARI of 1 percent as being an appropriate target for low-income agriculturally based countries. However, the ARI is influenced by several factors that need to be studied in depth at the country level. The ARI can be decomposed into four components: i) priority for research in agricultural expenditure; ii) priority for agriculture in total public expenditure; iii) fiscal capacity measured as the ratio of public expenditure to GDP; and iv) the (inverse of the) share of agriculture in GDP. Analysis of each of these elements in a country's effort highlights the importance of strategy and priorities, institutions and incentives, public sector finance and public expenditure management, and the role of global partners.

Emerging challenges, such as adaptation to climate change and increasing variability of weather, water scarcity and price volatility in global markets, will confront many of the countries that are least able to adapt to existing stresses. This lends increasing importance to the development of human and institutional capacity in agricultural research at the national level, to enable interaction with regional and global efforts. A systemic approach to planning will bring universities and research institutes closer together.

References

Akroyd, S. & Smith, L. 2007. *Review of public spending to agriculture: Main study and country case studies.* A joint DFID/World Bank Study. Oxford, UK, Oxford Policy Management.

Alston. J.M. 2002. Spillovers. *Australian Journal of Agriculture and Resources Economics,* 46: 315–346.

Alston, J.M., Pardey, P.G. & James, J.S. 2009. Setting agricultural science strategy in tumultuous economic time. *California Agriculture*, January–March 2009.

Alston, J.M., Pardey, P.G. & Smith, V.H. 1999. *Paying for agricultural productivity.* Baltimore, Maryland, USA, and London, Johns Hopkins University Press.

Alston, J.M., Chan-Kang., C. Marra., M.C., Pardey, P.G. & Wyatt, T.J. 2000. *A meta-analysis of rates of return to agricultural R&D:* Ex pede herculem? IFPRI Research Report No. 113. Washington, DC, IFPRI.

Beintema, N.M. & Di Marcantonio, F. 2009. *Female participation in African agricultural research and higher education: New insights.* Washington, DC, and Nairobi, IFPRI and CGIAR Gender and Diversity Program.

Beintema, N.M., Pardey, P.G. & Roseboom, J. 1998. *Educating agricultural researchers: A review of the role of African universities.* EPTD Discussion Paper No. 36. Washington, DC, IFPRI.

Beintema, N.M. & Stads, G.J. 2006. *Agricultural R&D in sub-Saharan Africa: An era of stagnation.* ASTI Background Report. Washington, DC, IFPRI.

Beintema, N.M. & Stads, G.J. 2008. *Diversity in agricultural research resources in the Asia-Pacific Region.* ASTI Synthesis Report. Washington, DC, and Bangkok, IFPRI and Asia-Pacific Association of Agricultural Research Institutions.

Beintema, N.M. & Stads, G.J. 2010. Public agricultural R&D investments and capacities in developing countries: Recent evidence for 2000 and beyond. Background note prepared for the Global Conference on Agricultural Research for Development (GCARD), Montpellier, France, 28–31 March 2010.

Beintema, N.M. & Stads, G.J. 2011a. *Agricultural R&D in the new millennium: Progress for some, challenges for many.* Food Policy Report. Washington, DC, IFPRI.

Beintema, N.M. & Stads, G.J. 2011b. *ASTI toolkit – Monitoring agricultural R&D capacity and investment indicators: A practitioner's guide to ASTI's methodologies and data collection standards.* Washington, DC, IFPRI.

Beintema, N.M., Koc, A., Anandajayasekeram, P., Isinika, A., Kimmins, F., Negatu, W., Osgood, D., Pray, C., Rivera-Ferre, M., Santhakumar, V. & Waibel, H. 2008. Agricultural knowledge, science and technology: Investment and economic returns. *In* B.D. McIntyre, H.R. Herren, J. Wakhungu and R.T. Watson, eds. *Agriculture at a crossroads: International assessment of agricultural science and technology for development global report*, chapter 8. Washington, DC, Island Press.

Bevan, D. 2001. *The budget and medium-term expenditure framework in Uganda.* Africa Region Working Paper Series No. 24. World Bank. www.worldbank.org/afr/wps/wp24.pdf.

Birner, R. & Palaniswamy, N. 2006. Financing agricultural development: The political economy of public spending on agriculture in sub-Saharan Africa. In *Proceedings of the German Development Economics Conference,* Berlin, Verein für Socialpolitik, Research Committee Development Economics.

CAADP. 2009. *A review of the Comprehensive Africa Agricultural Development Programme (CAADP): a focus on achievements.* Presentation by Prof. Richard Mkandawire. www.nepad.caadp.net.

Casas, J., Solh, M. & Hafez, H. 1999. NARS in the WANA region: An overview and a cross-country analysis. *In* J. Casas, M. Solh and H. Hafez, eds. *The National Agricultural Research Systems in the West Asia and North Africa region*, chapter 9. Aleppo, Syrian Arab Republic, International Centre for Agricultural Research in the Dry Areas, FAO, Association of Agricultural Research Institutions in the Near East and North Africa, and International Centre for Advanced Mediterranean Agronomic Studies.

CGC. 2008. *Creating an environment for the successful commercialization of Canadian crop innovation. Final report. Outcomes of a collaborative review by Canada Grains Council (CGC) and industry stakeholders.* www.canadagrainscouncil.ca.

CGIAR. 2009. *CGIAR strategy and results framework (SRF): Technical design and implementation meeting.* Washington, DC.

Easterly, W. 2007. *Walking up the down escalator: public investment and fiscal stability.* World Bank Policy Research Working Paper No. 4158. Washington, DC, World Bank.

Echeverria, R.G. & Beintema, N.M. 2009. *Mobilizing financial resources for agricultural research in developing countries: Trends and mechanisms.* Rome, Global Forum for Agricultural Research.

Eicher, C. 2006. The evolution of agricultural education and training: Global insights of relevance for Africa. Washington, DC, World Bank, AFTHD. (unpublished paper)

Elliott, H.E. 1995. The financing dilemma for agricultural research systems: New evidence and policy challenges. *In* ISNAR, ed. *Proceedings of the roundtable on research policy and financing in an era of adjustment.* Pretoria, South Africa, 27 June 1995. The Hague, International Service for National Agricultural Research.

Evenson, R.E. & Westphal, L.E. 1995. Technological change and technology strategy. *In* J. Behrman and T.N. Srinivasan, eds. *Handbook of development economics.* New York, Elsevier Science.

Fan, S. 2009. Agricultural public expenditure reviews: A synthesis. Presentation at World Bank/DFID Public Expenditure Workshop. 11–12 May 2009, Addis Ababa.

Fan, S., Mogues, T. & Benin, S. 2009. *Setting priorities for public spending for agricultural and rural development in Africa.* IFPRI Policy Brief No. 12. Washington, DC, IFPRI.

Fan, S. & Rao, N. 2003. *Public spending in developing countries: Trends, determination and impact.* Discussion Paper No. 99. Washington, DC, IFPRI.

FARA. 2006. *Agricultural research delivery in Africa: An assessment of the requirements for efficient, effective and productive National Agricultural Research Systems in Africa: Main report and Strategic recommendations.* Accra, Forum for Agricultural Research in Africa (FARA).

Fuglie, K.O. & Heisey, P.W. 2007. *Economic returns to public agricultural research.* Economic Brief No. 10. Washington, DC, Economic Research Service, United States Department of Agriculture.

Gray, R. & Weseen, S. 2008. *The economic rationale for public agricultural research in Canada.* Saskatchewan, Canada, University of Saskatchewan, Canadian Agricultural Innovation Research Network.

Gurib-Fakim, A., Smith, L., Acikgoz, N., Avato, P., Bossio, D., Ebi, K., Goncalves, A., Heinemann, J., Hermann, T., Padgham, J., Penarz, J., Scheidegger, U., Sebastian, L., Teboada, M. & Viglizzo, E. 2008. Options to enhance the impact of AKST on development and sustainability goals. *In* B.D. McIntyre, H.R. Herren, J. Wakhungu and R.T. Watson, eds. *Agriculture at a crossroads: International assessment of agricultural science and technology for development global report,* chapter 6. Washington, DC, Island Press.

Harberger, A.C. 2009. *Benefit-cost analysis: An overview.* Policy Research Initiative, Government of Canada. www.policyresearch.gc.ca.

Kreimer, A., Collier, P., Scott, C.S. & Arnold, M. 2000. *Uganda: post-conflict reconstruction.* Washington, DC, World Bank, Operations Evaluation Department.

IAC. 2004. *Inventing a better future: A strategy for building worldwide capacities in science and technology.* Amsterdam, Netherlands, InterAcademy Council (IAC).

Lewis, W.A. 1996. *Development planning: The essentials of economic policy.* London, George Allen and Unwin.

Mogues, T., Morris, M., Freinkman, L., Adubi, A. & Ehui, S. 2008. *Nigeria agriculture public expenditure review.* Nigeria Strategy Support Program Brief No. 2. Washington, DC, IFPRI.

OECD. various years. *OECD S&T indicators.* Paris.

Omamo, W.S., Diao, X., Wood, S., Chamberlin, J., You, L., Benin, S., Wood-Sichra, W. & Tatwangire, A. 2006. *Strategic priorities for agricultural research for development in Eastern and Central Africa.* Research Report No. 150. Washington, DC, IFPRI and Association for Strengthening Agricultural Research in Eastern and Central Africa.

Omilola, B., Yade, M., Karugia, J. & Chilonda, P. 2010. *Monitoring and assessing targets of the Comprehensive Africa Agricultural Development Programme (CAADP) and the First Millennium Development Goal; (MDG) in Africa.* ReSAKSS Working Paper No. 31. Washington, DC, IFPRI.

Oxfam International. 2007. *Adapting to climate change: What's needed in poor countries, and who should pay.* Oxfam Briefing Paper No. 104. Oxford, UK, Oxfam International Secretariat. www.oxfam.org/en/policy/briefingpapers/bp104_climate_change_0705.

Palaniswamy, N. & Birner, R. 2006. Financing agricultural development: The political economy of public spending on agriculture in sub-Saharan Africa. In *Proceedings of the German Development Economics Conference,* chapter 4. Berlin, Verein für Socialpolitik, Research Committee Development Economics.

Pardey, P.G. 2009. Putting US agricultural R&D and productivity developments in perspective. Presentation at the Farm Foundation Conference: Agricultural Research and Productivity for the Future, April 2009, Washington, DC, National Press Club.

Pardey, P.G. & Alston, J.M. 1995. *Revamping agricultural R&D.* 2020 Vision Brief No. 24. Washington, DC, IFPRI.

Pardey, P.G., Alston, J.M. & James, J.S. 2008. *Agricultural R&D policy: A tragedy of the international commons.* Staff Papers No. 43094. Minneapolis, Minnesota, USA, University of Minnesota, Department of Applied Economics.

Pardey, P.G., Kang, M.S. & Elliott, H. 1989. Structure of public support for national agricultural research systems: a political economy perspective. *Agricultural Economics*, 3(4) 261–278.

Pardey, P.G., Beintema, N.M., Dehmer, S. & Wood, S. 2006. *Agricultural research: A growing global divide?* IFPRI Food Policy Report. Washington, DC, IFPRI.

Pingali, P. 2009. Enhancing developing world agricultural performance: Getting beyond the current plateau through R&D. Keynote address to Conference on Integrated Assessment of Agriculture and Sustainable Development. Egmond aan Zee, Netherlands, 10–12 March 2009.

Pray, C.E. & Fuglie, K.O. 2001. *Private investments in agricultural research and international technology transfer in Asia.* ERS Agricultural Economics Report No. 805. Washington, DC, Economic Research Service, United States Department of Agriculture.

Roseboom, J. 2002. *Underinvestment in agricultural R&D revisited.* The Hague, International Service for National Agricultural Research.

Roseboom, J. 2004. Agricultural research and extension funding levels required to meet the anti-hunger programme objectives. Paper prepared for FAO. Rijswijk, the Netherlands.

Ruttan, V. 1982. *Agricultural research policy.* Minneapolis, Minnesota, USA, University of Minnesota Press.

Stads, G.J. & Beintema, N.M. 2006. Women scientists in sub-Saharan African agricultural R&D. Brief prepared for the USAID meeting on Women in Science: Meeting the Challenge. Lessons for Agricultural Sciences in Africa, Washington, DC, 21 June 2006.

Stads, G.J. & Beintema, N.M. 2009. *Public agricultural research in Latin America and the Caribbean: Investment and capacity trends.* ASTI Synthesis Report. Washington, DC, IFPRI and Inter-American Development Bank.

Stern, N. 2007. *The economics of climate change: The Stern Review.* Cambridge, UK, United Kingdom Treasury.

Valdés, A. & Foster, W. 2005. Reflections on the role of agriculture in pro-poor growth. Paper prepared for the Research Workshop: The Future of Small Farms. Wye, UK, June 2005.

Veeman, T., Unterschultz, J. & Stewart, B. 2007. *Canadian and prairie agricultural productivity: measurement, causes and policy implications.* Saskatoon, Saskatchewan, Canada, CAES Conference on Food and Fuel.

Widmer, L., Fox, G.C. & Brinkman, G.L. 1988. The rate of return to agriculture in a small country: The case of beef cattle research in Canada. *Canadian Journal of Agricultural Economics*, 36: 23–35.

World Bank. 1981. *Agricultural Research Sector Policy Paper.* Washington, DC.

World Bank. 2007. *Cultivating knowledge and skills to grow African agriculture.* Washington, DC.

World Bank. 2008a. *World Development Indicators 2007.* Washington, DC. (CD-ROM)

World Bank. 2008b. *World Development Report 2008: Agriculture for development.* Washington, DC.

CHAPTER 10

CAN TECHNOLOGY DELIVER ON THE YIELD CHALLENGE TO 2050?

Tony R. Fischer
Derek Byerlee
Gregory Owen Edmeades

Projecting crop yields, especially 40 years ahead, is fraught with uncertainty. However, three stylized facts emerge from several recent studies of world food needs. First, given land and water scarcity, climate change and rising energy prices on the supply side, and growing markets for food, feed and fuel on the demand side, global grain markets will be tighter in the future than over the past 40 years. Second, area expansion will at best be small, so future agricultural growth will be more reliant than ever on raising crop and animal yields. Third, the growth rate of cereal yields has been falling since the green revolution years. A major question for this chapter is whether this decline means that crop yields have reached a technological plateau, or there are still large unexploited sources of yield gains either on the shelf or in the research pipeline.

This chapter addresses these questions through the analysis of cereal yields and productivity. It does so by tracing recent sources of growth and identifying future technological opportunities for raising potential yields and closing the gaps between existing yields and those that could be economically attainable by farmers. It focuses on the big three cereals: rice, wheat and maize. Cereals account for 58 percent of annual crop area and provide about 50 percent of food calories. Rice and wheat alone have accounted for about half of the increased per capita energy intake in developing countries since 1960 (Figure 10.1). Maize has been the major source of energy supporting the rapid increase in consumption of animal products (Figure 10.2), accounting for more than 60 percent of energy in commercial animal feeds, and becoming a major feedstock for biofuels in recent years. Together, these three cereals will provide about 80 percent of the increase in cereal consumption to 2050 (Rosegrant *et al.*, 2008). However, the chapter also recognizes that diversification of food production is needed, and a comprehensive review would include relevant data for roots and tubers, pulses and oilseeds. Some of these crops show declining trends, but remain critical to the food security of millions, while others – such as potatoes, sugar cane, soybeans, canola and oil-palm – are booming commercial crops serving multiple uses for food, feed and fuel.

The chapter uses a bottom-up approach that reviews farm survey and experimental evidence on yields and yield gaps in the world's breadbaskets. This allows the discussion to go beyond the estimation of yield growth by simple extrapolation of aggregate trends, to explore the most likely sources of increased yields, including proximate factors, such as higher-yielding varieties, input use and reduced losses from biotic and abiotic stresses, and broader policy and institutional factors that influence crop management and include input market efficiency, risk management, and the information and skills of farmers. The chapter suggests some of the critical investments and institutional changes that will be needed to realize these changes.

Ultimately the chapter is about the potential for sustainable productivity growth, as the effects of productivity on food prices have major welfare implications for poor people. This leads from a discussion of yields *per se* to an assessment of input use and efficiency, and an analysis of trends in total factor productivity. In addition, sustainability is essential, to ensure that productivity can be maintained in the face of depleting non-renewable resources and that production systems do not degrade the environment.

The chapter employs both a global and a local approach to assessing crop yields. Changes in global yields are important for global food security. In a globalizing world, many countries will increasingly depend on trade for provisioning their food needs, and this should encourage production in the lowest-cost regions, if there are no significant trade barriers. However, there are many situations where trade will be inadequate to ensure food supplies. The "megacountries", China and India, have little choice but to produce most of their staple foods, especially rice, given the relatively small, thin world markets in relation to their huge domestic markets. In Africa too, poor infrastructure, land-locked locations and lack of foreign exchange necessitate the production of much of the food near where it is to be consumed. The high population growth in some of the more densely populated African countries adds urgency to accelerating domestic production (e.g., Ethiopia's projected population of 185 million in 2050). The 2008 food price spike, induced partly by export bans and by rising energy costs for long-distance transport, is likely to lead many other countries to put a premium on local supplies.

Figure 10.1
Sources of increased per capita calorie consumption in developing countries, 1961 to 2003

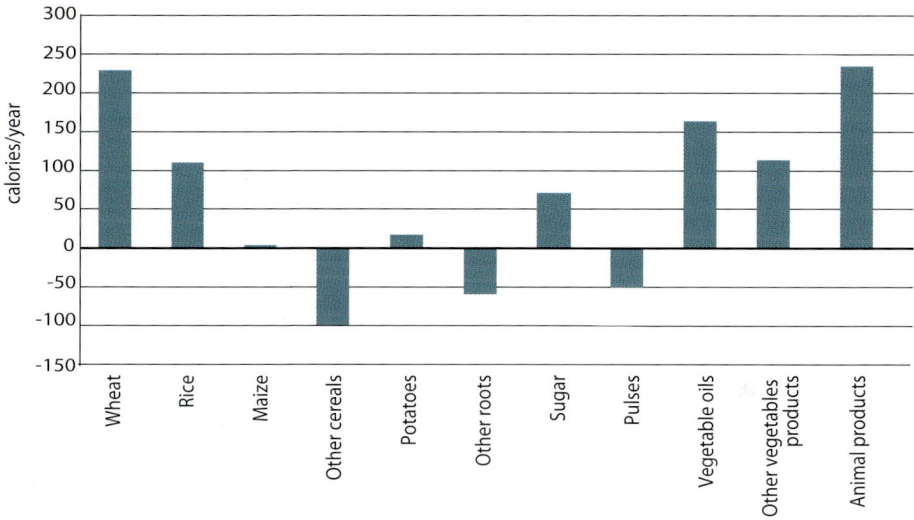

Source: FAOSTAT.

Figure 10.2
Sources of increased per capita protein consumption, developing countries, 1961 to 2003

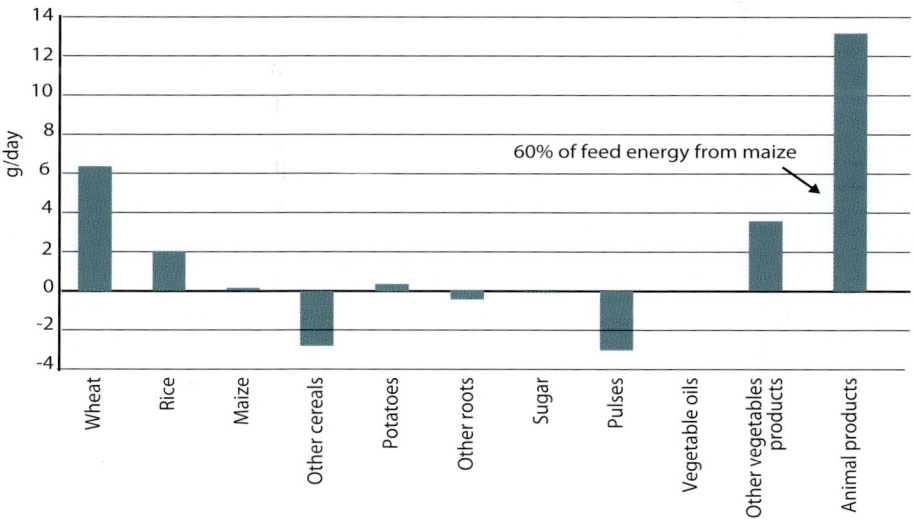

60% of feed energy from maize

Source: FAOSTAT.

391

Defining key concepts

There is a rich and evolving literature on various measures used for yields and efficiency gaps, but these terms are often used very loosely. This section defines the measures used in this chapter and their interpretation, relying largely on the work of Ali and Byerlee (1991), Loomis and Connor (1992) and Evans and Fischer (1999).

There are a number of measures used for crop yield, which here means the weight of grain harvested per unit of field area at a standard moisture content (Table 10.1). The starting point is average farm yield (FY), which forms the basis for calculating the gaps to attainable yield (AY) and then potential yield (PY). Water-limited potential yield (PY_W) is included as a sensible yardstick where crops receive on average only low to moderate water supplies (say < 75 percent of potential evapotranspiration). For increasing FY, which is the objective of this chapter, both increasing PY (or PY_W) and closing the yield gap are important, and somewhat different interventions operate on each. The overall gap PY to FY is considered in some detail because it is often easier to measure, but the key gap is the economically recoverable yield gap under current economics, and it is less, being AYa – FY (Figure 10.3; Table 10.1). Another gap, AYb – AYa, is the gap between attainable yield under efficient institutions and markets (AYb, which is ultimately linked to world prices), and that under current economics (AYa): because current economics are often less favourable to farmers, this gap is often positive, but can be negative where prices are subsidized to help farmers. Throughout this chapter, yield gaps are expressed as percentages of FY, for better comparability with the basis on which demand growth is estimated.

Progress in PY (or PY_W) through genetic and agronomic research is an important source of yield growth because raising the yield frontier lifts other yields as well – a rising tide that lifts all boats. This chapter's section on Sources of yield gains in the breadbaskets presents considerable evidence that $\Delta FY/FY \approx \Delta PY/PY$. However, much also depends on the interactions between genotype and management (Fischer, 2009). Generally, PY progress has exploited positive interactions between the genetic and agronomic routes for improvement in yield. For example, the increase in yields of semi-dwarf wheat and rice varieties at higher levels of management is significantly more than that of the tall varieties they replaced. In advanced systems however, yield increases from agronomy alone, and from these positive interactions, appear to be slowing, although the ongoing synergy between increased maize PY and higher plant population is an exception (Evans and Fischer, 1999).

Table 10.1
Definitions of yield measures

Yield	Symbol	Definition	Estimation
Average farm or on-farm yield	FY	Average yield achieved by farmers in a defined region over several seasons	Regional or national statistics, ground or satellite surveys of fields
Economically attainable yield given current markets and institutions	AYa	Optimum (profit-maximizing) yield given prices paid/received by farmers, taking account of risk and existing institutions	On-farm experiments, varying inputs, sometimes crop models, disaggregated farm surveys
Economically attainable yield assuming efficient markets/institutions	AYb	Optimum yield given prices that would prevail in efficient markets with well-functioning risk insurance markets	As for AYa, but adjusting to the price and risks of efficient markets, etc.
Potential yield	PY	Maximum yield with latest varieties, removing all constraints including moisture, at generally prevailing solar radiation, temperature and day length	Highly controlled on-station experiments or crop models calibrated with latest varieties, well-monitored crop contests
Water-limited potential yield	PY_W	Maximum yield under normal rainfed conditions, removing all constraints except for moisture	Highly controlled on-station experiments, crop models or crop contests
Theoretical yield		Maximum theoretical yield for prevailing solar radiation based on prevailing knowledge of crop physiology and photosynthetic efficiency	Accepted estimate given by the initial slope of the photosynthesis versus solar radiation response curve, discounted for dark respiration

Both farmer characteristics and system-wide constraints explain these various yield gaps and suggest how they may be closed. In general, yield gaps at the lower end, such as AYa – FY, are explained mainly by farmers' access to information and technical skills, while higher-order yield gaps reflect opportunities for research and broader policy and institutional constraints. Figure 10.3 depicts these overlapping sources of yield gaps.

These various definitions assume that underlying site characteristics, soil, climate and seasonal conditions that are beyond the control of farmers are uniform across a defined area. In reality, regional surveys reveal large variation in yields across farmers and fields, around the average FY, in part caused by site and season differences.[1] Often the distribution is negatively skewed (e.g., Lobell *et al.*, 2003),

---

1. This can be called the non-manageable natural resource base of the site. However, it depends on the time scale. Drainage, liming and terracing can be considered long-term investments to improve an initially deficient natural resource base.

but it is not clear how to relate such distributions to the prevailing AYa and PY. It might be expected that a proportion of farmers will always reach AYa, and a few reach PY. Crop contests that measure crop yield properly on sufficient field size (say > 4 ha) usually give very high yields, which can sometimes be taken as the prevailing PY when better sources are lacking. However, it is important to know whether the natural resource base of the winning fields (the part that cannot be changed with good management) is representative of the region. Similarly, experimental stations may be in more favourable sites, and the PY they estimate can be inflated by these site characteristics. In addition, optimum management is partly a function of seasonal conditions that are not known at the time of decision-making, so part of any yield gap is unpredictable and arises from the interaction between management (including variety choice) and variable seasonal conditions; risk aversion exaggerates this gap in rainfed situations.

As with site differences, the prices and institutions faced by farmers can vary, even within small areas. These differences may relate to farm size, to education, aspiration and skill differences, to differential access to credit and input markets, and to local power structures. Thus part of the gap between good and average FYs may be due in part to site characteristics (some of which might vary at random across years), and in part to differences among farmers in characteristics, resource constraints and prices.

Figure 10.3
Schematic view of key yields and yield gaps for a hypothetical favourable cereal region, and ways of closing them

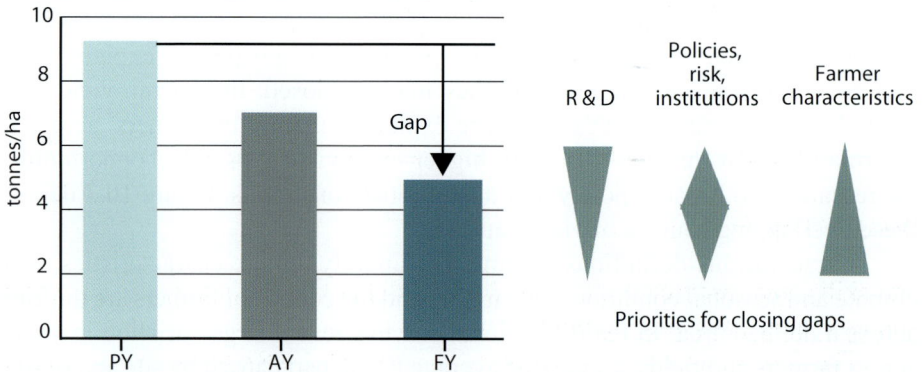

Source: Authors.

For reasons of both productivity and sustainability, this chapter is also interested in efficiency and the prospects of closing efficiency gaps. Put simply, efficiency is measured as the average cost for producing a given yield, relative to the lowest-cost option.

Economists generally distinguish between technical and allocative efficiency. Technical inefficiency refers to a failure to operate on the yield frontier: i.e., the same yield could be produced by using proportionally less of all inputs. Allocative inefficiency refers to failure to meet the marginal conditions for profit maximization where the marginal value of applying an additional unit of input is equal to the price of the input.

In green revolution settings – from Iowa to the Punjab – a useful framework for identifying these inefficiencies with considerable empirical support is given in Figure 10.4.[2] During the green revolution, farmers adopted modern varieties that shifted their production function from traditional varieties (TV) to modern varieties (MV). At the same time, they adopted modest levels of fertilizer and other inputs to reach point B. Initially, however, owing to risk, lack of knowledge

Figure 10.4
Measures of efficiency gaps

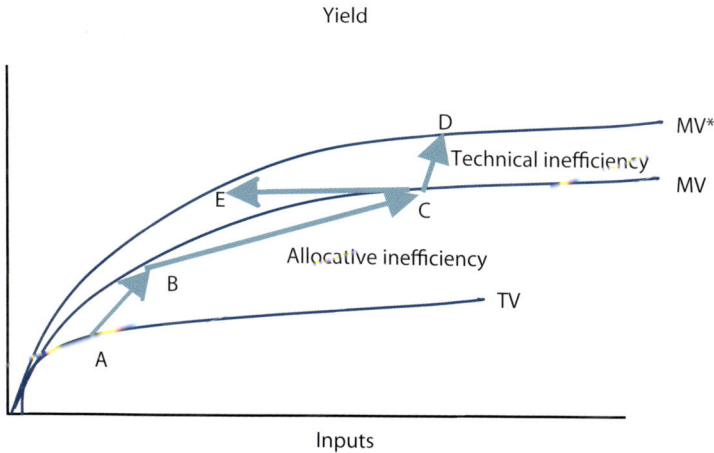

Source: Byerlee, 1992.

and skills, and resource constraints, farmers did not fully exploit the technology and used inputs at sub-optimum levels.

The first post-green revolution phase was characterized largely by input intensification, moving from B to a point C that is closer to the allocative optimum.

---

2. For simplicity, these efficiency measures are shown in one dimension with one input. Technically, their strict definition requires at least three-dimensional space with two or more inputs.

However, farmers still tended to operate considerably below the production frontier, implying a measure of technical inefficiency. In the second post-green revolution period, the emphasis has been on improving technical efficiency, substituting improved information and managerial skills for higher input use, and moving towards a point D, or – with appropriate incentives or regulations (e.g., on input pollution) – a point E, by reducing input use without sacrificing yields. The yield frontier MV* may be defined in terms of the highest production achieved from a given level of inputs in a population of farmers, or by reference to a potential frontier based on experimental data. In both cases, similar issues of site specificity and seasonal conditions that influence the measurement of yield gaps also affect the efficiency estimate. Most studies by economists have ignored these site and seasonal conditions, and therefore tend to overestimate inefficiency (Ali and Byerlee, 1991; Sherlund, Barrett and Akinwumi, 2002). Of course, MV* is not static but shifts upwards with the release of new technologies, especially newer generations of varieties. It may also shift downwards if there are serious long-term problems of resource degradation.

Yield gaps and efficiency gaps often measure the same things. However, efficiency gaps may exist even where there are no yield gaps. Farmers may be achieving the economically attainable yield, AYa, but using above-optimum input levels. Variation in efficiency across farmers and fields is also explained by factors related to farmer characteristics and system-wide constraints. Technical efficiency relates largely to timing and technical skills in using inputs, and is often explained by farmer-specific knowledge and skills. However, system-level factors such as management of irrigation systems can also explain technical inefficiency. Allocative inefficiency can be due to similar factors, as well as differential risks of using inputs, input market failures and financial constraints.

Ultimately, this chapter focuses on gains in total factor productivity (TFP) as a major determinant of long-term price trends – most productivity increases are ultimately passed on to consumers through lower prices. TFP is a measure of output in relation to the aggregate of all inputs, whereby changes in agricultural production are decomposed into a component relating to changes in inputs and the change due to productivity growth. The primary driver of productivity growth is investment in research and development (R&D) that raises PY. However, research and other factors contribute to TFP growth, such as extension and education that help farmers close yield gap AYa – FY; institutional change or better infrastructure and policy that close yield gap AYb – AYa; or related interventions that narrow efficiency gaps by reducing input costs. Thus TFP is a composite measure of gains in closing gaps, and is referred to later, in the subsection on Prospects for TFP growth.

Setting the scene: recent trends and the challenge to 2050

Much of the concern about feeding the world in 2050 relates to the slowing of yield growth in the major cereals over the past three decades (World Bank, 2007). This section briefly reviews global trends in key inputs and cereal yields, and summarizes available evidence on the growth in yields required to meet the world's food, feed and fuel needs in 2050. Yield trends and causes are disaggregated in the following section.

Recent changes in crop area, key inputs and yields

Land and water inputs are examined fully in other chapters, but being critical to this chapter's analysis they are mentioned briefly here. In recent decades, area growth has been a significant source of production growth in only Latin America and sub-Saharan Africa. Wheat area has fallen in industrial countries, while rice area has increased by only about 0.3 percent annually since 1990, and is actually falling in China, the Republic of Korea and Japan. However, maize area has expanded consistently at more than 1 percent per year in both developing countries (driven by livestock feed) and industrial countries (driven by biofuel demand, mainly in the United States of America). Even so, yield growth has also been the dominant source of production increases in maize (Figure 10.5).

Other crops have also been dynamic. Potatoes – traditionally a staple food in much of Europe – are now grown more extensively in developing countries. Because of both area and yield growth, China is the world's largest potato producer. Soybean has been the fastest growing crop, especially in Latin America, driven by demand for feed (Figure 10.5).

The growth of irrigated area slowed sharply in the 1980s and early 1990s (Rosegrant and Pingali, 1994). However, over the past decade irrigated area has expanded steadily at 0.6 percent per annum in developing countries. Given a productivity differential between irrigated and rainfed areas of 130 percent (Fuglie, 2008), irrigation alone accounted for about 0.2 percentage points in the overall annual yield growth of 1.1 percent for cereal yields from 1991 to 2007.

Increased use of fertilizer has been a major factor explaining perhaps one-third to one-half of yield growth in developing countries since the green revolution (FAO, 2003; Heisey and Norton, 2007). Developing countries now account for 68 percent of total fertilizer use, which has continued to increase by 3.6 percent per year over the past decade, so still accounts for a significant share of yield growth.[3]

---

3. With average rates of fertilizer use on cereals in developing countries of at least 100 kg of nutrients per hectare (Box 10.1), current growth in fertilizer use and a grain-to-nutrient response of 5:1 would add 18 kg/ha additional yield annually, or 0.6 percent.

Figure 10.5
Contributions of area and yield to production growth, 1991 to 2007

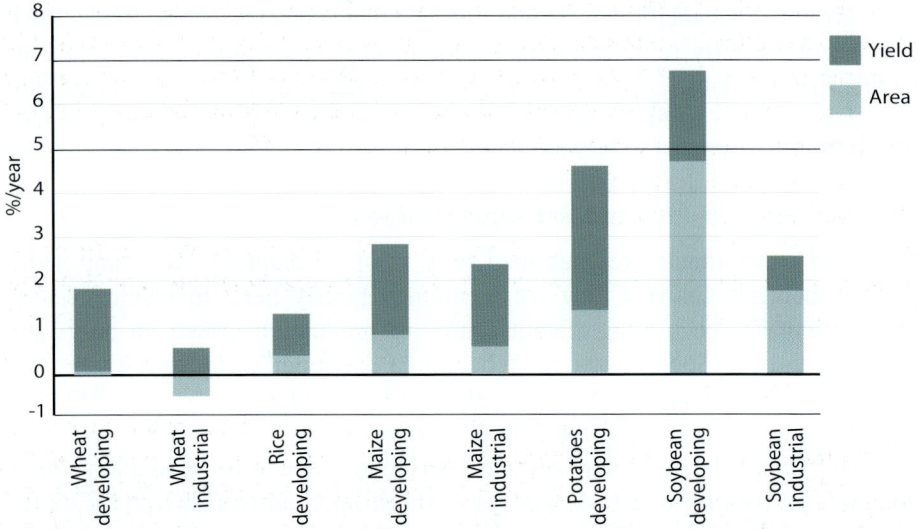

Source: FAOSTAT.

Figure 10.6
Trends in fertilizer use (nutrients per irrigated equivalent area)

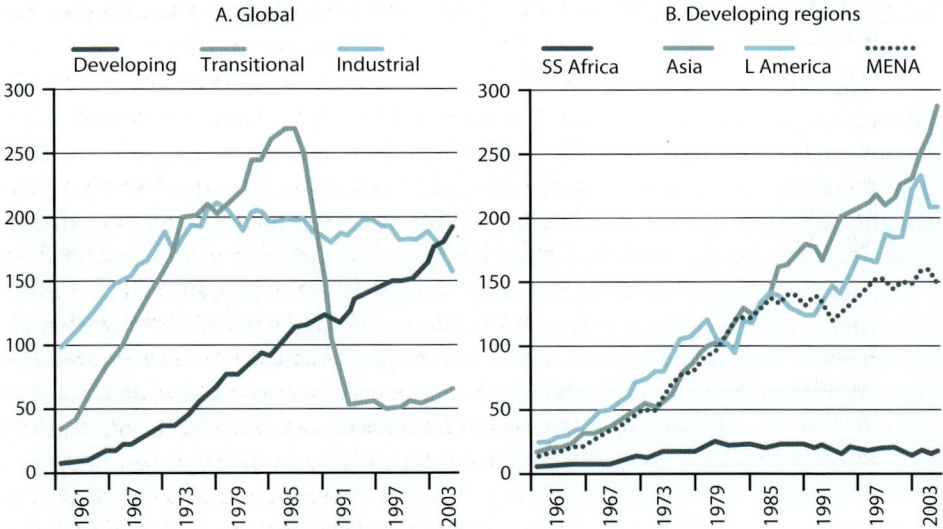

Sources: Nitrogen, phosphorus pentoxide and potassium oxide (N + P_2O_5 + K_2O) from FAOSTAT.
Computation of irrigated-equivalent area from Fuglie, 2008.

Using a measure of agricultural area standardized for land quality (Fuglie, 2008), fertilizer use per irrigated-equivalent hectare is also now higher in developing than in industrial countries (Figure 10.6).[4] Globally, fertilizer use has plateaued, owing to a decline in fertilizer use in industrial countries and a dramatic fall in the countries of the former Soviet Union after they moved towards a market economy.

The increase in fertilizer use has been surprisingly consistent across most developing regions. Asia still has the highest and fastest increase, but fertilizer use intensity in Latin America and the Near East and North Africa is comparable. However, fertilizer use per hectare in sub-Saharan Africa is abysmally low, for reasons such as high prices and poor markets, which have been well documented (Morris *et al.*, 2007). Low fertilizer use explains a large part of the lagging productivity growth in that region.

Box 10.1 - Fertilizer use on cereals

Wheat, rice and maize account for about half of all the fertilizer consumed globally. The following table provides data on fertilizer use for some countries and some years. The very high rates in countries such as China suggest little scope for further intensification, and huge scope for improved efficiency. Environmental pressures are likely to lead to pressure to reduce fertilizer use in many countries in Asia.

Estimated fertilizer use for wheat, rice and maize, selected countries

Country/region	Total nutrients *(kg/ha)*			Nitrogen *(kg/ha)*		
	Wheat	Rice	Maize	Wheat	Rice	Maize
Bangladesh		140			100	
China	296	310	213	197	192	180
India	164	160	67	117	106	45
Indonesia		108	146		93	109
Pakistan	182	190	161	140	146	123
Philippines	53	47		46	39	
Iran, Islamic Republic	118			84		
Argentina	77		79	44		46
Brazil	101	95	127	40	29	49
USA	129	250	269	86		152
EU15	186		373	135		227
Poland	142			90		
Sub-Saharan Africa		10	38			
World	128	155	153	87	101	98

Sources: Heffer, 2008; sub-Saharan Africa data from Heisey and Norton, 2007, for the late 1990s.

---

4. The quality-adjusted agricultural area weights land quality by irrigated, rainfed and pasture, based on relative productivity, to arrive at a rainfed equivalent area (Fuglie, 2008).

Figure 10.7
Long-term trends in cereal and wheat yields

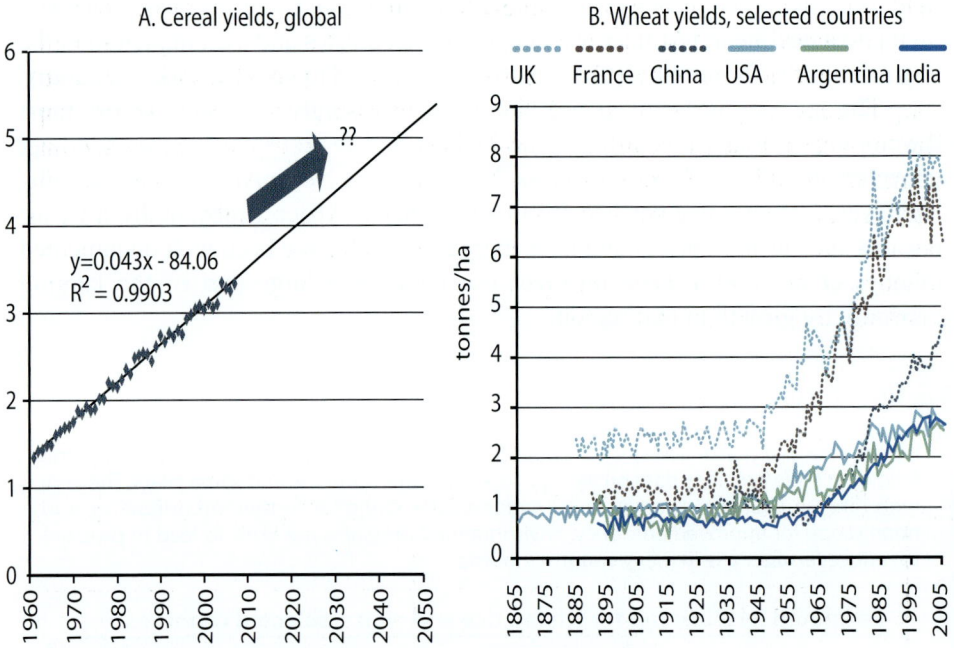

A. Cereal yields, global

B. Wheat yields, selected countries

UK   France   China   USA   Argentina India

$y=0.043x - 84.06$
$R^2 = 0.9903$

??

tonnes/ha

Sources: FAOSTAT; wheat yields updated from Pardey *et al.*, 2007.

Growth through intensification of fertilizer and irrigation use is no longer important in industrial countries. Fertilizer use and irrigation are also already high in some Asian countries, especially China, so their future contribution to yield growth will be modest at best (Box 10.1). However, there are still major regions of the developing world, especially sub-Saharan Africa, where input intensification is at an early stage. In addition, the Russian Federation, Ukraine and other transitional countries are already reversing the collapse of input use, providing scope for more rapid yield growth in the future.

Over the past five decades, global cereal yields have grown linearly at a constant rate of 43 kg per hectare per year, and with very low variability around the trend (Figure 10.7). However, this is a sharp departure from relatively stagnant yields in earlier periods. Note that linear growth in Figure 10.7 implies declining exponential growth, from 3.2 percent per year in 1960 to 1.5 percent in 2000. Projecting the same linear trend to 2050 would deliver annual growth of only 0.8 percent. Global atmospheric carbon dioxide (CO_2) has also increased approximately linearly over recent decades, and it can be estimated from Tubiello *et al.* (2007) that this is contributing about 0.2 percent of current annual yield growth of C_3 crops (such as wheat and rice, but not maize);[5] however C_3 crop yield sensitivity to CO_2 increase is likely to decline as CO_2 increases further.

The aggregate global picture disguises important differences by region and crop, as illustrated for wheat (Figure 10.7). Developing countries experienced a sharp increase in yield growth with the green revolution, and then a sharp drop. The ten-year moving average of growth rates for wheat and rice in developing countries has declined from the mid-1980s to about 1 percent annually in the most recent decade (Figure 10.8). Yield growth of wheat in industrial countries has also slowed, and fell to zero in the most recent decade. The trends for maize, although showing some decline in growth rates in both developed and developing countries, are not nearly so pronounced.

Figure 10.8
Ten-year moving average exponential yield growth rates for wheat, rice, maize and soybean

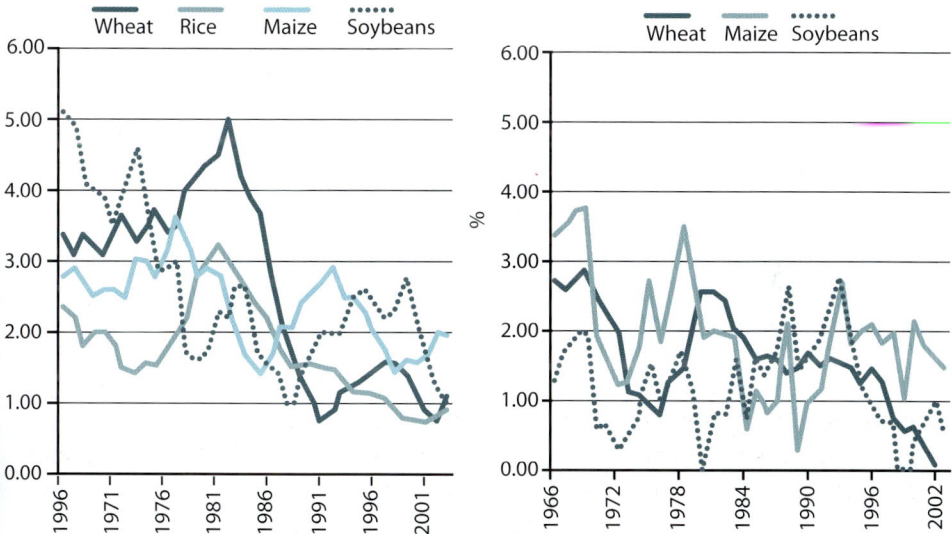

Growth rates estimated by log linear trend regression. Year refers to the mid-year of the decade.
Source: Computed from FAOSTAT.

---

5. C_3 and C_4 refer to two systems of photosynthesis found in common crop plants, and relate to the number of carbon atoms (three or four) in the primary molecule formed when CO_2 is first absorbed. C_3 species include wheat, rice, soybeans and barley, while common C_4 species are maize, sorghum and sugar cane. The C_4 photosynthetic system has an enzymatic and a morphological adaptation (the so-called "Kranz anatomy") that provides a CO_2-concentrating mechanism in the leaf. This gives it a higher photosynthetic efficiency under high light conditions and a generally greater photosynthetic output under higher temperatures than the C_3 system. Outcomes are usually higher yields and higher water, nitrogen and radiation use efficiencies under warm conditions, compared with C_3 crops.

At the regional level, Latin America has had the best yield performance for all cereals since 1991, averaging 2.5 percent per annum. The lowest average increases have been in sub-Saharan Africa and, surprisingly, East and Southeast Asia, each with about 1.2 to 1.3 percent a year. However, there is some good news in both these regions: sub-Saharan Africa has had a sustained period of modest yield growth from a very low base; and East and Southeast Asia already has high yields of 4.8 tonnes per hectare, so even this modest growth rate represents an achievement.

There is also evidence of a slowdown in absolute yield growth for rice and wheat. The coefficient c of the quadratic term of absolute yield trends was therefore tested by fitting the equation $y = a + bt + ct^2$, where y is national average yield, and t is year. To reduce the impact of the green revolution, the period analysed was 1980 to 2007, after modern varieties were widely adopted. The results indicate a clear slowing of the rate of absolute yield gains in rice and wheat. For wheat, this pattern prevails in most regions, and no region shows an accelerating trend. For rice, the declining trend is very evident in South and Southeast Asia, but Latin America shows an increasing rate of gain.

Again, the results for maize are different, showing a linear trend at the global level and an accelerating trend (positive and significant coefficient c) in the developing world. Both South Asia and Latin America show accelerating trends in absolute gains, while only Western Europe shows a declining trend.

The close linear trend in yield growth at the global level hides considerable heterogeneity in performance by crop and region. Maize has been most dynamic, and Latin America has been the star among regions, partly because maize is the most important grain in that region. As well as exponential growth rates, looking at absolute growth aids the interpretation of trends.

Scenarios to 2050 and the future yield challenge

Against this background, what rate of yield growth is needed to meet the food needs of the projected 9.2 billion people in the world's population in 2050? Studies by Rosegrant *et al.* (2008) at the International Food Policy Research Institute (IFPRI) and Tweeten and Thompson (2008) provide recent analyses of this challenge, while Hubert *et al.* (2010) provide a more accessible version of the IFPRI study.

Global demand and supply prospects are examined in depth in other chapters. Demand for grains is largely determined by population and income growth, with the recent addition of demand for biofuels. At the global level, per capita demand for cereals for food is projected to fall in all regions except sub-Saharan Africa, as increasingly affluent consumers diversify diets to higher-value products, including livestock ones. Livestock in turn will drive demand for feedgrain, especially maize. In addition, maize and some wheat will be used as feedstocks for biofuels. IFPRI projects that this demand for grain for biofuels will continue to increase to 2020/2025 before levelling off as second-generation technologies based on biomass conversion become available (Rosegrant *et al.*, 2008). Still, by 2020 industrial countries will consume about 150 kg per capita of mostly maize for biofuels, which is similar to today's per capita consumption of cereals for food in developing countries.

Tweeten and Thompson (2008) provide a simple analysis of what might happen by 2050 with linear growth in yields of major product groups, including cereals. They project an increase in cereal supply of 71 percent over 2000, or a total increase of 1.4 billion tonnes. This derives from projecting the linear annual yield growth of 43 kg per hectare suggested in Figure 10.7 over the whole period (the initial growth of 1.4 percent becomes 1.07 percent for the whole period).[6] Their middle estimate of demand growth gives an increase of 79 percent by 2050 (1.17 percent exponential over the whole period, with a world population of 9.1 billion in 2050). Thus, there will be a projected supply deficit in relation to demand, which implies an increase in weighted real agricultural prices of 44 percent by 2050 to "clear the market".

Using mid-range (baseline) estimates of population (9.2 billion by 2050), income growth and biofuel demands, Rosegrant *et al.* (2008) project an overall increase in cereal demand of 1.048 billion tonnes (56 percent) by 2050, from a 2000 base. This implies an average annual growth of 0.9 percent over the period, but the authors see demand growth declining from 1.4 percent in the first 25 years to 0.4 percent in the second. Fully 41 percent of this increase is for feed, especially in developing countries. As a result, maize accounts for 45 percent of the increase in cereal demand, wheat for 26 percent, and rice for only 8 percent.

On the supply side, Rosegrant *et al.* (2008) see land and water become increasingly constraining. Area devoted to cereals declines globally by 28 million ha, as loss of cropland and crop diversification in industrial countries and Asia cancels area expansion in Latin America and sub-Saharan Africa. Water available for agriculture also hardly increases, owing to competition from non-farm sectors, declining groundwater tables in the breadbaskets of India and China, and likely higher energy costs for irrigation (Molden, 2007; Tweeten and Thompson, 2008). Some 60 percent of global cereal production is now from irrigated areas, and with competition within these areas for higher-value production, projected irrigated area for cereals falls. Maize is the only cereal expected to show modest area expansion.

---

6. Tweeten and Thompson (2008) assume no change in area, so yield growth is equal to production growth.

The IFPRI projections also take account of climate change. However, climate change in the medium projection of the Intergovernmental Panel on Climate Change (IPCC) is not expected to have a significant effect on global yields by 2050 (IPCC, 2007), as yield gains in some regions (mostly temperate) balance losses in others (mostly tropical). The impacts of climate change are addressed in more depth in other chapters.

The IFPRI yield projections are based on the FAO expert opinions disaggregated by country and agro-ecological zone (FAO, 2003). Overall yield growth in the baseline projection for cereals is 1.0 percent per annum. Averaged for irrigated and rainfed production, the gains are 1.0 percent for wheat, 0.7 percent for rice and 0.9 percent for maize. FAO projections for 2030 are quite similar (FAO, 2003).

The global average annual absolute rate of yield gain to 2050 in the Rosegrant *et al.* (2008) projections (made more accessible by Hubert *et al.*, 2010) is 37 kg per hectare, 14 percent lower than the linear projection of past performance used by Tweeten and Thompson (2008). Given lower yield growth, the IFPRI baseline projects higher real price increases, of 91 percent for wheat, 60 percent for rice and 97 percent for maize from a 2000 base. Developing countries will increasingly depend on imports of cereals (and oilseeds) from industrial countries, Eastern Europe (including the Russian Federation), Brazil and Argentina.

Projections are only estimates, and the overall results are quite sensitive to the assumptions. In particular, Rosegrant *et al.* (2008) show that with an increase in public investment in agriculture of 13 percent over the baseline, especially in R&D, producing a 0.4 percentage point increase in annual yield growth, to 1.43 percent, world grain prices would resume their downwards trend characteristic of much of the past century, and could result in an almost halving of the number of malnourished children by 2050. By contrast, a yield growth of 0.4 percentage points lower (at 0.61 percent) would lead to a more than doubling of real cereal prices, to about USD 600 per tonne (in 2000 dollars) and stagnation in the number of malnourished people.

These studies have two major implications for the analysis of future yield perspectives. First, a continuous linear increase in yields at the global level, following the pattern established over the past few decades, will not be sufficient to meet food, feed and fuel needs – i.e., future demands at or below today's real prices. The world will need to do better in the next 40 years. Second, the outcome is quite sensitive to yield projections. An increase in yield growth of 0.4 percent percentage points can reverse projected price trends. Although this sounds like a relatively modest goal, these are exponential growth estimates (which must be maintained throughout the whole period) and require an increase of more than one-third in the current absolute yield growth rate. This cannot be taken for granted, especially because aggregate growth rates in both percentage and absolute terms are clearly in a declining phase (except for maize), and input growth may make a much smaller contribution than in the recent past. It should also be noted that the increase in demand for grains will be much greater to 2025 than for the following 25 years, so supply responses are needed relatively soon.

Sources of yield gains in the breadbaskets

This section reviews recent progress in grain yields through a series of case studies in some of the world's major breadbaskets. The full details of the case studies are reported elsewhere, in forthcoming work by Fischer and others, and only summary statistics are provided here.

The case studies indicate the depth of analysis that is necessary for understanding what is currently happening to crop yield on the farm (FY), which in turn is driven by: i) progress in potential yield (PY) arising from new agronomy and, increasingly, from new varieties; and ii) the adoption of new technologies that narrow the gap between FY and PY (expressed as a percentage of FY). The studies reveal considerable diversity among cases, based largely on crop species, agro-ecology and stage of economic development.

In all cases, PY and its rate of change were difficult to estimate, especially for crops under low to moderate rainfall (i.e., PY_W), because it is important that the PY or PY_W for a region comes from crops with the same natural resource endowment as the regional average. The estimates of current PY come from the latest breeders' trials, from simulation models calibrated using the latest cultivars, and sometimes, as a last resort, from yields in crop contests. Estimates of recent PY progress come from comparisons of historic sets of varieties grown inevitably under high inputs, preferably with disease and pest protection, as older varieties often become more susceptible over time. Progress is calculated simply by plotting yield against year of release for varieties released in the last 20 years or so; over this release period, relationships were always closer to linear than any other response shape. Note that this represents PY progress under advanced agronomy, and hence includes the genetic gains plus the usually significant gains from genotype-management interactions (Fischer, 2009). PY gains from agronomic innovation alone are thus not included. In advanced cropping systems, these are becoming a smaller factor in recent gains, although agronomic innovation remains very important for input use efficiency. In less developed systems, the lack of adoption of modern agronomy is often the major cause of the yield gap.

Finally, FY is usually obtained from official statistics, and sometimes from surveys. Yield progress for FY is not corrected for the effect of global CO_2

increase on C_3 crops mentioned earlier. However, PY growth estimated from trials of side-by-side comparisons of varieties of different vintages is not inflated by increased CO_2, and vintage-CO_2 interactions appear to be small where they have been studied.

Several cases from each major crop environment and stage of economic development should be examined to obtain a proper sample and full understanding of what is behind the aggregate numbers on FY, and to project with some confidence. Some researchers are using high-resolution Geographic Information System (GIS) and crop modelling approaches to deal with the challenge of bringing together all of the world's cropping regions (e.g., the Harvest Choice programme that includes IFPRI). However, although more extensive sampling would bring benefits, the approach adopted in this chapter is an appropriate way forward, and case study numbers are bolstered from other sources of data wherever possible. For illustrative purposes, some key case studies are described more fully in the following subsections. This provides the basis for discussion of the two paths for increasing FY further: reducing the gap between FY and PY, and increasing PY.

Wheat

Figure 10.9 illustrates two of the better-documented case studies with wheat: the Yaqui Valley in Mexico is irrigated low-latitude spring wheat (S1, irrigated or high-rainfall spring wheat environment 1), which represents 22 percent of the world's wheat area, found almost entirely in the developing world; and the United Kingdom is a well-watered winter wheat environment (W1, winter wheat environment 1), representing 31 percent of the world's wheat area, three-quarters of which is in industrial nations (Heisey, Lantican and Dubin, 2002). The Yaqui Valley has been a major target for the wheat breeding programme of the International Maize and Wheat Improvement Center (CIMMYT) and its predecessor for more than 50 years; its environment is similar to that for wheat in Pakistan, northwest India, southern China and Egypt, all of which experienced a green revolution in wheat yields associated with improved varieties, irrigation and fertilizer. In the Yaqui Valley, variety turnover is rapid, and nitrogen (N) rates have now reached 260 kg per hectare. Despite this, FY progress has slowed to about 49 kg/ha/year over the last 30 years (Figure 10.9A), but this should be corrected downwards for a significant and surprising decline in average minimum temperatures over the period, giving progress of only 18 kg/ha/year, or 0.3 percent per year. This is exactly the rate of progress seen in PY determined at an experimental station in the centre of the valley. Thus the yield gap is fairly steady at 50 percent of FY, somewhat surprising for a region of moderately sized farms in a reasonably well developed agricultural system; current FY is at 6 tonnes/ha and PY at 9 tonnes/ha.

The United Kingdom has one of the highest national wheat yields (just over 8 tonnes/ha), with modern agriculture and an active private (breeding) and public research base. Excellent records of the Home Grown Cereal Authority (HGCA) from its protected variety experiments across the country give a good indication of PY. The rates of FY and PY progress have been fairly steady over the last 20 years, at 0.7 and 0.6 percent respectively; N use has been steady at 190 kg/ha for most of the period, and the yield gap is also steady (currently 25 percent of FY, and probably close to AYa, with little or no further gap to AYb in the United Kingdom today).

Figure 10.9
Changes in wheat PY and FY in the Yaqui Valley of Mexico and the United Kingdom

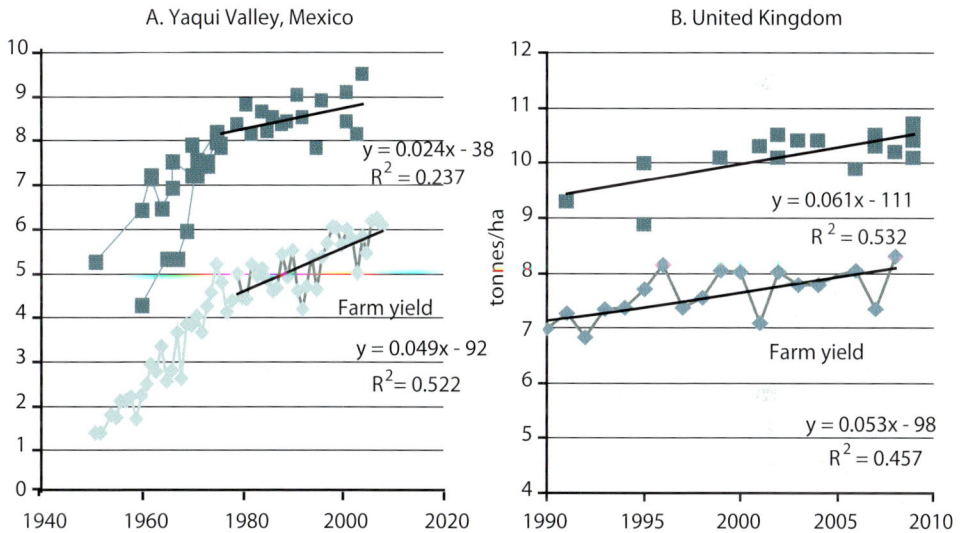

A. Yaqui Valley, Mexico

$y = 0.024x - 38$
$R^2 = 0.237$

Farm yield
$y = 0.049x - 92$
$R^2 = 0.522$

B. United Kingdom

tonnes/ha

$y = 0.061x - 111$
$R^2 = 0.532$

Farm yield

$y = 0.053x - 98$
$R^2 = 0.457$

Potential yield is plotted against the year of variety release.
Sources: A – FY from Cajeme District Statistics; PY from numerous unpublished CIMMYT experiments collated by R.A. Fischer. B – FY from FAOSTAT; PY from HGCA.

Results for the Yaqui Valley, the United Kingdom and all other wheat cases are summarized in Table 10.2. In addition to S1 and W1, three other important wheat mega-environments are included.

Table 10.2 shows a diversity of combinations of key parameters for wheat growing regions. Two key observations are that average PY progress is only about 0.6 percent, and that only some yield gaps are closing. The actual gaps given (averaging 43 percent) can be compared with those in the review by Lobell, Cassman and Field (2009). For wheat, these authors were able to summarize

12 estimations from developing countries in the 1990s, showing a FY range of 40 to 95 percent of PY, averaging 65 percent: expressing the gap as a percentage of FY it averaged 55 percent, somewhat larger than the estimate for developing countries in Table 10.2. The difference with Table 10.2 could easily arise from both lower estimates for FY (understandable give the earlier dates to which FY refers, and the inclusion of less-advanced regions) and higher estimates for PY in the Lobell, Cassman and Field (2009) study.

Table 10.2
Summary statistics[a] from case studies of wheat yield change

| Region and mega-environment[b] | Wheat area (million ha) | Yield and gap, 2007 or 2008 | | | Current rate of change relative to 2008 yield or gap | | | Comments |
		FY (tonnes/ ha)	FY (tonnes/ ha)	Gap (% FY)	PY (%)	PY (%)	Gap[c]	
Yaqui Valley S1	0.16	6.0	9.0	50	0.3	0.3	0	Case study
Punjab, India S1	3.9	4.3	6.25	45	0.2		0	Case study
Haryana, India S1	2.4	4.2	5.75	35	0.6		-	Case study
Egypt S1	1.2	6.5			1.6		--	High FY progress
Brazil S1	1.7	2.0			1.6		--	High FY progress
Western Australia S4	4.5	1.8	2.6d	45	1.4	0.5d	--	Case study
North Dakota, USA S6	3.4	2.5	3.7d	50	0.9	1.0d	0	Case study
UK W1	1.8	8.2	10.4	25	0.7	0.6	0	Case study
Eastern China W1	16	4.7?	7.0?	50		0.7		Zhou *et al.*, 2007a; 2007b
Kansas, USA W4	3.6	2.6	3.9d	45	0.6	0.4d	0	Case study

[a] All rates of FY change are from linear trends over last 20 to 30 years; 2008 yields are from the linear trends; no curvilinear fits were superior, unless noted. Where possible, FY trends have been corrected for secular weather change, but not for increasing CO_2. Blanks mean no data yet available.
[b] Mega-environments: S4 = low to moderate-rainfall spring wheat at low latitude, about 16 percent of world wheat area, equally distributed between industrial and developing countries; S6 = low to moderate-rainfall high-latitude spring wheat, 21 percent of world wheat area, mostly in industrial countries; and W4 = low to moderate-rainfall winter wheat, about 10 percent of world wheat area, equally distributed.
[c] + Increasing; 0 no change; -- decreasing.
[d] Actually PY_W.
Sources: FY and its change from FAOSTAT or United States Department of Agriculture (USDA) National Agricultural Statistics; PY from forthcoming work by Fischer and others, supplemented by reports from the literature.

S1 (irrigated and high-rainfall) is the most important wheat environment for the developing world. About 78 percent of the crop is irrigated, and was the first target of the green revolution. Several examples are given in Table 10.2. Progress in FY and PY have slowed markedly in Mexico and India (and South Asia in general), but Egypt, now exceeding the Yaqui Valley in yield, shows remarkable FY progress (discussed in the following subsection on rice), and high-rainfall countries such as Brazil also have good FY progress; acid soil tolerance and conservation tillage have been important factors in Brazil's progress.

The S4 environment characterizes rainfed wheat in the Mediterranean region, North Africa, West Asia, Australia and Argentina; it is probably the driest major wheat environment, with Western Australia – shown in Table 10.2 – providing an excellent example. It is the only mega-environment in which the yield gap has clearly closed lately, largely because of the adoption of many advances in wheat agronomy.

S6 is the high-latitude spring-sown wheat environment of the Northern Hemisphere, comprising 30 percent of the United States area, most of Canada, eastern parts of the Russian Federation and northern Kazakhstan, along with northeastern China. It is almost entirely rainfed and moderately dry. North Dakota, in the United States of America, fits S6 and shows modest progress and a yield gap fairly typical of rainfed wheat in the industrial world.

The United Kingdom, as already discussed, is probably reasonably representative of the favourable cool winter-habit W1 environment, comprising Europe, Ukraine, southern parts of the Russian Federation, the north China plain and the eastern United States.[7] In contrast, W4 refers to the drier cool wheat environments, dominated by the Great Plains of the United States, the Anatolian Plateau of Turkey and western China. It is represented in Table 10.2 by Kansas, whose low PY_W progress and modest yield gap are similar to those in rainfed Western Australia. FY progress would likely be similar or better in the W4 regions of Turkey and China because of the lower yield base; these regions would also have good scope for FY gains from increasing the currently low adoption of conservation tillage in this erosion-prone environment.

Rice

Figure 10.10 shows two case studies for rice, a crop that is almost entirely grown in developing countries (except for Japan, the Republic of Korea and the United States of America). Central Luzon in the Philippines includes the irrigated wet-season (I1, low-radiation) and dry-season (I2, high-radiation) tropical environments that dominate rice production, comprising about 54 percent of world rice area. Egypt represents irrigated rice in the very favourable intermediate-latitude high-radiation environment (I3), although this accounts for only 1 percent of the world's rice area, found equally in industrial and developing countries.

---

7. Although the eastern Chinese portion may have lower PY because of warmer grain filling.

The International Rice Research Institute (IRRI) has regularly surveyed FY in Central Luzon over the last 50 years; variety turnover has been rapid, and over the last 30 years rice area has been entirely planted to modern varieties, reaching high levels of fertilizer application (150 kg/ha of N, phosphorus [P] and potassium [K]). After greater initial FY progress with the first modern varieties, yield progress since the late 1970s has been a steady 0.6 percent, and large gaps (60 percent wet season, 100 percent dry season) persist compared with PY at IRRI (Figure 10.10). The yield gap is smaller (about 35 percent) for wet-season crops in provinces adjacent to Central Luzon and at PhilRice (Laguna and Neuva Ecija), where FY progress has almost ceased. PY progress has been very slow (estimated at zero percent per annum) in Central Luzon, although varietal disease and insect resistance, earliness and quality have improved markedly (Peng *et al.*, 1999). The current dry-season PY of 9 tonnes/ha is corroborated by dry-season yields of 9 to 10 tonnes/ha for optimally managed irrigated rice in tropical America under the Latin American Fund for Irrigated Rice programme (G. Zorrilla, personal communication). These estimates do not include the new tropical hybrid varieties just reaching farmers in the Philippines and showing 11 to 14 percent increases in PY in the dry season (Yang *et al.*, 2007).

Figure 10.10
Changes in rice FY and PY in Central Luzon, dry season and wet season, and in Egypt

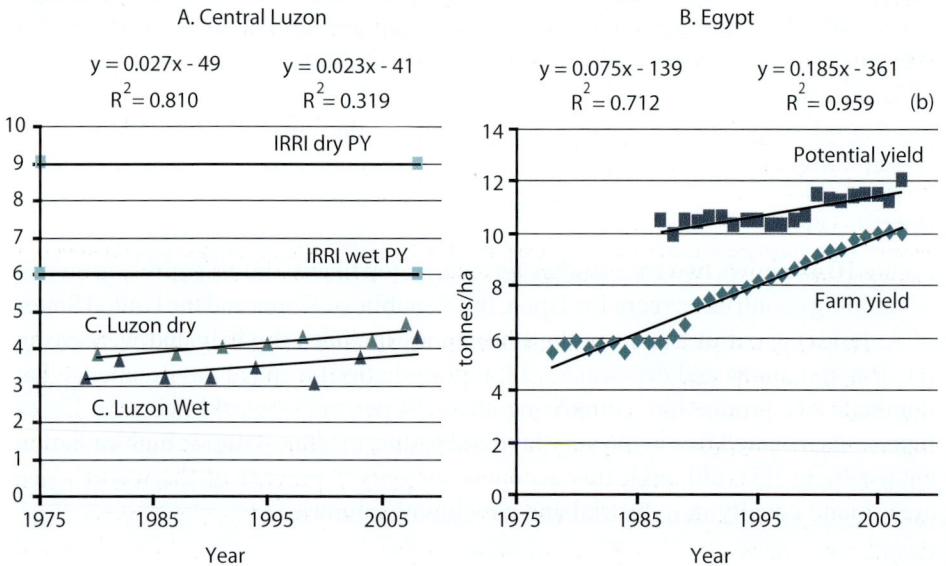

A. Central Luzon

$y = 0.027x - 49$
$R^2 = 0.810$

$y = 0.023x - 41$
$R^2 = 0.319$

B. Egypt

$y = 0.075x - 139$
$R^2 = 0.712$

$y = 0.185x - 361$
$R^2 = 0.959$ (b)

Sources: A – FY from IRRI surveys; PY from IRRI trials. B – FY from FAOSTAT; PY from on-farm demonstrations (A.E. Draz personal communication).

410

Egypt is noteworthy because of the contrast it represents: it has the highest FY in the world (10.1 tonnes/ha), exceeding that of California (of 9.4 tonnes/ha). FY has shown 1.8 percent annual growth in the last 20 years or so, while area has increased at 2 percent. PY is growing at only about 0.7 percent, meaning that there has been a marked closing of the yield gap, which is now about 15 percent of FY. It is suggested that the situation in Egypt reflects a strong research and extension effort; in addition, price reform in the late 1980s removed price disincentives for most crops, including rice. These and other case studies are summarized in Table 10.3.

Table 10.3
Summary statistics[a] from case studies of rice yield change in key regions

Region and mega-environment[b]	Wheat area (million ha)	Yield and gap, 2007 or 2008			Current rate of change relative to 2008 yield or gap			Comments
		FY (tonnes/ ha)	FY (tonnes/ ha)	Gap (% FY)	PY (%)	PY (%)	Gap[c]	
Central Luzon wet I1	0.8	3.8	6	60	0.6	0.2	0.0	
Punjab I1	2.4	3.8	8	110	0.9			
China I1	29.0	6.2			0.0			FY growth ceased 1996
Japan I1	3.0	6.5	10	55	0.3	0.4	-	Area decrease 1.7%
Central Luzon dry I2	0.4	4.5	9	100	0.6	0.2	0.0	
Egypt I3	0.7	10.1	11.6	15	1.8	0.7	--	Area increase 2%
California I3	0.2	9.4			0.0			
South Asia R1, R2, R3	28.5	1.8	3.6	100				IRRI, 2008

[a] All rates of FY change are from linear trends over the last 20 to 30 years; 2008 yields are from linear trends; no curvilinear fits were superior, unless noted. Blanks mean no data yet available.
[b] Mega-environments: R1 = rainfed lowland, 25 percent of global rice area; R2 = rainfed upland, 13 percent of global rice area; and R3 = deep-water, 7 percent of global rice area.
[c] + Increasing; 0 no change; -- decreasing.
Sources: FY and its change are average farm yield from FAOSTAT or USDA National Agricultural Statistics; PY from forthcoming work by Fischer and others, supplemented by reports from the literature.

The irrigated rice environment is well represented in Table 10.3. It was not possible to obtain reliable numbers for the other main rice ecologies – rainfed lowland (R1), rainfed upland (R2) and deep-water (R3) – but Table 10.3 attempts to cover aspects of these for South Asia. Notable in the table is the low FY growth, except for in Egypt, particularly the low or zero FY growth in China and Japan (and the Republic of Korea, but that is not shown). In China, this situation prevails, despite the 50 percent adoption of indica hybrids and the reporting of hybrid yields of up to 12 tonnes/ha in the rice bowl of the eastern China plains (Peng *et al.*, 2008). In Japan, eating quality requirements constrain FY. Also notable are the slow PY growth rates, and yield gaps are generally larger than with wheat, except for rice in Egypt.

The yield gaps in Table 10.3 can be compared with Lobell, Cassman and Field (2009), who summarize 41 estimates from developing countries of rice FY relative to PY: these range from 30 to 85 percent, with an average of 60 percent. This converts to a FY – PY gap of 65 percent. The authors supplement these numbers with results from a modelling exercise for irrigated rice PY across Asia, concluding that for northeast Asia, FY is about 75 percent of PY (gap = 35 percent of FY), but for northwest India it is only about 45 percent (gap = 120 percent).

For rice, where irrigated environments are fairly distinctive and dominant, another estimate of yield gaps can be generated by simply comparing regional or national yields for similar crop agro-ecologies, and assuming that the highest yield represents the current global attainable yield (AY), or at least a conservative estimate of it. For example, based on Egypt, where the current national average yield for I3 is 10.1 tonnes/ha, 9 to 10 tonnes/ha can be seen as the appropriate AY for intermediate-latitude countries with relatively cloud-free summers and an absence of chilling at meiosis, such as experienced in the Islamic Republic of Iran (current yield 4.9 tonnes/ha), Uzbekistan (3.4 tonnes/ha) and Chile (5.5 tonnes/ha).

Maize and related crops

CIMMYT has defined useful mega-environments for maize in the developing world, to which the industrial countries were added for the case studies. The Kenya case study encompasses all the low-latitude maize environments: tropical-lowland (M1), accounting for 32 percent of world maize area; subtropical and mid-altitude (M2), with 13 percent of area; and highland (M3) with 4 percent. All of these are found in developing countries. Generally, these are relatively humid environments with maize tailored to fit the wet season, but – as maize is relatively susceptible to water shortage – drought stress is not uncommon. Heisey and Edmeades (1999) estimate 21 percent of the area in the tropics and 14 percent in the subtropics to be "often stressed". The second case study is of Iowa in the United States of America, representing the relatively humid (or supplementally irrigated) favourable temperate environment (M4), which contains 51 percent of the world's maize area, equally distributed between industrial and developing nations (with China dominating the latter).

Maize in Kenya is complicated because of the diverse environments, but 75 percent is in the more favourable M2 and M3 environments at more than 1 100 m above sea level. Kenya was a pioneer in hybrid maize and other farmer support, but this declined in the early 1980s, and yield growth ceased or even fell after 1980 (Fig 10.11A). In the 1990s, fertilizer supply was privatized, and fertilizer use has slowly grown to reach about 45 kg/ha (N + P + K); after falling in the early 1990s for no clear reason, FY appeared to start growing in the mid-1990s, averaging 38 kg/ha/year since 1996, to give an impressive 2.1 percent current growth rate (before the problems of 2008 when yield fell to 1.4 tonnes/ha). Regardless of whether the recent trend is cause for optimism, many factors still constrain maize yield in Kenya, including degraded soils; insufficient nutrient supply from both fertilizer and manure; risk associated with drought, especially in the marginal areas to which maize is spreading; weeds such as Striga; and intercropping, which is not in itself a constraint. Thus PY in the favoured M2 and M3 areas is still so far above FY (the yield gap nationally is at least 200 percent) as to seem irrelevant. However PY_W in less-favoured parts of Kenya is currently the focus of intensive conventional breeding efforts by CIMMYT and the International Institute of Tropical Agriculture (IITA), which have shown good progress in trials throughout Southern Africa (Bänziger *et al.*, 2006). Recently, genetic modification approaches for drought tolerance have been included.

Figure 10.11
Changes in maize farm yields in Kenya, and Iowa State, United States of America

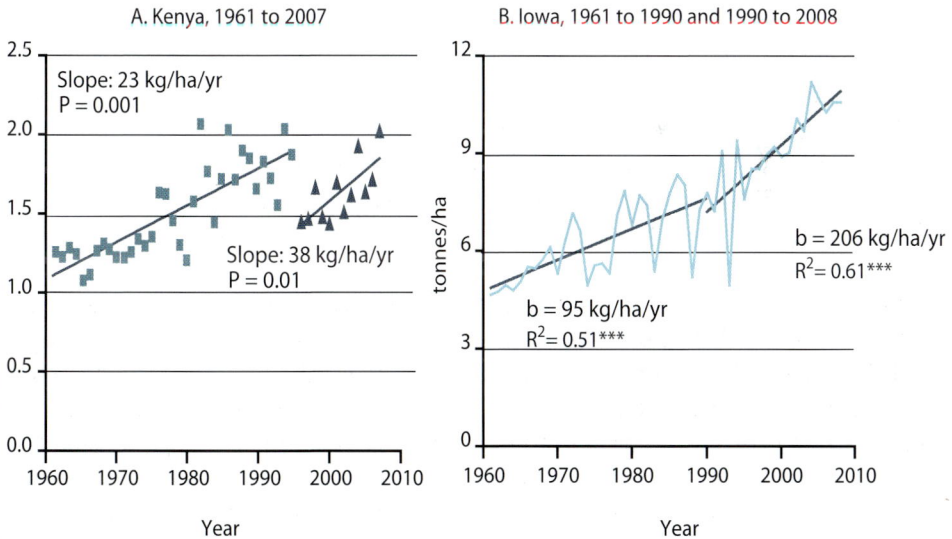

Sources: A – yields from FAOSTAT. B – Iowa grain yields (14 percent moisture) from USDA National Agricultural Statistics www.nass.usda.gov/quickstats/pulldata_us.jsp.

Iowa State grows 5 million ha of maize, largely in one-crop per year rotation with soybeans. FY progress has been impressive for many years (Figure 10.11B); it accelerated around 1990, and from 1990 to 2007 averaged 206 kg/ha/year, or 2.0 percent from the impressive 10.5 tonnes/ha projected FY in 2009. This reflects a large investment in private sector breeding and public sector research, combined with modern farming and a favourable climate: it is also suggested that the recent spurt in progress commenced with the arrival of genetically modified maize varieties. Certainly, herbicide-resistant maize favours conservation tillage and, perhaps, earlier sowing, and *Bacillus thuringiensis* maize may be giving resistance against yield losses not even recognized in the past (e.g., root worm resistance). Estimates of PY are few, and estimates of its rate of change even fewer: farmer contests suggest that PY is currently about 17 tonnes/ha, which would give a yield gap of 60 percent, perhaps surprising for advanced farming. The best hybrids in breeders' and agronomists' trials appear to be reaching about 15 tonnes/ha. These same breeders indicate gains in PY of about 100 to 200 kg/ha/year, or about 1.0 percent per annum (e.g., Hammer *et al.*, 2009; Edgerton, 2009), but this important number merits further study.

Table 10.4
Summary statistics[a] for case studies of maize yield change in key regions, and for related crops

Region and mega-environment[b]	Wheat area (million ha)	Yield and gap, 2007 or 2008			Current rate of change relative to 2008 yield or gap			Comments
		FY (tonnes/ ha)	FY (tonnes/ ha)	Gap (% FY)	PY (%)	PY (%)	Gap[c]	
Kenya M1, M2, M3	1.75	1.8	6[b]	200+[b]	2.1	++	--	FY growth in last 12 years only
Sub-Saharan Africa M1, M2, M3		1.6	4.1[c]	193[c]	0.8			Area increases
Brazil M2	12.5	3.6			2.6			
Iowa, USA M4	5.3	10.5	15	43	2.0	1.0	--	PY from trials versus contests
USA M4	32	9.7			1.5			
China M4	27	5.3			1.0			Area growth 1.4%
Egypt M4	0.8	8.4			2.0			
Other crops								
Sorghum Africa M2	27	1.0			0.4			Area growth 1.7%
Millet Africa M2	22	0.8			1.0			Area growth 1.3%
Millet India M2	11	0.9	1.8	100	1.7			Area decline -2.0%
Soybeans Brazil M2	21	2.7			1.7	0.7		Area growth 4.4%
Soybeans USA M4	31	2.8	3.6	30	1.0	0.7		Area growth 1.5%

[a] All rates of FY change are from linear trends over the last 20 years; 2008 yields are from linear trends; no curvilinear fits were superior, unless noted.
[b] Conservative expert opinion for PY across all environments.
[c] AY from on-farm with best-bet technologies (Sasakawa global 2000 reports).
[d] + Increasing; 0 no change; -- decreasing.
Sources: FY from FAOSTAT or United States Department of Agriculture (USDA) National Agricultural Statistics.

These maize case studies and other useful maize data are summarized in Table 10.4, which also includes sorghum, millet and soybean data. Sorghum and millet are the poor cousins of maize, tending to grow on the margins of maize areas where it is too dry for maize. Soybean, on the other hand, is a unique leguminous oilseed that, unlike cereals, has shown strong area growth in the last decades.

Notable in Table 10.4 are the relatively high maize FY growth rates compared with wheat and rice, not only in Brazil, the United States of America, China and Egypt, where hybrids dominate, but also with some growth in sub-Saharan Africa. Growth in sub-Saharan Africa is from a very low base, as yield gaps remain huge. Egypt since the early 1990s shows what can be achieved in a well-endowed environment with good policy on research, extension and prices.

Again the maize gaps in Table 10.4 can be compared with those in the extensive review by Lobell, Cassman and Field (2009), who cite nine tropical and subtropical maize cases (with FY ranging from 16 to 46 percent of PY, and averaging 33 percent) and two reports from Nebraska: irrigated (56 percent) and rainfed (40 percent). These convert into gaps of 200 percent of FY in the tropics and subtropics, 85 percent in Nebraska irrigated, and 150 percent in Nebraska rainfed. These numbers are quite comparable with those in Table 10.4, and suggest that yield gaps are larger for maize than for wheat and rice. However, the Nebraska data are surprising, and come originally from Duvick and Cassman (1999). Lobell, Cassman and Field (2009) later cite unpublished simulations of maize PY, which indicate that FYs in Nebraska are 75 percent (irrigated) and 65 percent (rainfed) of PY, amounting to gaps of only 35 percent (irrigated) and 55 percent (rainfed) of PY.[8]

In another approach, the poor yields in M1, M2 and M3 environments in sub-Saharan Africa in Table 10.4 can be contrasted with yields in relatively similar environments in Southeast Asia, averaging more than 3 tonnes/ha across 8 million ha, and Brazil, of 3.6 tonnes/ha.

Yields of sorghum and millet in sub-Saharan Africa are even poorer than those of maize, probably partly reflecting area expansion into more marginal areas. In India, millet is the target of the International Crop Research Institute for the Semi-Arid Tropics' (ICRISAT's) research effort: yield grows but area declines, while recent simulation modelling and on-farm demonstrations indicate PY to be 1.8 tonnes/ha, suggesting a gap of 100 percent (Murty *et al.*, 2007).

---

8. The discrepancy with the Duvick and Cassman (1999) report comes from the lower values of PY, derived from simulation in the later report (e.g., ranging from 12.2 to 17.6 tonnes/ha across Nebraska, irrigated). In addition, compared with the estimate of PY from contests in adjacent Iowa (15 tonnes/ha; Table 10.4) and current yields of contest-winning crops in Nebraska, these simulations seem unrealistically low, so the view that there is a moderate yield gap even in Nebraska, and even with irrigated maize, holds.

Soybean is showing remarkable yield and area growth globally, exemplified by Brazil and the United States of America; it is grown in maize environments, often in rotation with maize.

Summary of yield progress and yield gaps

In the wheat and rice examples, FY progress is generally below 1.5 percent, and usually below 1.0 percent. PY progress from breeding is no more than 1.0 percent, and often much less for wheat and rice, crops where breeders must give more attention to grain quality traits and disease resistance than for maize. In most situations, there is a gap exceeding 30 percent between FY and PY, but this reaches 100 percent in several rice cases. The rate of gap closing has been slow, except in the case of rice in Egypt. For maize, FY progress is often 1.5 percent or better. It has been difficult to obtain good estimates of PY progress for maize, but it likely exceeds that of wheat and rice, probably reflecting fewer selection constraints and the high involvement of hybrids and the private sector. The gap between FY and PY in maize is large in sub-Saharan Africa, where it easily exceeds 100 percent, but is only moderate and is closing in Iowa.

Closing existing yield gaps

Yield gaps exist because known technologies that can be applied at the local experiment station are not applied in farmers' fields with the same natural resource endowments. There are many reasons for this, but the first to consider are economics and risk aversion, about which there is a rich literature. Farm yields (FYs) that are constrained only by such considerations have usefully been defined as the attainable yield (AY, see section on Defining key concepts), but it must be borne in mind that AY is driven by farm-gate prices, which may be distorted from world prices by subsidies, taxes or poor infrastructure and institutions. Of the examples studied in the previous section, wheat yields in the United Kingdom – which has modern farmers, institutions and infrastructure, and minimal subsidies – should approach AY: the 25 percent yield gap between FY and PY (Table 10.2) is therefore a useful estimate of the minimum gap to be expected due to economics and risk. Another approach to calculating AY is to look at the distribution of field yields within a region and assume that some proportion of the higher yields indicates the AY, for example, the ninth decile (Yaqui Valley case study). However, this has problems: it is hard to obtain a large unbiased sample of field yields; and yield variation may be due to variation in the natural resource base of the fields, not solely to that in exploitable factors.

In the case studies, only one yield gap smaller than that for wheat in the United Kingdom was found – 15 percent for rice in Egypt. In Egypt, there appears to be no large price subsidy, but there is an especially strong and focused research and extension effort for rice, which is highly concentrated in the Nile Delta region. It is interesting that Lobell, Cassman and Field (2009) also suggest that a gap of 25 percent of FY may represent the economically optimum level of production, while recognizing that risk and uncertainty in farmers' decision-making (especially in rainfed situations) may raise the estimate of this yield gap somewhat. Taking a conservative 30 percent as the minimum above which there is scope for economic exploitation, 14 of the 17 cases outlined in the previous section appear to have exploitable gaps, some being quite large, notably maize in sub-Saharan Africa. As might be expected, there is also a strong tendency for smaller gaps in industrial countries. Other things being equal, PY increases might be expected to be important for the future where the gap is small, and gap closing possibilities to increase as the size of the gap increases. This section looks at gap closing.

Constraints contributing to yield gaps

Poor infrastructure, weak institutions and bad farm policy can create huge obstacles to the adoption of improved technologies. These obstacles are exhibited particularly in price disincentives at the farm-gate, expensive credit and increased risk in general; for example, the N-to-grain fertilizer price ratio in much of Africa is on average double that in other regions, and higher still in inland land-locked regions (Morris *et al.*, 2007). Solutions lie with public investment in infrastructure and institutions, and with sound policy, the lack of which has been a major contributor to the large yield gap in places such as sub-Saharan Africa (e.g., Table 10.4). These are widely canvassed in other chapters; this section focuses on those other (non-market) constraints that contribute to the exploitable yield gap (the Agronomic column of Table 10.5).

The Breeding column in Table 10.5 points to ways in which targeted breeding can help close the yield gaps arising from given constraints, not by raising PY or PY_W, but essentially by making varieties more resilient: new varieties are generally adopted more readily than new management techniques, often because they are a less expensive option for the farmer and the extension organization, so this is always a favoured route if the required genetic variation exists. In contrast to breeding, there is nothing new in the other two Resolution columns of Table 10.5; these technologies and policies already exist in many parts of the world (although some might be refined with further research, such as information technology for smallholders, or seasonal forecasting) and all have had or should have positive impacts on FY where appropriate.

Table 10.5

Constraining factors contributing to the FY – PY gap, and ways of resolving them

| Constraint | Resolution[a] | | |
	Argonomic	Breeding	Institutional/ infrastructural
General farmer constraints			
Lack of farmer awareness, conviction or skill	On-farm demonstration	On-farm testing and selection	Education, media campaigns, extension
Farmer risk aversion	Forecasts, tactical decision-making, e.g., for N top-dressing	Tolerance of extreme weather events, e.g., drought, flooding, hail, frost, wind	Insurance, favourable credit terms
Inadequate labour supply	Mechanization, reduced tillage, herbicides	Selection for uniform maturity to favour mechanical harvesting	Facilitated labour migration; credit for mechanization
Technical constraints			
Lack of major long-term soil amelioration	Drainage, land levelling, liming, deep tillage, gypsum	Waterlogging and salt tolerance	Long-term credit
Excess tillage and loss of moisture, soil compaction	Conservation tillage options and suitable machinery, controlled traffic	Suitable varieties; disease and herbicide tolerance	Credit for new machinery
Manageable topsoil toxicities	Amelioration, e.g., lime for acidity	Acid tolerance	Input suppliers, credit
Sub-optimal nutrient supply	Diagnostics, application of nutrients, tactics	Some scope for improved N, P and zinc uptake and utilization	Input suppliers, quality control
Soil variation within and between adjacent fields	Diagnostics for adjustment of application rates	Greater tolerance of soil stresses	
Use of old varieties or poor seed	Better on-farm seed management and storage	F1 hybrids and licensed traits to encourage strong seed industry	Strong seed industry and regulation, credit
Incorrect time of sowing	Mechanization and reduced tillage to accelerate sowing	Varieties with a range of maturities; herbicide-tolerant varieties	Policy for favouring mechanization, contract seeding
Poor plant population	Better drilling procedures and machines, quality seed storage	More robust varieties, e.g., long coleoptile in wheat, more tillage	Strong seed industry
Diseases and pests, above and below ground	Biocides, sanitation, crop rotation	Host plant resistance	Input suppliers, quality control
Weeds	Herbicides, cultivation, sanitation, crop rotation	Enhanced crop plant competitiveness, herbicide tolerance	Herbicide quality control, release regulation
Poor water management in irrigated systems	Improved water application techniques and skills	Greater tolerance to water shortage and excess	Efficient supply systems to farms
Long-term soil degradation	Crop rotation, fertilizer, green manuring, farmyard manure, conservation tillage, zero tillage	Varieties adapted to biotic and abiotic stresses of high plant residue levels, and with good residue production	Regulations ensuring farmers' landownership

[a] Resolution to allow FY to approach the AY corresponding to current PY with realistic economics.

Without doubt, plant breeding's major role in gap closing lies in host plant resistance. Oerke (2006) presented a meta-analysis of actual global yield losses due to biotic stress (weeds, insects, fungi, bacteria and viruses), which averaged more than 23 percent of estimated AY (hence a greater percentage of FY) across the major cereals (without any controls, potential losses were estimated to average 32 percent) (Table 10.6). This is part of the exploitable yield gap, and its reduction is the aim of host plant resistance breeding. Conventional breeding is protecting progress by maintaining resistance levels in the face of evolving pest agents, while aiming to make progress by strengthening resistances, especially through exploiting durable sources of resistance. This has recently been documented globally for the case of wheat rusts (Dubin and Brennan, 2009). Others have pointed to the growing impact of transgenic insect resistance, particularly with maize (and cotton), and linked it to yield gains as more effective, less expensive host plant resistance replaces insecticides, which were often not 100 percent effective. It would seem that the scope for using better host plant resistance to halve a portion of the global yield gap – which is about 30 percent of FY and due to biotic stresses – is good in the medium term (15 years), especially if transgenic resistance to fungal diseases, which currently exists in a few cases, can be delivered.

Table 10.6
Global estimates of potential crop losses without physical, biological or chemical protection, and actual crop losses

	Potential losses *(% of AY)*			Actual losses *(% of AY)*		
Biotic stress agent	Wheat	Rice	Maize	Wheat	Rice	Maize
Weeds	23.0	37.1	40.3	7.7	10.2	10.5
Animal pests	8.7	24.7	15.9	7.9	15.1	9.6
Pathogens	15.6	13.5	9.4	10.2	10.8	8.5
Virus	2.5	1.7	2.9	2.4	1.4	2.7
Total	49.8	77.0	68.5	28.2	37.4	31.2

Source: Oerke, 2006.

The Oerke (2006) meta-analysis also estimates actual losses due to weeds at 10 percent (with potential losses of 33 percent). Modern varieties tend to be more susceptible to weed competition, so breeding did not help until the advent of herbicide-tolerant cultivars, first using natural resistance, and then in the last 15 years resistance based on genetic modification. Gyphosate- ("round-up") and glufosinate-resistant genetically modified varieties have been very successful in maize, soybean and canola in the Americas, facilitating weed control, conservation tillage and often earlier planting, all leading to somewhat higher yields. Genetically modified herbicide resistance will undoubtedly spread into the rest of the world, but integrated weed management, employing a suite of agronomic and breeding approaches will remain vital for sustainable weed control, and will be a special challenge for extension in developing countries.

Prioritizing constraints and reaching farmers

There are usually multiple constraints in any situation, and it is a challenge determining which constraints are more critical and more amenable to change while recognizing that interventions often interact positively and are thus most effective when adopted together (de Wit, 1992). This can only be achieved by on-farm survey and experimentation, which started many years ago with farming systems research, farm management clubs and rapid rural appraisal, and continues in many guises in the industrial world, especially influenced by the privatization of agricultural extension and the use of advances in remote sensing and information and communication technology. It is noteworthy that large commercial maize seed companies in the industrial world, such as Monsanto and Pioneer, employ more agronomy extensionists than breeders to ensure that new varieties reach their potential in farmers' fields.

In the developing world, more traditional approaches remain, although with growing emphasis on farmer participation (Paroda, 2004). Lobell, Cassman and Field (2009) recount how IRRI conducted on-farm rice experiments in Asia in the 1970s to test high inputs, and learned that FY varied greatly, as did responses to inputs, especially fertilizer and insecticide, which were often uneconomic. This pointed to the importance of field-to-field variability, and the need to adjust inputs accordingly and throughout the season, whether by site-specific nutrient management, which reached maturity some 20 years later (Dobermann *et al.*, 2002), or via field-level pest monitoring as part of integrated pest management packages. Another lesson is that this is scientist-intensive and expensive research, often taken over by farmers and their advisers in the industrial world, and explaining why large yield gaps often persist in developing countries, where circumstances demand innovative approaches to reach the billion small farmers (e.g., Paroda, 2004). However, there are also cases of unique progress, as demonstrated by the almost instantaneous delivery of field-specific recommendations to small farmers in the Philippines through the ubiquitous mobile phone (Roland Buresh, personal communication).

Very recently, IRRI re-examined rice yield gaps, this time using expert knowledge to assess constraints and possibilities for irrigated rice in South Asia (IRRI, 2008). For this crop, FY is currently 5.1 tonnes/ha on 34.3 million ha; it was estimated that yield was constrained to an average of 1.9 tonnes/ha

(37 percent) by yield-limiting factors including nutrients (10 percent), diseases (7 percent), weeds (7 percent), water shortage (5 percent) and rats (4 percent). IRRI predicted that the adoption of existing technology and ongoing breeding for robustness would reduce the total loss by about one-third over the next 15 years, increasing FY by 35 kg/ha, or 0.7 percent, per year. The exercise was repeated for the 28.5 million ha of rainfed lowland and upland rice in South Asia, with a current FY of 1.8 tonnes/ha. Yield-limiting factors amounted to 68 percent of FY, including nutrients (23 percent), disease (15 percent) and weeds (12 percent); about one-quarter of these losses are predicted to be eliminated by research for development, including extension, over the next 15 years, adding 19 kg/ha, or 1.0 percent, per year to FY. With this background on South Asia, IRRI – along with National Agricultural Research Systems (NARS), the Consultative Group on International Agricultural Research (CGIAR) and private sector partners – has recently embarked on a new extension approach, the Cereal Systems Initiative for South Asia. This primarily involves hubs staffed by experts mandated to adapt and deliver existing technologies to local farmers.

For wheat in the Yaqui Valley, a recent concerted effort has been made to understand the yield gap PY – FY (currently 50 percent) (Table 10.2), using the latest high-resolution satellite imagery to estimate field-level yields (Lobell *et al.*, 2003) and supplement a long history of farm surveys. From images over several years, it was estimated that wheat yields were constrained by late planting (Ortiz-Monasterio and Lobell, 2007), delays in the first post-plant irrigation (Lobell and Ortiz-Monasterio, 2008) and summer fallow weeds (Ortiz-Monasterio and Lobell, 2007). Improved institutions and farm management decisions could largely eliminate these constraints (which averaged a total of about 10 to 15 percent of FY a year), and would bridge about half of the gap to estimated AY in the valley. It is interesting that N nutrition was only a very minor limitation, surveys and on-farm fieldwork pointing to considerable scope for better N management to improve N fertilizer use efficiency, if not increase yield (Ortiz-Monasterio and Raun, 2007).

The persistence of large yield gaps, especially in the developing world, draws attention to situations where these gaps have been closed. As already mentioned, rice in Egypt is an obvious example. A second example of dramatic technology adoption, albeit with less immediate implications for FY than for sustainability of the whole cropping system, relates to the uptake of conservation tillage for wheat, maize and soybeans in southern South America (Argentina, Brazil and Paraguay), which rose from zero in 1970 to 24 million ha in 2000. This was very much driven by farmers' groups and farmers themselves, faced with the threat of serious soil degradation and the opportunity provided by knock-down herbicides and knowledge spill-over from the North (e.g., Ekboir, 2002). This revolution has yet to reach other developing continents, but is beginning in northwest South Asia. A third success story among poor small farmers has recently emerged from winter maize in northeastern India and Bangladesh.

Conclusions on gap closing

Despite individual success stories such as rice in Egypt, yield gaps generally appear to be persistent and to close only slowly; this is the case even for gaps that are well above those expected for the economics and risk aversion concerned, and even when PY progress has slowed so that catch-up through eliminating excessive lags in varietal adoption is not a major issue. The problem is that gap closing on the large scale needed requires massive investments in rural infrastructure and institutions, as well as technology transfer, and these are not forthcoming, as exemplified by maize in sub-Saharan Africa.[9] Elsewhere, public sector agencies, particularly those reaching the billion small farmers in Asia (Paroda, 2004), aided by the private sector, particularly in Latin America, have made some inroads on the yield gap; they should continue to do so in proportion to the investments made, but there is also scope for innovation, for example based on modern information and communication technologies (see previous subsection and the section on Prices, efficiency, productivity and R&D investment). The employment of agronomists by private seed companies is a pattern that is bound to be followed in the developing world as its seed industry grows in strength and competitiveness. With gap closing, there are no spill-ins as there are in the case of PY advances through R&D; developments need to be made locally, but it can be argued that the Internet and mobile phones are relevant spill-in technologies whose role could greatly expand. Given the persistence of yield gaps, it remains critically important to continue to lift FY through improved PY, the subject of the next section.

Increasing yield potential

As described in previous sections, PY has grown substantially in the past, through breeding backed by improved agronomy, and this has driven FY growth. Earlier discussion suggests that in the future, growth in PY is probably going to depend more on breeding than on new developments in crop agronomy. New management through breeding synergies will certainly be discovered but is difficult to anticipate. There is a sense that genetic variation for yield must, at some time, become exhausted, and that the relatively easy improvements, such as

---

9. The first comprehensive report on the African Millennium Village Project (Nziguheba *et al.*, 2010) offers useful and encouraging insights into this key issue.

increases in harvest index (HI)[10] and adaptation of phenology, have already been made. Progress will probably depend increasingly on molecular and physiological knowledge of plant growth processes to improve the targeting of breeding efforts for PY, although empirical breeding continues to make some yield progress. This section considers the prospects and avenues for increased PY, under conditions of adequate water and of water constraint (PY_W). Brief mention is also made of PY under N limitations.

Components of PY

Crop physiologists have developed useful analytical frameworks for exploring potential grain yield and its components under radiation- or water-limited conditions (Monteith, 1977; Passioura, 1977):

$$PY = \text{total above-ground dry weight (TDW)} \times HI \tag{1}$$
$$PY = \int PAR_i \times RUE \times HI \tag{2}$$
$$PY_W = \text{transpiration (T)} \times TE \times HI \tag{3}$$

where $\int PAR_i$ is the integral of photosynthetically-active radiation (PAR, MJ)[11] intercepted by green tissue over the life of the crop; and RUE, or radiation use efficiency, is the efficiency with which PAR_i is converted into above-ground biomass (in grams per megajoule [g/MJ]). For PY_W, T is the amount of water taken up and transpired by the plant (in millimetres [mm]);[12] and TE is transpiration efficiency for creating dry weight (milligrams per gram, or kilograms per hectare per millimetre). A parallel to equation (3) for PY_N, N-limited potential yield, can be written as N absorbed and NUE (nitrogen use efficiency). There are many variations of these identities (Mitchell, Sheehy and Woodward, 1998), but they all point towards the efficiency with which a limiting input (radiation, water, N) is captured and used to create dry weight, and how efficiently the biomass is converted to grain (HI). The concept of PY per day is also important; in tropical rice, for example, PY has remained static, while varieties have become earlier, resulting in a gain in PY/day (Peng *et al.*, 1999).

Progress in PY through agronomy has largely come through better crop nutrition, especially N nutrition, giving greater leaf area of longer duration, hence increased PAR_i, and modest increases in RUE (Muchow and Sinclair, 1994;

---

10. The HI is the ratio of the grain yield to total biomass at maturity.

11. Crop physiologists work with either total solar radiation or PAR, the latter being close to 0.5 times the former wherever sunshine is involved (Mitchell, Sheehy and Woodward, 1998); this chapter uses PAR throughout.

12. Throughout this chapter, mm refers to rainfall or water use in depth of water over the land surface; thus for 1 ha, 1 mm equals $10m^3$.

Bange, Hammer and Rickert, 1997). Altered planting date, especially earlier planting, can also give small gains in PY and PY_W through better crop timing with respect to expected weather patterns. Altered planting configuration can give earlier full radiation capture and more even radiation distribution among plants, both important for PY. Progress in breeding for increased PY over the past 50 years has been very significant, and is generally attributed to increases in HI, often via shorter stature in wheat, rice and tropical maize (e.g., Johnson *et al.*, 1986). An exception is temperate maize adapted to the United States of America or Argentina, where PY has increased because TDW has increased, while HI has remained relatively high and stable (Duvick, 2005). Typical values of HI are 0.5 to 0.55 under good conditions for modern winter wheat, rice and temperate maize varieties, but only 0.4 to 0.45 for spring wheat and modern tropical maize varieties (Johnston *et al.*, 1986; Duvick, Smith and Cooper, 2004). There appears little scope for further increase in HI beyond 0.5 because the crop needs a stable structure to distribute its leaf area, support its seeds and prevent lodging. However, there seems to be scope for a 20 percent increase in HI in spring wheat and tropical maize.

The increase in TDW in temperate maize appears to be related to a number of small changes: more erect leaves, which should give higher RUE; more grains per square metre at high planting density, meaning greater sink strength and RUE during grain filling; greater "stay-green", meaning more PAR_i in late grain filling; and a general improvement in tolerance to minor stresses such as cool nights, sudden changes in radiation, high plant density and oxidative chemicals (Tollenaar and Wu, 1999; Duvick and Cassman, 1999).[13] More recently, early cold tolerance, permitting earlier planting, has been highlighted (Kucharik, 2008), and Hammer *et al.* (2009) have made the very novel proposition, supported largely by modelling, that modern hybrids are apparently generating more biomass by capturing and transpiring about 270 mm of additional water from deeper in the soil than their counterparts of 70 years ago. In the case of wheat and rice, however, TDW has increased relatively little through breeding, although there are some reports of increased RUE (see following subsection).

A key aspect of gains in PY in the past has been increased numbers of grains per square metre of land area, rather than changes in weight of individual grains (e.g., Bolaños and Edmeades, 1996; Fischer, 2007). Seed number per square

---

13. Duvick and Cassman (1999) argue that even under irrigation and excellent management, apparently minor but common stresses such as cool nights, sudden changes in radiation as clouds move over the sun, and occasional high temperature are important. They conclude that yield gains with selection have come about because of better tolerance to these "minor stresses", rather than because of increase in yield potential *per se*. At modern densities (about 100 000 plants/ha), plants are also under substantial stress from crowding.

(e.g., Bolaños and Edmeades, 1996; Fischer, 2007). Seed number per square metre is related to crop growth rate from 20 to 30 days before flowering to ten days after flowering in all three cereals (see later), and to the variety's ability to partition assimilate to the developing ear (Andrade, Otegui and Vega, 2000; Shearman *et al.*, 2005). Rice and wheat varieties with the highest PY appear also to accumulate and later translocate larger amounts of temporarily stored pre-anthesis carbohydrate to the grain (Shearman *et al.*, 2005; Katsura *et al.*, 2007). Grains that are set at flowering must be filled adequately from current assimilate plus stored carbohydrate, and adequate water and N nutrition are essential (Wolfe *et al.*, 1988).

In summary, the likeliest routes for further increases in PY are through increases in RUE or PAR_i by boosting photosynthetic activity and/or extending the active life of leaves, while for PY_W, preventing the common decline in HI when crops are under stress, especially around flowering, is also an important possibility. The challenge of RUE and its constituent components attracts many plant scientists. To quote Duvick (2005) "Finally … maize breeders can always hope for the Holy Grail of plant physiologists, major [increases in RUE], effected without disrupting the rest of the infinitely complicated network of interacting genetic systems".

Increasing radiation use efficiency

RUE is the ratio of gross photosynthesis minus (crop respiration + root dry matter) to radiation intercepted over periods that range from a few days to the crop's complete lifetime. RUE was initially found to be a relatively stable number and a useful integrator across leaf positions and radiation levels (Mitchell, Sheehy and Woodward, 1998). Crops differ in their photosynthetic systems. Maize has a C_4 photosynthetic system that allows its leaves to respond to higher levels of irradiance than the C_3 system of wheat and rice, but performs poorly in cool conditions. The C_4 system has a CO_2-concentrating mechanism in bundle sheath cells (the so-called Kranz anatomy) that sharply reduces CO_2 losses from the photorespiration observed in C_3 crops. As irradiance of the leaf increases, the photosynthetic rate of C_3 species reaches a maximum (P_{max}) at a lower irradiance and a lower value of photosynthesis than that of C_4 species; a C_3 species therefore has lower RUE (Figure 10.12), TE and NUE than a C_4 species. However, C_3 species are generally better adapted to cooler conditions.

The main source of variation in RUE is among the species themselves, and P_{max} and RUE are positively associated. Although RUE increases less than a given relative increase in P_{max}, the exact relationship depends on how light is distributed down into the canopy. Mitchell, Sheehy and Woodward (1998) found were 2.7 g/MJ for wheat, 2.2 g/MJ for rice, 3.3g/MJ for maize, and 1.9 g/MJ for soybean, and varietal differences in RUE within crops are quite small. More recent evaluations of RUE in modern maize hybrids result in a value of 3.8 g/MJ, suggesting a possible increase in RUE had occurred with selection (Lindquist *et al.*, 2005). However, selection specifically for higher leaf photosynthetic rate in several past studies, although sometimes successful, has failed to raise crop yield (Crosbie and Pearce, 1982; Austin, 1989; Evans, 1993). Nevertheless, Long *et al.* (2006) suggest theoretical maximum limits to RUE of 5.8 g/MJ for C_3 crops, and 6.9 g/MJ for C_4 crops.

Figure 10.12
Response of leaf net photosynthetic rate to radiation as a proportion of full sunlight for C_3 and C_4 species

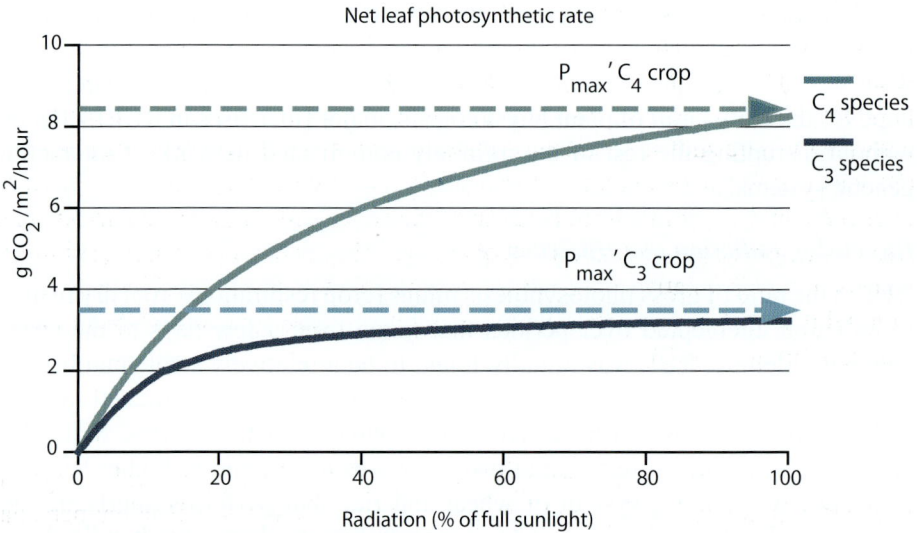

Source: Loomis and Connor, 1992.

As leaves spend much of their lives in shade, the likely route to improving PY is to increase RUE under radiation levels of 10 to 50 percent full radiation (Figure 10.12). RUE values of 3.9 g/MJ for rice (Katsura *et al.*, 2007) and 7.6 g/ MJ for maize (Tollenaar and Wu, 1999) grown under low radiation conditions support this contention. Most modern cereal varieties have erect leaves and a high ratio of leaf area to ground area. This results in lower irradiance at the leaf surface, and hence a higher RUE, but there is little scope for further improving RUE via canopy structure in these crops, because all the modern varieties have very erect leaves. Loomis and Amthor (1999) also conclude that crop respiration is very efficient, with only modest prospects of improvement through targeted selection for low respiration rates.

Future increases in RUE via breeding are therefore likely to be through increases in P_{max}; recent evidence suggests that P_{max} is higher in modern varieties of wheat (Fischer *et al.*, 1998) and rice (Horie *et al.*, 2003), while it has been shown in the United Kingdom that modern varieties of winter wheat have higher RUE; this progress in photosynthesis was measured during the critical period determining seed number. What kind of additional progress could be made by focusing on P_{max} itself? One opportunity for dramatic changes in P_{max} lies in genetic engineering of the leaf photosynthetic system, especially the central photosynthetic enzyme, Rubisco, by increasing its efficiency in capturing CO_2, or increasing the supply of CO_2 or other limiting substrates to the enzyme. A very ambitious project under way at IRRI involves genetic engineering of the C_4 pathway into C_3 crop rice to improve CO_2 supply to Rubisco. Long *et al.* (2006) predict RUE increases at annual rates of 1 to 4 percent over the next ten to 20 years through mechanisms such as these. Other strategies include reducing photorespiration in C_3 crops or reducing the thermal sensitivity of Rubisco activase by gene shuffling so that Rubisco remains active at higher temperatures (Salvucci, 2008). However, these transgenic approaches may have a low chance of success in the medium term because of the complexity of the tasks involved. A less challenging approach could involve a search, for example among primitive wheats and wild relatives, for more efficient photosynthetic machinery, bearing in mind that such wheats have already exhibited higher P_{max} levels than modern varieties (Evans, 1993).

Projections of potential yields

Wheat: A well-researched estimate of wheat PY for the United Kingdom (Sylvester-Bradley, Foulkes and Reynolds, 2005; R. Sylvester-Bradley personal communication) – based on reasonable assumptions, including an RUE of 2.8 g/ MJ and an HI of 0.6, while deploying stem dry matter as efficiently as possible to minimize lodging risk – resulted in 19 tonnes of grain per hectare under well-watered conditions; this could result in a 50 percent increase in average farm yields to about 13 tonnes/ha by 2050.

Rice: Mitchell, Sheehy and Woodward (1998) predicted that conventional selection could result in a tropical and subtropical rice PY of 11.3 tonnes/ha for IR72 maturity. On the other hand, application of IRRI's New Plant Type principles in the large Chinese "super rice" breeding programme has already given a 10 to 20 percent jump in PY, to 12 tonnes/ha, in hybrids grown in lowland eastern

China (Peng *et al.*, 2008). Sheehy *et al.* (2007) predict yields of 50 percent greater than the present 9 tonnes/ha if C_4 photosynthesis could be engineered into rice; the relative advantage could rise as global temperatures increase.

Maize: It is difficult to find consistent PY projections for maize. Chile has the world's highest national maize yield (11.5 tonnes/ha from 130 000 ha in 2005 to 2007) and yields of more than 20 tonnes/ha have been observed under irrigation in Chile's central valley (unpublished data), but this may reflect the more favourable climate compared with that in the United States Corn Belt. This is obviously an issue of great interest in the mid-west of the United States, given the huge maize research investments there. On the one hand, Cassman *et al.* (2003) argue that the limit of PY has already been reached under irrigation in Nebraska, as reflected in a stable average yield of contest winners of 18.8 tonnes/ha. Higher yields have been observed in contests since 1975 (of 21 to 23 tonnes/ha)[14] but the Nebraska number is an average for the period 1983 to 2002. At the other extreme, Monsanto, a leading seed company,[15] has set a goal of doubling United States maize FY by 2030, based on 2000 yields of 8.5 tonnes/ha, resulting in a FY target of 17 tonnes/ha (Edgerton, 2009). This would be unprecedented breeding progress (2.3 percent exponential, or 3.3 percent linear at the outset, 1.7 percent by 2030); but can it be sustained over time, and what would it imply for PY increase to 2030? United States yields for 2007/2009 averaged 9.82 tonnes/ha, with a 2 percent increase per year from 2000 to 2008, already well behind the goal. Monsanto breeders claim they will achieve these record gains in equal measure through conventional breeding, molecular-aided marker selection and genetic engineering for yield. FY in Iowa is about 12 percent above the United States average FY for maize, so Monsanto's claim translates into an Iowa FY of about 19 tonnes/ha in 2030. The PY – FY yield gap in Iowa is currently thought to be about 45 percent (Table 10.4). If this gap is sustained, it would imply a PY of 27 tonnes/ha across Iowa by 2030, somewhat higher than the theoretical maximum yield of 25 tonnes/ha cited by Tollenaar and Lee (2011). Of course the yield gap could close further, but even at 25 percent, PY would be 24 tonnes/ha. A further complication for these projections are the recent findings by Hammer *et al.* (2009), which imply that yield and water use are more tightly coupled than previously believed and that there may not be enough water from rainfall to support much higher yields in Iowa, a region usually considered to be relatively free of water stress and operating under PY not PY_W.

---

14. The highest yield reported in the United States National Corn Growers' Association yield contests in 2007 was 23.9 tonnes/ha: www.ncga.com/files/pdf/2008cycnationalwinners.pdf. Contest yields (rainfed) in Iowa and Nebraska also show steady yield progress at levels about double the state averages.

15. www.monsanto.mediaroom.com.

Water-limited potential yield

Equation (3) underlies understanding of PY_W progress, despite it limitations (Blum, 2009). There has been breeding progress for PY_W, but generally at lower absolute and even relative rates than that for PY. Initially, progress has derived from better fitting of the crops' phenological development to the particular rainfed environments, usually meaning selection for earliness – whether for wheat in a Mediterranean environment or maize in a tropical one. This brings the growth of the crop into a moister period, when TE is higher[16] and reduces the risk of exhausting available moisture before grain filling (maintaining HI). PY_W progress has also derived from spill-over of progress in PY; for example, when higher intrinsic HI is maintained under stress, yield improves in both equations (1) and (3), and higher RUE may also deliver higher TE. Recent analysis of old versus new maize hybrids shows that progress in a dry year in Iowa matches that under wetter conditions (Duvick and Cassman, 1999), although the authors claim that this is spill-over of improved micro-stress tolerance with higher PY, not spill-over of PY *per se*. Such is the importance of variation in flowering date and PY that attempts to study other factors in PY_W variation usually correct for them (Fischer and Maurer, 1978; Bidinger, Mahalakshmi and Rao, 1987), but the picture is less clear for rice, with marked specific adaptation to flooded or rainfed conditions limiting spill-over from favourable environments.

Numerous other factors may influence performance under rainfed conditions, including early vigour, to cover the soil and enhance T at the expense of soil evaporation (a special advantage of proper soil nutrition under rainfed conditions); osmotic adjustment; leaves with waxiness and low epidermal water conductance; and deeper roots (Blum, 2009). For example, for maize in Iowa, it has been suggested that selection has increased tolerance to stress at flowering (Campos *et al.*, 2004) and significantly increased deep-soil water uptake (Hammer *et al.*, 2009). Modest gains in PY_W of wheat have also been made by selecting for TE directly (Richards, 2004). However, many putative drought tolerance traits have not proved useful as selection criteria, or carry a significant yield penalty under well-watered conditions.

One area of opportunity derives from cereals – especially rice and maize – being sensitive to drought at flowering, when a sharp reduction in the numbers of kernels set can occur (Fischer, 1973; 1985; Bruce, Edmeades and Barker, 2001), inevitably reducing HI. Maize ovaries starved for carbohydrate grow slowly, and the ovary's ability to be successfully fertilized can be severely reduced.

---

16. TE is inversely related to the prevailing vapour pressure deficit (vpd) of the air.

Pollen is also directly affected by water stress at meiosis in rice and wheat, and carbohydrate starvation does not explain all of the damage. Selection gains occur when stress is managed to coincide with these critical periods. Indirect selection for rapid ear growth rates in maize under managed drought stress has resulted in improved tolerance (Edmeades *et al.*, 2000). Useful genetic variation (not genetically modified) in the sensitivity of grain set in wheat to water stress around meiosis has recently been demonstrated (Ji *et al.*, 2010).

Water-limited potential yield projections

A variation of equation (3) used in Australia (French and Schultz, 1984) states that $PY_W = k (ET - 110)$, where ET is water used in mm, and 110 mm estimated soil evaporation, while $k = 20$ kg/ha/mm is essentially an average TE across the season multiplied by a good value for HI. This defines an upper limit to PY_W of wheat for a given level of ET; for example if average ET for wheat in southern Australia is 300 mm^3, PY_W is 3 800 kg or 3.8 tonnes/ha (c.f., current national average is about 2 tonnes/ha). This approach is an oversimplification, but has proved a very useful practical guide to PY_W in Australia (Fischer, 2009) and for discussing PY_W increase (Passioura and Angus, 2010). Yield increase through breeding or agronomy can only come from increases in T (e.g., by storing more water, developing a more efficient root system or reducing losses through evaporation from soil or by weeds), or from increases in TE or HI. These generally appear to be modest in extent, but added together may lift PY_W by 25 percent (Passioura, 2002; Passioura and Angus, 2010).

Revisiting equation (3), the largest differences in TE are seen between C_4 and C_3 crops, which average 159 and 83 g of biomass per kilogram of water transpired, respectively (Loomis and Connor, 1992). In a warmer and water-limited world, this provides another strategic reason for developing C_4 versions of rice, wheat and other crops, although C_4 crops are not necessarily more drought-tolerant than C_3 ones (Ghannoum, 2009). There is probably continued scope for PY_W increase through increasing HI, particularly via lessening water shortage-induced reductions in grain number. There is no sign of slowing in recent striking PY_W progress of about 100 kg/ha (or 5 to 8 percent) per year that has been achieved under managed drought stress in the field in tropical maize over a ten-year period, mainly through increases in HI. This selection methodology is also currently delivering useful gains in farmers' fields in Africa (Bänziger *et al.*, 2006). Progress for drought tolerance in rice is also encouraging, with a single large-effect chromosomal region adding 47 percent to yield under severe drought (Bernier *et al.*, 2007), and pedigree selection under managed stress reporting gains of 4 to 10 percent per year (Venuprasad *et al.*, 2008). Genetic engineering possibilities abound in the literature and are discussed in a later subsection.

Exploiting heterosis

Heterosis, present in hybrids and obtained by crossing two genetically dissimilar parents, is considered a form of stress tolerance, and is often greater for PY_W than for PY. In general, hybrids offer about 15 percent yield advantage over open-pollinated parents in maize, and about 10 percent over inbred parents in wheat and rice (e.g., Bueno and Lafarge, 2009). Hybrids have been widely used in maize for 80 years, and are deployed on about 70 percent of the global cultivated area. In rice and wheat, both normally self-pollinated crops, the limitation is the poor yield of the female parent line when it is forced to out-cross, resulting in expensive seed production. Adoption of hybrids in rice is still quite low, except in China where indica hybrids account for 60 percent of the planted area. In wheat, technical issues in seed production have prevented any large-scale adoption. Seed yield constraint is likely to be resolved in the next ten to 20 years, probably using genetic modification technology, thus permitting hybrids to take over most of the world's rice and wheat area. CIMMYT is not optimist about wheat hybrids (Dixon, Braun and Crouch, 2008), but recently it has launched a new hybrid wheat project, while at IRRI there is now strong confidence regarding the viability of indica hybrids for tropical latitudes. Thus wheat, rice and maize yields could rise in one-off yield increases of 10, 8 and 5 percent, respectively, as the proportion of hybrids under cultivation approaches 100 percent. Because there is an on-farm advantage to growing fresh F1 hybrid seed every year, hybrids foster a viable commercial seed industry and a superior level of intellectual property (IP) protection, thereby creating a positive environment for private investment in crop improvement.

Genetic modification using transgenes

Prospects for augmenting PY by increasing P_{max} and RUE through genetic modification are currently based mainly on engineering C_4 photosynthesis into rice, and possibly wheat, or on modifying Rubisco and Rubisco activase enzymes or other enzymes close to Rubisco. These are formidable technical challenges. Other promising genetic modification routes to higher PY have been proposed, but few have been demonstrated in the field, and the compensatory response among yield components is often overlooked. Engineering better abiotic stress resistance (greater PY_W) may be easier, although many putative drought tolerance genes reduce yield unacceptably in well-watered conditions, or simply fail to deliver in the field. In 2012, Monsanto aims to launch commercial maize hybrids carrying the cold shock protein gene *cspA* from *Bacillus subtilis*, which functions under drought stress as a protein that protects RNA from degradation and for which there are some credible published field plot data (Castiglioni *et al.*,

2008).[17] Reports suggest that this transgene is active throughout the life of the maize crop, rather than affecting stress tolerance only at flowering, and will lift yields by 6 to 10 percent under a moisture stress that reduces yields to about 50 percent of irrigated yield levels.[18] This may mark a breakthrough for genetic modification breeding targeting abiotic stress and crop yield. Of particular interest is Monsanto's intention to release this technology for use in adapted maize in sub-Saharan Africa on a royalty-free basis, through the Water Efficient Maize for Africa Project, in an exciting private-public sharing of cutting-edge technology to benefit those who need it most. Preliminary results in Southern Africa suggest the gene is very background-specific, meaning that it has little or no effect in some conversions, and has its greatest effect around flowering. Several other recent studies point to possibilities of greater stress tolerance in rice, which is the common candidate crop for published work on genetic modification for PY because the genome is sequenced and widely available. However, there are few convincing published reports of yield effects due to transgenes in either wheat or rice (but see Xiao *et al.*, 2009).

Engineering for biotic stress and herbicide resistance has already been hugely successful. It has had a significant environmental benefit through reduced pesticide applications, and has lifted yields of crops under insect attack (Brookes and Barfoot, 2009), but has had little effect on PY *per se*. Engineered herbicide tolerance in soybeans, maize and canola has facilitated conservation tillage and permitted more timely planting, with modest benefits for yield. Transgenic resistance to corn root worm in maize has improved yield under water-limited conditions where the insect infestation is severe by retaining more roots and increasing water uptake; before genetic modification it was very difficult to control this pest. The benefits of genetic modification for maize yield are probably reflected by the rate of increase in maize yields in Iowa, which have been significantly greater than those in France and Italy since 1996, the year transgenic maize was first introduced to farmers' fields (Figure 10.13). Transgenic technologies are not used in the field in France and Italy, but an estimated 90 percent of Iowa maize carries at least one transgene for herbicide or insect resistance. It is unlikely that less favourable weather in Europe than in Iowa accounts for all of this difference.

In conclusion, further yield increase via genetic modification for biotic stress resistance and herbicide tolerance is a good possibility; this yield gap is closing. Whether increase will also come from increased PY and PY_W *per se* is less

---

17. An earlier genetically modified maize from Monsanto incorporating an *Arabidopsis* transcription factor and showing improved field drought tolerance (Nelson *et al.*, 2007) appears to have been allowed to lag.

18. www.monsanto.com.

certain. However, the likelihood of transgenic options for stable and long-lasting disease resistance in rice and wheat in the next 15 years or so has the advantage of sharply reducing the need for maintenance breeding in these two crops, an activity that currently consumes about 30 to 50 percent of the breeding effort at IRRI and in the CIMMYT Wheat Program – a much larger proportion than in maize. This would release considerable additional breeding resources for focusing on PY in rice and wheat.

Figure 10.13
Maize yields in Iowa and in France and Italy

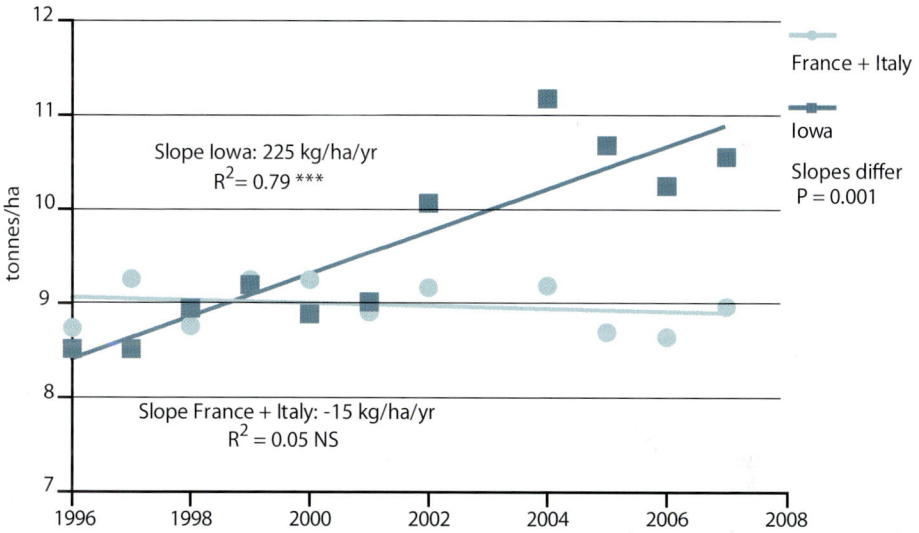

Slope Iowa: 225 kg/ha/yr
$R^2 = 0.79$ ***

France + Italy

Iowa

Slopes differ
P = 0.001

Slope France + Italy: -15 kg/ha/yr
$R^2 = 0.05$ NS

2003 is excluded because of severe drought in Europe.
Sources: USDA and FAOSTAT, 2009.

New tools, efficiency and structures for yield breeding

Conventional plant breeding is a relatively slow, somewhat empirical but very successful process resulting in genetic gains in raised PY and PY_W that have matched the demand for grains over the past century. It has depended on large investments in empirical yield testing, and has been driven by genetic diversity supplemented by effective wide crossing. Progress has been aided by developments in genetics, population theory, crop and genetic modelling, plot mechanization, robotics, remote sensing, biometry, computing and environmental characterization. Despite this, yield progress through breeding, as a percentage of current yield and in absolute terms, has been declining over the past decades for rice and wheat (see section on Sources of yield gains in the breadbaskets), but apparently not for maize, although gain per unit of investment has probably been declining for some time in maize too (Duvick and Cassman, 1999).

Molecular breeding technologies offer real hope of accelerated progress, provided useful genetic variation continues to be available. These technologies, such as marker-assisted selection (MAS) and marker-assisted recurrent selection (MARS), are now being integrated with conventional breeding approaches, but have not been widely adopted outside industry leaders in the private sector because of capital constraints. As noted previously, Monsanto has set a goal of doubling maize yields between 2000 and 2030, claiming accelerated gains in yield (2.5 times their historical rates) partly via more efficient MAS.

Are such yield gains probable, or even possible? Leading private seed companies are investing considerable resources in maize breeding, blending conventional breeding with MAS, MARS and transgenics, coupled with extensive multilocation testing. Early MARS studies using association mapping[19] suggest that gains in yield in elite germplasm of 4 percent per year are possible (Crosbie *et al.*, 2006) in favourable and stressed environments, effectively doubling the rate of yield gain compared with conventional breeding (Eathington *et al.*, 2007; Edgerton, 2009). Association mapping is based on dense marker maps, usually using single nucleotide polymorphisms, a full-genome marker scan, accurate yield assessment, and statistical algorithms that develop many gene-to-phenotype associations (Heffner, Sorrells and Jannick, 2009). Again, the biggest unknown is how useful transgenic variation will be in creating novel variation to supplement natural variation for grain yield traits, such as RUE and functional stay-green that tolerates drought, for root growth that explores the soil volume more thoroughly, and for some types of drought tolerance. If maize was engineered to tolerate light frosts, this would extend its effective season length in temperate environments and increase its PY. The same applies to rainfed wheat at intermediate latitudes, where frost resistance at flowering would likely bring earlier flowering and significant yield benefits. These additional genetic modification gains appear technically feasible, but far less certain.

Realizing these additional gains requires that genetic variation (natural or transgenic) is present and that genotypic (laboratory assays of genes and markers) and phenotypic data (field measures of plant performance) can be brought together in the tight time frame demanded by large breeding programmes today. Physiological understanding will be critical to yield increase via genetic modification, but maybe less so for MAS, MARS and genomic selection, which will depend more on whether methods for detection of gene-phenotype associations and their use within a routine pedigree breeding system, such as "mapping-as-

---

19. This is now more commonly referred to as "whole of genome selection" or "genomic selection".

you-go" (Podlich, Winkler and Cooper, 2004), deliver on their early promise. Phenotyping capability in the field and greenhouse is expanding far more slowly than the ability to genotype huge arrays of germplasm in the laboratory, and cost per phenotypic data point is declining far more slowly that cost per genotypic data point, but both classes of data are critical to future success in crop improvement. Improvements in phenotyping efficiency will depend largely on a combination of carefully managed stress levels in the field, and remote sensing of large numbers of plants, again with a greater role for physiology than in the past. Such changes will likely require significant advances in agronomy, especially in N nutrition, if they are to be fully exploited in the farmer's field.

Intellectual property (IP) considerations are a constraint to the widespread use of molecular breeding techniques, but offer the protection that ensures continued private sector investment. Coupled with the use of hybrids, IP protection where farmers and companies benefit from annual purchase of seeds provides a powerful incentive for investment in crop improvement, as is reflected partly in the greater genetic gain seen in maize than in rice and wheat. There are advantages of scale in global breeding, seen initially in the international breeding programmes of CGIAR centres such as CIMMYT and IRRI and currently in the global operations of multinationals such as Monsanto, Dupont, Syngenta and Bayer. Among small and medium enterprises (SMEs), CGIAR centres and multinational seed companies, research alliances for addressing the needs of national or niche markets have generated viable business models for seed SMEs, and are needed to maintain a healthy competitive environment in the seed industry.

Transformation and marker-aided back-crossing are now relatively cheap and routine. However, the search for appropriate candidate transgenes, IP agreements and royalties, regulatory compliance and commercialization is an expensive undertaking, perhaps costing USD 50 to 70 million per gene in industrial countries. The scale of these costs excludes many developing countries and SMEs from the technology, and the recent agreements to waive IP restrictions on the use of technologies associated with high pro-vitamin A "Golden Rice" and the Water Efficient Maize for Africa Project are welcome signs of corporate social responsibility and public-private collaboration. However, regulatory compliance costs remain high, and have increased greatly in recent years. This reflects societal unease with genetic modification technology, which should reduce in time as experience reveals the true level of risk. In the meantime, with very few exceptions, unease has prevented the commercial use of transgenes in major food staples. It is safe to assume that by 2050 transgenic technology will still be monitored, but will be cheaper, far more widely available, and used to a much greater extent to improve the PY and yield stability of staple food crops.

Yield potential toward 2050

Prophecy is an uncertain business, and can only be based on extrapolation of existing trends. An accelerated and sustained gain in cereal yield progress on the farm is needed, with an increase from less than 1 percent to about 1.5 percent per annum: this will come largely from new varieties with increased PY, helped by the development of agronomic practices that exploit the new capability while conserving agriculture's natural resource base. New varieties will also need to be able to cope with climate change. The following areas call for increased research investment:

- Conventional breeding, increasingly aided by genome analysis and other molecular marker-aided breeding focused on increasing PY and PY_W, and possibly underlying key mechanisms. This will involve sequencing genomes of a diverse but representative array of rice, wheat and maize genotypes, and must be linked with high-throughput precise protected phenotyping facilities, as well as representative production fields with managed input levels (e.g., water supply). Physiology, remote sensing, informatics and biometrics are critical tools in this.

- Increased photosynthetic rates, using conventional but targeted approaches, as well as longer-term transgenic ones, such as developing C_4 options for rice and wheat, or otherwise increasing the efficiency of net photosynthesis in warmer environments by modifying Rubisco, Rubisco activase and the enzymes that modulate photorespiration in C_3 plants. Because crop plants have a finely balanced source – sink interrelationships (Denison, 2007) – a major change in source will take several decades of adaptive breeding to deliver its full benefits as grain yield.

- Eliminating out-crossing barriers for successful hybrid production in rice and wheat.

- Crop genetic enhancement, through the use of wild species (for wheat, see Ortiz *et al.*, 2008).

- Ongoing focus on stress tolerance as well as PY in all crops. This will continue the trend towards higher yields, enhanced yield stability and improved input use efficiency that is already evident in the temperate maize crop.

- Continued strong investment in protecting genetic and agronomic gains through pest resistance, because climate change will bring changes in the pest-predator balance. The global soil resource must also be protected from erosion, which is a huge unfulfilled role for conservation tillage, and from degradation caused by nutrient depletion, providing an inescapable role for efficient use of chemical fertilizers.

A suitable *policy framework* is needed to attract private investment, develop technology and guide its benefits to those most in need. Such a framework should include:

- a strong but balanced emphasis on IP protection for molecular and varietal products and on F1 hybrid production in maize, wheat and rice;

- societal acceptance of transgenic food products, and reduced costs of transgene deregulation, which will greatly increase the range of tools at the breeder's disposal;

- development of a win-win social contract for sharing technology outcomes with resource-poor countries and encouraging more private-public partnerships in the developing world; both private and public sectors are key components of efficient international agricultural research, and strengthening of the CGIAR system and of regional and global commercial activities are essential complements.

Prices, efficiency, productivity and R&D investment

The ultimate concern is not with yields *per se*, but with improving productivity and reducing the prices of food staples. Declining real prices of food staples for 1961 to 2006 – by annual average rates of 1.8 percent for wheat, 2.6 percent for rice and 2.2 percent for maize in world markets – have been a major source of poverty reduction, given that food staples account for a large share of expenditures of the world's poor (for a review of the evidence, see World Bank, 2007). This decline in real prices has been driven by growth in TFP, averaging 1.0 percent globally for all agriculture for the period 1961 to 2006, and 1.7 percent for the industrial countries that provide most grain exports (Fuglie, 2008). A distinguishing feature of this period has been that TFP has risen faster than prices have declined, so both farmers and consumers have benefited (Lipton, 2005).

This section reviews the prospects for sustainable productivity growth and food prices. In particular, it briefly analyses three major determinants of future prices: i) pressure from rising prices of non-renewable resources and the need for more sustainable systems; ii) opportunities for closing efficiency gaps; and iii) prospects for continuing gains in TFP.

Prices of non-renewables

Looking ahead to 2050, the potential for sharply increasing prices of non-renewable resources that have no close substitutes could have major implications for crop yields and food prices. The two resources of most concern are fossil fuels for the manufacture of nitrogenous fertilizers and the provision of farm power, and reserves of phosphates as an essential macroelement for soil fertility.

Fossil fuels: All indications are that fossil fuels have entered a new era of higher and more volatile prices, with an expected upwards trend. Modern agriculture uses an estimated 12.8 EJ[20] of fossil energy, or about 3.6 percent of global fossil fuel consumption. This is roughly divided as 7 EJ for fuel and machinery, 5 EJ for fertilizer (90 percent of which is for N), and the rest for irrigation and pesticides (Smil, 2008). The intensity of commercial energy consumption (nearly all from fossil fuels) varies widely, from about 0.14 to 0.16 GJ[21] per tonne of grain for rice in the Philippines and maize in Mexico in traditional systems, to 2.4 GJ/tonne for improved rice in the Philippines, 2.5 GJ/tonne for wheat in Germany, and 5.9 GJ/tonne for irrigated maize in the United States of America (FAO, 2000; Langreid, Bockme and Kaarstad, 2004). Both machinery and fertilizer costs account for growing shares of production costs in developing countries (World Bank, 2007).

Nitrogen: Current global consumption of around 100 million tonnes of N fertilizer provides more than two-thirds of the N taken up by crops (Socolow, 1999). Although N fertilizer use is now steady or falling in industrial countries, it continues to rise in developing ones (see the section on Setting the scene). Future projections of N fertilizer consumption vary widely, from a relatively modest increase to 121 million tonnes in 2050 (Wood, Henao and Rosegrant, 2004) to 180 million tonnes in 2070 (Frink, Waggoner and Ausubel, 1999), depending on assumptions, including N use efficiency change.

Fossil energy (usually natural gas) accounts for 70 to 80 percent of the cost of manufacturing N fertilizer.[22] Increased efficiency in manufacturing allowed N fertilizer prices to fall until the 1980s. For example, the energy for manufacturing ammonia using the best technology at the time declined from 80 GJ per tonne of ammonia in 1950, to 50 GJ/tonne in 1980, and about 40 GJ/tonne in 2000 (Smil, 2008).[23] However, the best plants are now approaching the stoichiometric limit

---

20. 1 EJ = 10^{18} joules.

21. 1 GJ = 10^9 joules; 1 litre of diesel contains 38 MJ of energy, 1 tonne of maize or wheat about 15 GJ.

22. The actual figure varies depending on the location and age of the manufacturing plant, the fertilizer product, and the costs of natural gas. Although natural gas is cheap in the Gulf States, fertilizer must still be transported to the point of consumption (A. Roy, personal communication).

23. The conversion of ammonia to urea adds 10 GJ per tonne of N to the energy costs of fertilizer, giving a final energy cost of urea of 55 to 58 GJ per tonne of N (Smil, 2008).

for energy efficiency. Since 1981, N prices have closely tracked energy prices, with 1 tonne of urea (46 percent N) costing about 40 times as much as 1 GJ of natural gas (Figure 10.14), although significant efficiency gains could still be made by abandoning older less efficient plants.

Figure 10.14
Real prices of urea in bulk in Eastern Europe (left axis), and natural gas in Europe (right axis)

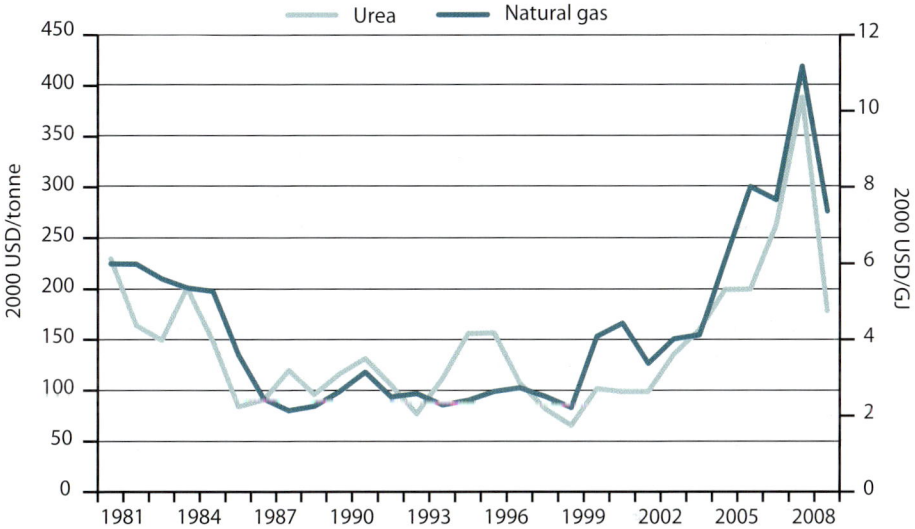

Source: World Bank data files.

As the major efficiency gains have already been made, it is likely that the price of N fertilizer will rise in line with energy prices. In addition, some high-income countries are now taxing N fertilizer use, as a disincentive to pollution. A tax on greenhouse gas emissions is also likely in the future. This would hit prices of N fertilizer particularly hard, owing to N fertilizer's fossil energy intensity and the fact that once applied it can become a significant source of nitrous oxide, an especially potent greenhouse gas that accounts for about one-third of all agricultural greenhouse gas emissions (Crutzen *et al.*, 2008).

Increasing the efficiency of on-farm use of N and the supply of biologically fixed nitrogen are the best options for confronting rising N prices. Numerous studies have documented low on-farm efficiency of applied N, with an average of only 33 percent being taken up by the crop, dropping to 29 percent in developing countries (Raun and Johnson, 1999). Many Chinese farmers may be using N at above-optimum levels (Buresh *et al.*, 2004). With better management and, in many cases, lower application rates, N use efficiency could be improved by 33 percent for irrigated maize to more than 100 percent for rainfed rice (Balasubramanian *et al.*, 2004) (Table 10.7). Improvement is already evident in United States maize, for example, where N use per hectare has declined through more site-specific application rates, even as yields have increased (see the section on Sources of yield gains in the breadbaskets). Precision agriculture provides new tools for improving efficiency further (discussed in the following subsection). New products such as controlled and slow-release fertilizer can also increase efficiency with rice (IFDC, 2009). In Bangladesh, more than half a million farmers have adopted Urea Super Granules, which are deep-placed at planting time, enabling N use to be cut by about one-third, with a corresponding increase in yields of almost 20 percent (IFDC, 2007). As plant breeding raises yields, it inevitably results in more efficient N use (Ortiz-Monasterio *et al.*, 1997; Bänziger, Edmeades and Lafitte, 1999; Echarte, Rothstein and Tollenaar, 2008); this general principle also applies to most other inputs, such as phosphorus and water (de Wit, 1992; Fischer, 2009).

Table 10.7
Mean recovery efficiency of N (REN) for harvest crops under current farming practices and research plots

Crop	Mean REN under current farming practice	Mean REN in research plots	Maximum REN of research plots
	(% of N applied)		
Rice			
Irrigated	31–36 (Asia)	46–49	88
Rainfed	20	45	55
Wheat			
Irrigated	33–34 (India)	45–57	96
Rainfed	17 (USA)	25	65
Maize			
Irrigated and rainfed	36–57	42–65	88

Sources: Balasubramanian *et al.*, 2004; Dobermann, 2007.

Biological N fixation is the other major opportunity for increasing the supply of N while reducing the dependence on fossil fuels. Biological fixation already accounts for about one-third of world N supply to agriculture, and more in some countries such as Australia. Legumes cover only about 11 percent of cropped land; although using generally lower-yielding legumes to replace more cereals would depress world food supplies, there are some opportunities for fitting legumes into gaps in even relatively intensive cropping systems, as shown by the adoption of 60-day mung beans on nearly 1 million ha in the rice-wheat system of the Indo-Gangetic plains, which reduced the cost of the following wheat crop by 23 percent

(Ali *et al.*, 1997). N fixation in cereals themselves is also being researched, but is unlikely to be a feasible technology by 2050, and the gain in N would have to be balanced against a possible yield penalty for energy diverted to N fixation (Ladha and Reddy, 2000).

Farm power: Conservation farming using zero tillage is a major opportunity to reduce fuel use for farm power in agriculture by an average of 66 to 75 percent, as well as helping to sequester soil carbon. Globally, no-tillage is now used on an estimated 100 million ha of about 1 170 million ha of total cropped land (FAO, 2008b), with a large concentration in the Americas where wide adoption of transgenic herbicide-resistant maize and soybeans has strongly accelerated the trend (Brookes and Barfoot, 2008) (Table 10.8). There are also good examples from irrigated South Asian systems, where small-scale farmers have adopted zero tillage on as much as 5 million ha of wheat in rice-wheat systems, with estimated savings in fuel costs of 60 to 90 percent and an increase in wheat yields of 11 percent (Erenstein *et al.*, 2008; FAO, 2008a).[24] Conservation tillage has also been suggested as a potentially important source of carbon sequestration in tropical soils (IPCC, 2007).

Table 10.8
Estimated area under no-tillage[a] in major adopting countries

Country	1988–1991 (million ha)	2003–2007 (million ha)	2003–2007 (%)
Argentina	0.5	19.7	67
Brazil	1.4	25.5	38
Paraguay		2.1	49
Canada	2.0	13.5	26
USA	6.8	25.3	14
Kazakhstan		1.8	8
Australia	0.4	9.0	36
Total[b]	11.4	99.9	≈9

[a] No-tillage is defined as a system of planting crops into untilled soil by opening a narrow slot, trench or band of sufficient width and depth to obtain proper seed coverage only. No other soil tillage is done (FAO, 2008a).

[b] Including countries with less than 1 million ha in 2003 to 2007.

Source: FAO, 2008b.

With less than 10 percent of the world's cropland under conservation tillage, wider adoption of the practice represents a major opportunity for improving the sustainability, energy efficiency and yield of cropping. However, conservation

---

24. This figure is not included in Table 10.8 because farmers practise tillage in the following rice crop, so do not meet the strict definition of zero tillage.

agriculture is knowledge-intensive and location-specific and will require sharply increased investment in research on suitable varieties, management practices adapted to specific sites, appropriate machinery, and advisory services and farmer networks. If successful, current discussion of payments for soil carbon sequestration will greatly add to the incentive for adopting conservation tillage, provided conservation tillage sequesters more C in soils – an issue that is currently under much scrutiny – and monitoring systems can be devised.

Phosphorus: Phosphorus (P) is the other major non-renewable resource for which scarcity could significantly affect crop yields by 2050. Recent work by Cordell, Drangert and White (2009) estimates that production of phosphates will peak by 2034, using the Hubbert curve that predicts declining production of oil and other mineral resources when half of reserves have been exploited (Figure 10.15). Production will also become more concentrated, especially in Morocco, as the United States of America has only 20 to 25 years of reserves remaining, and China has a high export tax. The quality of deposits is also declining, raising the cost of extraction of remaining reserves. A recent report (IFDC, 2009), however, casts considerable doubt on the imminent P scarcity predicted by Cordell, Drangert and White (2009), estimating the global supply of phosphate rock reserves of reasonable grade to be several hundred times annual consumption.

Figure 10.15
Projection of peak global phosphorus extraction

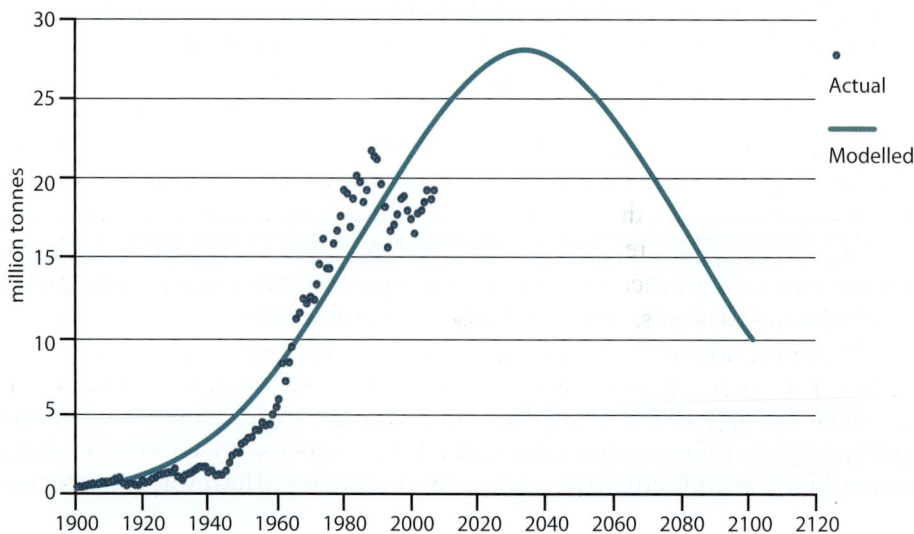

Source: Cordell, Drangert and White, 2009.

As with N, there is much room for enhancing the efficiency of P use. Of the 14.9 million tonnes of P mined for agriculture, only 6.1 million tonnes is removed in crop biomass. On-farm efficiency can be improved through application of many of the same site-specific management practices as for N, although the big difference is that N is a mobile element that can be leached, while P remains in the soil, slowly building up (in advanced agriculture, more P is applied than removed in biomass) in forms that are less available to most plants; microbial additives and genetic engineering of crop roots may improve the accessibility of these unavailable forms of soil P. It is also likely that increased recovery of P from human and animal excreta for use as fertilizer will become common as the technology for recycling is developed and P prices rise (Cordell, Drangert and White, 2009).

Agricultural price policies

Price policies can also be important in achieving high yields and efficiency. Historically, developing countries have taxed their agriculture sectors heavily, in part to provide cheap food, penalizing overall growth rates in the sector. This situation has largely been resolved under the liberalization policies of the 1990s, and the average tax on agriculture is now low (Anderson, 2009). This has provided a one-off opportunity to spur productivity growth. However, yields of food crops are generally quite inelastic with respect to prices, at least in the short term (Binswanger, 1989; Rosegrant *et al.*, 2008). Progress in dismantling price distortions has been much slower in industrial countries, where farm subsidy programmes have favoured a few crops and discriminated against the adoption of more sustainable cropping systems, especially crop rotations.

Subsidies on many inputs, and outmoded pricing structures, especially for water, are still common in Asia. These policies played a role in stimulating adoption of green revolution inputs in the 1970s and 1980s, but given the current high levels of input use, they undermine incentives to use inputs more efficiently. Supporting institutional reforms will also be important, such as greater devolution of water management decisions to users, and a gradual shift to market-determined water allocation systems.

In Africa, where yields and input use are still very low, there is a case for "market-smart" input subsidies to promote the adoption of fertilizers and stimulate the development of input markets. Several countries have reintroduced such subsidies (World Bank, 2007), but high fiscal costs and the displacement of commercial sales threaten their long-run sustainability and effectiveness.

The production efficiency gap

Many areas could produce the same or higher yields with lower input costs through practices designed to enhance input efficiency. Over the past two decades, economists have carried out hundreds of studies to estimate farm-level efficiency in relation to the production frontier reached by the best farmers. A meta-analysis of 167 such studies concluded that average technical efficiency is 72 percent, with a high of 82 percent in Western Europe and a low of 70 percent in Eastern Europe (Bravo-Ureta *et al.*, 2007).

While most of these studies fail to account adequately for site and season characteristics specific to plots and farms, they find efficiency is most closely related to farmer characteristics, especially education, location and access to information (Ali and Byerlee, 1991). A further finding is that education has a significant impact on productivity in most post-green revolution settings where management is increasingly knowledge-intensive.

Information and communication technologies in what is often termed "precision agriculture" have much potential to enhance productivity, as well as contributing to more sustainable production systems. These new tools include yield mapping, leaf testing to time N application, remote sensing, crop modelling and expert systems, improved weather forecasting and wireless in-field monitoring; they aim to improve input use efficiency by allowing inputs to be calibrated more precisely to within-field variability and seasonal conditions (Sudduth, 2007). They are also being applied in small farm agriculture; for example, very small-scale farmers are using the leaf colour chart to time N application on rice (Islam, Bagchi and Hossain, 2007). With the spread of mobile phones and village information kiosks, farmers can increasingly tap external sources of information on prices and crop management, and identify pests and diseases remotely.

However, to achieve its full potential, this type of precision farming will require greatly improved knowledge transfer systems, additional equipment, and skilled and educated farmers. To date, the potential of such an information technology revolution has received far less attention than the biotechnology revolution.

Prospects for TFP growth

What does all of this mean for TFP growth? In general, the share of TFP growth in agricultural output growth grows as agricultural economies develop (Pingali and Heisey, 1999). TFP growth was responsible for half of output growth after 1960 in China and India, and for 30 to 40 percent of the increased output in Indonesia and Thailand (World Bank, 2007). There is little evidence that growth in TFP is slowing (Box 10.2).

Box 10.2 - Is TFP growth slowing?

Recent work by Fuglie (2008) provides an up-to-date and comprehensive overview of TFP growth (see following table). While these estimates are for all agriculture and not just cereals, the general conclusion is that TFP growth has accelerated in the most recent period since the green revolution, 1991 to 2006, in spite of slower output growth. Input growth has slowed in all regions, and in developed countries is now negative, especially in the former Soviet Union, where inputs were used very inefficiently before the transition to markets.

In developing countries, total output growth has not slowed, implying that growth from diversification to higher-value products has cancelled slower growth in cereals. High growth in both output and TFP is led by large countries, especially Brazil and China, with TFP growth of more than 3 percent per year. Nonetheless, Fuglie (2008) recognizes that cereal growth has slowed significantly and that TFP growth for individual commodity groups may show diverse patterns. A recent review by Kumar, Mittal and Hossain (2008) suggests some slowing of TFP growth in cereals in South Asia, with negative growth in rice in the Punjab. This supports earlier evidence of slowing TFP growth in rice-wheat systems in India and Pakistan (Murgai, Ali and Byerlee, 2001).

In developing countries overall, the share of growth accounted for by TFP has risen from one-third in the period 1970 to 1990 to nearly two-thirds in 1991 to 2006. In line with the earlier analysis, sub-Saharan Africa is the outlier, with growth dependent on land expansion rather than TFP – land area has expanded more rapidly than output, although there is evidence of recent acceleration of productivity growth in some countries such as Ghana (Fuglie, 2009).

Growth of total output, input and TFP in agriculture

Region	Output (%/year) 1970–1990	Output (%/year) 1991–2006	Input (%/year) 1970–1990	Input (%/year) 1991–2006	TFP (%/year) 1970–1990	TFP (%/year) 1991–2006
Sub-Saharan Africa	2.03	2.67	1.72	1.81	0.31	0.86
Latin America	2.69	3.03	1.68	0.59	1.02	2.44
Asia	3.36	3.57	1.85	0.95	1.51	2.62
Near East and North Africa	3.15	2.54	2.02	1.01	1.14	1.53
North America	1.49	1.61	0.00	-0.30	1.49	1.91
Europe	1.10	-0.15	-0.16	-1.66	1.26	1.52
Russian Federation, Ukraine and Central Asia	0.99	-1.57	1.17	-3.95	-0.17	2.38
Developed	1.35	0.87	-0.27	-1.18	1.61	2.05
Transitional[a]	0.95	-1.48	0.94	-3.28	0.00	1.79
Developing	3.16	3.41	2.08	1.22	1.08	2.19
World	2.16	2.13	1.37	0.57	0.79	1.56

[a] Countries of the former Soviet Union.
Source: Fuglie, personal communication, recalculated from Fuglie, 2008.

TFP growth is largely explained by investments in research, extension, education, irrigation and roads, and by policy and institutional changes (Pingali and Heisey, 1999; Binswanger, 1989; World Bank, 2007; Kumar, Mittal and Hossain, 2008). Decompositions of productivity gains consistently point to investment in research, often associated with extension, as the most important source of growth. Improved varieties alone contributed as much as half of TFP gains in Pakistan and China in the post-green revolution period (Rozelle *et al.*, 2003; Ali and Byerlee, 2002). Even in sub-Saharan Africa, the impact of R&D has been identified as important in the region's (limited) productivity growth (Lusigi and Thirtle, 1997).

The key role of R&D investments

Considerable uncertainty surrounds the question regarding what level of investment in R&D will be needed to realize the gains in yields and productivity necessary to secure global food security to 2050. For example, IFPRI's high R&D investment scenario reverses an upwards trend in real prices of grain to 2050 relative to the baseline, and boosts yield growth from 1.0 to 1.4 percent (see subsection on Scenarios to 2050 and the future yield challenge) and involves an approximate doubling of agricultural R&D, along with increases in other key areas of agriculture (Hubert *et al.*, 2010). Von Braun *et al.* (2008) estimate that a doubling of investment in R&D in developing countries would increase R&D's contribution to overall output growth by 1.1 percentage points (i.e., an approximate doubling of current rates), sufficient to ensure a continued decline in poverty (and presumably food prices) through 2020. However, there remains a wide margin of uncertainty in estimates of the quantitative relationship between R&D investments and yield and productivity growth, especially regarding the time lags involved, even though *ex-post* analyses of research impact have invariably yielded very attractive rates of return.

These scenarios do not consider investment in R&D in industrial countries, which will continue to play a major role in global food security as developing countries urbanize and are likely to increase their dependence on food imports. Spill-overs from R&D in industrial countries are also important for developing countries. Combined public and private agricultural R&D investment in industrial countries is double that in developing countries. There are worrying signs of reduced public investment in R&D in industrial countries, and a reallocation to non-productivity issues such as food safety and the environment could reduce resources for long-term strategic research of relevance to developing countries, such as efforts to push out the yield frontier (Pardey *et al.*, 2007). Meanwhile, private investment in R&D has increased rapidly in industrial countries. A conservative estimate puts private sector spending on maize research in the United States of America at about USD 1 billion a year, compared with USD 181 million in 1990 (in 2008 dollars) (Byerlee and Lopez-Pereira, 1994). This huge increase is a likely explanation for the continuing impressive yield gains in maize in the United States, and in similar environments where these companies and their subsidiaries operate.

Nonetheless, there are worries about the sustainability of recent trends in private R&D spending, which has been increasing exponentially while yields have been increasing linearly (Duvick and Cassman, 1999). The large jump in private spending may have finally driven returns on investment in R&D down from their very high levels of more than 50 percent to rates closer to a risk-adjusted cost of capital. If so, the era of rapid growth in private investment in maize and soybean research may be over, although the spread of hybrid rice could result in a similar burst of investment in rice. Unpublished data from USDA indicate a levelling of private spending in the United States of America since 2000. One factor that may trigger a new round of private investment in food crops would be if transgenics become accepted by the public for major food staples such as rice and wheat.

It is likely that over the long term, productivity-enhancing investments will be driven by prices. There is evidence that public investment in rice research and irrigation in Asia was negatively affected by the long-term fall in real rice prices (Hayami and Morooka, 1987; Rosegrant and Pingali, 1994). Private research is likely to be even more responsive to prices, and the recent increases in food prices may have already led to a resurgence of R&D spending. Thus, over the long term, yields may be much more elastic with respect to prices than they are in the short to medium term.

Conclusions

It is common that when world grain prices spike, as they did in 2008, a small enclave of world food watchers raises the Malthusian spectre of a world running out of food. The original demon of an exploding population has evolved to include the livestock revolution and, most recently, biofuels. However, since the 1960s, the global application of science to food production has maintained a strong track record of staying ahead of growing demands. Even so, looking to 2050 new demons on the supply side, such as water and land scarcity and climate change, provoke claims that "this time it is different". Even so, after reviewing what is happening in the breadbaskets of the world and what is in the technology pipeline, there is cause for cautious optimism about the world's ability to feed itself to 2050; this optimism was shared by Evans (1998) at the end of his long excursion through these same issues.

First, despite impressive gains in yields over the past 50 years in most of the world, large and economically exploitable yield gaps remain in many places, especially in developing countries, and nowhere more so than in sub-Saharan Africa, where food supply is the most precarious.

Second, in the short to medium term, many of the newer technologies that are in their early stages of adoption promise a win-win combination of enhancing productivity and managing natural resources sustainably. These include conservation farming approaches based on no-tillage and the genetic modification technology revolution – both still used on less than 10 percent of the world's cropland – and information and communication technologies for more efficient and precise management of modern inputs, which are still at an even earlier adoption phase.

Third, yield gains are not achieved by technology alone, but also require complementary changes in policies and institutions. This is now recognized in much of the developing world, and policies are becoming more favourable to rapid productivity growth, while a range of innovations in risk management, market development, rural finance, farmers' organizations and the provision of advisory services show considerable promise for making markets work better and providing a conducive environment for technology adoption. In sub-Saharan Africa, these innovations are a necessary condition for wider adoption of critical technologies such as fertilizer. The recent progress in cereal yields in Egypt reflects attention to technology, policy and institutions together.

Fourth, plant breeders continue to make steady gains in PY and PY_W, more slowly than in the past for wheat and rice, but with little slackening in the case of maize; there is no physiological reason why these gains cannot be maintained, but progress from conventional breeding is becoming more difficult. Genomics and molecular techniques are now being regularly applied to speed breeding in the leading multinational seed companies and elsewhere, and their costs are falling rapidly. Transgenic (genetic modification) technology has a proven record of more than a decade of safe and environmentally sound use, and its potential to address critical biotic and abiotic stresses in the developing world – with positive consequences for closing the yield gap – has yet to be tapped. The next seven to ten years are likely to see its application to major food crops in Asia and Africa, and after its initial adoption, the currently high regulatory costs will begin to fall. However, this will require significant additional investment, not least in the areas of phenotyping on a large scale, and it still takes ten to 15 years from the initial investment before the resulting technologies begin to have major impacts on food supply. Transgenics for greater PY_W may also appear by then, but trangenics for greater PY arising from significant improvements in photosynthesis may take longer than even the 2050 horizon.

These are broad generalizations, and there are important variations by crop and region. This review of the big three cereals has shown that maize is the dynamic crop, with no evidence of slowing yields and with huge potential in the developing world. It is also the crop experiencing the most rapid increase in demand, largely for feed and fuel, and the crop attracting the largest R&D research budget. Wheat demand and yield growth appear to be intermediate, the latter perhaps because of disease resistance, industrial quality constraints on breeding, and the greater role of water stress in wheat's production environment. Yield gains in rice are more problematic, but demand growth is also lower for rice, although it is a particularly important food staple for the poor of Asia (where rice area is shrinking) and increasingly of Africa. Although increases in food production in Asia over the past 50 years have been impressive, no country in sub-Saharan Africa has yet experienced a green revolution in food crops in a sustained manner, despite generally better overall performance of the agriculture sector in the past decade.

However, a number of cautions should be raised. First, this chapter has not (yet) reviewed other food crops: sorghum and millet, roots and tubers, pulses and oilseeds. Many of these crops are not globally important, but are critical to local food security, such as cassava in Africa. Others are commercial crops for an urbanizing population – potatoes for fast foods, and oilseeds for oil and feed.

Second, the future of biofuels is the new wild card in the world food economy. To some extent the need to accelerate global cereal yield trends beyond the historic annual rate of 43 kg/ha for 1961 to 2007 relates to this new demand. By 2020, the industrial world could consume as much grain per capita in its vehicles as the developing world currently consumes per capita directly for food.

Third, many countries face huge challenges in achieving food security, even from the narrow perspective of food supply. There is less cause for concern about China and India, as they should continue to be largely self-sufficient for food needs (although dependent on imports for part of their feed needs), but much depends on investments in R&D and management of natural resources. However, many other countries do not have the capacity to import large amounts of grain, or find it prohibitively costly to do so, but still have very high population growth. Most of these countries are in Africa, but even Pakistan, with an estimated 335 million people in 2050, faces a potential food crisis. Climate change will also be a major challenge for many of these countries, adversely affecting yields and diverting R&D resources towards adaptation rather than yield improvement – adding a new dimension to maintenance research.

449

Past agricultural success has been achieved partly by mining non-renewable resources, such as fossil energy, phosphate and underground water. This chapter's review of the impact of looming limitations for this strategy raises major concerns, and places a premium on improved efficiency in using these resources, which must be at the top of the agenda for feeding the world in 2050. Contrary to popular opinion, however, generally increased yield through breeding and modern agronomy is lifting resource use efficiency.

The history of agriculture in the twentieth century teaches that investment in R&D will be the most important determinant of whether this chapter's optimistic view is realized. There are indications that major developing countries such as China, India and Brazil are poised to close their gaps in research intensity with the industrial countries. CGIAR is also revamping its efforts, aiming to double its budget in the coming years. However, many technological orphans are falling behind in R&D spending (Chapter 9 in this volume). The private sector must also be encouraged to make a major impact beyond its mainstays of maize and soybeans, especially in rice. Innovative partnerships will be needed to ensure access to and adaption of technologies for the world's 800 million small farmers.

Resilience, flexibility and policies that favour R&D investment in staple food research and efficient input use will be the pillars on which future food security depends. Darwin, whose 200th birthday was celebrated in 2009, leaves two relevant statements: "If the misery of the poor be caused not by the laws of nature, but by our institutions, great is our sin," and, "It is not the strongest of the species that survives … [but] the one that is the most adaptable to change."

References

Ali, M. & Byerlee, D. 1991. Economic efficiency of small farmers in a changing world: A survey of recent evidence. *Journal of International Development*, 3: 1–27.

Ali, M. & Byerlee, D. 2002. Productivity growth and resource degradation in Pakistan's Punjab: A decomposition analysis. *Economic Development and Cultural Change*, 50: 839–863.

Ali, M., Malik, I.E., Sabir, H.M. & Ahmad, B. 1997. *The mungbean green revolution in Pakistan*. Taiwan Province of China, Asian Vegetable Research and Development Center.

Anderson, K., ed. 2009. *Distortions to agricultural incentives: A global perspective, 1955–2007*. London, Palgrave Macmillan, and Washington, DC, World Bank.

Andrade, F.H., Otegui, M.E. & Vega, C. 2000. Intercepted radiation at flowering and kernel number in maize. *Agronomy Journal*, 92: 92–97.

Austin, R.B. 1989. Genetic variation in photosynthesis. *Journal of Agricultural Science*, 112: 287–294.

Balasubramanian, V., Alves, B., Aulakh, M., Bekunda, M., Cai, Z., Drinkwater, L., Mugendi, D., van Kessel, C. & Oenema, O. 2004. Crop, environment, and management factors affecting nitrogen use efficiency. *In* A.R. Moiser, R.K. Syers and J.R. Freney, eds. *Agriculture and the nitrogen cycle: Assessing the impact of fertilizer use on food production and the environment.* Washington, DC, Island Press.

Bange, M.P., Hammer, G.L. & Rickert, K.G. 1997. Effect of specific leaf nitrogen on radiation use efficiency and growth of sunflower. *Crop Science*, 37: 1201–1207.

Bänziger, M., Edmeades, G.O. & Lafitte, H.R. 1999. Selection for drought tolerance increases maize yields across a range of nitrogen levels. *Crop Science*, 39: 1035–1040.

Bänziger, M., Setimela, P.S., Hodson, D. & Vivek, B. 2006. Breeding for improved drought tolerance in maize adapted to southern Africa. *Agricultural Water Management*, 80: 212–224.

Bernier, J., Kumar, A., Venuprasad, R., Spaner, D. & Atlin, G. 2007. A large-effect QTL for grain yield under reproductive-stage drought stress in upland rice. *Crop Science*, 47: 507–518.

Bidinger, F.R., Mahalakshmi, V. & Rao, G.D.P. 1987. Assessment of drought resistance in pearl millet (*Pennisitum americanum* (L.) Leeke). II Estimation of genotype response to stress. *Australian Journal of Agricultural Research*, 38: 49–59.

Binswanger, H. 1989. The policy response of agriculture. In *Proceedings of the World Bank Annual Conference on Development Economics 1989*. Washington, DC, World Bank.

Blum, A. 2009. Effective use of water (EUW) and not water-use efficiency (WUE) is the target of crop yield improvement under drought stress. *Field Crops Research*, 112: 119–123.

Bolaños., J. & Edmeades, G.O. 1996. The importance of the anthesis-silking interval in breeding for drought tolerance in tropical maize. *Field Crops Research*, 48: 65–80.

Bravo-Ureta, B.E., Solís, D., Moreira López, V.H., Maripan, J.F., Thiam, A. & Rivas, T. 2007. Technical efficiency in farming: a meta-regression analysis. *Journal of Production Analysis*, 27: 57–72.

Brookes, G. & Barfoot, P. 2008. Global impact of biotech crops: socio-economic and environmental effects, 1996–2006. *AgBioForum*, 11: 21–38.

Brookes, G. & Barfoot, P. 2009. *GM crops: global socio-economic and environmental impacts 1996–2007*. UK, PG Economics. www.pgeconomics. co.uk/pdf/2009globalimpactstudy.pdf.

Bruce, W.B., Edmeades, G.O. & Barker, T.C. 2001. Molecular and physiological approaches to maize improvement for drought tolerance. *Journal of Experimental Botany*, 53: 13–25.

Bueno, C.S. & Lafarge, T. 2009. Higher crop performance of rice hybrids than of elite inbreds in the tropics: 1. Hybrids accumulate more biomass during each phenological phase. *Field Crops Research*, 112: 229–237.

Buresh, R., Peng, S., Huang, J., Yang, J., Wang, G., Zhong, X. & Zou, Y. 2004. Rice systems in China with high nitrogen inputs. *In* A.R. Moiser, R.K. Syers and J.R. Freney, eds. *Agriculture and the nitrogen cycle: Assessing the impact of fertilizer use on food production and the environment*, pp. 143–153. Washington, DC, Island Press.

Byerlee, D. 1992. Technical change, productivity, and sustainability in irrigated cropping systems of South Asia: Emerging issues in the post-green revolution era. *Journal of International Development*, 4: 477–496.

Byerlee, D. & Lopez-Pereira, M.A. 1994. *Technical change in maize production: A global perspective*. Economics Working Paper No. 94-02. Mexico, CIMMYT.

Byerlee D. & Traxler, G. 2001. The role of technology spillovers and economies of size in the efficient design of agricultural research systems. *In* J.M. Alston, P.G. Pardey and M.J. Taylor, eds. *Agricultural science policy: Changing global agendas,* Chapter 9, pp. 207–250. Baltimore, Maryland, USA, Johns Hopkins University Press.

Campos, H., Cooper, M., Habben, J.E., Edmeades, G.O. & Schussler, J.R. 2004. Improving drought tolerance in maize: a view from industry. *Field Crops Research*, 90: 19–34.

Cassman, K.G., Dobermann, A., Walters, D.T. & Yang, H. 2003. Meeting cereal demand while protecting natural resources and improving environmental quality. *Annu. Rev. Environ. Resour.*, 28: 315–358.

Castiglioni, P., Warner, D., Bensen, R.J., Anstrom, D.C., Harrison, J., Stoecker, M., Abad, M., Kumar, G., Salvador, S., D'Ordine, R., Navarro, S., Back, S., Fernandes, M., Targolli, J., Dasgupta, S., Bonin, C., Luethy, M. & Heard, J.E. 2008. Bacterial RNA chaperones confer abiotic stress tolerance in plants and improved grain yield in maize under water-limited conditions. *Plant Physiology*, 147: 446–455.

Cordell, D., Drangert, J. & White, S. 2009. The story of phosphorus: global food security and food for thought. *Global Environmental Change*, 19: 292–305.

Crosbie, T.M. & Pearce, R.B. 1982. Effects of recurrent phenotypic selection for high and low photosynthesis on agronomic traits in two maize populations. *Crop Science*, 22: 809–813.

Crosbie, T.M., Eathington, S.R., Johnson, G.R., Edwards, M., Reiter, R., Stark, S., Mohanty, R.G., Oyervides, M., Buehler, R., Walker, A.K., Dobert, R., Delannay, X., Pershing, J.C., Hall, M.A. & Lamkey, K.R. 2006. Plant breeding: past, present, and future. *In* K.R. Lamkey and M. Lee, eds. *Plant breeding: the Arnel R. Hallauer International Symposium*, pp. 3–50. Ames, Iowa, USA, Blackwell.

Crutzen, P.J., Mosier, A.R., Smith, K.A. & Winiwarter, W. 2008. N_2O release from agro-biofuel production negates global warming reduction by replacing fossil fuels. *Atmospheric Chemistry and Physics*, 8: 389–395.

Denison, R.F. 2007. When can intelligent design of crops by humans outperform natural selection? *In* J.H.J. Spiertz, P.C. Struik and H.H. Laars, eds. *Scale and complexity in plant systems research: Gene-Plant-Crop Relations*, pp. 287–302. Berlin, Springer.

de Wit, C.T. 1992. Resource use efficiency in agriculture. *Agricultural Systems*, 40: 125–131.

Dixon, J., Braun, H.-J. & Crouch, J.H. 2008. Overview: Transitioning wheat research to serve the future needs of the developing world. *In* J. Dixon, H.-J. Braun and P. Kosima, eds. *Wheat facts and futures 2007*. Mexico City, CIMMYT.

Dobermann, A. 2007. Nutrient use efficiency – measurement and management. In *Fertilizer best management practices. General principles, strategy for their adoption and voluntary initiatives vs. regulations*, pp. 1–28. Paris, IFA.

Dobermann, A., Witt, C., Dawe, D., Abdulrachman, S., Gines, H.C., Nagarajan, R., Satawathananont, S., Son, T.T., Wang, G.H., Chien, N.V., Thoa, V.T.K., Phung, C.V., Stalin, P., Muthukrishnan, P., Ravi, V., Babu, M., Chatuporn, S., Sookthongsa, J., Sun, Q., Fu, R., Simbahan, G.C. & Adviento, M.A.A. 2002. Site-specific nutrient management for intensive rice cropping in Asia. *Field Crops Research*, 74: 37–66.

Dubin, H.J. & Brennan, J.P. 2009. *Combating stem and leaf rust of wheat: historical perspective, impacts, and lessons learned.* IFPRI Discussion Paper No. 910. Washington, DC, IFPRI.

Duvick, D.N. 2005. The contribution of breeding to yield advances in maize (*Zea mays* L.). *Advances in Agronomy*, 86: 83–145.

Duvick, D.N. & Cassman, K.G. 1999. Post-green revolution trends in yield potential of temperate maize in the North-Central United States. *Crop Science*, 39: 1622–1630.

Duvick, D.N., Smith, J.C.S & Cooper, M. 2004. Long-term selection in a commercial hybrid maize breeding program. *Plant Breeding Reviews*, 24: 109–151.

Eathington, S.R., Crosbie, T.M., Edwards, M.D., Reiter, R.S. & Bull, J.K. 2007. Molecular markers in a commercial breeding program. *Crop Science*, 47(S3): S154–S163.

Echarte, L., Rothstein, S. & Tollenaar, M. 2008. The response of leaf photosynthesis and dry matter accumulation to nitrogen supply in an older and a newer maize hybrid. *Crop Science*, 48: 656–665.

Edgerton, M.D. 2009. Increasing crop productivity to meet global needs for feed, food and fuel. *Plant Physiology*, 149: 7–13.

Edmeades, G.O., Bolaños, J., Elings, A., Ribaut, J.-M., Bänziger, M. & Westgate, M.E. 2000. The role and regulation of the anthesis-silking interval in maize. *In* M.E. Westgate and K.J. Boote, eds. *Physiology and modeling kernel set in maize*, pp. 43–73. Crop Science Society of America Special Publication No. 29. Madison, Wisconsin, USA, CSSA.

Ekboir, J., ed. 2002. *CIMMYT 2000–2001 world wheat overview and outlook: Developing no-till packages for small-scale farmers.* Mexico City, CIMMYT.

Erenstein, O., Farooq, U., Malik, R.K. & Sharif, M. 2008. On-farm impacts of zero tillage wheat in South Asia's rice–wheat systems. *Field Crops Research*, 105: 240–252.

Evans, L.T. 1993. *Crop evolution, adaptation and yield.* Melbourne, Australia, Cambridge University Press.

Evans, L.T. 1998. *Feeding the ten billion: Plants and population growth.* Cambridge, UK, Cambridge University Press.

Evans, L.T. & Fischer, R.A. 1999. Yield potential: its definition, measurement and significance. *Crop Science*, 34: 1544–1551.

FAO. 2000. *The energy and agriculture nexus.* Environment and Natural Resources Working Paper No. 4. Rome.

FAO. 2003. *World agriculture towards 2015/2030: An FAO perspective*, edited by J. Bruinsma. Rome.

FAO. 2006. *World agriculture towards 2030/2050: Interim report.* Rome.

FAO. 2008a. *Global overview of conservation agriculture adoption*, by R. Derpsch and T. Friedrich. Rome.

FAO. 2008b. *Investing in sustainable agricultural intensification: The role of conservation agriculture.* Rome.

Fischer, G., Shah, M., van Velthuizen, H. & Nachtergaele, F. 2001. *Global agro-ecological assessment for agriculture in the 21st century.* Vienna, IAASA.

Fischer, R.A. 1973. The effect of water stress at various stages of development on yield processes in wheat. *In* R.O. Slatyer, ed. *Plant responses to climate factors*, pp. 233–241. Paris, UNESCO.

Fischer, R.A. 1985. Number of kernels in wheat crops and the influence of solar radiation and temperature. *Journal of Agricultural Science*, 105: 447–461.

Fischer, R.A. 2007. Understanding the physiological basis of yield potential in wheat. *Journal of Agricultural Science*, 145: 990–113.

Fischer, R.A. 2009. Exploiting the synergy between genetic improvement and agronomy of crops in rainfed farming systems of Australia. *In* V. Sadras and D. Calderini, eds. *Crop physiology*, pp. 235–254. Amsterdam, Netherlands, Elsevier.

Fischer, R.A. & Maurer, R. 1978. Drought resistance in spring wheat cultivars. I. Grain yield responses. *Australian Journal of Agricultural Research*, 29: 897–912.

Fischer, R.A., Rees, D., Sayre, K.D., Lu, Z.-M., Condon, A.G. & Larque-Saavedra, G. 1998. Wheat yield progress associated with higher stomatal conductance and photosynthetic rate, and cooler canopies. *Crop Science*, 38: 1467–1475.

French, R.J. & Schultz, J.E. 1984. Water use efficiency of wheat in a Mediterranean-type environment. I. The relation between yield, water use and climate. *Australian Journal of Agricultural Research*, 35: 743–764.

Frink, C.R., Waggoner, P.E. & Ausubel, J.H.1999. Nitrogen fertilizer: retrospect and prospect. *Proceedings of the National Academy of Sciences*, 96: 1175–1180.

Fuglie, K. 2008. Is a slowdown in agricultural productivity growth contributing to the rise in commodity prices? *Agricultural Economics*, 39: 431–441.

Fuglie, K. 2009. Agricultural productivity in sub-Saharan Africa. Paper presented at the Cornell University Symposium on The Food and Financial Crises and their Impacts on the Achievement of the Millennium Development Goals, 1–2 May 2009, Ithaca, New York.

Ghannoum, O. 2009. C_4 photosynthesis and water stress. *Annals of Botany*, 103: 635–644.

Gollin, D. 2006. *Impacts of international research on inter-temporal yield stability in wheat and maize: An economic assessment*. Mexico. Mexico City, CIMMYT.

Hammer, G.L., Dong, Z., McLean, G., Doherty, A., Messina, C., Schussler, J., Zinselmeier, C., Paszkiewicz, S. & Cooper, M. 2009. Can changes in canopy and/or root system architecture explain historical maize yield trends in the US Corn Belt? *Crop Science*, 49: 299–312.

Hayami, Y. & Morooka, K. 1987. *The market price response of world rice research*. Agricultural Economics Department Paper No. 87-21. Los Baños, Philippines, IRRI.

Heffer, P. 2008. *Assessment of fertilizer use by crop at the global level*. Paris, International Fertilizer Industry Association.

Heffner, E.L., Sorrells, M.E. & Jannick, J.-L. 2009. Genomic selection for crop improvement. *Crop Science*, 49: 1–12.

Heisey, P.W. & Edmeades, G.O. 1999. *Maize production in drought-stressed environments: technical options and research resource allocation*. CIMMYT 1997/98 World Maize Facts and Trends Part 1. Mexico City, CIMMYT.

Heisey, P.W., Lantican, M.A. & Dubin, H.J. 2002. *Impacts of international wheat breeding research in developing countries, 1966–1997*. Mexico City, CIMMYT.

Heisey, P. & Norton, G.W. 2007. Fertilizer and other chemicals. *In* R. Evenson and P. Pingali, eds. *Handbook of agricultural economics*, Volume 3, pp. 2747–2786. Amsterdam, Netherlands, Elsevier.

Horie, T., Lubis, I., Takai, T., Ohsumi, A., Kuwasaki, K., Katsura, K. & Nii, A. 2003. Physiological traits associated with high yield potential in rice. *In* T. Mew, D.S. Brar, S. Peng, D. Dawe and B. Hardy, eds. *Rice science: Innovations and impacts for livelihood*, pp. 117–146. Manila, IRRI.

Hubert, B., Rosegrant, M., van Boekel, M.A.J.S. & Ortiz, R. 2010. The future of food: scenarios for 2050. *Crop Science*, 50(Suppl.): S33–S50.

IFDC. 2007. *Mitigating poverty and environmental degradation through nutrient management in South and Southeast Asia*. http://betuco.be/compost/urea%20 briquettes%20machine%20bangladesh.pdf.

IFDC. 2009. *Controlled release fertilizers – an emerging technology for food security*. IFDC Focus on Food Security Issue No. 13. Muscle Shoals, Alabama, USA.

IFDC. 2010. *World phosphate rock reserves and resources*. Technical Bulletin No. 75. Muscle Shoals, Alabama, USA.

IPCC. 2007. *Climate change 2007 synthesis report*. Geneva.

IRRI. 2008. Investment returns opportunities in rainfed/irrigated environments of South Asia. Unpublished paper by A. Dobermann and D. Mackill. Los Baños, Philippines.

Islam, Z., Bagchi, B. & Hossain, M. 2007. Adoption of leaf color chart for nitrogen use efficiency in rice: Impact assessment of a farmer-participatory experiment in West Bengal, India. *Field Crops Research*, 103: 70–75.

James, C. 2009. *Global status of commercialized biotech/GM crops: 2008*. ISAAA Brief No. 39. Ithaca, New York, USA, ISAAA. 243 pp.

Ji, X., Shiran, B., Wan, J., Jenkins, C.L.D., Condon, A.G., Richards, R.A. & Dolferus, R. 2010. Importance of pre-anthesis sink-strength for maintenance of grain number during reproductive stage water stress in wheat. *Plant, Cell and Environment*, 33: 926–942.

Johnson, E.C., Fischer, K.S., Edmeades, G.O. & Palmer, A.F.E. 1986. Recurrent selection for reduced plant height in lowland tropical maize. *Crop Science*, 26: 253–260.

Katsura, K., Maeda, S., Horie, T. & Shiraiwa, T. 2007. Analysis of yield attributes and crop physiological traits of Liangyoupeiju, a hybrid rice recently bred in China. *Field Crops Res.*, 103: 170–177.

Kucharik, C.J. 2008. Contribution of planting date trends to increased maize yields in the central United States. *Agronomy Journal*, 100: 328–336.

Kumar, P., Mittal, S. & Hossain, M. 2008. Agricultural growth accounting and total factor productivity in south Asia: a review and policy implications. *Agricultural Economics Research Review*, 21: 145–172.

Ladha, J.K. & Reddy, P.M. 2000. *The quest for nitrogen fixation in rice*. Los Baños, Philippines, IRRI.

Lal, R. 2009. Soil degradation as a reason for inadequate human nutrition. *Food Security*, 1: 45–57.

Langreid, A., Bockme, O.C. & Kaarstad, O. 2004. *Agriculture, fertilizers and the environment*. Wallingford, UK, CAB International.

Lantican, M.A., Pingali, P.L. & Rajaram, S. 2003. Is research on marginal lands catching up? The case of unfavourable wheat growing environments. *Agricultural Economics*, 29: 353–361.

Lindquist, J.L., Arkebauer, T.J., Walters, D.T., Cassman, G. & Dobermann, A. 2005. Maize radiation use efficiency under optimal growth conditions. *Agronomy Journal*, 97: 72–78.

Lipton, M. 2005. *The family farm in a globalizing world: The role of crop science in alleviating poverty, 2020*. Discussion Paper No. 40. Washington, DC, IFPRI.

Lobell, D.B., Cassman, K.G. & Field, C.B. 2009. Crop yield gaps: their importance, magnitudes, and causes. *Annual Review of Environmental Resources*, 34.

Lobell, D.B. & Ortiz-Monasterio, J.I. 2008. Satellite monitoring of yield responses to irrigation practices across thousands of fields. *Agronomy Journal*, 100: 1005–1012.

Lobell, D.B., Asner, G.P., Ortiz-Monasterio, J.I. & Benning, T.L. 2003. Remote sensing of regional crop production in the Yaqui Valley, Mexico: estimates and uncertainties. *Agricultural Ecosystems and Environment*, 94: 205–220.

Long, S.P., Zhu, X.-G., Naidu, S.L. & Ort, D.R. 2006. Can improvement in photosynthesis increase crop yields? *Plant, Cell and Environment*, 29: 315–330.

Loomis, R.S. & Amthor, J.S. 1999. Yield potential, plant assimilatory capacity, and metabolic efficiencies. *Crop Science*, 39: 1584–1596.

Loomis, R.S. & Connor, D.J. 1992. *Crop ecology. Productivity and management in agricultural systems*. Cambridge, UK, Cambridge University Press.

Lusigi, A. & Thirtle, C. 1997. Total factor productivity and the effects of R&D in African agriculture. *Journal of International Development*, 9: 529–538.

Mitchell, P.I., Sheehy, J.E. & Woodward, F.I. 1998. *Potential yields and the efficiency of radiation use in rice.* IRRI Discussion Paper Series No. 32. Manila, IRRI.

Molden, D., ed. 2007. *Water for food, water for life. A comprehensive assessment of water management in agriculture.* Colombo, Sri Lanka, International Water Management Institute.

Monteith, J.L. 1977. Climate and the efficiency of crop production in Britain. *Philosophical Transactions of the Royal Society, London Series B*, 281: 277–294.

Morris, M., Kelly, V.A., Kopicki, R.J. & Byerlee, D. 2007. *Fertilizer use in African agriculture. Lessons learned and good practice guide.* Washington, DC, World Bank.

Muchow, R.C. & Sinclair, T.R. 1994. Nitrogen response of leaf photosynthesis and canopy radiation use efficiency in field-grown maize and sorghum. *Crop Science*, 34: 721–727.

Murgai, R., Ali, M. & Byerlee, D. 2001. Productivity growth and sustainability in post-green revolution agriculture: the case of the Indian and Pakistan Punjabs. *World Bank Research Observer*, 16: 199–218.

Murphy, D.J. 2007. *Plant breeding and biotechnology: Societal context and the future of agriculture.* Cambridge, UK, Cambridge University Press.

Murty, M.V.R., Singh, P., Wani, S.P., Khairwal, I.S. & Srinivas, K. 2007. *Yield gap analysis of sorghum and pearl millet in India using simulation modelling.* Global Theme on Agroecosystems Report No. 37. Patancheru, Andra Pradesh, India, ICRISAT.

Nelson, D.E., Repetti, P.P., Adams, T.R., Creelman, R.A., Wu, J., Warner, D.C., Anstrom, D.C., Bensen, R.J., Castiglioni, P.P., Donnarummo, M.G., Hinchey, B.S., Kumimoto, R.W., Maszle, D.R., Canales, R.D., Krolikowski, K.A., Dotson, S.B., Gutterson, N., Ratcliffe, O.J. & Heard J.E. 2007. Plant nuclear factor Y(NF-Y) B subunits confer drought tolerance and lead to improved corn yields on water-limited acres. *Proceedings of the National Academy of Sciences*, 104: 16450–16455.

Nziguheba, G., Palm, C.A., Berhe, T., Denning, T., Dicho, A., Diouf, O., Diru, W., Flor, R., Frimpong, F., Harawa, R., Kaya, B., Manumbu, E., McArthur, J., Mutuo, P., Ndiaye, M., Niang, A., Nkhoma, P., Nyadzi, G., Sachs, J., Sullivan, C., Teklu, G., Tobe, L. & Sanchez, P.A. 2010. The African green revolution: results from the Millennium Villages Project. *Advances in Agronomy*, 109: 75-115.

Oerke, E.C. 2006. Crop losses to pests. *Journal of Agricultural Science*, 144: 31–43.

Ortiz, R., Braun, H.J., Crossa, J., Crouch, J., Davenport, G., Dixon, J., Dreisigacker, S., Duveiller, E., He, Z., Huerta, J., Joshi, A.K., Kishii, M., Kosina, P., Manes, Y., Mezzalama, M., Morgounov, A., Murakami, J., Nicol, J., Ortiz-Ferrara, G., Ortiz-Monasterio, I., Payne, T.S., Peña, R.J., Reynolds, M.P., Sayre, K.D., Sharma, R.C., Singh, R.P., Wang, J., Warburton, M., Wu, H. & Iwanaga, M. 2008. Wheat genetic resources enhancement by the International Maize and Wheat Improvement Center (CIMMYT). *Genetic Resources and Crop Evolution*, 55: 1095–1140

Ortiz-Monasterio, J.I. & Lobell, D.B. 2007. Remote sensing assessment of yield losses due to sub-optimal planting dates and fallow period weed management. *Field Crops Research*, 101: 80–87.

Ortiz-Monasterio, J.I. & Raun, W. 2007. Reduced nitrogen and improved farm income for irrigated spring wheat in the Yaqui Valley, Mexico, using sensor based nitrogen management. *Journal of Agricultural Science*, 145: 215–222.

Ortiz-Monasterio, J.I., Sayre, K.D., Rajaram, S. & McMahon, M. 1997. Genetic progress in wheat yield and nitrogen use efficiency under four N rates. *Crop Science*, 37: 892–898.

Pardey, P.G., Alston, J., James, J., Glewwe, P., Binenbaum, E., Hurley, T. & Wood, S. 2007. Science, technology and skills. Background paper for *World Development Report 2008: Agriculture for development*. Washington, DC, World Bank.

Paroda, R.S. 2004. Scaling up: how to reach a billion resource-poor farmers in developing countries. Plenary paper in *Proceedings of the 4th International Crop Science Congress*, Brisbane, Australia. www.cropscience.org.au/icsc2004/plenary/4/223_paroda.htm.

Passioura, J.B. 1977. Grain yield, harvest index, and water use of wheat. *Journal of the Australian Institute of Agricultural Science*, 43: 117–120.

Passioura, J.B. & Angus, J.F. 2010. *Advances in Agronomy*, 106: 37–75.

Peng, S., Cassman, K.G., Virmani, S.S., Sheehy, J. & Khush, G.S. 1999. Yield potential trends of tropical rice since the release of IR8 and the challenge of increasing rice yield potential. *Crop Science*, 39: 1552–1559.

Peng, S., Khush, G.S., Virk, P., Tang, Q. & Zou, Y. 2008. Progress in ideotype breeding to increase rice yield potential. *Field Crops Research*, 108: 32–38.

Pingali, P.L. & Heisey, P.W. 1999. *Cereal crop productivity in developing counties: Past trends and future prospects*. Economics Working Paper No. 99-03. Mexico City, CIMMYT.

Pingali, P., Hossain, M. & Gerpacio, R.V. 1997. *Asian rice bowls: The returning crisis.* Wallingford, UK, CAB International.

Podlich, D.W., Winkler, C.R. & Cooper, M. 2004. Mapping as you go: an effective approach for marker-assisted selection of complex traits. *Crop Science,* 44: 1560–1571.

Raun, W.R. & Johnson, G.V. 1999. Improving nitrogen use efficiency for cereal production. *Agronomy Journal,* 91: 357–363.

Richards, R.A. 2004. Physiological traits used in the breeding of new cultivars for water-scarce environments. In *Proceedings of the 4th International Crop Science Congress,* Brisbane, Australia. www.cropscience.org.au.

Rosegrant, M. & Pingali, P. 1994. Policy and technology for rice productivity growth in Asia. *Journal of International Development,* 6: 665–688.

Rosegrant, M.W., Huang, J., Sinha, A., Ahammad, H., Ringler, C., Zhu, T., Sulser, T.B., Msangi, S. & Batka, M. 2008. *Exploring alternative futures for agricultural knowledge, science and technology (AKST).* ACIAR Project Report ADP/2004/045. Washington, DC, IFPRI.

Rozelle, S., Jin, S., Huang, J. & Hu, R. 2003. The impact of investments in agricultural research on total factor productivity in China. *In* R.E. Evenson and D. Gollin, eds. *Crop variety improvement and its effect on productivity: The impact of international agricultural research.* Wallingford, UK, CAB International.

Salvucci, M. 2008. Association of rubisco activase with chaperonin-60β: a possible mechanism for protecting photosynthesis during heat stress. *Journal of Experimental Botany,* 59: 1923–1933.

Shearman, V.J., Sylvester-Bradley, R., Scott, R.K. & Foulkes, M.J. 2005. Physiological processes associated with yield progress in the UK. *Crop Science,* 45: 175–185.

Sheehy, J.E., Ferrer, A.B., Mitchell, P.L., Elmido-Mabilangan, A., Pablico, P. & Dionora, M.J.A. 2007. *In* J.E. Sheehy, P.L. Mitchell and B. Hardy, eds. *Charting new pathways to C_4 rice,* pp. 3–26. Manila, IRRI.

Sherlund, S.M., Barrett, C.B. & Akinwumi, A.A. 2002. Smallholder technical efficiency controlling for environmental production conditions. *Journal of Development Economics,* 69: 85–101.

Smil, V. 2008. *Energy in nature and society.* Boston, Massachusetts, USA, MIT Press.

Socolow, R.H. 1999. Nitrogen management and the future of food: lessons from the management of energy and carbon. *Proceedings of the National Academy of Sciences,* 96: 6001–6008.

Sudduth, K.A. 2007. Current status and future directions of precision agriculture in the USA. In *Proceedings of the 2nd Asian Conference on Precision Agriculture*, 2–4 August 2007, Pyeongtaek, Republic of Korea.

Sylvester-Bradley, R., Foulkes, J. & Reynolds, M. 2005. Future wheat yield: evidence, theory and conjecture. In *Proceedings of the 61st Easter School in Agricultural Sciences*, pp. 233–260. Nottingham, UK, Nottingham University Press.

Tollenaar, M. & Lee, E.A. 2011. Strategies for enhancing grain yield in maize. *Plant Breeding Reviews*, 34.

Tollenaar, M. & Wu, J. 1999. Yield improvement in temperate maize is attributable to greater stress tolerance. *Crop Science*, 39: 1597–1604.

Tripp, R., Louwaars, N. & Eaton, D. 2007. Plant variety protection in developing countries. A report from the field. *Food Policy*, 32: 354–371.

Tubiello, F.N., Amthor, J.S., Boote, K.J., Donatelli, M., Easterling, W., Fischer, G., Gifford, R.M., Howden, M., Reilly, J. & Rosenzweig, C. 2007. Crop response to elevated CO_2 and world food supply. *European Journal of Agronomy*, 26: 215–223.

Tweeten, L. & Thompson, S.R. 2008. *Long-term agricultural output supply-demand balance and real farm and food prices*. Working Paper AEDE-WP 0044-08. Columbus, Ohio, USA, Ohio State University.

Venuprasad, R., Sta. Cruz, M.T., Amante, M., Magbanua, R., Kumar, A. & Atlin, G.N. 2008. Response to two cycles of divergent selection for grain yield under drought stress in four rice breeding populations. *Field Crops Research*, 107: 232–244.

von Braun, J., Fan, S., Meinzen-Dick, R., Rosegrant, M.W. & Pratt, A.N. 2008. *International agricultural research for food security, poverty reduction, and the environment: What to expect from scaling up CGIAR investments and "best bet" programs*. Washington, DC, IFPRI.

Weinberger K. 2003. *Impact analysis of mungbean research in South and Southeast Asia*. Final Report of GTZ Project. Shanhua, Taiwan Province of China, AVRDC.

Wolfe, D.W., Henderson, D.W., Hsiao, T.C. & Alvino, A. 1988. Interactive water and nitrogen effects on senescence of maize. I. Leaf area duration, nitrogen distribution, and yield. *Agronomy Journal*, 80: 859–864.

Wood, S., Henao, J. & Rosegrant, M.W. 2004. The role of nitrogen in sustaining food production and estimating future nitrogen needs to meet food demand. *In* A.R. Moiser, K. Syers and J.R. Freney, eds. *Agriculture and the nitrogen cycle: Assessing the impact of fertilizer use on food production and the environment*, pp. 245–265. Washington, DC, Island Press.

World Bank. 2007. *World Development Report 2008: Agriculture for development.* Washington, DC.

Xiao, B., Chen, X., Xiang, C., Tang, N., Zhang, Q. & Xiong, L. 2009. Evaluation of seven function-known candidate genes for their effects on improving drought resistance of transgenic rice under field conditions. *Molecular Plant*, 2: 73–83.

Yang, W., Peng, S., Laza, R.C., Visperas R.M. & Dionisio-Sese, M.L. 2007. Grain yield and yield attributes of new plant type and hybrid rice. *Crop Science*, 47: 1393–1400.

Zhou,Y., He, Z.H., Sui, X.X., Xia, X.C., Zhang, X.K. & Zhang, G.S. 2007a. Genetic improvement of grain yield and associated traits in the northern China winter wheat region from 1960 to 2000. *Crop Science*, 47: 245–253.

Zhou, Y., Zhu, H.Z., Cai, C.B., He, Z.H., Zhang, X.K., Cia, X.C. & Zhang, G.S. 2007b. Genetic improvement of grain yield and associated traits in the southern China winter wheat region: 1949 to 2000. *Euphytica*, 157: 465–473.

CONCLUSIONS

CHAPTER 11

CRITICAL EVALUATION OF SELECTED PROJECTIONS

Nikos Alexandratos

The purpose of this chapter is to summarize in a coherent manner the different views embodied in the projections concerning possible futures for world food and agriculture to 2050. The preparation of this chapter started when syntheses of the expert meeting (EM) papers of June 2009 were being prepared for the High-Level Expert Forum on "How to feed to the world in 2050". It soon become apparent that not only views were diverse, but also that it was difficult to understand why they differed. To illustrate this problem, it suffices to take as an example some results on the climate change impacts on world market prices reported in the EM papers, whose revised versions have become chapters of this volume. Percentage differences over baselines without climate change in 2050 are as follows.
Chapter 3, price index for all cereals:

- climate model Hadley without CO_2 fertilization: + 10 percent;

- climate model Hadley with CO_2 fertilization: -1 percent;

- climate model Commonwealth Science and Industrial Organization (CSIRO) with;

- CO_2 fertilization: + 2 percent.

Chapter 2, Table 2.3[1]; models National Center for Atmospheric Research (NCAR) and CSIRO:

- wheat: NCAR + 111 percent; CSIRO + 94 percent;

- maize: NCAR + 52 percent; CSIRO + 55 percent;

- rice: NCAR + 37 percent; CSIRO + 32 percent.

---

1. Chapter 2 reports results which are also found in a recent paper of the International Food Policy Research Institute (IFPRI) (Nelson *et al.*, 2009).

A similar degree of diversity is observed also in the projection of other key variables. It is therefore important to compare the main results reported on different themes, and, as far as possible, to understand the origin and the reasons for such differences.

The rest of the chapter is organized along five topics: projections on world prices and consumption volumes; impact of climate change; impact of biofuels; economic growth, global inequality and poverty; and expected developments in sub-Saharan Africa. For each of these themes projections are analyzed and compared, and reference is made to the associated food and nutrition outcomes. Focus is mostly on the results offered by Alexandratos in Chapter 1, by Msangi and Rosegrant in Chapter 2, by Fischer in Chapter 3, by Hillebrand in Chapter 4, and by van der Mensbrugghe, I. Osorio-Rodarte, A. Burns and J. Baffes in Chapter 5. Both Chapter 2 and this chapter draw also on more recent IFPRI work (Nelson *et al.*, 2009)[2].

Prices and quantities in the baseline scenarios

Prices

A frequently asked question is whether the price surges of recent years were a harbinger of things to come or, as in the past, a temporary occurrence (Alexandratos, 2008; Mitchell, 2008). Do the price projections reported in previous chapters provide answers? When the first versions of these chapters were prepared, in June 2009, cereal prices had fallen by 30 percent from their peaks of spring 2008. Nevertheless, they were still well above averages for the pre-surge period.

Chapter 5 reports price projections only for total agriculture to 2030. They show that with a productivity growth of 2.1 percent per annum in agriculture there is a small negative trend over the long term (Figure 5.10).[3] Chapter 3, shows detailed price projections to 2050 for several commodity groups and total agriculture (Table 3.4). Chapter 2 reports percentage price changes from 2000 to 2050 for the three major cereals, drawing on Nelson *et al.* (2009) (Table 2.3). Chapter 1 does not provide price projections.

For comparisons, the different price projections need to be rebased on a common base-year denominator. Figure 11.1 shows that, at least for cereal and in terms of annual averages, the price surge started in 2006. Therefore, a good base for comparing projections vis-à-vis "the present" is the three-year average

---

2. Clarifications were received from the authors of Chapter 2, 4 and 5. The author of this chapter thanks all the authors for their help, and particularly Siwa Msangi who provided detailed background material.

3. Presumably "overall agriculture" corresponds to "total agriculture" used in construction of the price indices in the World Bank's commodity price data.

2006/2008. For comparing future prices with those of the pre-surge period, instead, the three-year average 2003/2005 is used. If the 2006/2008 average were used, it would be concluded that prices will fall. If a pre-surge three-year average is used, the opposite general conclusion is reached. Figure 11.2 shows price projections of the 2009 OECD/FAO agricultural outlook to 2018 (OECD/FAO, 2009), after rebasing them on the two three-year averages: prices are higher in 2018   by 6 to 19 percent for the three cereals   with respect to those of the pre-surge years, but much lower than those of 2006/2008.

Figure 11.1
Cereals and agriculture real price indices

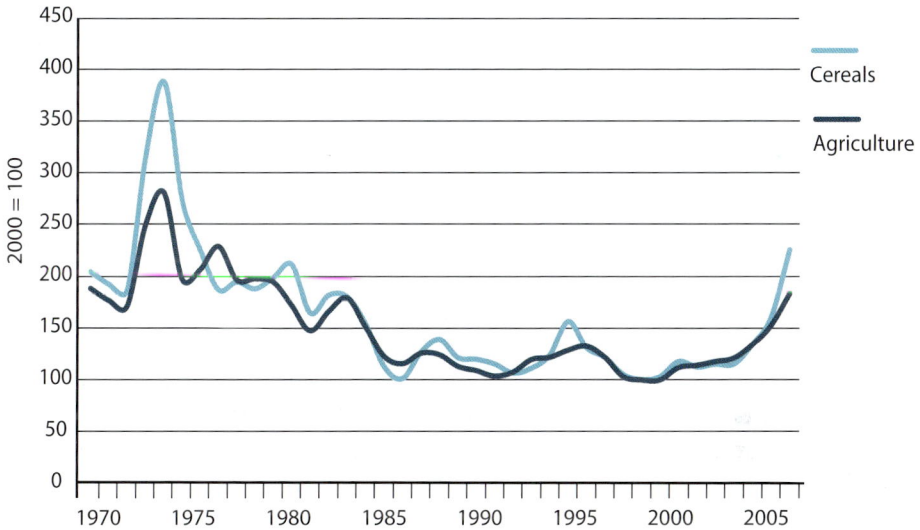

Source: World Bank commodity price data. Indices are from price data at constant 2000 dollars.

Figure 11.3 compares price projections to 2030 from Chapters 3, 5 and from IFPRI[4], computed from both the 2003/2005 and the 2006/2008 bases. The main message seems to be that over the next two decades real prices will be much lower than the average of surge years 2006/2008. The picture is more mixed when viewing the price projections in relation to those of the pre-surge period.

Concerning total agriculture to 2030, Chapters 3 and 5 suggest that the average price level in 2030 will be not very different from that of the pre-surge period 2003/2005. The same applies to the IFPRI projections of wheat and maize

---

4. Chapter 2 only reports projections to 2050. The background data for Figure 11.3 was supplied separately by the authors of Chapter 2.

Figure 11.2
OECD-FAO 2009 projections: cereals price indices to 2018

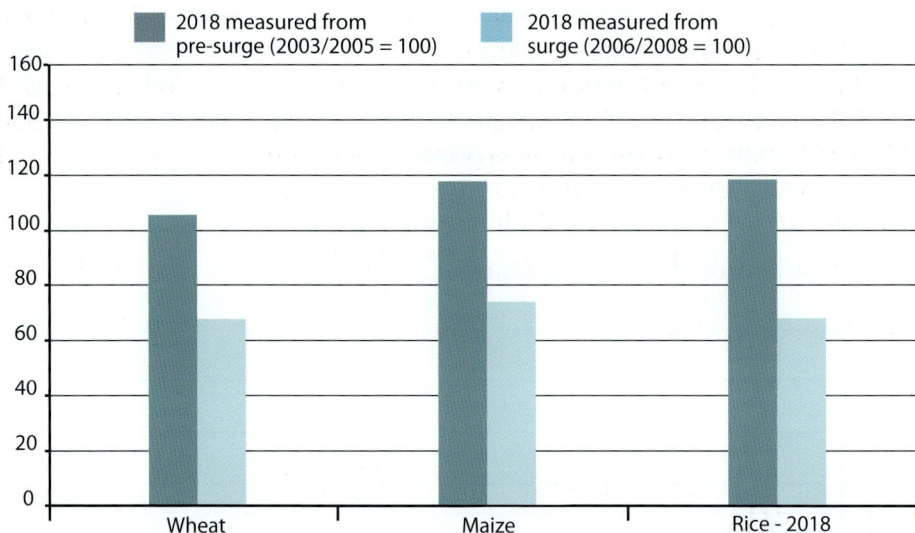

Source: OECD/FAO, 2009: Figures 6.8 to 6.10.

Figure 11.3
Price indices to 2030: projections from chapters 2, 3 and 5

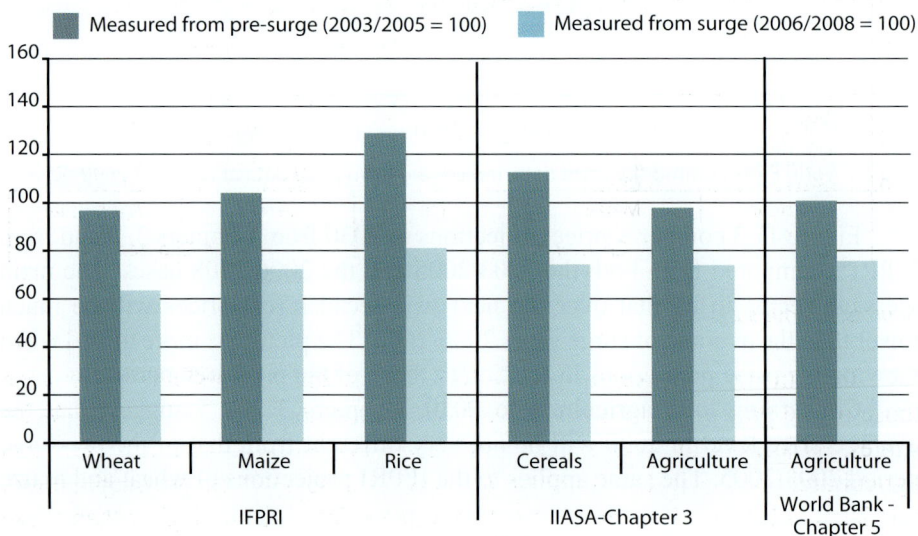

Sources: IFPRI: detailed tables for Chapter 2 (supplied separately by the authors), Chapters 3, 4 and 5.

prices. The only significant differences are in rice prices, which could be 29 percent higher than the pre-surge 2003/2005 average according to IFPRI; and the average cereals price index, which may be 13 percent higher than the pre-surge 2003/2005 average, according to Chapter 3.

Figure 11.4 reports baseline price projections to 2050 from Chapters 2 and 3. The average level of agricultural prices according to the IIASA projection (Chapter 3) may be just 10 percent higher than in the pre-surge period, and well below "present levels". However, Chapter 3 projects that cereal prices will rise faster than the average for all agriculture. Prices of the other agricultural products increase by less than those of cereals (Table 3.4). The cereals price index implied for wheat, maize and rice in Chapter 2 drawing on Nelson *et al.*, 2009 is broadly in line or somewhat higher than that projected by IIASA in Chapter 3 (Figure 11.4).

Figure 11.4
Price indices to 2050: baseline projections from chapters 2 and 3

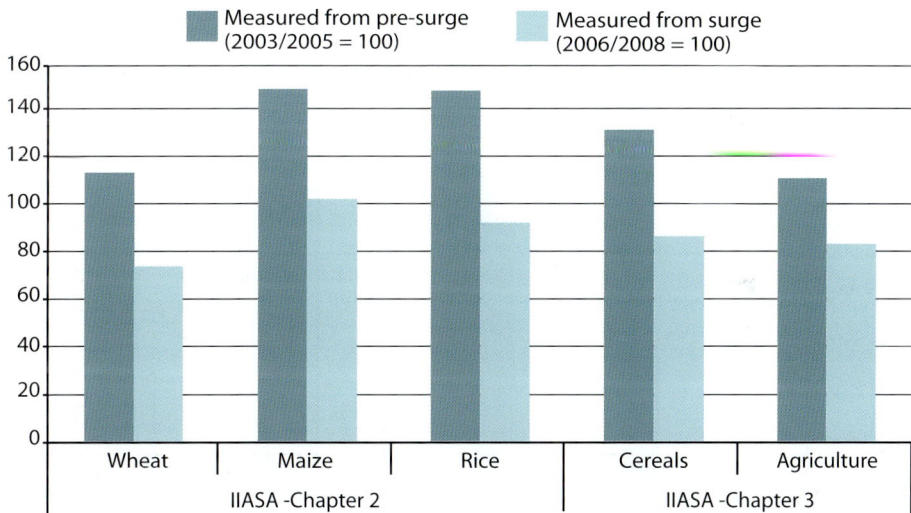

Sources: Chapters 2 and 3.

To sum up, 2050 cereal prices projected in the baseline are higher than those of the pre-surge period, but much lower than those of the period of price surges. Projections to 2030 of Chapters 3 and 5 generally suggest that real world agricultural prices will tend to revert to levels near those of the pre-surge period. For 2050, all projections broadly indicate that prices will be higher than those of the pre-surge years but lower than those reached in the years of price surges.

They will not revert to the long-term trend of decline, but this is not really a novel element: the path of decline had already been largely halted from about the mid-1980s.

Consumption volumes

Projected world cereals consumption in 2050 differs widely among the different authors (Figure 11.5). Chapter 3 projects 3 388 million tonnes in 2050[5]; and Chapter 2 has 2 739 million tonnes. As noted earlier, the price projections for individual cereals in Chapter 2 are only slightly higher than those implied by Chapter 3 for all cereals. This raises the question regarding why, if their price projections are similar, the papers' projections of cereals consumption are so different from each other?

Figure 11.5
Cereals projections: world consumption in 2050 and growth rates – Chapters 1, 2 and 3

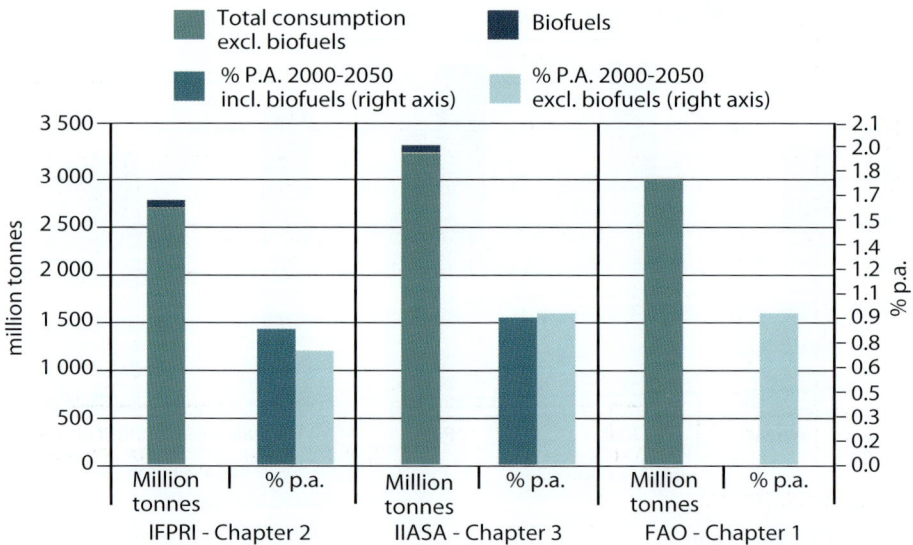

Sources: Chapter 1; Chapter 2, Table 2.1 (biofuels are from the detailed files); Chapter 3, Table 3.3.

The answer is to be found partly in the wide differences in the historical data used in the two analyses. Chapter 2 (Table 2.1) starts with 2000 world consumption

---

5. Use of cereals for biofuels is maintained at the 2008 level throughout the projection years in this IIASA scenario (Scenario WEO-01 in Chapter 3).

of 1 818 million tonnes, while Chapter 3 has 2 144 million tonnes for the same year (Table 3.3). Both are different from the FAO historical data (1 900 million tonnes average for 1999/2001). The figure for 2000 in Chapter 3 is even higher than the increased world consumption of recent years, i.e. the 2006/2008 average. These discrepancies are depicted in Figure 11.6. They prevent a comparison of the absolute values of projected world cereals consumption. A better idea of the differences in projected world cereals consumption can be obtained by observing growth rates of consumption, which are also shown in Figure 11.5. The growth rate in Chapter 3 is higher than that of Chapter 2, and the difference is even more pronounced if biofuels are excluded from total cereals consumption.

Figure 11.6
World cereals consumption historical data versus data for 2000 used in chapters 2 and 3

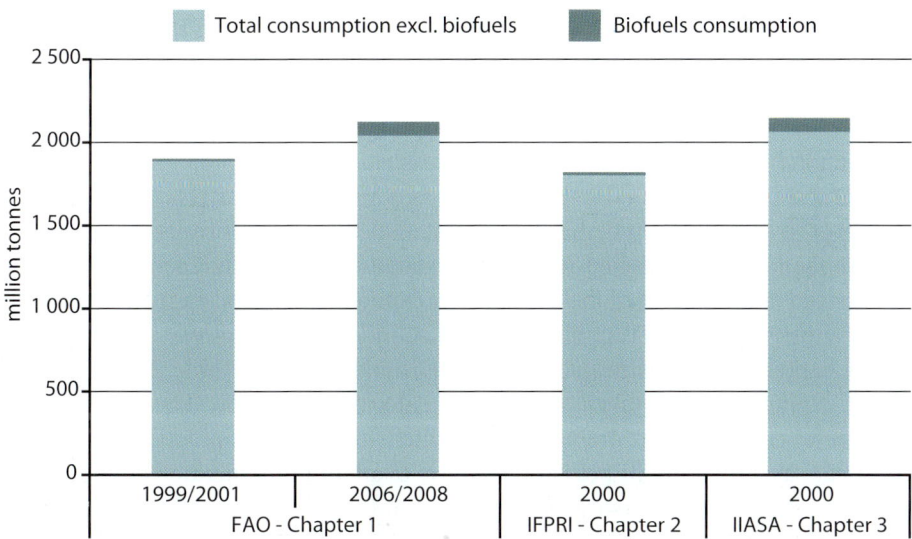

Sources: OECD/FAO.

Differences in historical data bedevil the attempt to form an idea concerning the volume of world cereals production that would be required to meet the growth of consumption in 2050. Figure 11.5 also shows the FAO projection to 2050 (FAO, 2006). The 3 billion tonnes projected for 2050 is based on historical data for up to 2001, which does not include biofuels. On this benchmark, the IFPRI projection in Chapter 2 looks too low, mainly because of the lower starting data for 2000 and to a lesser extent because of the lower growth rate; and the one of IIASA in Chapter 3 definitely too high, again because of the starting 2000 figure.

However, is the FAO projection any better? It certainly has two advantages. Firstly, it is grounded on reliable historical data; and, secondly, it is subjected to a reality check in Chapter 1, where the trajectory of world production/consumption for 2000 to 2050 was compared with outcomes to 2008 and medium-term projections to 2018 from OECD/FAO (2009) (Table 1.1 and Figure 1.1). The FAO projection trajectory was close enough to both actual outcomes to 2008 and the subsequent ten-year OECD/FAO projection, both not including cereals use for biofuels.

Of primary interest for the issue at hand are the food and nutrition outcomes associated with the different prices and quantities projected. FAO, and Chapter 1 in this volume, estimate changes in the incidence of undernourishment based on per capita food consumption expressed in kcal/person/day (FAO, 2006: Box 2.2). In Chapter 3, IIASA provides projections of population at risk of hunger, which presupposes the availability of projections in terms of kilocalories per person per day for all commodities, not only cereals. The IIASA World Food System model used in that chapter computes the population at risk of hunger based on the correlation between the share of undernourished in total population and the ratio of average per capita food supply to the average per capita food requirements. However, it is doubtful whether this is a statistically valid correlation, given that the numbers undernourished in the FAO statistics are not independent data, but are derived as a function of the average per capita dietary food supply, the average national per capita food requirements and an index of inequality.

Chapter 2 contains projections of child malnutrition (Table 2.4), for which a key explanatory variable is per capita calorie availability. The IFPRI model "computes changes in the prevalence of malnutrition in the population aged zero to five years as a function of per capita calorie availability (generated endogenously by the model), as well as exogenous projections of schooling rates among females of secondary school age, the share of population with access to clean water, and the ratio of female-to-male life expectancies" (Chapter 2). The projections of this variable are not given in Chapter 2, but are in Nelson *et al.* (2009: Table 5). Comparisons on food security outcomes are in the next sections of this chapter, and especially in the last one, that discusses prospects for Sub-saharan Africa.

Climate change impact on agriculture

This issue is analyzed in most detail in Chapter 3. Chapter 2, drawing on Nelson *et al.* (2009), presents results for scenarios incorporating climate change effects and Chapter 5 addresses the issue of climate change and agriculture, but only for the sector as a whole[6]. The FAO projections in Chapter 1 do not address climate change.

Conclusions of the different chapters about climate change impacts are disparate. Differences are not primarily due to the many uncertainties regarding what climate changes are in store; they persist even after controlling for such uncertainties. It is therefore worth starting by outlining what is involved in building scenarios of possible agricultural and food security outcomes under climate change. The following steps may be distinguished:

1. Define future levels of greenhouse gas (GHG) concentrations in the atmosphere from some concept of the future rate and pattern of socio-economic development, demographics, energy use, etc. for the globe. These are called "development pathways" in the terminology of the Intergovernmental Panel on Climate Change (IPCC).

2. Use climate models – general circulation models (GCMs) – to translate these GHG concentrations into deviations of key climate variables of relevance to agriculture, such as temperature, precipitation, etc., from those assumed in 2050 in the baseline scenario. The baseline usually assumes prevalence of present climate in the future, although this may be an impossibility.

3. Superimpose deviations in climate variables on biophysical data and models (e.g., agro-ecological attributes, crop growth models) to portray how resource characteristics may be altered (e.g., changes in the length of growing periods, soil moisture, incidence of pests/diseases), and how plants may respond to the changed conditions. Resulting estimates refer to the impact of climate change on the potential productive capacity of resources, not on projected production.

4. From this interface, derive information to modify the model of agriculture used to project production, consumption, trade and prices. This means changing the way in which models recognize the altered biophysical base of agricultural production. Revised estimates of land/water constraints should be included, along with assumptions of alternative paths to adaptation, i.e., how people may respond by changing crop calendars or introducing crop varieties better suited to the altered climate.

Steps 1 and 2 need not be an integral part of the agricultural modeller's task: changes in climate variables necessary for estimating impacts can be taken

---

6. This section is also based on personal communication from the authors of Chapter 5, and evidence reported in Nelson *et al.* (2009).

from the work of others specializing in GCMs. However, caution is required: the socio-economic pathway generating the GHGs that will affect the climate, and especially assumptions on GDP and population growth, must not be too different from those assumed in the baseline. This is necessary if the projected results of the scenarios with and without climate change are to be compared. Moreover, the definition of a scenario without climate change is not unambiguous: assuming that present climate will prevail in the future is almost tantamount to assuming that present GHG concentrations will remain constant. But could such a situation exist? Even ignoring the possibility that climate variables can continue to change even if current GHG concentrations do not change, there is no possible development and demographic path that will not increase GHG concentrations. The debate is entirely about how much emissions could be reduced, and there is no option of reducing them to zero. GHG concentrations will increase anyway, bringing climate change; so there seems to be no point in estimating a scenario with present climate that, by definition, cannot exist.

Step 3 requires that the modeller has access to, and can manipulate, detailed biophysical data describing land resources and the physiology of crops in relation to climate variables. For this purpose, IIASA uses the Global Agroecological Zones Study (GAEZ) that it produced with FAO and is reflected in Chapter 3. IFPRI uses "a hydrology model and links to the Decision Support System for Agrotechnology Transfer (DSSAT) crop-simulation model, with yield effects of climate change at 0.5-degree intervals aggregated up to the food-production-unit level" (Nelson *et al.*, 2009: 1); this is reflected in Chapter 2.

Step 4 is the crucial one. It can be speculated that in previous chapters this is mostly carried out by modifying the values of exogenous variables of models, most likely intercepts and rates of productivity growth. IFPRI's International Model for Policy Analysis of Agricultural Commodities and Trade (IMPACT) "simulates growth in crop production, determined by crop and input prices, externally determined rates of productivity growth and area expansion, investment in irrigation, and water availability" (Nelson *et al.*, 2009: 1). Similar observations may apply to the IIASA model[7].

Climate change effect on prices and quantities

The results described in Chapter 5 include climate change effects, but only for agriculture. In the model, climate change influences agriculture by lowering the growth rate of agricultural productivity. In the case of no climate change,

---

7. "Results of [Agro-Ecological Zone] agricultural production potential assessments, under various climate change/CO_2 emission scenarios and obtained for different GCM-based climate experiments, were input to IIASA's system of national agricultural models to further assess world food system and trade implications" (Fischer, Shah and van Velthuizen, 2002: 22).

productivity growth is exogenously assumed to be 2.1 percent per annum. The extent to which climate change affects this growth rate is derived from Cline (2007), assuming that the 2.5 oC temperature increase is reached in 2050, rather than in 2080 as in Cline's work.[8] The impact on the rate of productivity growth is the average of Cline's scenarios with and without CO_2 fertilization effects. In this scenario, temperature rises by 1.7 oC in 2030, which translates into a reduction in the rate of productivity growth from 2.1 percent per annum to 1.76 percent per annum. The result is that the baseline projects the price index of agriculture in 2030 to be a little below that of the base year 2005. As noted (Figure 11.3), this translates into 2030 prices nearly equal to those of the pre-surge period (average 2003/2005) and well below those reached in the surge years (average 2006/2008). There is no scenario without climate change, so price outcomes with and without climate change cannot be compared. However, it is reasonable to assume that without the reduction in agricultural productivity growth brought about by climate change, projected prices would be lower, at least up to 2030. This is an interesting finding, which is in sharp contrast to the predictions of other analyses, particularly that of Chapter 2.

Chapter 3 estimates a number of climate change scenarios. The analysis assumes that climate change is driven by GHG emissions generated by the socio-economic development/demography path of the IPCC Special Report on Emissions Scenarios (SRES) A2.[9] These projected concentrations are combined with the GCMs of the Hadley Centre for Climate Prediction and Research (United Kingdom) and CSIRO (Australia) to generate the projected changes in temperature, precipitation, etc. As noted, changes in climatic variables are defined for each individual grid point of the GAEZ, and the implied changes in the biophysical base of the analysis are derived.[10]

Estimates of impacts on potential production are derived from such changes in the biophysical base, and adaptation options are restricted to currently existing varieties. Such restriction is justified when estimating changes in production

---

8. Note that Cline's estimates refer to climate change effects on "agricultural capacity" in the 2080s, not on projected production.

9. This scenario implies that GHG concentrations in the atmosphere will rise from their current 354 parts per million (ppm) to 536 ppm in 2050 (IPCC, 2007b: Table 1). A more recent IPCC report indicates that "Atmospheric CO_2 concentrations were 379 ppm in 2005. The best estimate of total CO_2-eq[uivalent] concentration in 2005 for all long-lived GHGs is about 455 ppm, while the corresponding value including the net effect of all anthropogenic forcing agents is 375 ppm CO_2-eq" (IPCC, 2007a: 67).

10. The author of Chapter 3 computes implications only for the impacts on potential production of rainfed agriculture for any particular crop, and only for land currently under cultivation. He does not provide estimates of the implications for irrigated and total land, whether currently under cultivation or not.

potentials under the assumption that climate change was occurring today. But estimates of future impacts of climate change should include the option that novel varieties may be developed that can better withstand the new climatic stresses. This is an important issue, especially when considering options for responding to climate change, such as optimal combination of mitigation and adaptation actions.

Impacts estimated in Chapter 3 refer to changes in production potentials. Climate change would have only a minimal impact on global quantities, lowering them by between 0.2 percent to 1.4 percent (Table 3.16). Effects on projected cereal prices would also be small, ranging from -1 percent to + 10 percent at worst, under the Hadley A2 scenario without CO_2 fertilization. These projected changes would occur with cereal prices in 2050 being 31 percent above pre-surge levels as per baseline (Figure 11.4, see also Figure 11.7).

Figure 11.7
Effects of climate change on cereal prices in 2050: Chapter 2 versus Chapter 3

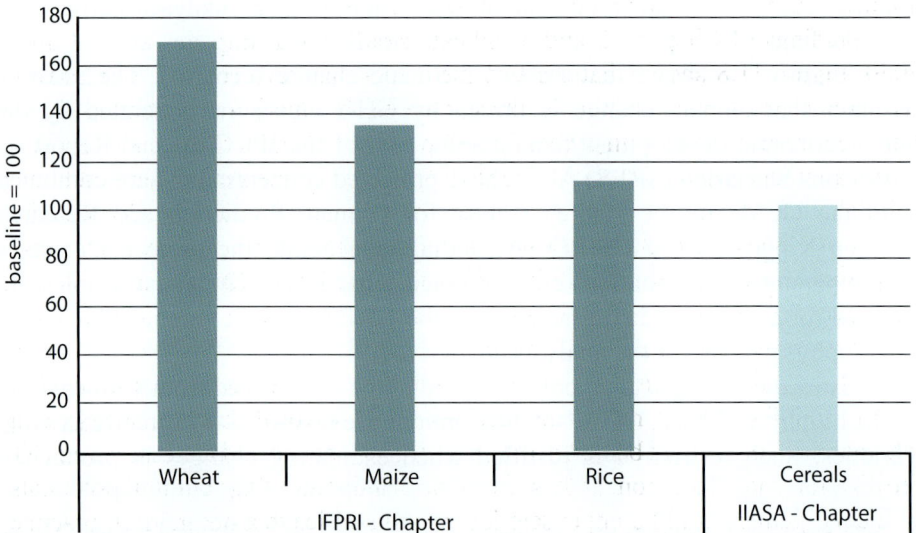

Sources: Nelson *et al.*, 2009: Table 2; Chapter 3, Table 3.15.

Findings on climate change impacts reported in Chapter 2 are those reported in Nelson *et al.* (2009). They are based on the IPCC SRES A2 scenario, which is combined with GCMs from NCAR (United States of America) and CSIRO to derive changes in the climate variables that affect production potentials. One of IFPRI's scenarios is "CSIRO with CO_2 fertilization" and "autonomous adaptation as farmers respond to changing prices with changes in crop mix and input use"

(Nelson *et al.*, 2009: 6).[11] Therefore, the results can be compared with those of Chapter 3 for the same scenario: prices in 2050 are projected to be 70 percent higher than the baseline for wheat, 35 percent higher for maize, and 12 percent higher for rice (Figure 11.7) (Nelson *et al.*, 2009: Table 2). This is far more than the 2 percent projected in Chapter 3; hence some structural fundamentals differ widely in the approaches of IFPRI and IIASA. One of these differences may be that in Chapter 2 the most severe declines in potential yields occur in irrigated wheat and rice, while Chapter 3 reports effects for rainfed production potentials only.

Impacts of climate change by region

Climate change impacts can differ widely among latitudes, regions, countries and agro-ecological zones. Outcomes for the world as a whole are made up of pluses and minuses from different regions, countries and country groups. Increased production potentials in the parts of the world benefiting from climate change are not necessarily available to make up losses suffered in other parts.

Findings of Chapters 3 and 5 indicate small global impacts, at least up to 2050. Figure 11.8 shows that the two most food-insecure regions, sub-Saharan Africa and South Asia, suffer declines in cereal production potentials of 7 and 6 percent respectively, with a peak of -32 percent in Southern Africa. With CO_2 fertilization and adaptation most countries/groups become net gainers, including sub-Saharan Africa and South Asia. Such gains disappear in later decades, but do not turn negative globally (Table 3.12). Full gains from CO_2 fertilization and adaptation are uncertain, hence the outcome could be less rosy than it appears. Moreover, as recalled these estimates refer to cereals production potentials only, and not on projected production; and they refer to rainfed land currently cultivated.

The possibility of a net gain in production potentials at the world level, at least for rainfed cereals, raises the interesting and provocative question regarding whether the world could be better-off with climate change and the associated CO_2 fertilization and adaptation, at least from the standpoint of agricultural potentials. In theory, gainers could compensate losers and still leave a net gain. In practice, however, this looks unlikely. Losers could only gain if the improvement in global potential could be translated into increases in their consumption. The most food-insecure region, sub-Saharan Africa, is projected to suffer a marginal decline in 2050 cereals consumption compared with the baseline, despite the finding that

---

11. This seems to be similar to the adaptation concept used in Chapter 3, but may not be. As noted, Chapter 3 speaks of farmers shifting to crop types that are better able to withstand the new climatic stresses. Chapter 2 speaks of farmers adapting their crop mixes and input use in response to changing prices.

its cereals production capacity could increase by 1 percent under this scenario (Hadley A2 with CO_2 fertilization and adaptation). A stronger negative impact is projected for the other food-insecure region, South and Southeast Asia (Tables 3.16 and 3.17), with production down by 3.7 percent and consumption by 1.0 percent.[12]

Figure 11.8
Chapter 3: effects of climate change on rainfed cereals production potential from currently cultivated land in 2050

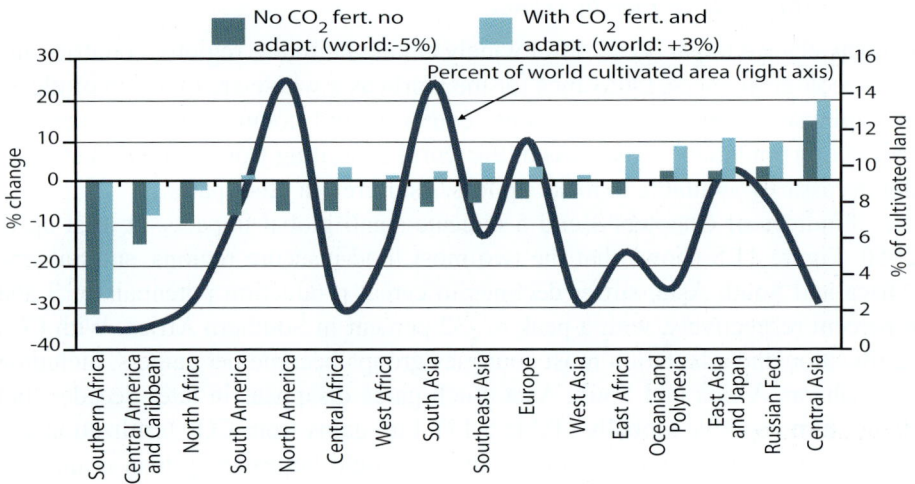

Source: Chapter 3, Table 3.11.

Projections in Chapter 2 indicate that global cereal production in 2050 could be 15 to 17 percent lower than in the baseline (Nelson *et al.*, 2009: Table 3). This implies substantial declines in some regions. For example, in Europe and Central Asia production is expected to decline up to 51 percent for wheat and 38 percent

---

12. Impacts on rainfed production potentials (Table 3.12) are given in greater detail (by country/country group) than the economic model projections (Tables 3.16 and 3.17). It is therefore not clear how changes in potentials are related to changes in projected production and consumption. The evaluation of impacts on production potentials refers only to rainfed production on currently cultivated land, so there need not be a close relationship between such potentials and the production projected by the economic model. In the latter, production also comes from irrigated land , that is particularly important in South Asia; and from expansion on to currently uncultivated land, that could be important in sub-Saharan Africa.

for maize.[13] This finding requires further investigation, as this region includes areas like the Russian Federation and Central Asia that in the estimates of Chapter 3 are shown to benefit the most from climate change, with increases of 3 and 14 percent, respectively, for rainfed cereals production potential (Table 3.11).

Figure 11.9
IFPRI: effects of climate change on per capita food availability in 2050

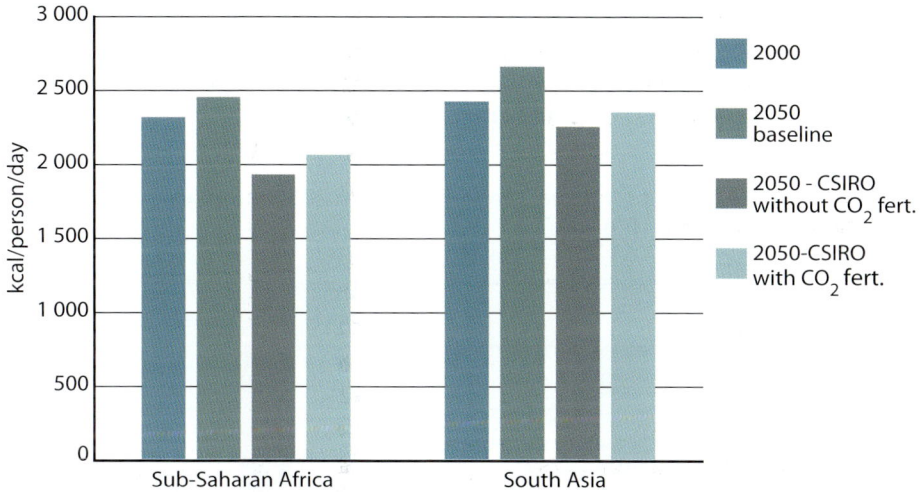

Source: Nelson *et al.*, 2009: Table 5.

Findings of Chapter 2 for the two most food-insecure regions are dire. Negative impacts on crop production are pronounced in sub-Saharan Africa and South Asia. In sub-Saharan Africa rice, wheat, and maize yield may decline compared to the baseline by 15 percent, 34 percent and 10 percent, respectively. And in South Asia climate results in a 14-percent decline in rice production, a 44- to 49-percent reduction in wheat production, and a 9- to 19-percent reduction in maize. (Nelson *et al.*, 2009: 6 and Table 3). For both regions, the modest gains projected in the baseline are more than reversed in the scenarios with climate change (Figure 11.9): food availability would be lower not only compared to the baseline, but also compared to the starting point in year 2000.[14] This is a

---

13. These findings refer to the NCAR and CSIRO scenarios without CO_2 fertilization but with adaptation in Chapter 2, as defined earlier.

14. "Calorie availability in 2050 will not only be lower than in the no-climate-change scenario – it will actually decline relative to 2000 levels throughout the developing world" (Nelson *et al.*, 2009: vii and Figure 5).

really catastrophic impact, given that in 2000 these regions had very low food availability in terms of calories per person per day. It raises the question as to how credible it is that India's star economic performance, which dominates that of South Asia, will go into reverse because of climate change impacts on its agriculture, as implied by the finding that South Asia may have lower per capita food consumption in 2050 than in 2000.

As seen, IIASA projections in Chapter 3 are less pessimistic. The only feasible direct comparison with Chapter 2 is between impacts on consumption for cereals under the CSIRO with CO_2 fertilization scenario for sub-Saharan Africa and South Asia (Figure 11.10). Impacts in sub-Saharan Africa are minimal in estimates from Chapter 3, but quite large in Chapter 2.

Figure 11.10
Effects of climate change on cereals consumption in 2050, sub-Saharan Africa and South Asia (baselines = 100)

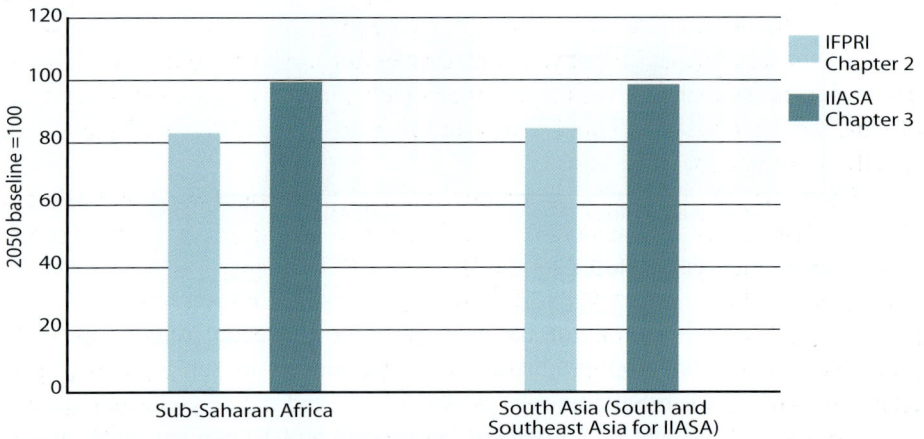

Effects are from scenarios with GHG from SRES A2 and climate from CSIRO with CO_2 fertilization. IFPRI data are for food consumption; IIASA data are for all uses. Non-food use of cereals is small in both regions, so the difference in definitions should not distort comparisons unduly.
Sources: Nelson *et al.*, 2009: Table 4; Chapter 3, Table 3.17.

Discussion of climate change impact

The reasons why climate change impact evaluations vary so much need to be better understood. The large differences among models suggest that much work is still needed before well-grounded views can be formed on the significance of climate change for agriculture and food security. The reasons for these differences lie in a combination of the data, coefficients, exogenous assumptions, databases and models. Differences in the representation of biophysical environments could play a role in explaining why results are so different. They affect the production potentials that are used as inputs into the economic models and therefore play a key role in the projections of production, consumption, prices.[15]

Easterling *et al.* (2007: Table 5.6) give an idea of the relative importance of biophysical versus economic models in determining the differences. They present projected climate change impacts from the application of the same economic model employed in Chapter 3 to two crop modeling systems: one application by Fischer, using the GAEZ and one by Parry using the DSSAT, which is the same crop model employed by IFPRI in the analysis reported in Chapter 2. Impacts are given only for the projected numbers of people "at risk of hunger" under alternative SRES scenarios and the Hadley climate (GCM) model with CO_2 fertilization. The impacts on the numbers at risk of hunger for 2050 are minuscule compared to the baseline, and nearly identical whether the IIASA economic model is coupled with the GAEZ or the DSSAT crop-modelling system. This suggests that the reasons why impacts of climate change differ so much between Chapters 2 and 3 are more likely to be found in the economic models. However, more needs to be understood about both economic and biophysical models before any firm conclusions can be drawn. It is noted that much of the decline in yield potentials in the IFPRI analysis of Chapter 2 occurs in irrigated wheat and rice, for which Chapter 3 reports no specific information.

There may be a bias in the comparisons of projections with and without climate change. Both Chapters 2 and 3 estimate scenarios assuming future GHG concentrations generated by the IPCC SRES A2 scenario which has world population of 11.3 billion in 2050[16]. Impacts are then derived by comparing results of such scenarios with those of the baselines which use a much lower world population in 2050 - 9.1 billion. This means that climate impacts originating in GHG concentrations of 536 ppm in 2050 are compared with baselines that, due to lower population, would generate GHG concentrations of less than 500 ppm.[17]

---

15. "The simulated crop-price changes in response to climate change are quite moderate due to the relatively small net global impact on crop-production potential" (italics added; Fischer, Shah and van Velthuizen, 2002: 108).

16. The identity commonly used to convey the relationship is: impact = population × affluence × technology (Nakicenovic and Swart, 2000: Chapter 3). SRES A2 assumes high population growth and slow technological change: both tend to magnify the climate impact.

17. An IPCC scenario with 2050 population of 9.3 billion (scenario SRES B2) generates GHG concentrations of 478 ppm (IPCC, 2007b: Table 1). Naturally, the lower GHG concentrations in SRES B2 are due not only to the lower projected population but also to differences in other scenario variables, such as developments in the energy efficiency and carbon intensity of growth, more equitable development, and a more environment-conscious policy context. (See scenario storylines in Nakicenovic and Swart, 2000: Chapter 4.).

Analyses may thus tend to overstate the impact of climate change because, if the baselines had been projected with the higher population, the demand would have been higher and so would the projected prices. As a consequence the difference in projected prices between the with- and the without climate change scenarios would have been smaller.

Indeed, in another IIASA study using a projected 2050 population of 11.1 million, the difference in cereal prices results as 14.4 percent for 2080 (Fischer, Shah and van Velthuizen, 2002: Tables 4.1, 4.10; it has projected prices only for 2080), i.e. smaller than the 23 percent estimated for the same year in the present IIASA paper (Table 3.15). In both cases, the climate change effects are those of the Hadley GCM with CO_2 fertilization. Fischer, Shah and van Velthuizen (2002: Tables 4.1,4.10) estimated also scenarios with a 2050 population of 9.3 billion and GHG of 478 ppm originating from the same 9.3 billion population of scenario SRES B2: they project 2080 prices with climate change to be 2.1 percent higher than the baseline without climate change. In conclusion, the discrepancy in the population assumptions between the baseline projections and the scenarios that generate GHG concentrations can be at the origin of an upward bias in the estimation of climate change effects in Chapter 3. Mutatis mutandis, the same may apply to the analyses of Chapter 2.

Concerning food security, the analyses presented in previous chapters analyze the effects of climate change mainly via its impact on potential production in agriculture, and the related changes in food consumption. This would be acceptable if changes in food consumption depended only upon changes in production potentials. But climate change may affect food consumption and food security through several channels. First, impacts on production may have implications for incomes generated both in agriculture and other sectors. Poverty, and hence consumption and food security can be affected by climate change via this feedback link, especially in countries where agriculture accounts for significant shares of GDP and employment. Such feedback effects may be accounted for in the projection generated with general equilibrium models, such as IIASA's economic model described in Chapter 3 and the World Bank model described in Chapter 5. IFPRI's IMPACT model, employed in Chapter 2, is partial equilibrium, and the consumption projections probably do not account for such feedback effects. If they did, projected climate change impacts on consumption may have probably been lower than reported (Figure 11.9).

Second, climate change may affect economic growth, incomes, poverty and food security via other phenomena, such as sea-level rise, population dislocations, destruction of infrastructure, natural catastrophes from extreme events, altered biodiversity, and growing incidence of disease.[18] These impacts may be particularly relevant for developing countries. Many of these countries are located in latitudes with temperatures close to thresholds beyond which even small increases can have large effects (Cline, 2007: 41). They are less equipped to respond, and their high dependence on agriculture enhances their vulnerability.[19] This suggests that projections of consumption and food security outcomes under climate change scenarios should account for the lower GDP growth, along with reductions in agricultural production potentials. Among those employed in previous chapter, only the World Bank model (Chapter 5) seems to have the potential for doing so. The GDP and poverty projections reported in Chapter 5 account for climate change damages, but only those operating via lower agricultural productivity. Should other damages be considered in the simulations, the macroeconomic and poverty projections for developing countries may become somewhat less optimistic.

In general, two visions of world futures with climate change seem to emerge from the preceding discussion. The first is one of a significantly more prosperous world in 2050, with reduced poverty and food insecurity, in which climate change affects the rate of development by reducing the growth of agricultural productivity. This position of Chapter 5 is also echoed in contributions such as Easterling *et al.* (2007) and Schmidhuber and Tubiello (2007). The second vision is of a world of falling per capita food consumption and growing food insecurity, as depicted in Chapter 2, in which climate change makes the world poorer than today, as implied by falling food consumption caused by sharp declines in crop yield potentials, and assuming no corrective action is taken.

FAO's long-term assessments and projections have not so far addressed climate change impacts. This is a gap that must be filled. Based on the above discussion, the following preliminary indications can be put forward to orient future work.

Firstly, the new GAEZ should provide a sound basis for estimating climate change impacts on production potentials. More efficient way should be sought, however, of using them in economic models.

Secondly, if the objective of FAO long term scenario work is to generate a depiction of the likely future, assuming that present climate will prevail in the baseline might not be a realistic approach, since this might well be an impossibility, given that the consensus is that temperatures are going to rise anyway.

---

18. On issues concerning total costs of climate change see Mendelsohn, 2007; Weitzman, 2007; Economist, 2009.

19. "Estimates are that they would bear some 75 to 80 percent of the costs of damages caused by the changing climate. Even 2 °C warming above preindustrial temperatures – the minimum the world is likely to experience – could result in permanent reductions in GDP of 4 to 5 percent for Africa and South Asia" (World Bank, 2010: 6, Box 1).

Thirdly, scenarios with climate change must be based on values for the exogenous variables (e.g. population growth) similar to those used in the baseline. Perhaps the main difference from the baseline could be specified in terms of volumes of emissions and the associated GHG concentrations in the future. The latter could be related to assumptions about mitigation policies.

Finally, given that climate change affects not only agricultural potentials but also the overall development prospects, income growth rates used to project food demand should not be maintained as exogenous variables. Feedbacks from climate change via effects on agriculture must be brought to bear on the GDP growth assumptions.

Impact of biofuels

In this volume, only Chapter 3 proposes detailed projections on biofuels. Chapter 2 makes references to biofuels, and shows projections of crops used for biofuels as part of IFPRI's baseline scenario, but does not indicate what difference biofuels can make to projected levels of other variables, such as prices and per capita food consumption. Chapter 5 underlines the potential impact biofuels development on agricultural prices, but the agricultural variables in the baseline projection, such as the productivity growth and prices, do not account for biofuels (van der Mensbrugghe *et al.* personal communication). FAO's long-term agricultural projections proposed in Chapter 1 do not include biofuel use of crops (FAO, 2006)[20] because the projections were prepared from 2003 to 2005 using historical data up to 2001, when such use was not a major issue. However, Chapter 1 does show biofuels projections to 2018 from FAO's more recent work (OECD/FAO, 2009).

Several scenarios of possible biofuels outcomes are analyzed in Chapter 3. These seems to be based on IIASA's work for the Organization of the Petroleum Exporting Countries Fund for International Development (OPEC/OFID) (Fischer *et al.*, 2009), and constitute variants around the reference projection of the International Energy Agency's (IEA's) 2008 World Energy Outlook (WEO) (IEA, 2008) (Box 11.1, based on Table 3.26).[21]

---

20. Unknown amounts of cereals, sugar cane and vegetable oils used for biofuels were included as part of the more general non-food industrial use category of these products (there were no separate biofuels use statistics at the time).

21. There is some confusion as to which scenario is the baseline. Chapter 3 indicates that it reports "a baseline projection without any use of agricultural feedstocks for biofuel production, as portrayed in the FAO-REF-00 scenario ..."; but in Table 3.3 it presents cereals projections based on scenario FAO-REF-00 while describing the projections as being from "baseline simulation without climate change and biofuel expansion" (scenario FAO-REF-01). As there is no cereals projection with zero biofuels, the comparisons in this chapter use as baseline the FAO-REF-01 scenario, which assumes that cereals use for biofuels remains constant at the 2008 level of 83 million tonnes.

In order to examine what previous chapters project for biofuels and what consequences this has for production, consumption and prices of agricultural products, we start from the projections to 2020, and compare them with those for 2018 reported in OECD/FAO (2009) (Figure 11.11). Cereals use for biofuels was 15 to 20 million tonnes in 1999/2001 and grew rapidly to 105 million tonnes in 2008, or 4.8 percent of world consumption of cereals. The OECD/FAO projections use these data in their starting years. Chapter 3 assumes that cereals use for biofuels in 2000 was 83 million tonnes, a level achieved only later.

Figure 11.11
World cereals use for biofuels, 2018 (OECD/FAO) and 2020 (Chapters 2 and 3)

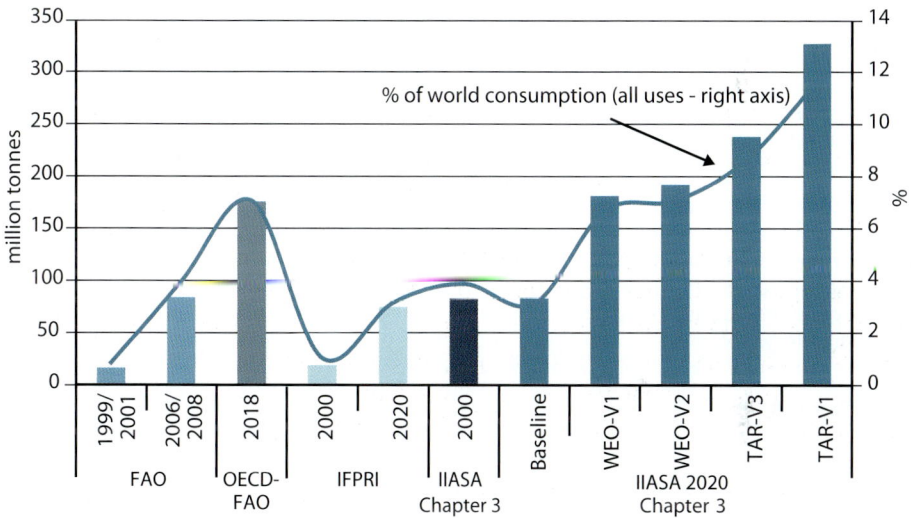

Sources: OECD/FAO, 2009; IFPRI: detailed tables for Chapter 2 (supplied separately by the authors); Chapter 3.

The OECD/FAO projection foresees further rapid growth in cereals-based ethanol production, although slower than in the recent past. This is driven mostly by policies, especially the United States Energy Independence and Security Act, and the European Union (EU) Renewable Energy Directive, which together account for the bulk of the projected cereals use for ethanol. Chapter 2 projects a much lower biofuel use in 2020, at 75 million tonnes, which is less than the actual quantities of 2008.[22] In the five scenarios of Chapter 3 (Box 11.1) biofuel use in 2020 ranges from 83 million tonnes to 327 million tonnes in scenario TAR-V1,

---

22. Quantities for biofuels are included in IFPRI's Baseline in the total demand for cereals shown in Table 2.1.

assuming full achievement of the mandates and targets and slow penetration of second-generation biofuels. The bulk of ethanol production in this scenario comes from cereals and sugar cane. OECD/FAO projection for 2018 is similar to those of two Chapter 3 scenarios for 2020. This probably reflects both studies' reliance on IEA's (2008) energy sector projections.

Figure 11.12 compares projections from Chapters 2 and 3. The projected share of biofuel in cereal use peaks in 2030 under the high-biofuel scenario of Chapter 2 (TAR-V1); but there is little further growth after 2030 even in this scenario, given that second-generation biofuels are assumed to kick in after 2030.

Box 11.1 - IIASA biofuels scenarios

There are five scenarios:

1. Baseline scenario: Cereals use for biofuels is kept constant in all projection years at the level used in 2008 (83 million tonnes). This level is assumed to have existed also in 2000, when actual use was only 15 to 20 million tonnes. So this is a problem when seeking to compare projections with historical data.

In the other four scenarios, the basis is the reference scenario projection of total demand for transport fuel of WEO 2008: demand for transport fuel is projected to increase by 40 percent from 2006 to 2030. IIASA extends the projection to 2050, implying a further rise of 18 percent from 2030 to 2050. From this WEO scenario, the other IIASA scenarios are generated by assuming alternative rates of: i) achievement of the biofuels mandates (mandatory, voluntary or indicative) in different countries; and ii) progress in second-generation biofuels (Chapter 3, Tables 3.26 and 3.33).

2. Scenario WEO-V1: Same as WEO reference scenario with extension of transport fuel to 2050 (as above). Biofuels mandates in the different countries are partly achieved (as in the WEO reference scenario) and the share of biofuels in total transport fuel rises from 1 percent in 2006 to 3.5 percent in 2020, 4.2 percent in 2030, and 6.0 percent in 2050 (total transport fuel is 2 830, 3 171 and 3 750 million tonnes of oil equivalent [TOE] in the three years; Chapter 3, Table 3.21). Second-generation biofuels have a slow start: they contribute 3 percent of total biofuels in 2020, 13 percent in 2030 and 30 percent in 2050. This scenario generates the lowest projection of cereals use for biofuels. This chapter uses the term "low-biofuels" scenario for easy reference in further discussions.

Impact on prices and quantities

The very rapid growth of cereals use for biofuels in recent years is thought to have been a key demand-side factor that contributed to the price surges (Alexandratos, 2008; Mitchell, 2008). It is therefore interesting to consider what the analyses in Chapters 2 and 3 imply in terms of prices. Figure 11.13 shows price effects of the high- and low-biofuels scenarios. The left side of Figure 11.13 shows that projected 2050 prices could be from 10 percent to 27 percent higher under the low-biofuels and high-biofuel scenarios respectively, compared to a baseline in

Figure 11.12
IFPRI and Chapter 3 projections of cereals use for biofuels, all years

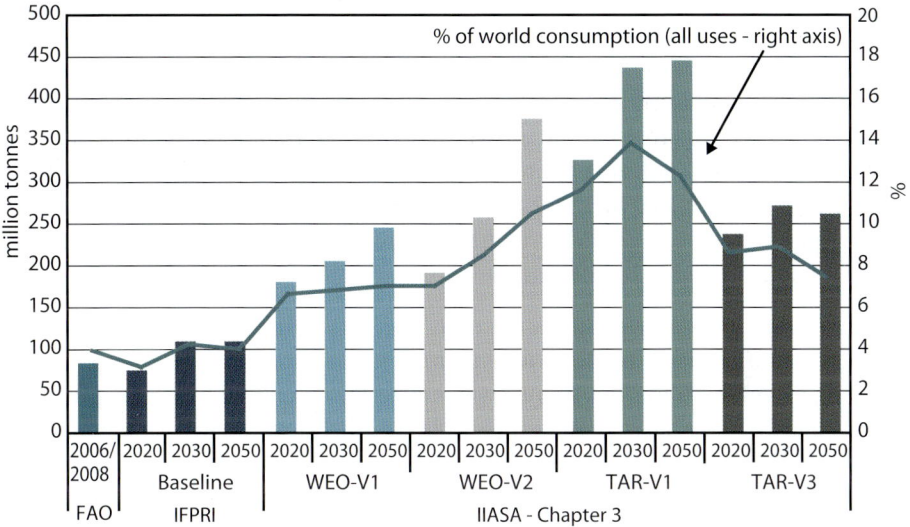

Sources: IFPRI: detailed tables for Chapter 2 (supplied separately by the authors), Chapter 3.

Figure 11.13
Effects of biofuels on cereal prices – Scenarios in Chapter 3

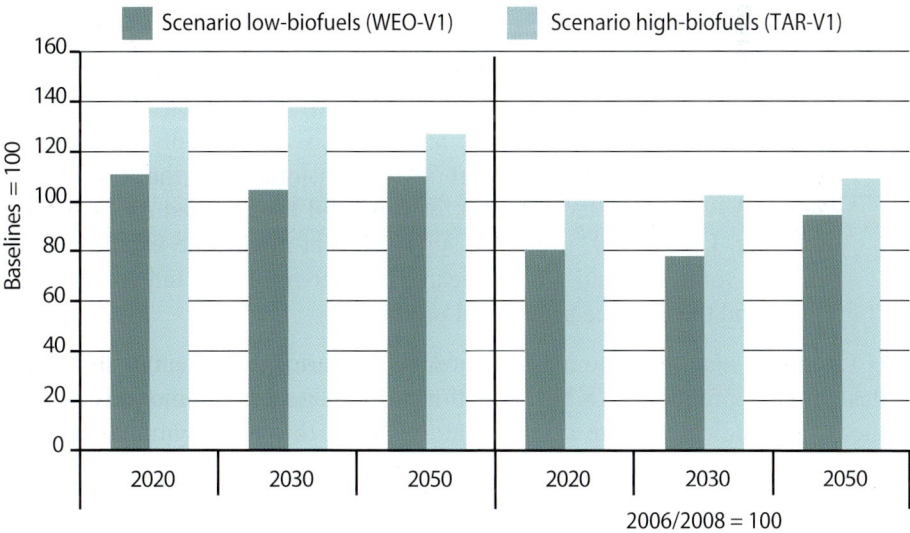

Source: Chapter 3, Tables 3.4 and 3.27.

which biofuels use of cereals is constant at the 83 million tonnes of 2008. The right side of Figure 11.13 rebases these price projections to the levels reached in the surge years (2006/2008 = 100). It shows that the high-biofuel scenario implies that prices in 2050 could be 9 percent above those reached in the surge years. Altogether, the impact of biofuels on cereal prices could be substantial, according to projections of Chapter 3.

Chapter 2 refers to what the growth rate of prices would have been if cereals use for biofuel had not accelerated from 2000 to 2007, and had grown at the same rate as between 1990 and 2000. The authors conclude that the growth rate of average grain prices would have been 30 percent lower than the actual ones. Based on data available today, this conclusion implies the following.

1. The actual growth rates of cereals use for biofuels were 6.0 percent per annum for 1990 to 2000, leading to 18 million tonnes of such use in 2000 (of which 16 million tonnes was maize in the United States of America); and 24.7 percent per annum for 2000 to 2007, leading to 85 million tonnes in 2007 (of which 77 million tonnes was maize in the United States).[23] If cereals use for biofuels from 2000 to 2007 had grown at 6.0 percent per annum as in the 1990s, as implied by IFPRI's counterfactual scenario, biofuel use in 2007 would have been 27 million tonnes, i.e., 58 million tonnes less than the actual 85 million tonnes.

2. If prices grew 30 percent less than the actual 7.0 percent from 2000 to 2007 per annum, they would have grown at 4.9 percent per annum. The World Bank cereals price index in 2007 would have been 140 instead of the actual 160, i.e., prices would have been 13 percent lower in 2007. The reduction of cereals use for biofuels by 58 million tonnes in this counterfactual scenario would have led to a reduction of aggregate demand for cereals of less than 58 million tonnes, as the lower prices would have stimulated demand for food and feed. Based on findings in Chapter 3, however, it can be surmised that about one-third of the 58 million tonnes would have appeared as increased demand for food and feed, and the balance as reduced production.

Chapter 2 also refers to another biofuels scenario, that assumes the United States target of producing 15 billion gallons of first-generation biofuels by 2022 is met.[24] No indications are given regarding what this would imply for cereal

---

23. The growth rate of the 1990s is for United States use of maize for ethanol, as there was little cereals use for biofuels outside the United States. Data from 2000 to 2007 are from the OECD/FAO (2009) files.

24. The 15 billion gallons is the target for 2015, which continues at same level until 2022.

quantities used for biofuels, and for prices. The authors argue that the additional quantity of maize going to biofuels would be considerable and the effects on food security would be negative. If such effects were to be offset by additional production through accelerated yield growth, they estimate that average world cereal yields would have to grow at 1.8 percent per annum from 2000 to 2030, up from the 1.3 percent per annum of the baseline.

The projected higher prices following increased use of cereal in biofuel production have the potential for depressing demand for food and feed. It is interesting to consider what previous chapters indicate on this matter. Chapter 3 shows projected levels of food, feed and biofuel use of cereals under different scenarios (Figure 11.14). In 2030, for instance, food/feed consumption would fall from 2 845 million tonnes in the baseline scenario (when biofuels use was assumed to be only 83 million tonnes) to 2 712 million tonnes in the high-biofuels scenario, or to 2 800 million tonnes in the low-biofuels scenario. The high-biofuels scenario implies that of the total increase in cereals use for biofuels in 2030, 38 percent will come from reduced food/feed consumption and 62 percent from increased production. Figure 11.15 illustrates these outcomes in more detail. Overall, more of the incremental use of cereals for biofuels would come from increased production.

Figure 11.14
Cereals consumption: biofuels scenarios in Chapter 3

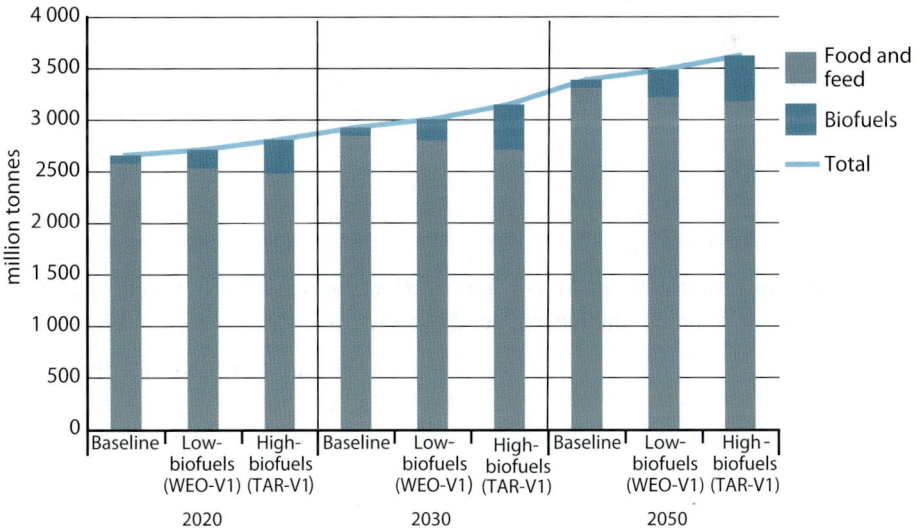

Source: Chapter 3.

489

Figure 11.15
Increments of cereals use for biofuels: scenarios in Chapter 3

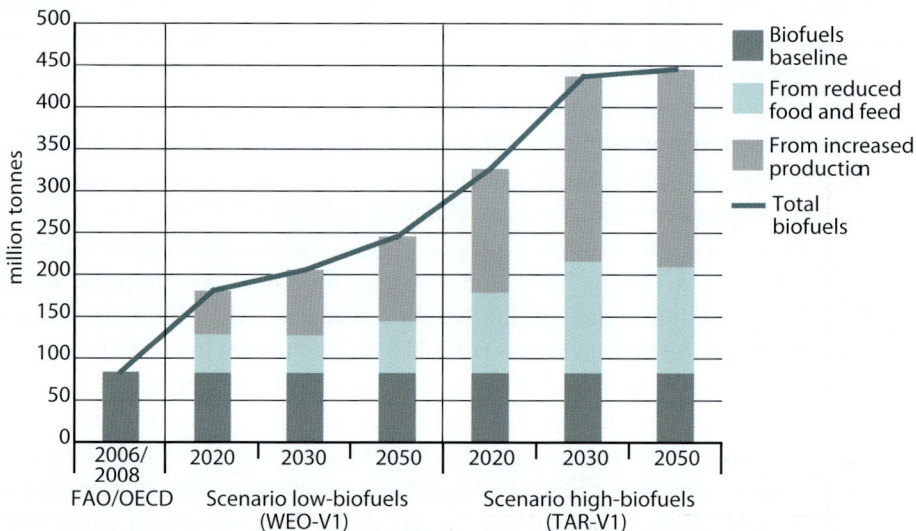

Source: Chapter 3.

It would be interesting to further analyze the implications of this scenario in terms of per capita food consumption and nutrition. Figures 3.6 and 3.7 imply that about 30 percent of the total decline in combined food/feed use comes from food, mostly in developing countries, and the rest from feed, mostly in developed countries. Chapter 3 implies but does not show an estimated effects on per capita food consumption in terms of kcal/person/day from all food, not only cereals, because it shows impacts on the numbers of people at "risk of hunger", which depend on the values of the kcal/person/day.

Biofuels and land use

Scenarios discussed in Chapter 3 provide estimates of the extent to which cultivated land – that is "arable land" and "land in permanent crops" in FAOSTAT terminology would need to expand beyond what is envisaged in the baseline to produce the crop feedstocks required by the biofuels industry. Figure 11.16 shows this under different scenarios, along with land-use projections from FAO (Chapter 6, Table 6.7). At first glance, biofuels seem to add little to total projected land use. However, increases in land for biofuels do not seem to include land required for the production of lignocellulosic biomass for second-generation biofuels embodied in projections. The author estimates that some 50 million ha of land would be required in 2030 for the production of second-generation feedstocks under scenario TAR-V1.

Figure 11.16
Cultivated land

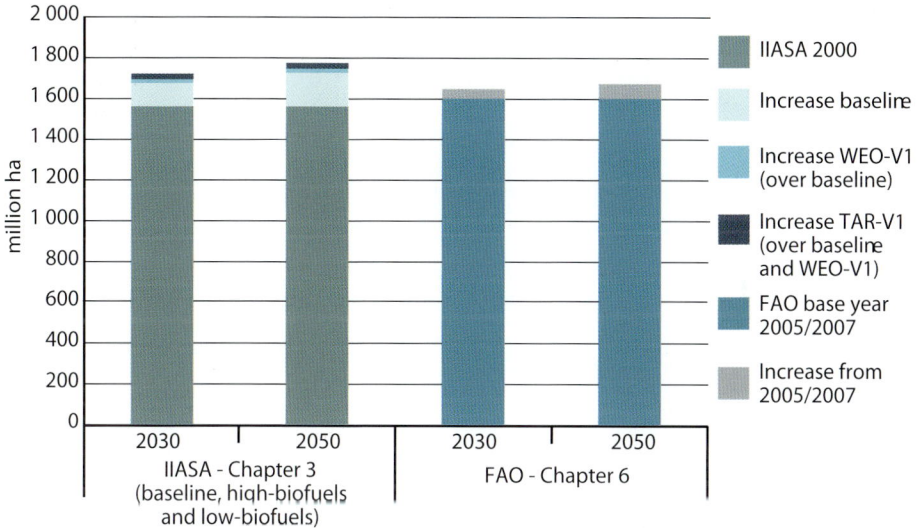

Sources: Chapter 3, Tables 3.7 and 3.31; Chapter 6, Table 6.6.

Are the projected land expansions in the baseline or for increased biofuels feasible? Chapter 3 does not address this issue directly by providing estimates of land with crop production potential into which the expansion could be made. In Chapter 6, the assessment of land with rainfed crop production potential in various agro-ecological suitability classes suggests that this may amount to 4.2 billion ha for the world as a whole (Table 6.6). This information comes from the old IIASA GAEZ, and includes all types of potentially cultivable land, including land under forest, in protected areas and in human settlements. The part that is really available for expansion without deforestation and disturbance of protected areas is much smaller. About 1.6 billion ha are currently used for crop production. This is approximately the same figure that IIASA uses for 2000 in Chapter 3. Projections in Chapter 6 show a small increase in land used for crop production by 2050, of 70 million ha. This contrasts sharply with the projection reported in Chapter 3, which points to a 166 million ha increase in the baseline scenario, implying smaller increases in yields and/or cropping intensities than Chapter 6.[25] Figure 11.17 shows the classification of all land surface globally according to Chapter 3.

Figure 11.17
Land areas by use and suitability for lignocellulosic feedstock crops (billion hectares)

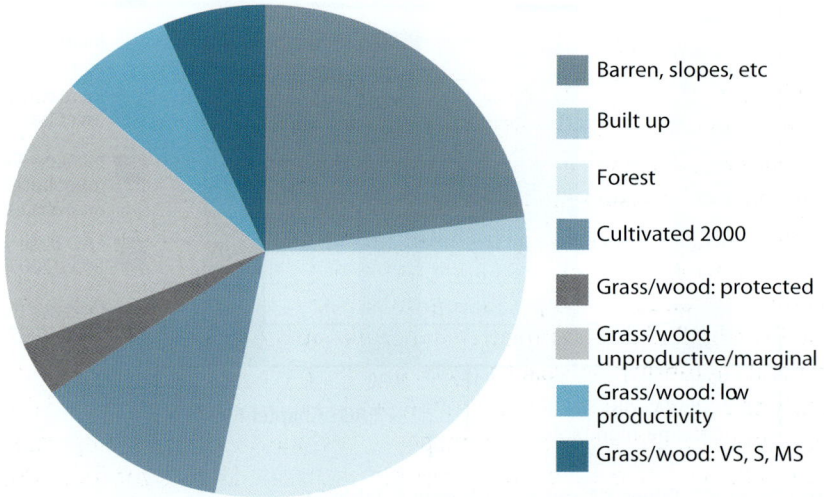

Suitability classes for lignocellulosic feedstock production: VS = very suitable; S = suitable; MS = moderately suitable.
Sources: Chapter 3, Table 3.37; Fischer *et al.*, 2009: Table 2.7-3e.

The same Chapter 3 does provide an estimate of land "potentially usable" for production of lignocellulosic biofuel feedstocks (Table 3.37). This is apparently based on the OFID work in Fischer *et al.* (2009: Table 3.6-7), using the updated 2009 GAEZ. The authors derive a figure of 1.75 billion ha of grassland and woodland with potential for lignocellulosic feedstocks; however, the OFID report (Fischer *et al.*, 2009: Table 2.7-3e) classifies only 860 million ha of this land as very suitable, suitable or moderately suitable for such feedstocks. Fischer *et al.* (2009) point out that much of this grass- and woodland may be used for livestock grazing, and only some 700 to 800 million ha may currently be available

25. Figures on cultivated land in Chapter 3 are similar to those of FAO and other studies, as they all come from FAOSTAT data for arable land and land under permanent crops; However cereals production in year 2000 looks higher in Chapter 3. This implies that the yields and/or cropping intensities derived from the interface between land and production may not provide a good basis for the projections on yield growth versus area expansion.

for lignocellulosic feedstock production. However, this is less comforting than it appears, as this same grass- and woodland may be the only area into which agriculture in general, and not only bio-fuel feedstocks, can expand in the future.

Finally, Chapter 3 suggests that competition for land may be limited, as production of feedstocks for second-generation biofuels are expected to be grown mainly outside cultivated land, and that some 100 million ha would be sufficient to achieve the target biofuel share in world transport fuels in 2050.

Discussion on biofuels

In analyzing biofuel development scenarios, increased attention should be devoted to the implications for food consumption and nutrition of a faster growth of agriculture generated by the expansion of biofuels. The potential negative impacts on food consumption, and by implication on nutrition, are not always discussed in sufficient detail. It is not always clear whether potential positive effects are accounted for, such as those associated with raising rural incomes and food consumption for the population groups who may benefit from higher prices arising from biofuel development. As noted also in the section on climate change, the general equilibrium approach may be more suitable than the partial equilibrium for capturing feedbacks from agricultural production to GDP and to food demand and nutrition. Future work on the impact of biofuels on food security cannot be limited to the negative effects of higher prices on the demand for food. Positive developmental impacts are an important aspect of the biofuels issue (ODI, 2009).

In terms of land resources, as seen Chapter 3 only addresses land-use and the suitability of an expansion to a limited extent. Even if there were plenty of land with production potential for lignocellulosic feedstocks, however, there can be no assurance that the cultivation of such feedstocks would expand only on to that land: it could also invade the land suitable for food crops, if economic realities so dictated. This means that the question as to whether the advent of second-generation biofuels will remove food versus fuel competition remains open.

Energy markets seem not to play a large role in the analyses of biofuels proposed in this volume; for example, the sensitivity of projections to oil prices is not addressed. The only link to energy markets is the one implied by adoption of the WEO 2008 reference scenario projection of total transport fuel to 2030, and the part of it to come from biofuels in IIASA's WEO-V1 scenario (Chapter 3). All other biofuels projections are derived from exogenous assumptions concerning the implementation of mandates and the shares of second-generation biofuels in total biofuels, with total transport fuel remaining constant[26].

If the future of biofuels is assumed to continue to depend predominantly on mandates, a gradual easing of competition with food and feed, and of the associated concerns about food security impacts, can be expected. This is because the role of mandates as the major driving force in biofuel development will tend to be gradually exhausted/weakened in the future: in the United States of America, the demand for ethanol created by the 10 percent blend mandate may soon be filled, and act as a constraint on further growth[27]; and with falling demand for transport fuel in the United States during the current crisis, the 10 percent mandate may not allow the absorption of the full 15 billion gallons of ethanol target for 2015 to 2022. Of more interest to long-term issues is the possible strengthening of links between the energy and agriculture markets, which may come about if the oil crunch that some foresee for the not-too-distant future were to materialize (Stevens, 2008; IEA, 2008).

Such an event would alter the economic fundamentals driving the biofuels sector, making it less dependent on mandates and other support policies, and more on market forces. In such circumstances, intensified competition for feedstock crops would tend to siphon off supplies and resources from the food sector, to the detriment of the food security of weaker population groups. Trade policies also have an important role, as they can help spread pressures on resources to countries with more resource endowments, such as in South America or sub-Saharan Africa. The eventual advent of second-generation biofuels may ease the impact, but will not eliminate it, because the production of lignocellulosic feedstocks may not be confined to land with no food crop production potential. In conclusion, analysis of the longer-term aspects of the food versus fuel issue requires that biofuel outcomes be explored in the context of alternative energy futures, and not only of alternative mandates and other support policies within a single energy future.

Demography, growth, inequality and poverty

Population data

Authors who use population projections in their quantitative analyses generally state that they are using those of the United Nations (UN), not always specifying which version. The UN has conducted four assessments over the last five years

---

26. There is no reference to the oil prices that are associated with such assumptions. Implicitly, the oil prices underlying the IIASA basic scenario (WEO-V1) are those of IEA's reference scenario: an average of USD 100/barrel over the period 2008 to 2015, rising to US 120/barrel by 2030 (in 2007 dollars; in nominal terms, 2030 prices are assumed to reach USD 200/barrel [IEA, 2008: 59]).

27. In jargon, it is about to hit the "blend wall".

between the one for 2002, published in 2004, and the latest of 2008. The world projected population, according to the medium variant projection, has changed in successive assessments, moving from 8.9 billion in the 2002 assessment, to 9.2 billion in the 2008 assessment.

At the global level, the differences in the projected 2050 population are not large enough to affect significantly the issues addressed in this chapter. However, it is important to note that the differences in the world totals reflect predominantly higher projected populations in sub-Saharan Africa (1 557 million in the 2002 assessment and 1 753 million in that of 2008). This is the region with the most severe problems of poverty and hunger. Obviously, 200 million more people in the region's population could have a significant impact on the prospects for improving food security. For example, the 2050 projected incidence of undernourishment of 5.8 percent (FAO, 2006: Table 2.3), when applied to a population that includes an additional 200 million people, would add nearly 12 million to the region's projected numbers of undernourished. Other factors also change when the demographic situation changes and the projected numbers of undernourished increase by more than the 12 million people, including changes resulting from revisions of the historical data and the parameters used in the estimation (Table 11.1).

Income growth and distribution

Economic growth can play different roles in projections exercises. Generally it enters the analyses in two distinct ways: i) as a result in economy-wide models; and ii) as exogenous assumption in partial equilibrium analyses. Two chapters in this volume adopt the first approach: these are Chapters 4 and 5; Chapter 2 adopts the second approach, and Chapter 3 is based on a general equilibrium model, hence in principle GDP is modified by feedbacks received from changes in the solution for the agriculture sector.

Chapters 4 and 5 specifically address the issue of how the world economy may grow to 2050, and how income differentials between developed and developing countries may evolve. The common theme is "convergence": per capita incomes in developing countries are projected to grow faster than those in developed ones, so by 2050 the income divide will have narrowed compared to the present, at least in relative terms, and at the level of the two large aggregates. The magnitude of the income divide, and how much it narrows, depends on how the starting situation is measured, as well as on the differential growth rates assumed for the two country groups (Figure 11.18).

Figure 11.18
GDP growth rates and ratios of per capita incomes between different country groups, 2005 to 2050

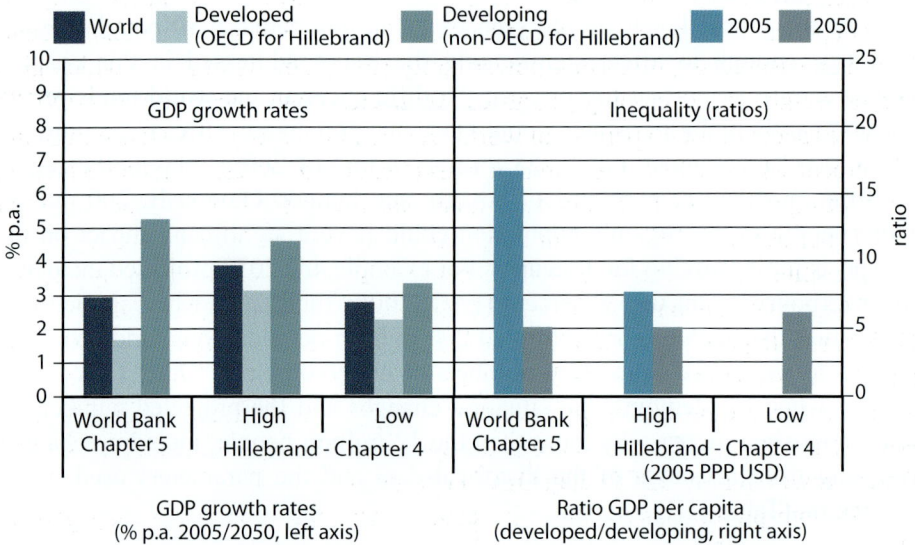

Sources: Chapter 4, Table 4.1; author's calculations from World Bank, 2009 and UN population data.

Chapter 5 uses market exchange rates to measure GDP. The low and middle-income countries (LMYs)[28] account for 21 percent of global GDP,[29] but for 84 percent of world population. The ratio of per capita GDP of the high-income countries (HICs) to that of the LMYs is 1:20. This is a huge divide, which has been narrowing, but very slowly: it was 1:22 in 1990. The working hypothesis in the World Bank baseline scenario is for GDP to grow at rates of 1.6 percent per annum in HICs, 5.2 percent for LMYs and 2.9 percent for the world. The ratio of

---

28. In the GDP scenarios, neither of the two chapters use the term "developed and developing countries". However, the authors of Chapter 5 use the term "developing" in its poverty estimates. The large country groups used in the GDP scenarios are largely overlapping but not identical. The World Bank classifies countries into "high-income" (accounting for 16 percent of world population) and "low and middle-income" groups (84 percent). (Country lists available from the World Bank's World Development Indicators website: http://data.worldbank.org/data-catalog/world-development-indicators). In Chapter 4, Hillebrand divides countries into OECD (14 percent) and non-OECD (86 percent) groups. (Hillebrand's OECD group comprises only those countries that were OECD members in 1981.) The UN uses the terms "more developed" (19 percent of world population in 2005 according to the 2006 UN population assessment) and "less developed" (81 percent). FAO (2006) uses the terms "developed" (22 percent) and "developing" (78 percent).

29. GDP data in constant 2000 dollars are from World Bank, 2010.

per capita GDP in LMYs'to that of HICs improves drastically: from 1:20 in 2005 to nearly 1:5 in 2050.

Chapter 4 addresses the same issue of economic growth to 2050, and presents two scenarios: a high one, called "market first", with world GDP growing by 3.8 percent per annum; and a low one, called "trend", with GDP growth of 2.8 percent. GDP is measured in purchasing power parity (PPP) dollars of 2005, which give higher weight to developing countries. With GDP thus measured, non-OECD countries account for 46 percent of world GDP in 2005, a huge difference from the 21 percent when GDP is measured at market exchange rates. As a consequence, the starting per capita GDP divide is much narrower: the ratio of non-OECD to OECD per capita GDP is 1:7.3 in 2005, falling to 1:5.3 or 1:6.1 by 2050.

Despite starting with huge differences in their global inequality measures, the two Chapters come to nearly identical conclusions: in 2050 the ratio of per capita GDP in rich to poor countries would be in the range of 1:5 or 6. This result reflects mainly the views embodied in the scenario hypotheses. The authors of Chapter 5 are upbeat regarding LMYs' prospects and rather conservative regarding those of HICs. The author of Chapter 4 are in the opposite direction, although both project faster growth in non-OECD than OECD countries.[30]

In both projections, declines in relative inequality imply widening gaps in absolute per capita incomes. People who are concerned about absolute gaps may see this as deterioration, but for those concerned with raising the income levels of low-income countries and fighting absolute poverty, the projections imply significant improvement, even if absolute gaps increase. In practice, achieving rising incomes for the poor often depends on the incomes of the rich growing in tandem, a point made by Hillebrand in Chapter 4. However, the link may be weaker in the future, following the emergence of growth poles in the non-OECD area.

As for poverty, in Chapter 5 this is measured as the number of people living on less than USD 1.25 per day in 2005 PPP dollars. Poverty is projected to virtually disappear in developing countries. Only sub-Saharan Africa and Latin America are projected to have measurable poverty rates in 2050, at 2.8 and 1.0 percent respectively. Hillebrand (Chapter 4) is less optimistic, but still projects significant declines in the high scenario. However, under his low scenario the poverty rate actually increases in sub-Saharan Africa, from the already very high 52 percent in 2005 (Figure 11.19).

---

30. Using market exchange rates rather than PPP to value and compare GDPs exaggerates the initial income gaps and can lead to erroneous conclusions regarding the implications of moving towards any given degree of convergence over the projection period. See "Hot potato" (Economist, 15 February 2003) and "Hot potato revisited" (Economist, 8 November 2003).

Figure 11.19
Poverty rates, population on less than USD 1.25/day

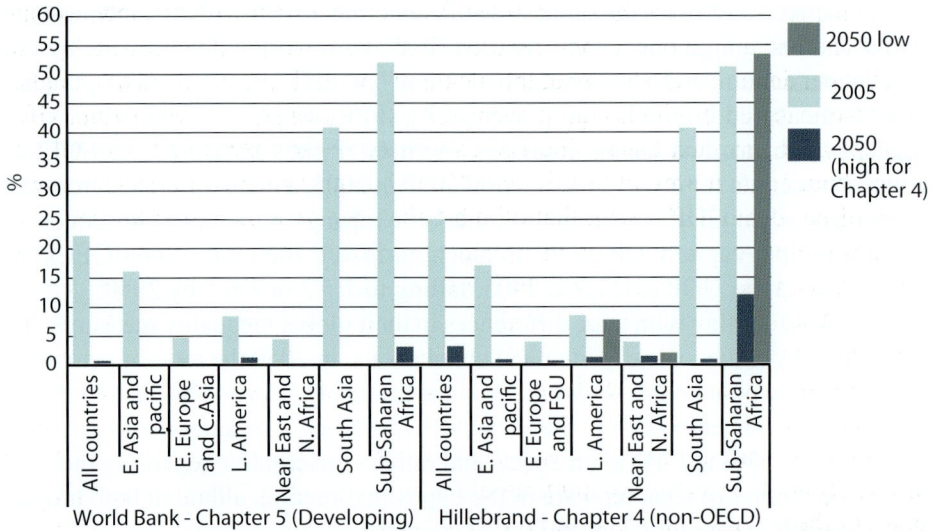

FSU = former Soviet Union.
Sources: Chapter 5, Table 5.8; Chapter 4, Tables 4.1 and 4.10.

Discussion on demography, growth and distribution

For the specific concern regarding food, agriculture and nutrition, the projected narrowing of the relative income divide should lead to an even sharper narrowing in the food consumption divide, because the latter, in contrast to income, is a bounded variable.[31] Chapter 4 explores the food and nutrition consequences of the projections of income, implied poverty and global inequality. In its "high" scenario, the non-OECD-to-OECD ratio of per capita food consumption (kcal/person/day) falls from 1:1.29 in 2005 to 1:1.16 (Table 4.9). This relatively modest change does not tell much in terms of food welfare and nutrition. Absolute levels of food consumption are the important factor. As reflected in Hillebrand's market first scenario, the outlook for 2050 is for food consumption in non-OECD countries to increase from an inadequate 2 662 kcal/person/day in 2005 to 3 135 kcal/person/day. This would be compatible with a significant reduction in undernourishment, but an adequate average across this large group of countries does not mean that every country, or even region, will be in the same happy position. Next section

---

31. A poor person must consume a minimum amount of food, while physiology limits the amount of food a rich person can consume.

points out that Hillebrand's food consumption projections for sub-Saharan Africa are probably too pessimistic.

This leads into the issue of likely rising inequalities within low-income countries. Chapter 5 states that the process of convergence between HICs and LMYs and declining inequality at the global level will be accompanied by higher within-country and within-region inequality. For example, the per capita GDP ratio of East Asia and the Pacific LMYs to that of sub-Saharan Africa will rise from about 1:2.5 at present to 1:5.7 in 2050.[32] Chapter 4 shows projected growth rates of per capita GDP, with that of sub-Saharan Africa being the lowest of all regions (Table 4.8). Moreover, its projected poverty rates and food consumption levels make it clear that there will be increasing differentiation, with sub-Saharan Africa making far less progress in these variables than the other regions in the non-OECD group.

A final point worth mentioning is the view, implicit in scenario projections, that world economic activity can continue to grow unimpeded by constraints imposed by natural resources and the environment. In Chapters 4 and 5 the world economy in 2050 is a multiple of that in 2005.[33] Authors do not explicitly address these constraints and the possible impacts on global growth. Chapter 4 refers briefly to the Club of Rome's Limits to Growth. Chapter 5 reports possible impacts of climate change on agriculture, and also speaks of exploring the impacts of rising energy prices on agriculture, without pursuing the matter. This can mean either that such constraints will not prove binding in the 45-year projection period, or – more likely – that in time-honoured fashion, a more prosperous world will be finding ways around such constraints, as and when they arise, reflecting a Julian Simon concept of economic progress (Simon, 1996). Hillebrand, in Chapter 4 explicitly recognizes that "Resource constraints, if not met by technological solutions, will surely make the poverty estimates shown here worse".

These broader issues cannot be addressed here, but it is worth considering the possibility of binding constraints emerging in the sphere of natural resources and the environment, when dealing with long-term processes involving huge increases in economic activity. The main binding constraints to solving the hunger problem may not be agricultural resources, but rather those that can stand in the way of ever-expanding global economic activity at rates sufficient to raise incomes and eliminate poverty in countries with food insecurity problems. This possibility was already being raised 15 years ago in FAO's 1995 edition of World agriculture: Towards 2010 (Alexandratos, 1995: 136–137).

---

32. "This ratio drops to six by 2050, but varies highly across regions, with a low of 3.5 in East Asia and the Pacific and a high of 20 in sub-Saharan Africa" (Chapter 5). Therefore, if YHIC/YSSA = 20 and YHIC/YEAP = 3.5, then YEAP/YSSA = 20/3.5 = 5.7.

33. Both chapters emphasize that their long-term projections are scenarios rather than forecasts.

Sub-Saharan Africa

The future of agriculture and food security of this region was the topic of separate sessions at the 2009 Expert Meeting on "How to feed to the world in 2050". Not all expert contributions provided quantitative projections on this region. Those that did had often disparate variables -- income growth rates, poverty rates, food consumption, numbers undernourished ¬ hence results cannot always be easily compared. The main variables on which results are available are shown in Table 11.1.

In Chapters 1 and 2 growth rates are exogenous, and Chapter 5 does not show GDP projections by region. Per capita growth rates reported in other chapters are shown in Table 11.1. Three are in the range 2.3 to 2.5 percent per annum for 2005 to 2050, and only Chapter 2 has a significantly lower rate, at 1.9 percent. Chapters 4 and 5 shows projections of poverty rates by region. Chapter 5 is upbeat, with absolute poverty virtually disappearing by 2050 in sub-Saharan Africa (Table 11.1). This implies that whatever GDP growth rate underlies the World Bank's poverty projections, must be fairly robust; moreover, these drastic poverty reductions occur despite the assumption that within-country income distribution becomes somewhat more unequal. Chapter 4 is less optimistic: the region's poverty rate is projected to decline to 11.7 percent by 2050 in the high GDP growth scenario when per capita GDP grows at 2.5 percent per annum. This assumes no changes in within-country income distribution and in the part of GDP devoted to private consumption, so here too it is economic growth that drives declines in the poverty rates, rather than internal redistribution. The lower-growth scenario of Chapter 4 is outright pessimistic: poverty actually increases slightly, to 53.1 percent of a much larger population, so the implied absolute numbers in poverty are huge.

Two main questions arise from these results. Firstly: does evidence support the proposition that sustained economic growth leads to declines in the poverty rates? The latest World Development Indicators data, reporting figures up to 2005 show negative correlation between the two variables (Figure 11.20).[34] The poverty rate started falling from 1999, when growth of per capita GDP accelerated. Macroeconomic projections assume a continuation of the GDP per capita growth of recent years, so the significant further reductions in the poverty rates do not seem out of reach, provided these high growth rates actually materialize. Secondly: is it realistic to expect that the relatively high growth rates of per capita incomes achieved in the decade to 2008 can be sustained for several decades? One reason for scepticism is the region's initial poverty conditions: starting with 52 percent

---

34. Assuming that the two data sets are independent, i.e., the poverty rate is not estimated as a function of income.

of the population below this very low poverty line   USD 1.25/day -- means that poverty rates will continue to be high for long periods in the intervening years. Would such a situation provide the social and political environment necessary for sustaining high economic growth rates for several decades?

Table 11.1
Key figures for sub-Saharan Africa

	Actual		Projection
GDP/capita (% per annum)	2000–2005	2000–2050	2005–2050[a]
Hillebrand (Chapter 4, Table 4.8, high scenario)	1.9		2.5
FAO (2006: Table 2.5)	1.9	2.3	2.4
IFPRI (personal communication)	1.9	1.9	1.9
IIASA (Chapter 3, Tables 3.1 and 3.2)	1.9	2.2	2.3
B. Poverty (% population with < USD 1.25/day in 2005 PPP dollars)	2005		2050
Hillebrand (Chapter 4, Table 4.10, high scenario)	51.2		11.7
Hillebrand (Chapter 4, Table 4.10, low scenario)			53.1
World Bank (Chapter 5, Table 5.8)	51.7		2.8
	Data used in projections		Projection
C. Food consumption (kcal/person/day)	2000	2005	2050
Hillebrand (Chapter 4, Table 4.9, high scenario)		2 256	2 588
FAO (2006: Table 2.1)	2 194		2 830
FAO (Chapter 1, Table 1.4)	2 128	2 167	2 708
IFPRI (Nelson *et al.*, 2009: Table 5, baseline)	2 316		2 452
IFPRI (Nelson *et al*, 2009: Table 5, climate change-CSIRO with CO_2 fertilization)			2 064
D. Undernourished (% population)[b]			
FAO (2006: Table 2.3)	33.3		5.8
FAO (Chapter 1, Table 1.3, actual and revised)	32	30.5	7
IIASA (Chapter 3, Tables 3.1 and 3.5, on risk of hunger)	29.9		12.6
E. Undernourished (million people)			
FAO (2006: Table 2.3)	201		88
FAO (Chapter 1, Table 1.3, actual and revised)	202	213	118
IFPRI (Nelson *et al.*, 2009: Table 6, baseline, children only)	33		42
IFPRI (Nelson *et al.*, 2009: Table 6, CSIRO with CO_2 fertilization, children only)			48
IIASA (Chapter 3, Table 3.5, at risk of hunger)	196		239

[a] Except for Chapter 4, the projected growth rates for 2005 to 2050 are derived from those for 2000 to 2050 and the actual rates for 2000 to 2005.
[b] Hillebrand's EM paper gave estimates for this variable (percentage of population malnourished), which have been removed in the version for this volume (Chapter 4).

Figure 11.20
Sub-Saharan Africa: per capita GDP, per capita food consumption and poverty rate

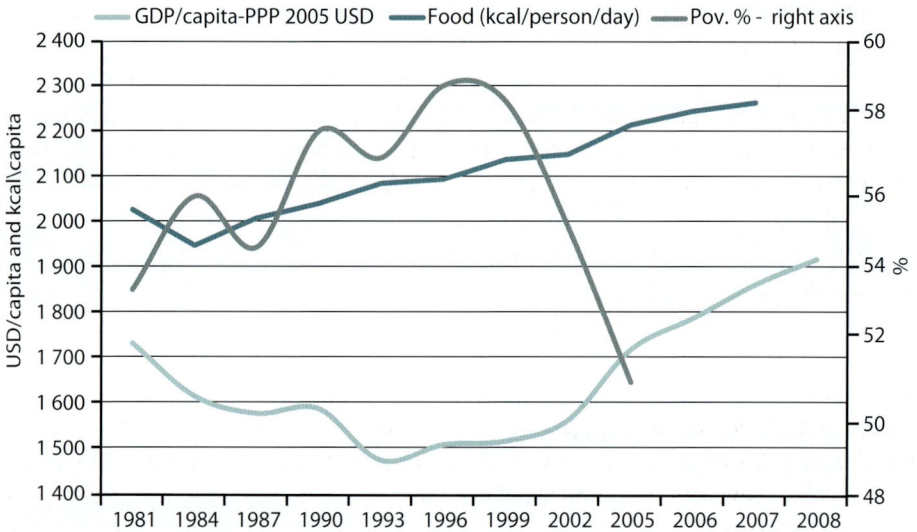

Sources: World Bank, World Development Indicators; FAOSTAT.

These questions cannot be answered, but it is worth considering them. It must also be kept in mind that reducing or even eliminating this kind of poverty does not imply a region free of deprivation for significant parts of the population: the poverty line used is very low, and 45 years is a long time to wait for such an outcome.

Food security in Sub-saharan Africa

Of prime interest here is what may happen to the food situation in Sub-saharan Africa, and particularly whether the projected reductions in poverty rates will be accompanied by commensurate declines in the rates of undernourishment. In principle this should be so, given that "almost all the national poverty lines use a food bundle based on prevailing diets that attains predetermined nutritional requirements for good health and normal activity levels, plus an allowance for non-food spending" (World Bank, 2009). In turn, the international poverty line of USD 1.25/person/day is based "on the mean of the poverty lines in the poorest 15 countries" in the sample of countries surveyed. Other countries have higher poverty lines (Haughton and Khandker, 2009: 185; Ravallion, Chen and Sangraula, 2009). Assuming the latter are also based on the "bundle of food" principle, the

USD 1.25 level will not be sufficient to ensure that "nutritional requirements for good health and normal activity levels" are met in all countries. Therefore, it is not surprising that studies using similar assumptions of per capita income growth find that undernourishment rates in 2050 are projected to be higher than the USD 1.25 poverty rates.

As mentioned, kcal/person/day is the main variable used by FAO to estimate the incidence of undernourishment (FAO, 2006: Box 2.2). The projection in Chapter 1 is the most optimistic among those reported in this volume: food consumption is projected to rise from 2 167 kcal/person/day in 2003/2005 to 2 708 kcal in 2050, and the percentage of population undernourished falls from 32 to 7 percent. The absolute numbers undernourished remain high, with 118 million in 2050, down from 212 million in 2003/2005. The target of the 1996 World Food Summit of halving the numbers undernourished by 2015 is not achieved, even by 2050, assuming the global halving target is applied to the region.[35] These developments in the absolute numbers undernourished, despite the significant decline in the percentage of population, reflect the very rapid population growth in sub-Saharan Africa, at 142 percent from 2005 to 2050, compared with 38 percent for the other developing countries. This raises the issue of how to compare the relative performances of different countries/groups regarding progress towards the target. Defining the target in absolute numbers rather than as a percentage of population tends to penalize countries with high demographic growth, by making progress look less than it would be in percentage terms.

In Chapter 4, Hillebrand's 2.5 percent per capita income growth rate in the "high" scenario is associated with an increase in food consumption, from 2 256 kcal/person/day in 2005 to 2 588 kcal in 2050. The implicit income elasticity is unrealistically low (at 0.12), given the very low starting level, even when food is expressed in calories.[36] How is this result derived? The model represents the crop sector in terms of physical quantities with all crops lumped together as one variable plus meat, dairy and fish, but "Calories consumed per capita are estimated as a function of GDP per capita and relative prices" (personal communication).[37]

Chapter 2, as recalled, provides projections in terms of kcal/person/day, and numbers of malnourished children in millions, but not as percentages of the population in the zero to five years age group. In these projections, sub-Saharan

---

35. Progress towards the target is measured from the 169 million people undernourished in 1990/1992 (FAO, 2008: Table 1), the base year used at the 1996 World Food Summit.

36. Estimates of calorie-income elasticities are provided by Skoufias (2002) and Ohri-Vachaspati *et al.*, (1998).

37. implicit income elasticity may be so low because projected food prices rise significantly. The author does not address this issue.

Africa's per capita food consumption increases only marginally in the baseline from 2000-2050 (Nelson *et al.*, 2009: Table 5) and suffers a sharp decline (to 2 064 kcal/person/day) in the climate change scenario. This is a very pessimistic outlook, given that it considers a 50-year horizon. As a consequence, the numbers of malnourished children in the baseline are projected to increase from 33 million in 2000 to 42 million in 2050 (Nelson *et al.*, 2009: Table 6). This may be the outcome of the small improvement in per capita food consumption projected by IFPRI, and the projected price increases (Figure 11.4). As in the case of cereals (Figure 11.6), part of the reason may be the historical data. Food consumption in the region is given as 2 316 kcal/capita/day in year 2000, while FAO records 2 128 kcal (Table 11.1).

In general, statements about food security in the future require a credible projection of per capita food consumption, which is generally derived as a function of GDP per capita and prices. However, at least for the regional average for sub-Saharan Africa, evidence suggests that per capita food consumption is more closely related to the evolution of domestic food production than to anything else. This was observed in FAO (2006: Figure 2.2, example of Nigeria) where the evolution of per capita food consumption seemed to bear little relation to per capita incomes. In most sub-Saharan African countries with limited shares of imported food in total supplies, high economic dependence on agriculture and little non-food use of food commodities, food consumption seems to follow the evolution of production and, to a lesser extent, food imports.[38] As explained in the methodology of the earlier FAO projections (Alexandratos, 1995), in such cases, the growth of food consumption is derived iteratively in the process of evaluating the scope for increasing production and trade. Much of the credibility of the food consumption projections therefore depends on the credibility of the food production projections.

Looking at cereal projections for Sub-saharan Africa, Chapter 2 has the smallest increase in per capita food consumption, and also the lowest growth rate for cereals production in its baseline projection, at 1.7 percent per annum from 2000 to 2050 (Nelson *et al.*, 2009: Table 3). Chapter 3 shows a higher growth rate for cereals production, at 2.5 percent per annum for the same period (Table 3.3). Projections in Chapter 1 are in between, at 2.1 percent per annum (FAO, 2006: Table 3). It is to be noted that sub-Saharan Africa's cereals production increased by 4.5 percent per annum between 2000 and 2008, and production grew by more

---

38. In the food balance sheets, food consumption is derived as part of the utilization of total supplies of each food product (production + net imports + net changes in stocks). It is therefore natural to expect a close correlation between production and consumption in countries with little net trade in food, and limited non-food uses (e.g., feed) of food commodities.

than 4 percent per annum in several major producers in the region, including Ethiopia, Nigeria and the Sudan (Chapter 1, Annex 1.1). Such high growth rates were partly due to recoveries and favorable weather; however, there is no reason to believe that the region's cereals production cannot grow at a somewhat higher rate than that of the population. This may not apply to all countries, given that, for example, the Niger has annual population growth of more than 3 percent to 2050.

At the same time, it is useful to note that cereals are a relatively small component of food production in sub-Saharan Africa, accounting for 17 percent of total food production. Potential links between production and consumption may be even stronger for the other major food groups in the regions, such as roots, tubers and plantains.

Discussion on Sub-saharan Africa

The chapters of this volume present contrasting views on the prospects for sub-Saharan Africa. At one extreme, Chapter 5 foresees a much improved future, judging from its projection of drastic falls in poverty, despite the assumption of no improvement or even deterioration of income distribution, and after accounting for climate change damage to agriculture. This view seems to reflect confidence in continuation of the upturn in the region's performance of the last decade, after a long history of stagnation and decline. The most recent World Bank assessment (World Bank, 2011: Table R6.3) foresees continuation of robust GDP performance to 2012. It must be kept in mind that the region now has ten oil-exporting countries and receives growing amounts of foreign private investment (World Bank, 2011: Table R6.1). A recent article in the Economist highlights sub-Saharan Africa's good economic growth record of the current decade, and its future prospects.[39] As already noted, agriculture has also improved its performance in the current decade.

At the other extreme, Chapter 2 paints a pessimistic future for the region, as deduced from the limited progress in raising per capita food consumption by 2050 in its baseline projection. The future is projected to be dismal, as a consequence of climate change effects on agriculture: in the best climate change outcome, per capita food consumption is projected to suffer a drastic fall from its already very low level. As noted in Figure 11.10, this is at great variance also with the findings of Chapter 3. However, catastrophic predictions are common, and the latest foresees doom already by 2025: "about two-thirds of arable land in Africa is expected to be lost by 2025" (Hisas, 2011). Climate change futures that seem to

---

39. "A more hopeful continent: The lion kings? Africa is now one of the world's fastest-growing regions", Economist, 8 January 2011.

imply such huge divergences in views about the region's long-term prospects are clearly in great need of further analysis.

References

Alexandratos, N., ed. 1995. *World agriculture: Towards 2010, an FAO study*. Chichester, UK, J. Wiley and Sons, and Rome, FAO.

Alexandratos, N. 2008. Food price surges: Possible causes, past experience, and longer-term relevance. *Population and Development Review*, 34(4): 663–697. www.fao.org/es/esd/foodpricesurges-alexandratos.pdf.

Cline, W.R. 2007. *Global warming and agriculture: Impact estimates by country*. Washington, DC, Center for Global Development and Peterson Institute for International Economics.

Easterling, W.E., Aggarwal, P.K., Batima, P., Brander, K.M., Erda, L., Howden, S.M., Kirilenko, A., Morton, J., Soussana, J.-F., Schmidhuber, J. & Tubiello, F.N. 2007. Food, fibre and forest products. In M.L. Parry, O.F. Canziani, J.P. Palutikof, P.J. van der Linden and C.E. Hanson, eds. *Climate change 2007: Impacts, adaptation and vulnerability. Contribution of Working Group II to the Fourth Assessment Report of the Intergovernmental Panel on Climate Change*, chapter 5, pp. 273–313. Cambridge, UK, Cambridge University Press.

Economist. 2009. Is it worth it? 3 December 2009.

FAO. 2006. *World agriculture: towards 2030/2050, interim report*. Rome. www.fao.org/es/esd/gstudies.htm.

FAO. 2008. *The State of Food Insecurity in the World 2008*. Rome.

Fischer, G. 2000. Simulating global agricultural development according to IPCC SRES scenario paradigms. Paper prepared for the Global Modelling Forum, New York, 2 to 4 February 2000.

Fischer, G., Shah, M. & van Velhuizen, H. 2002. *Climate change and agricultural vulnerability*. Special report for the World Summit on Sustainable Development. Laxenburg, Austria, IIASA.

Fischer, G., Shah, M., Tubiello, F.N. & van Velhuizen, H. 2005. Socio-economic and climate change impacts on agriculture: an integrated assessment, 1990–2080. *Phil. Trans. R. Soc. B*, 360: 2067–2083.

Fischer, G., Hizsnyik, E., Prieler, S., Shah, M. & van Velthuizen, H. 2009. *Biofuels and food security*. Study prepared by IIASA for the OPEC Fund for International Development (OFID). Laxenburg, Austria, IIASA.

Haughton, J. & Khandker, S. 2009. *Handbook on poverty and inequality*. Washington, DC, World Bank.

Hisas, L. 2001. The food gap: The impacts of climate change on food production: A 2020 perspective. Alexandria, Virginia, USA, Universal Ecological Fund.

IEA. 2008. *World energy outlook 2008*. Paris, OECD/IEA.

IPCC. 2007a. *Climate change 2007: Synthesis report*. Nairobi.

IPCC. 2007b. *General guidelines on the use of scenario data for climate impact and adaptation assessment*. Nairobi.

Mendelsohn, R. 2007. A critique of the Stern Report. *Regulation*, Winter 2006–2007.

Mitchell, D. 2008. *A note on rising food prices*. Policy Research Working Paper No. 4682. Washington, DC, World Bank.

Msangi, S. & Rosegrant, M. 2009. World agriculture in a dynamically-changing environment: IFPRI's long-term outlook for food and agriculture under additional demand and constraints. Paper presented at the FAO Expert Meeting on How to Feed the World in 2050, 24 to 26 June 2009, Rome, FAO. www.fao. org/wsfs/forum2050/background-documents/expert-papers/en/.

Nakicenovic, N. & Swart, R., eds. 2000. *Special report on emissions scenarios*. Cambridge, UK, IPCC and Cambridge University Press. 570 pp.

Nelson, G.C., Rosegrant, M.W., Koo, J., Robertson, R., Sulser, T., Zhu, T., Ringler, C., Msangi, S., Palazzo, A., Batka, M., Magalhaes, M., Valmonte-Santos, R., Ewing, M. & Lee, D. 2009. *Climate change: Impact on agriculture and costs of adaptation*. Washington, DC, IFPRI.

ODI. 2009. *Biofuels: Could the South benefit?* Briefing Paper No. 48. London, Overseas Development Institute (ODI).

OECD/FAO. 2009. *Agricultural outlook, 2009–2018*. Paris.

Ohri-Vachaspati, P., Rogers, B.L., Kennedy, E. & Goldberg, J.P. 1998. The effects of data collection methods on calorie–expenditure elasticity estimates: a study from the Dominican Republic. *Food Policy*, 23(3/4): 295–304.

Ravallion, M., Chen, S. & Sangraula, P. 2009. Dollar a day revisited. *World Bank Economic Review*, 23(2): 163–184.

Rosegrant, M. 2008. Biofuels and grain prices: Impacts and policy responses. Testimony for the United States Senate Committee on Homeland Security and Governmental Affairs, 7 May 2008. Washington, DC, IFPRI.

Schmidhuber, J. & Tubiello, F. 2007. Global food security under climate change. *Proceedings of the National Academy Science*, 104(50).

Simon, J. 1996. *The ultimate resource II*, second edition. Princeton, New Jersey, USA, Princeton University Press.

Skoufias, E. 2002. *The sensitivity of calorie-income demand elasticity to price changes: Evidence from Indonesia*. FCND Discussion Paper No. 141. Washington, DC, IFPRI.

Stevens, P. 2008. *The coming oil supply crunch*. London, Chatham House.

Weitzman, M. 2007. The Stern Review of the economics of climate change.

World Bank. 2009. *World Development Indicators 2009.* http://data.worldbank.org/data-catalog/world-development-indicators.

World Bank. 2010. *World Development Report 2010: Development and climate change.* Washington, DC.

World Bank. 2011. *Global economic prospects 2011.* Washington, DC.

CHAPTER 12

CHALLENGES AND POLICIES FOR THE WORLD AGRICULTURAL AND FOOD ECONOMY IN THE 2050 PERSPECTIVE

Piero Conforti
Alexander Sarris

Previous chapters have presented a number of scenario analyses on global agriculture, its connection with the expected evolution of the world economy, the availability of natural resources, investment requirements, and prospects for research and innovation. Chapter 11, by Alexandratos, has reviewed projections for main variables, highlighting differences, and their origins in the data and analysis. This last chapter takes a broader look at the main results, and proposes insights into how this wealth of information can be utilized. The need to shape policy action is currently one of the fundamental reasons for taking interest in long-term projection exercises such as the one presented in this volume. Another need is for looking beyond short-run phenomena, to understand what may be long-lasting phenomena, and disentangle these from what is likely to be short-run noise. The following discussion is based on the chapters of this book; hence it is less comprehensive than recent ambitious analyses of agriculture and the global food system, such as the one by the World Bank (2010) on climate change impacts, and the Foresight project report (Foresight, 2011).

This chapter is divided into three sections. The first section summarizes key conclusions from the four parts of the volume, highlighting the challenges that they pose. The second section highlights information and analytical gaps that may hinder understanding of the likely evolution of world agricultural and food economies over the coming decades. Lessons from the rest of the volume are identified in terms of areas where more information and knowledge would be useful. The third section uses conclusions from the four parts of the volume to outline policy directions and identify intervention areas that could be explored today, given the outlook to 2050. Some concluding remarks close the chapter.

Challenges and opportunities

The first part of this volume shows how challenging and controversial projection exercises for world agriculture can be. Insights presented in the three chapters by Alexandratos, Msangi and Rosegrant, and Fischer indicate that such projections are the result of a complex combination of data, assumptions and modelling, drawing on different subject areas. In Chapter 1, Alexandratos concludes that the last comprehensive projection to 2050, undertaken a few years ago by FAO (2006) without accounting for the impact of biofuels, is still to a large extent valid. This implies that the turbulence observed in food markets during recent years is not necessarily going to affect the long-term picture, at least as regards global quantities of consumption and production. The presence of potentially disruptive forces, including further development of the bioenergy sector and climate change, is acknowledged, but uncertainties surrounding these phenomena warrant a cautious attitude: the emphasis is on what is not yet known. Chapter 3, by Fischer, addresses these two aspects – bioenergy and climate change – to conclude along similar lines. Climate change, it is argued, will not necessarily imply large changes in the world food system until 2050, and there may be room for farmers to adapt and mitigate. The combined influence of climate change and a fast development of biofuels may make the situation more critical and generate significant stress on resources and markets, but this result is judged to be the upper extreme of a range of possible outcomes.

While taking a similarly cautious attitude, in Chapter 2 Msangi and Rosegrant emphasize the likelihood of extreme conditions materializing in the 2050 horizon. Changes in climate variables are expected to impose considerable pressures on land and water resources. Combined with the dynamics of major crop yields, it is argued, this could lead to very high increases in real prices of key agricultural commodities by 2050. Depending on the climate scenario, Msangi and Rosegrant project reductions of up to 30 percent in yields of rainfed maize in developed countries, up to 18 percent in yields of irrigated rice in developing countries, and up to 34 percent in the yield of irrigated wheat. In sub-Saharan Africa, the number of malnourished children in 2050 is projected to be up to 24 percent higher under climate changes scenarios. The decline in malnutrition prevalence expected in sub-Saharan Africa in the baseline projection becomes contingent on significant investment to improve knowledge, science and technology. This is deemed necessary to dampen the increases that are otherwise expected in the prices of key staples, such as cassava and maize.

Even without climate change, Msangi and Rosegrant project meat demand to generate substantive increases in feed demand for cereals, following fast per capita income growth, especially in East Asia. Alexandratos's comparison of global cereals consumption (without biofuels) projected for 2050 (Chapter 11, Figure 11.5) shows that there are wide differences. Broadly speaking, the projections of Msangi and Rosegrant are on the low side, those of Fischer are on the high side and those of FAO are in between. Alexandratos suggests that a major cause of these differences is to be found in the different historical data used by the three projection studies.

The expected development of biofuels, according to Msangi and Rosegrant, may require developing countries to increase their yield growth by 1 percent per year up to 2030, over and above the increases foreseen in the authors' scenario without biofuels. Biofuels are projected to result in substantive increases in market prices and a worsening of food security conditions. In Chapter 1, Alexandratos instead argues that the quantities by which biofuels would increase world aggregate cereals utilization would be a relatively modest 7 percent of world consumption in 2015 – based on the FAO/OECD outlook figures – and much of it will likely come from increased production over and above what it would have been without biofuels.

The three projection exercises of the first part of the volume disagree on the extent and likelihood of a global stress in food markets. However, they all seem to point towards two main challenges. First is the need to improve efforts on the supply side of the market, towards increasing productivity. This is especially urgent in poor developing countries, given that they may be vulnerable to changes in the natural environment, and that their consumption levels are more directly related to production. Second is the need to improve the functioning of agricultural markets, again especially in poor developing countries, where transaction costs and lack of infrastructure may hinder the transmission of scarcity signals and contribute to distorting incentives. The three outlook exercises attribute different emphases to the generation and adoption of technology, but there seems to be less controversy regarding the importance of market incentives and policies to promote increased productivity as one way of addressing undernutrition. As noted in Chapter 11, the three projection exercises pay less attention to opportunities; in fact, higher prices over the coming decades, although a potential problem for poor consumers, may also serve to drive investment and technical change, as well as an improved functioning of production chains.

The second part of the volume presents less controversial evidence. This seems to be mostly the consequence of one key assumption – made in Chapters 4 and 5 and described in Chapter 11 – of faster GDP growth in non- OECD countries compared with OECD ones. In Chapter 4, Hillebrand indicates that global per capita income is expected to grow at an average annual rate of 1.9 to 3 percent from 2005 to 2050, corresponding to 2 to 2.9 percent in high-income countries, and 2.4 to 3.6 percent for developing countries. The faster dynamic of the latter group is expected to generate significant reductions in absolute poverty and undernourishment. If Hillebrand's "high" scenario GDP projections materialize, which is an optimistic scenario, the incidence of global poverty could fall from about 21 percent in 2005 to 2.6 percent in 2050, and the number of people living in absolute poverty could decline by 1.1 billion. The absolute gap between per capita incomes in OECD and non-OECD countries, and the global Gini coefficient are still expected to remain high. That is, relative inequality is still expected to remain high between today and 2050.

However, Hillebrand also shows that if regions that have been lagging do not achieve growth rates higher than those observed in the last 25 years, the picture may be far different, and poverty could still be considerable in 2050, especially in sub-Saharan Africa. The picture may also look less optimistic if resource constraints are taken into account, or if the assumption that technological solutions will become available is relaxed.

Agriculture can contribute significantly to economic growth, especially in developing countries, where the sector's share in GDP is relatively high. However, if growth prospects materialize, one challenge in these countries will be the adjustment that is likely to be associated with growth. Agriculture will be called on to adapt and diversify its productive role. In this respect, bioenergy could constitute an opportunity for farmer communities to obtain access to larger and possibly more dynamic income sources. Price conditions, so far, do not seem viable: the use of agricultural feedstock for energy production is mostly policy-driven, apart from in few cases, such as Brazil. More opportunities may be found in the area of environmental services, provided that private and public demands combine to create viable markets.

The potential impacts of climate change on economic growth, analysed by van der Mensbrugghe and his colleagues in Chapter 5, appear altogether limited. As also Alexandratos notes in Chapter 11, the highest impacts would occur in South Asia and sub-Saharan Africa, where income in 2030 may be, respectively, 4.5 and 3.5 percent lower. Globally, agriculture is the largest user of natural resources, accounting for about 14 percent of total greenhouse gas emissions without considering deforestation, and for up to 30 percent when deforestation is included in the computation. The characteristics of agricultural production processes therefore have an important impact on carbon emissions and mitigation.

Based on this picture, one challenge for global agriculture is to be able to counteract its own potential negative impact on climate change, while contributing to climate change mitigation. As FAO (2009) shows, there are many potential synergies between certain agricultural practices and climate change mitigation.

The challenge is therefore to enhance the latter while reducing the former. This entails the adoption of specific farming practices, such as crop rotations, more efficient fertilizer use, low-energy irrigation and conservation techniques, which are often site-specific. Opportunities may be explored for providing farmers with incentives for adopting these practices, so that they can assume the short-run costs of adapting production systems while improving the long-run sustainability of their activity.

The third part of the volume quantifies the amounts of capital, investment and natural resources required to produce the foods, feeds and fibres that are expected to be demanded and consumed in 2050. The resource outlook proposed by Bruinsma in Chapter 6 makes the point that at the global level, the amounts of land and water available is probably sufficient to support the increase in production needed to satisfy projected consumption. The projected growth in crop production to 2050, following the FAO baseline, would result in a net expansion of arable land of about 70 million ha, with an increase of 120 million ha in developing countries being offset by a decline of about 50 million ha in developed countries. Globally, available water resources are also judged to be most probably sufficient to support expected yield increases, but they are unevenly distributed: there is already considerable scarcity in the Near East and North Africa and in South Asia, and the situation in these regions is projected to worsen. More in general, one interesting conclusion in Chapter 6 is that yield increases that may materialize are likely to bring about increased environmental pressures. Considerable efforts are therefore required to manage resources more efficiently. Public interventions and private participation will be necessary to prevent and mitigate damages.

These requirements have to be added to an already large bill. In Chapter 7, von Cramon-Taubadel and colleagues argue that fixed capital stock in agriculture, at the global level, has been growing steadily over the last three decades, although at declining rates. Government expenditure on agriculture is correlated with capital formation in a sample of developing countries, and is shown to have a significant positive impact on total factor productivity (TFP), as well as on foreign direct investment. According to Schmidhuber, Bruinsma and Boedeker in Chapter 8, the amount of gross investment required for developing countries' agriculture to produce a volume of supply consistent with FAO's long-term outlook is projected at USD 9.2 trillion over the next 44 years, corresponding to an annual gross investment of about USD 210 billion, from both private and public sources. Primary agriculture accounts for about 46 percent, while the rest is projected to be absorbed by processing, transportation, storage and other downstream activities. Within primary agriculture, mechanization and irrigation would together account for about 54 percent. Net investment is projected to show a declining trend over the coming decades, owing to various reasons, including the expected decline in the rate of production increase. This tendency should complement the shift towards more efficient input use and more capital-intensive production, in which capital is expected to replace labour at an increasing pace. Sub-Saharan Africa and Latin America show incremental capital output ratios that are lower than those in other parts of the world. This suggests that large additional supply can be obtained with relatively less capital investment in these regions.

Over the last few years, agriculture has attracted considerable attention in high-level political fora, following the sense of urgency generated by the turbulence of world prices. A number of international initiatives have been launched recently, accompanied by resource mobilization commitments: the Aquila Food Security Initiative, the European Union's (EU's) food facility, and FAO's Soaring Food Prices Initiative; the establishment of the United Nations (UN) Secretary-General's High-Level Task Force on the Global Food Security Crisis, and the reform of the UN Committee on World Food Security. Coherence in resource mobilization for agriculture and food security is sought through the Global Agriculture and Food Security Programme. National policies are also evolving fast. Recent initiatives by the governments of large countries such as China, the Russian Federation, India and Indonesia indicate a shift of emphasis from emergency measures, such as export restrictions, towards increasing productivity and improving risk management, which are becoming prominent medium- to long-term objectives. Translating this renewed attention to agriculture into increased income opportunities for poor farmers and increased participation of smallholders in national and global production chains in developing countries is a major challenge. This depends on how specific policies are shaped, and where resources are invested. This chapter's section on Policy directions provides some insights on this.

The fourth part of the volume deals with the more specific and technical aspect of what the research system can do to foster productivity in world agriculture. Research capacity is increasingly concentrated in a few countries, and efforts to build collaboration at the global, regional and national levels still appear insufficient. Investment in public agricultural research declined globally during the 1990s, especially in the regions where it may have had the strongest impact on poverty, such as sub-Saharan Africa. However, the challenges are increasing, owing to climate change and possible increase in weather variability and water scarcity. It is not easy for governments to increase expenditure, because of fiscal constraint and the fact that benefits from investment in agricultural research materialize only in the long term: hence the condition of generalized underinvestment.

In Chapter 9, Beintema and Elliot show that the share of expenditure on agricultural research in total expenditure devoted to agriculture is fairly similar across national income levels, and that while total expenditures on agriculture are low in absolute terms, they represent a higher share of total public expenditure in low-income agriculture-based economies than in wealthier countries. It therefore appears that low-income countries as a group are not underspending on agriculture compared with higher-income countries. However, the share of agricultural research expenditure in agricultural GDP, which the authors call the "agricultural research intensity", is lower in developing countries. There does not appear to be any useful benchmark in this domain, although the International Food Policy Research Institute (IFPRI) suggests that research is still the most productive investment for supporting agriculture, followed by education, infrastructure and input credits. Moreover, research would imply a larger impact on productivity than non-research spending (Fan and Rao, 2003). An interesting development of this analysis could be the comparison with alternative policies, in terms of cost, feasibility and coverage. This would provide insights into the costs and benefits of more than one alternative.

In fact, yield increases are slowing down in developing countries, despite the existing potential. In Chapter 10, Fischer, Byerlee and Edmeades analyse this point in detail. They show that wheat has an average yield gap of about 40 percent, while for rice the gap averages 75 percent, and for maize it ranges from 30 to more than 200 percent. Reducing these gaps, especially that for maize in sub-Saharan Africa, is a primary goal. Chapter 10 also reveals that there is considerable scope for pushing the agricultural technology frontier outwards in the short to medium term, as many new technologies at an early stage of adoption promise to enhance productivity sustainably. These include conservation farming approaches and biotechnology – each still used on less than 10 percent of the world's cropland – and information and communication technology, which is at an even earlier adoption phase and seems to be particularly promising in terms of efficient inputs management. Plant breeders continue to make steady gains in potential yields and water-limited potential yields, more slowly than in the past for wheat and rice, but with little slowdown in the case of maize. New opportunities are also presented by genomics and molecular techniques, which are now regularly applied to speed up breeding processes in seed companies.

It takes considerable time and resources to develop new technologies and bring them to the field. Despite the increase in private sector research and development (R&D), most of the world's resources devoted to agricultural research are still public. A few large countries account for the bulk of agricultural R&D spending, but the social returns on such spending have been and still are quite high. Some of the large countries that have invested heavily in agricultural research, such as China and India, have experienced rapid yield and general productivity growth, with subsequent general income growth. However, the rate of growth in agricultural research investment has been declining globally, and a large number of developing countries experienced negative growth rates during the 1990s, especially in sub-Saharan Africa. In the last decade, while a number of mostly larger African countries have increased their commitments to agriculture and agricultural research, many countries still experience declining agricultural spending.

Existing synergies among technologies should be taken into account and enhanced. Reducing yield gaps is likely to require improved efficiency throughout entire production processes, including better information and skills for input use. In other words, TFP – which measures the value of output over and above what can be accounted for by physical inputs – needs to improve. R&D appears to be one of the major drivers of TFP; other factors that contribute significantly to enhancing TFP are extension services and education, which allow farmers to close gaps between farm and attainable yields. Improved infrastructure is also important for reducing yield gaps, especially those between attainable yields at undistorted prices and attainable yields.

Can data and analyses be improved?

In recent decades, the technical possibilities for researchers to exchange and share data and resources have improved significantly. However, comparisons of projection exercises in similar subject areas highlight basic controversies, if not inconsistencies. Starting points, such as historical information, can be different, and account for many of the differences in results. Such differences may also prevent comparison of the impacts of the different theoretical assumptions that inform different projections. For example, this is the case of demand and supply projections to 2050, and of the use of feedstock in biofuel production, as shown by Alexandratos in Chapter 11.

The tools now available for analysing long-term developments of world agriculture are extremely wide-ranging and complex. To mention just a few, general equilibrium models, integrated assessment models including climate and economic variables, and geo-referenced databases are today relatively accessible. This richness is an advantage in terms of the potential information that can be produced, but it also requires that analysts maintain access to and knowledge of several large data sets, and make assumptions that range over a large number of subjects, encompassing both natural and social sciences. This is seldom feasible for individual researchers or even research groups.

Communication, networking and exchange of data and analytical tools among concerned institutions, working groups and individual analysts are among the most promising ways of addressing such complexity and the need to integrate several subject areas, and of improving the transparency and comparability of results. A methodological discussion of the land market – land is just one aspect of global agriculture, and encompasses biophysical, social and economic information and analysis – concludes that "there remain significant barriers to entry in this field of study. ... [and] ... there could be tremendous returns to public investments in open-source, publicly available data base infrastructure for explicitly spatial, global analysis of long-run issues related to agriculture and its links with the environment" (Hertel, 2010: 46). Similar considerations apply to more complex phenomena, such as climate change, the interaction between agricultural and energy markets, and development of the biofuel industry.

Equally important, and related to this issue, is the need to establish better dialogue between scientists dealing with the natural environment and social scientists, such as economists. Analyses in Chapters 2 and 3 – where Msangi, Rosegrant and Fischer study climate change and biofuel development scenarios – demonstrate how results are driven, *inter alia*, by how the interaction between biophysical and economic variables is modelled. In this area, biophysical analyses seem mainly to emphasize the limitations imposed by expected changes in the natural environment, and the finite dimension of resources. Economists, on the other hand, may emphasize more how agents can adapt to changes in the natural environment, and the role that incentives can play in adaptation. To date, the role of incentives seems to be less clearly identified – at least in analyses of the prospects for global agriculture. A lot of emphasis is placed on mitigation and adaptation, and on what could be promoted by policies, but projections for global agriculture seem to make less effort to model technology adoption and the latent demand for innovation. The expected changes in the natural environment may well affect these mechanisms, and this aspect of adaptation to and mitigation of climate change impacts is also highlighted by the World Bank in its most recent World Development Report (World Bank, 2010: Chapter 7). Incentives for entrepreneurs to adopt innovations, as well as existing technologies, depend largely on the environment in which they operate, and the possibilities for obtaining finance for the investment and risk management involved in more efficient technologies are key elements of this environment. Adaptation to changes in the production potential of different areas is included in Chapter 2, through the International Model for Policy Analysis of Agricultural Commodities and Trade (IMPACT) and a wide survey of crop models; in the Agro-Ecological Zone (AEZ) model, which is behind Fischer's work in Chapter 3; and in the ENVironmental Impact and Sustainability Applied General Equilibrium (ENVISAGE) in Chapter 5. These seem to be based on a wealth of engineering parameters, but seem less equipped to represent explicitly the economics of technology adoption. Given the size of the gaps between potential and actual yields, this seems to be an area where additional efforts could be useful. More needs to be understood about the reasons for technical inefficiencies, if any, and also the reasons for wide discrepancies between economic and technical efficiency, which may explain yield gaps. This point is further explored by Fischer, Byerlee and Edmeades in Chapter 10.

Other areas that could generate useful results are the sensitivity of projections to assumptions regarding income and population. The analyses presented in this volume typically assume population as an exogenous influence, with data sourced from the UN. The evolution of population, as noted in Chapter 11, is among the major drivers of several projections. Population is one of the few variables for which projections are available for distant years into the future, given that its dynamic is considered relatively predictable. In fact, most results presented in the volume are sensitive to population assumptions. One interesting topic, which seems not to have been explored in depth at the global level and in connection with agriculture, is migration. As well as the sensitivity to assumptions about the overall global population level – the 9.2 billion people expected in 2050, according to the 2008 UN medium variant – it would be interesting to study the long-run impact that potential changes in the distribution of population across regions could have on agricultural and food demand and supply. Qualitatively, the migrations observed today come from some of the regions where the ratio of population to resources is reported to be high, or where per capita incomes are low. For instance, results in Chapter 6 indicate that the ratio of population to resources is already high in the Near East and North Africa and in South Asia, and is expected to increase further in the future. It would probably be worth exploring how assumptions about migration affect projections and analysing scenarios in which population is endogenous with respect to economic variables or changes in climate variables.

Nutrition is another area where projections of global agriculture could benefit from more information. Projections of changes in the number of undernourished people undertaken by FAO and IFPRI – and referred to in Chapters 1, 2 and 3 – are computed mostly on the basis of total available calories, albeit using different methodologies. Calories are an efficient proxy of the overall nutrition status, especially when the focus in on undernutrition. However, nutrition and its adequacy are typically multidimensional: proteins, fats and micronutrients are of great importance in determining an adequate nutrition status, at all income levels. Furthermore, health and nutrition status interact with the ability to absorb and use foods correctly. Malnutrition is notoriously a growing phenomenon in many countries, including many developing ones, especially where calorie intake is increasing rapidly. Evidence from large long-term surveys shows that obesity has increased substantially over the past decades (Finucane *et al.*, 2011) and is mounting in developing countries, following fast population growth (Shetty and Schmidhuber, 2006). Recent provisional projections from FAO – not included in this volume – show that obesity may become an even more pressing issue than undernourishment in the coming decades. The interactions among quality of nutrition, poverty and agriculture are attracting increasing attention. Imbalances in diets associated with poverty in developing countries may in some cases be as problematic as insufficient amounts of calories. It is well known that nutritional imbalances may be a cause of low labour productivity on the one hand, and a consequence of limited production possibilities within subsistence agriculture on the other. Associated with this is another important information gap, on food waste. Food consumption data are still largely computed from total availability (production + net trade + stock changes) in the food balance sheets, after deductions are made for non-food uses (industrial non-food use, seed and waste). Estimates of waste rates are in great need of improvement, but can only be assessed through costly sample-based statistical analysis.

More detailed information on nutrition – based on multiple parameters and more accurate food composition data – and consumption may contribute to improving the projections for food insecurity and malnourishment. These are areas where substantive benefits may arise from the type of collective efforts described for geo-referenced data: sharing resources across the key national and international institutions that produce, maintain and use them can improve data quality.

Two additional areas should be considered as potential generators of interesting scenario analyses on global agriculture: international political economy, and gender analysis. The way in which nation states and firms shape their relations interacts with markets and, more generally, contributes to determining economic outcomes. International trade is one area that can be directly affected by international relations, through both rules, such as trade agreements, and transaction costs. Foreign investments and financial flows can also play a role in shaping economic perspectives. Some contributions in the area of international political economy produce scenario analyses (e.g., Patomäki, 2008) that could be introduced into projection exercises. Regarding gender analysis, the economic impact of existing gaps in opportunities between men and women is attracting increasing attention. A recent comprehensive analysis of the available evidence on this subject (FAO, 2011) shows that this economic impact may be substantive in agriculture. Changes in productivity brought about by increased gender equality could also be factored into future scenario analyses, to gain relevant policy indications.

Policy directions

This section is divided into three subsections devoted to policies directed towards addressing market failures in demand, supply and global public goods.

The demand side

In recent years, the volatility of world food prices has placed renewed emphasis on a number of phenomena that affect the global food system from the demand side. Policies affect the demand for food and agricultural products from a variety of entry points. In the coming decades, the demand for raw homogeneous agricultural products is expected to slow down, due to the small income elasticity of demand for these products, coupled with increasing global incomes and growing urbanization. This result is highlighted in several chapters in this volume, particularly in the first part. It suggests that the share of basic food in total food consumption is likely to decline in the coming decades, and this may result in a decline of the global price elasticity of the demand for basic food products. Under such conditions, in which demand will be relatively stable, any cause of supply disruption is likely to determine large price swings in the global food system. Hence price volatility may increase. Although this topic is not directly addressed in this volume, it is probably going to become a major concern for policy-makers, and increasingly so in the future, as the share of urbanized population potentially affected by volatility grows larger. Especially in developing countries that depend on imports, policies for dealing with unpredictable price volatility are likely to attract increasing attention.

A related development that is bound to shape demand-related policies is the degree of price transmission from international to domestic markets. Transmission is affected by policies, such as trade policies, infrastructure and their efficiency, and by the marketing system. Prices are the most important signals affecting demand and supply responses in both the short and long runs. It appears that the downwards trend in real agricultural basic commodity prices, which was a feature of the global food system for many decades, has stopped. The issue is whether policy-makers should allow the changing price signals to be reflected fully in domestic prices, and hence guide producers and consumers, or should manage domestic prices and allow imperfect pass-through, to achieve specific domestic objectives.

Deviation of domestic from global equilibrium market prices for long periods is a risky proposition, even though the latter may be influenced by policies enacted by other trading countries. As prices affect both domestic demand and domestic supply, it is always prudent to follow global price developments to project proper signals for domestic agents. However, a case could be made for preventing the full brunt of short-term unpredictable international price changes from being fully reflected in domestic prices. Whatever the domestic issues that may dictate such a policy stance, the relevant price deviations should not be allowed to persist for long periods, to avoid creating permanent distortions in domestic markets.

More than one chapter in this volume highlights the growing reliance of a number of countries on imports for the fulfilment of domestic food needs. This leaves them more exposed to global market variations, and highlights the need for them to be able to afford the necessary food imports. It also highlights a policy dilemma that many governments may contemplate: whether to increase the degree of their countries' self-sufficiency in food, to reduce their exposure to trade. The feasibility of this depends on the domestic production possibilities, supply conditions and government's ability to affect domestic demand. Policies can influence demand in a variety of ways. Irrespective of the feasibility of such a policy stance from the production perspective, however, the issue here is whether a country should use demand-side policies to change its exposure to international trade. While such policies may be feasible in the short run, for instance via trade controls, they may not be sustainable in the medium to longer run, owing to their impact on other countries and to existing international trade regulations.

As observed in Chapters 4 and 5, strong economic growth is the key to future poverty reduction. Nevertheless, poverty will remain in some parts of the world, especially sub-Saharan Africa. And, as seen, the decreasing price elasticity of demand for most agricultural products may make food price spikes more frequent. These two points highlight the importance of safety nets relating to food security for vulnerable populations, especially in the perspective of less predictable international market prices.

In Chapter 2, Msangi and Rosegrant indicate that in terms of social protection of the most vulnerable sections of the population, much can be accomplished through policy-driven strengthening of national social "safety net" programmes, which provide relief for those who are most threatened by escalating food prices, while avoiding blanket policies such as price controls, which are easier (and cheaper) for governments to enact, but which have the perverse effect of reducing the producer response that could otherwise soften the price rises through increased output. As Msangi and Rosegrant highlight, the main challenge of policy in this case is to balance the need to maintain producer incentives and avoid distorting self-correcting signals from markets, while supporting human welfare through protecting the most vulnerable.

Major issues for such programmes are how to build on households' existing coping strategies, and how to achieve efficient targeting. Programmes need to be wide in scope, to support households beyond the short term. They may include measures to protect the short-term nutrition levels of those most in need, along with measures that can restore assets, to enable households to maintain productive capacity. Regarding targeting, although general food subsidy schemes often benefit many food-insecure households, they may also result in large leakages to the non-poor. The ideas and methods employed in recently implemented conditional cash transfer programmes provide a fertile area for adapting traditional food safety net programmes (European Commission, 2010). Food safety net programmes can also make a significant contribution to growth by partially alleviating the need to self-insure. However, these programmes may discourage poor households from investing. A large share of their savings, for instance, may be used to constitute food reserves. Social protection can also promote growth by creating physical assets through employment schemes. Public works programmes can make particularly relevant contributions – in both supporting the entitlements of the vulnerable and improving the access to markets and basic services that is essential for longer-term economic growth – through the provision of appropriate public infrastructure goods.

Another set of policy tool that may be worth exploring and enhancing is food-related global safety nets. These may assist countries that become unable to pay for increased food import bills in periods of high food prices. Global safety nets can be based on different measures, including those aimed at insuring food import bills. Countries could consider insuring their financial needs, to reduce the impact of high food import prices. Schemes may be conceived for providing relief through physical or monetary means, aimed at partially compensating for higher-than-expected costs of food imports.

In Chapter 5, van der Mensbrugghe and colleagues highlight the role of what they term the "emerging global middle class". This is a large population group that is projected to grow to about 2.1 billion people, or 28 percent of the global population, and will be found almost totally in developing countries, especially China and India. The food needs of the global middle class will affect global food demand. Hence policies will need to anticipate its changing preferences, as well as its food demands.

A major demand-related area where policy can be influential is the demand for food commodities for biofuel production. As seen, the likely size of this market varies across the projections: Msangi and Rosegrant (Chapter 2) and Fischer

(Chapter 3) indicate that the use of agricultural feedstocks for the production of biofuels is likely to grow in the medium to long term. Alexandratos (Chapter 11), instead, argues that at least the cereals involved in this type of use are likely to remain small compared with world aggregate consumption. Large persistent increases in oil prices – arising from phenomena unrelated to agriculture – can alter the economic fundamentals and make biofuels less dependent on mandates and more on market forces. In such circumstances, intensified competition for feedstock crops (and the underlying resources) will tend to siphon off supplies and resources from the food sector, to the detriment of the food security of weaker population groups. The eventual advent of second-generation biofuels may ease the impact, but will not eliminate it. Msangi and Rosegrant make the point that it may be useful for policy interventions to limit the use of agricultural feedstocks in first-generation ethanol and biodiesel production. Direct support to biofuel producers and blenders, and blending targets are currently applied in developed countries, along with trade barriers. These measures are likely to have significant impacts on global agricultural markets. If developing countries adopt similar measures, demand-side pressure in the feedstock markets will increase. Such policies currently present challenges in terms of fiscal affordability in developing countries, and their monitoring and coordination at a global level seem to be a fertile area for future policy coordination.

A consequence of a highly interdependent world food system is the establishment of global supply chains, such as exist today. A main driving force behind this phenomenon was the development of supermarkets and large food processors. Consumers have played a willing role in this development and, with the increasing size of the global middle class, these tendencies may become more manifest. In developed countries, the successful attempt to package foodstuffs as having attributes of health and environmental responsibility, along with animal welfare and fair labour conditions in some cases, has transformed the economics of food production and trade. In developing countries, middle-income consumers have embraced the availability of non-local foods and the better reliability and quality control that can come with company size and management expertise. Most of these tendencies have given rise to private standards that are additional to or different from those applied under the Sanitary and Phytosanitary Agreement of the World Trade Organization (WTO), which sought to control governments' ability to set import standards that were not justified by risk assessment and not based on scientific evidence. The Sanitary and Phytosanitary Agreement has been useful, particularly in the area of animal and plant diseases, but it has been less effective in the area of private standards. In terms of policy, the importance attached to food safety and quality point out the need to expand the international discussion on these topics. It would be useful to improve the regulation of non-safety-related standards in traded foods, and to review health and safety standards periodically, in light of changing scientific evidence.

In sub-Saharan Africa – as argued in Chapter 11 – food consumption is likely to be directly related to the evolution of domestic food production, more than to income growth. With limited shares of imported food in total supplies, high economic dependence on agriculture and little non-food use of food commodities, food consumption follows the evolution of production. This implies that policies for dealing with consumption in this region are largely intertwined with those for production.

Despite the slowing rate of growth in demand for cereals and meat, increases in income are still bound to lead to fast growth of meat consumption, especially in middle-income developing countries. Hence demand for feeds will also increase rapidly. This raises the issue of whether demand policies should target the dietary patterns of the growing global middle class, through nutrition education aimed at slowing the growth in demand for meat and meat products and reducing diet-related diseases. This policy issue would be more relevant in developed countries, where meat consumption and diet-related diseases are high (on a per capita basis, almost three times as high as in developing countries), but it also applies to developing countries with fast growth. The policy issue in this case is how to orient consumer behaviour towards more healthy habits. While large-scale nutrition education campaigns could assist, appropriate consumption taxes to discourage the consumption of specific types of product could also be envisioned. Mazzocchi *et al.* (2009) provide a wide discussion of these topics, mostly with reference to developed countries. The limited evidence available on campaigns shows that they may have a low impact, mainly on consumers who are already well informed. Labelling can be effective, to some extent. Taxes and subsidies for healthy foods can be effective in orienting consumers towards healthy foods to some extent; but as Schmidhuber and Traill (2006) observe, where the income elasticity of food is small, taxes can hardly affect consumers' behaviour. Hence education campaigns still need to be considered a key tool in this field.

An interesting demand-related policy direction may be the use of consumption taxes to transmit to consumers the resource and scarcity values of various raw materials that enter the food they eat. For instance, if a food product uses a water-intensive raw material and water has a high scarcity value, which is not reflected in the product's commercial price, consumption could be discouraged by the imposition of relevant resource-related consumption taxes. Such taxes could be imposed nationally or, if there are global resource sustainability issues, globally, by international agreement. The matter, however, is quite complex, as some of these taxes may imply considerable side-effects. This would be the case, particularly, of taxes imposed on energy-intensive production.

The supply side

One major conclusion of the analyses in Chapters 1 and 6 is that the world will need to increase agricultural production by about 70 percent between 2005 and 2050, to satisfy the growing demand for food and agricultural products. About 80 percent of this increase will need to come from increases in productivity, such as higher crop yields. To ensure that production prospects materialize with minimum negative side-effects, adjustments are likely to be necessary, especially to allow the transmission of appropriate scarcity signals and to correct market failures through policy interventions.

Agriculture is a part of almost every country's economy, and is a potential contributor to growth, but countries differ in their factor endowments and institutional settings, and there are large differences in how agriculture can contribute to growth and poverty alleviation. Based on their stocks of natural and human resources, and their existing potentials for agricultural growth, countries should design policy intervention with the aim of promoting production that generates higher actual and potential comparative advantages. There are no unique recipes for agricultural development. Instead, countries could identify their specific binding constraints and market failures, and address these. Such strategies, moreover, should not be static. They should take into account past experiences, developments in technology, changes in external and internal constraints, and emerging concerns.

A major shift in the focus of agricultural policies is required, and is partly taking place, from a product production approach to a value chain approach. Supply improvements through the implementation of production-enhancing measures alone may not be sufficient. Supply depends on a variety of downstream and upstream activities. For instance, unless a fertilizer supply system is in place, enhanced technology that is based on increased fertilizer applications may not be adopted, or may be adopted inefficiently. Similarly, unless the marketing side of any new volume of production is assured, efforts to increase supply may be thwarted.

To understand the various policy issues relating to supply it is helpful to adopt the yield definitions of Fischer and colleagues in Chapter 10. The gap between yields attainable under actual conditions – i.e., economically attainable under profit-maximizing conditions – and farm yields observed under current practices is largely due to improper practices, which may be improved through extension. The second gap, between yields attainable under efficient markets and actual yields, is due largely to factors beyond farmers' control, but within the scope of government policies. The third gap, between potential yields and yields attainable under efficient markets, is largely due to knowledge gaps. Policies can contribute to narrowing all these gaps, and this results in an increase of TFP.

Information on the size of the different yield gaps identified in Chapter 10 would assist policy-makers in prioritizing interventions. However, such information is rarely, if ever, available. What is more frequently available is information on the technical and allocative efficiencies of the agriculture sector, which relate to input use to achieve specific output levels or yields. Technical and allocative inefficiencies can be dealt with by policies relating to knowledge, through extension, and policies for removing factor distortions that prevent efficient factor use, such as inputs and credit.

A major policy issue for the coming decades appears to be how many resources to devote to increasing potential yields, and how many to closing the various yield gaps. For individual countries too, the major policy issue concerning agricultural supply – apart from expanding the resource supply – seems to be whether to emphasize the increase of potential yields or the closing of yield gaps. This must be decided at the country level, as the policies needed for each sphere are different. Unfortunately, some gaps may be easier to close than others. For instance, increases in potential yields may be derived from improved seed varieties, which may be relatively simple to distribute among and take up by farmers under existing input and market settings. Reducing factor market distortions, such as those affecting the supply of credit, may be more difficult. Given the large yield gaps in many developing countries, particularly in sub-Saharan Africa, policies aimed at closing yield gaps would seem to be preferable to policies aimed at raising potential yields, and therefore potentially making the yield gap even larger.

Governments' commitments to agriculture depend on their perceptions of the sector's potential for overall growth. In developed countries, where the yield gap is smaller, the best strategy seems to be to devote available public R&D resources to pushing the technology frontier outwards. A growing share of total agricultural R&D expenditures in these countries is financed by private companies. It is known that the private sector can capture only a share of the benefits of technology, and that some innovations produce hardly any appropriable benefit; hence the need for public-private partnerships. Concerns have been expressed because the total public resources devoted to agricultural R&D in developed countries are not increasing in line with either the likely social returns or the importance of future challenges. As indicated in Chapter 9, the private sector is unlikely to make up the difference, and the private sector also benefits from the basic research results that originate in publicly financed research. It is therefore important to keep up the momentum of public funding for agricultural R&D in developed countries, concentrating it on more basic research with longer-term benefits.

Among developing countries, large ones such as China and India that have no option but to supply a large share of their food needs from domestic production, will need to continue devoting resources to pushing out the production frontier, and they are already doing so. For smaller countries, supply-side policy options regarding technology may be related more to the issues of closing yield gaps and improving productive efficiency than to pushing out the yield frontier. In this context, resources should be devoted to maintaining the capacity to assess a country's potential in terms of water resources, soils and climate; where to obtain the necessary knowledge, science and technology to realize its potential; and where partnerships need to be negotiated and technology needs to be purchased.

R&D needs to become more responsive to farmers' needs, and one way of achieving this is via partnerships with farmers' organizations. Such partnerships can take many forms, ranging from simple consultation, to formal representation in research bodies, and financing via levies on specific products. More farmer participation in extension services can also be envisioned, and is crucial in bringing new technologies to the production level. Extension governance can be decentralized to the regional and local levels, with farmers' representatives being directly involved in the decentralized governance of technology and extension agencies.

A major policy issue for most developing countries is the private sector's role in providing some of the services that are traditionally deemed as being in the public sector's sphere. A major pathway for smallholders' integration into larger national and international markets involves contractual agreements between farmers or farmers' groups and a company with the technology and marketing network to take up production. For many developing countries, therefore, policies should be put in place to facilitate private companies' partnering or subcontracting of local producers to increase the production of certain products. Such companies can bring technical expertise, extension and ancillary services, such as credit and marketing, to increase production in specific areas and products. The environment for their operation should therefore be a focus for policy in many developing countries. Technology adaptation can also be enhanced by partnerships between the public and private sectors.

In Chapter 10, Fischer and colleagues point out that a key agricultural technology is irrigation, which is drastically underinvested in some regions, including sub-Saharan Africa. Chapter 6 shows that the global area equipped for irrigation in developing countries could expand by 32 million ha by 2050, especially in land-scarce regions such as East Asia, South Asia and the Near

East and North Africa. However, as Chapter 10 highlights, increases in irrigation would have to be accompanied by corresponding investments in installing adequate drainage facilities, to avoid problems of salinity. The policy issues in this area include the types of irrigation to be promoted, the institutional setting for managing irrigation systems, and the financing of irrigation schemes. Irrigation policy also needs to focus on minimizing waste and improving water efficiency.

A major way of closing the yield gap in many developing countries, as pointed out in Chapter 10, is to improve the functioning of markets. In this context, policies may need to target the reduction of margins between prices at the farm-gate and at central market locations. Large margins lead to disincentives for farmers and other stakeholders along value chains. At the same time, supply response also depends on the availability of credit, insurance, information and inputs. If agricultural supply in developing countries is to be made more responsive to market signals, improvements will be needed in all of these areas. Of crucial importance is the supply of financial services to farmers; the range of innovations in this area has included microfinance. However, apart from policies to support such institutions, there is also large scope for policy to assist the establishment of better financial services, by reforming public agricultural banks and linking them more closely with innovative institutional arrangements.

Agricultural production depends on the availability of inputs such as seeds, fertilizers and agrochemicals. The provision of many of these involves economies of scale, and most developing countries have relied on the public sector for this. However, given the well-known inefficiencies of the public sector, the role of the private sector needs more attention.

A new area for policy is the introduction of so-called "smart subsidies" to increase the use of crucial inputs such as fertilizer. Apart from its obvious budgetary cost, the use of subsidies needs to be carefully examined, and applied when there are good reasons to believe that such subsidies can overcome substantial market failures. For instance, if there are insufficient volumes of marketed products in some parts of a country, owing to low production, this may lead to an absence of permanent private marketing networks for the provision of inputs, as well as the marketing of outputs. In such cases, the use of time-bound input subsidies may lead to the expansion of marketed production at sufficient volumes to create the scale for the private sector to enter and provide the services and products cost-effectively.

Chapters 7 and 8 highlight the crucial role of agricultural capital in the production process, and the large amounts of new and, especially, replacement capital needed to meet the 2050 global targets. Imperfect as these estimates may be, they raise three major policy issues. The first concerns the choice of appropriate locations for new investments in agricultural production. Policies should target regions and countries where the incremental capital output ratios are relatively low, to facilitate the productive use of underutilized or unutilized resources, especially land and water.

The second policy issue, related to the result presented in Chapter 7, is that government expenditure on agriculture is correlated with capital formation and has a significant positive impact on TFP. This confirms the decisive role of public expenditure in creating an enabling environment in terms of infrastructure and sustainable access to natural resources. Adequate incentives for the private sector, particularly farmers, to invest in productive assets largely depend on the possibility of establishing such an environment. Governments in developing countries may change priorities in budget allocations and avoid, or at least reduce, discrimination against agriculture. However, it must be noted that the nature of public expenditures for agriculture is important. Expenditures on infrastructure and extension may have different effects compared with expenditures for salaries in ministries of agriculture. Foreign direct investment is also strongly correlated with productivity growth, and the presence of an efficient bureaucracy, the lack of corruption and democratic political structures; policy should focus on improving these areas.

The third policy issue concerns the source of finance. The bulk of investment finance for agricultural capital needs to come from domestic savings, and the major sources of these are lending from the domestic banking system and own savings from farmers and other stakeholders along production chains. Policy needs to make sure that these savings are mobilized and recirculated to finance capital needs directed at productive investments, rather than, for instance, keeping them at the household level to finance self-insurance against unpredictable shocks.

In Chapter 6, Bruinsma points out that arable land for agriculture in developing countries needs to expand by about 120 million ha by 2050. This expansion would benefit from policy actions ensuring that any new land is developed and cultivated sustainably and equitably. As mentioned, this volume does not emphasize this aspect; but it is clear that focus should be on preventing the conversion of forest land into arable land; policies should discourage such conversion in environmentally sensitive areas and regions, while providing alternatives for people who encroach on forests in the absence of other livelihood choices. Another area for policy focus is the adoption of conservation farming using zero tillage. Chapter 10 indicates that this is a major opportunity for reducing fuel use in agriculture and sequestering soil carbon. As conservation agriculture is knowledge- and location-specific, there is ample scope for public policy to encourage such practices via adaptive research and knowledge management.

An important potential avenue for encouraging the adoption of such practices in developing countries is payment for soil sequestration schemes – an area on which both national and international policy should focus.

One way in which land expansion can be combined with sustainable agricultural practices is through the use of a cluster approach. In this approach, certain areas or regions within a country are designated as priorities for agricultural development, and infrastructural and other facilities are built to facilitate private companies' investments in a cluster of activities related to agricultural production. The advantage of such an approach is that public investment and policy are combined with private sector capital and expertise.

Global public goods: trade policies, domestic support and the environment

It has long been recognized that agricultural policies have distorted not only the domestic agricultural markets of most countries, but also international agricultural markets. The agriculture sector was the main cause of delay in concluding the WTO Uruguay Round of trade negotiations, and is one of the main causes of delay in concluding the current Doha Round. In addition, analysis of the recent food crisis revealed that a large share of the global price rises during 2007/2008 could be attributed to short-term trade policies (Headey, 2010).

Increased investment in agriculture and adequate incentives to farmers are required to meet the global challenge to 2050, as highlighted in the third part of this volume. A key question is how to shape and design support to farmers in both developed and developing countries, while minimizing global market distortions that are potentially harmful to developing countries, and promoting global food supply adequacy, food security for the undernourished, and poverty-reducing and growth incentives for farmers in low-income, food-deficit countries.

Agricultural support in OECD countries is costly and distorts international commodity markets. It also disproportionately benefits the wealthier households that own large amounts of agricultural land, while raising food costs, which disproportionately reduces the real incomes of lower-income households. However, agricultural support is not uniformly distorting. While overall OECD support to farming has been remarkably stable over time, periodic reforms undertaken since the onset of the Uruguay Round (Skully, 2009) have resulted in declining levels of market distortion, owing to a reinstrumentation of policies. Subsidies directly attached – or "coupled" – to production have gradually been reduced and substituted by measures that support farmers' incomes and reduce their risk exposure. Among trade policies, tariffs are the predominant form of border measure. Market price support and payments based on output have decreased, and export subsidies and foreign surplus disposal are now relatively minor, having been heavily used in

OECD countries in the 1980s. Since the Uruguay Round, many OECD countries have introduced direct payments to producers as partial compensation for reduced tariff protection and lower product-specific support. One form of direct payment, decoupled support, has been prominent in the United States of America and the EU. Direct payments still distort output and trade, but to a lesser degree than tariff protection. However, estimates suggest that import tariff barriers represent 81.4 percent of total support to agriculture in all countries: tariffs accounted for USD 691 billion, direct domestic subsidies for USD 97 billion, and export subsidies for only USD 61 billion (Anderson, Martin and Valenzuela, 2006). All of the 70 countries most penalized by agricultural protectionism are developing countries (Bouët and Laborde, 2009), which suggests that the Doha negotiations' market access agenda should focus on the cutting of market access provisions.

As OECD farm support has shifted from commodity-based to decoupled measures, farm incomes have become more variable, and safety nets in the form of risk mitigation measures, such as revenue or weather insurance, are increasingly being relied on to provide protection from unpredictable income swings. Agricultural insurance has been widely subsidized in OECD countries, and legally so for WTO, which classifies it as a "green box" or "minimally distorting" policy. For instance, in the United States of America, farm insurance and payments under crop and weather insurance are projected to reach USD 22 billion in the 2008 to 2012 period, a substantial share of total United States farm support. Nevertheless, such measures tend to distort incentives, especially regarding investments, and can be a partial substitute for direct support policies. OECD (2009) reports that risk-related policies account for a significant share of the producer support estimate in OECD countries, averaging about 51 percent in the EU and 63 percent in the United States for the 2002 to 2007 period.

In developing countries, the period before the recent food crisis saw steadily reduced spending and investments in agriculture, with the latter receiving a disproportionately small allocation of public resources (Bezemer and Headey, 2006). Foreign aid to agriculture also contracted during this period. Developing countries' farm policies have been driven largely by the need to accelerate the transition from low-income agrarian structures to more developed industrialized and service-oriented economies. The overall effect of such policies has mainly been taxes on producers. In the process, the agriculture sectors in many countries have faced negative policy biases and low growth, while import dependence has increased. However, when average incomes grow (typically to levels of at least USD 8 000 per capita per year), the type of farmer support in developing countries seems to turn positive and to follow a pattern similar to that of developed countries. The results from a recent World Bank study estimating agricultural distortions for

75 developing countries from 1955 to 2007 (Anderson, 2009) bore this out, by showing that broadly developing countries taxed agriculture via price and trade policies from the early 1960s to the late 1970s/early 1980s, before gradually reducing taxation and switching to slightly positive assistance to agriculture, in aggregate, by the mid-1990s.

Table 12.1
Non-distorting farm support for developing countries' agriculture

Policy goal	Interventions
Maintain or improve productive capacity	R&D: new varieties
	Better management techniques
	Efficient use of inputs: water, fertilizer, pesticides
	Development of input market systems
	Improved storage, processing, product quality
	"Hard" infrastructure: irrigation, land restoration
	"Soft" infrastructure: information systems, lower transaction costs, extension of best practices
Correct market failures	Facilitation of exchange between producers and buyers
	Provision of credit: subsidized
	Technology dissemination, farmers' training
	Support to producers' organizations/inter-professional agreements
	Promotion of value chain development
Reduce income and price risks/ uncertainty	Support to information for insurance markets
	Market information systems for exchange
	Investments in post-harvest storage
	Veterinary services for livestock
	Insurance/safety nets against crop failures, droughts, etc.
Improve food security and reduce hunger	Fostering of rural employment
	Targeted input subsidies: fertilizer, seeds
	Storage/safe processing for staple foods
	Subsidized credit for farm and off-farm activities
	Promotion/creation of demand for staple food/cash crops
	Increased R&D in staple food varieties, improved techniques
	Investments/subsidies for post-harvest storage
	Quality control for stored grain
	Improved processing for perishable staples
Preserve natural resources and environment	Soil fertility management
	More efficient use of water: proper pricing
	R&D in varieties adapted to climate change
	Best practices for lower levels of pesticides

Source: Elbehri and Sarris, 2009.

The issue in developing countries is how to support farmers, to promote production, productivity and food security, without generating large domestic and international distortions. Elbehri and Sarris (2009) list a variety of public interventions that could serve several developing country goals and be deemed non- or minimally distorting. These are reproduced in Table 12.1. Many of these policies have already been reviewed or mentioned in various parts of the book; Table 12.1 outlines those that are not likely to raise issues with WTO. Implementing the interventions indicated would be a tall order for any developing country, but they give a menu of possible non-distorting policies.

As well as domestic policies, many other events are likely to shape future agricultural trade and trade policies. The past 30 to 35 years, since the food crisis of the mid-1970s, have seen the emergence of a more globalized food system and the shifting of national and global policy concerns to issues of growth in non-agriculture and more open trade. WTO and the debates surrounding agricultural trade have tended to neglect food security concerns. Nevertheless, the recent global food market events have refocused many policy-makers' views back on to food security. In addition, there have been a series of developments that are likely to impinge considerably on global food markets and trade.

Projections presented in the second part of this volume suggest that growth in the next few decades, whether fast or slow, will be faster in developing countries, especially in Asia. As seen, this will increase demand for the most income-elastic food products, such as livestock products, fruits and vegetables. If most of the growth in many of the faster-growing economies occurs outside agriculture, the demand for imports will increase faster than overall demand. Concerns about how to satisfy this growing domestic demand for food will be a major factor in shaping developing countries' agricultural trade policies and their attitudes towards WTO in the years to come. Fast growth in non-agricultural sectors may induce the familiar (in developed countries) political pressure to ease the adjustment via subsidies to rural areas. This will bring pressures for protection or domestic support. If WTO restricts countries' freedom to apply relevant policies, conflicts may arise between WTO commitments and domestic adjustment pressures. WTO commitments may therefore need to allow policy space for countries that are at different stages of development.

Perhaps more worrying for the world trade system as a whole is whether the aftermath of the recent financial and economic crisis and the attendant slowdown in global economic growth will create pressure for trade disruptions. Of particular concern are oil supplies, which depend on a relatively small group of countries. Periods of inflation and slow growth in the past have been associated with sharp increases in the price of crude oil. The issue is whether the global trade system as it has emerged since the Kennedy Round can survive self-preservation policies that may destroy laboriously established mechanisms. Fortunately, the world trading system embodied in WTO agreements seems to have survived the recent financial crisis quite well, and the main consequence seems to be the further stalling of the Doha Round.

The period since 1985 has seen a paradigm shift in management of the agricultural economy in both developed and developing economies, towards deregulation and more focus on preserving market signals and incentives and promoting risk management. In light of political demands generated by the increased food price volatility observed in recent years, an issue for the coming decades is whether the reform process will continue along the same path: in other words, whether the tendency will continue to be towards less market intervention and more risk management. Such a trend would be consistent with a more open trade system and removal of the impediments that developing countries face in supplying food to industrial country markets. However, the pace of reforms could stall if the Doha negotiations are delayed, or even abandoned.

A fundamental question is whether developing countries will follow the same pattern of protection for domestic markets and producers as developed countries. Much of the impetus for public intervention in developed country agricultural markets came as a reaction to different adjustment patterns in agriculture and non-agriculture sectors, and these pressures are already apparent in several developing countries that are going through a process of transition. Pressures from developing countries for relaxation of WTO rules pertaining to protection and domestic support to agriculture could become an issue in the next few decades.

The historical pattern of agricultural protection suggests that agriculture is first unprotected or even taxed at early stages of development, then goes through a cycle of protection and support while the country achieves middle-income status, when it is liberalized. If developing countries follow this pattern in the future, attempts to bind in WTO the current levels of protection and support may deny some developing and least-developed countries the policy flexibility needed to pass through the middle-income phase of their development. It is not clear whether developing countries will need to follow the historically traditional pattern and rate of protection of agriculture. However, if they do, and if the new WTO rules on agriculture do not allow it, pressures may be created for other types of support that are deemed compatible with WTO; in the worst case, this may threaten WTO itself. To prevent this, it may be appropriate to allow policy space for developing countries' agricultural trade-related policies.

Recent research has demonstrated that world trade in most products, including food products, is dominated by a few large multinational firms. Although this has resulted in more diverse and cheaper food and provided more consumer choice, especially in developed countries, a side-effect is that corporate decisions can affect millions of farmers and consumers. Concern has grown that the concentration of economic power could, at some stage, constrain rather than empower farmers and consumers.

In manufactures, much trade moves within the same firm, as supply chains lengthen. The same trend is noticeable in the food trade. While many countries, especially developed ones, apply anti-monopoly and anti-trust laws within national borders, such rules are non-existent in international trade. Competition issues are among the so-called "Singapore issues" that many countries deemed undesirable as part of the current Doha agenda. One of the main problems that hamper developments in this area is the lack of appropriate information, as well as the legal vacuum. For instance, if a multinational company is monopolizing a market, which national or international authority should be responsible for disciplining it? Whether and when global competition policy will re-emerge remains to be seen.

Will the global food market begin to fragment as more regional and bilateral trade agreements are concluded? Or will these regional and bilateral agreements effectively merge to create global free trade? Large countries and trading blocs such as the EU, the United States of America and Japan have already concluded many bilateral and regional trade agreements, and more are under negotiation. Agriculture is usually included cautiously, if at all, to avoid upsetting the status quo, and its inclusion entails many exemptions. There is an inherent asymmetry in such agreements, as the larger country with a larger market has an advantage over the smaller one. Preferential access to the larger market is usually bought at the cost of freer entry of the developed country partner's product into the smaller country's market. A major obstacle to taking advantage of such agreements is adherence to the rules of origin, which can place undue costs and other burdens on many administratively weaker economies, with the consequence that the agreements' potential benefits are not realized.

Finally, both the expansion of cultivated areas and the agricultural intensification indicated by projections to 2050 (particularly those by Bruinsma in Chapter 6) have significant implications in terms of global environmental policy. As mentioned, these are not emphasized in this volume, but it is worth recalling that agriculture emits about 30 percent of all greenhouse gases in the atmosphere (including through deforestation), and can therefore contribute to a reduction of such gases through environmentally friendly production practices. Key questions in the current debate are: How can society motivate farmers to reduce the negative environmental side-effects of agricultural production, while continuing to meet the increasing demand for farm products and enhancing the positive environmental impacts? and How can countries agree on common policies to address global externalities?

Areas for future policy and analytical focus

Previous chapters, as well as the discussion in this one, have highlighted a very wide set of subject matters, policies and actions that will need to be explored to promote the global agrifood system's ability to respond to the needs of a growing and wealthier population, such as that expected in the 2050 perspective. Bearing in mind that there are still more questions than answers in each of these areas, this last section proposes a list of points that seem to follow from the analyses conducted in other chapters, and on which analysts and policy-makers may focus their attention. The following six such areas seem to follow from the analyses conducted in the rest of the volume.

- *Growing consideration of the demand side along with supply policies*: Until the end of the 1970s, the main focus of agricultural and food policies was production and the supply side. From the mid-1980s to more recent years, the emphasis was mostly on reducing distortions in world markets. The recent food crises have brought a renewed emphasis on the need to invest in agriculture and increase productivity. In the coming decades, the demand side is also likely to require an increased policy focus, to provide the proper signals for the supply side, and also to anticipate and even influence patterns of global food needs and the use of agricultural products. Increasing attention will likely need to be devoted to food-related policies aimed at orienting and informing consumers, to prevent malnutrition, reduce the incidence of diet-related diseases and ensure food safety. The outlook exercises suggest that poverty and undernourishment may become smaller problems compared with their current levels. However, food-related safety nets aimed at improving food access are likely to continue to play a key role, at both the national and international levels, at least in areas such as sub-Saharan Africa.

- *Production chains rather than individual producers should become the target and the focus of analysis and policy-making on the supply side*: Relations along production chains tend increasingly to shape incentives and outcomes for individual stakeholders, especially for poor farmers, whose ability to participate in this type of setting should be enhanced. Policies should increasingly focus on market failures that prevent production chains from working effectively, by simultaneously considering input and output markets, from the farmer to the final consumer.

- *Agricultural production expansion will need to rely on new technologies, but a lot can be achieved by simply promoting the adoption and adaptation of existing technologies*: As seen in several chapters of this volume, there are large gaps in productivity between developed and developing countries. This is clearly a major area for policy focus, and will necessitate major adaptations to existing institutional and incentive environments in many countries. However, it has also been pointed out that closing yield gaps alone will not be sufficient for meeting the 2050 production challenges. New technologies will most probably be required, and it appears that several R&D directions already promise to expand the production frontier considerably.

- *More sustainable production systems need to be promoted*: Current agricultural production practices in both developed and developing countries seem to have created a growing burden on the world's resources, such as land, water and the environment. While different conclusions may be drawn about the probability of a global Malthusian scenario, it is clear that the resource basis will undergo significant stress in some of the countries that are already more sensitive in terms of poverty and food insecurity. The thrust of agriculture in the future will need to be oriented towards far more resource-sustainable practices and technologies.

- *Climate change will need to be an integral part of agriculture and food policies*: It has become clearer in recent years that agriculture is both part of the climate change problem and a potential part of the solution. It will require a major shift of current policy thinking to acknowledge this in many countries, and to shape policies to make agriculture an integral part of the climate change adaptation and mitigation landscape. As seen in this volume, opportunities exist, and policies should assist in involving agriculture more in the overall international climate change debate and policy arena. Once more, emphasis should mostly be on those areas that are already more sensitive in terms of potential impacts on poverty and food insecurity.

- *The world trading system will need to be flexible to accommodate both old and new agricultural concerns*: The outlook exercises presented indicate a likely perspective of growing interaction among countries through trade in the coming decades. WTO has proved remarkably resilient and accommodating to changes in policies and views among its growing number of members. However, as the now developing countries become a larger part of the international trade sphere, rules related to agriculture will have to be adapted to meet their concerns, and give them some policy space in the period of transition to a more developed status. As part of the future global trading system, a global mechanism should be considered for addressing competition issues at the international level. Although this is an overly ambitious proposition, the growing influence of multinational companies and the diffusion of international production chains seem to call for more attention in this area.

References

Anderson, K. 2009. *Distortions to agricultural incentives: A global perspective, 1955–2007.* London, Palgrave Macmillan, and Washington, DC, World Bank.

Anderson K., Martin, W. & Valenzuela, E. 2006. The relative importance of global agricultural subsidies and market access. *World Trade Review*, 5(3): 357–376.

Bezemer, D. & Headey, D. 2006. Something of a paradox: The neglect of agriculture in economic development. Paper presented at the IAAE Conference, 12–18 August 2006, Gold Coast, Australia.

Bouët A. & Laborde, D. 2009. Market access versus domestic support: Assessing the relative impacts on developing countries agriculture. *In* A. Elbehri and A. Sarris, eds. *Non-distorting farm support to enhance global food production.* Rome, FAO.

Elbehri, A. & Sarris, A. 2009. Introduction and overview. *In* A. Elbehri and A. Sarris, eds. *Non-distorting farm support to enhance global food production.* Rome, FAO.

European Commission. 2010. *Social protection for inclusive development: A new perspective in EU cooperation with Africa.* European Report on Development. Brussels.

Fan, S. & Rao, N. 2003. *Public spending in developing countries: Trends, determination and impact.* Discussion Paper No. 99. Washington, DC, IFPRI.

FAO. 2006. *World agriculture: towards 2030/2050. Prospects for food, nutrition, agriculture and major commodity groups. Interim report.* Rome, Global Perspective Studies Unit.

FAO. 2007. *The State of Food and Agriculture: Paying farmers for environmental services.* Rome.

FAO. 2009. *Food security and agricultural mitigation in developing countries: Options for capturing synergies.* Rome.

FAO. 2011. *The State of Food and Agriculture. Women in agriculture. Closing the gender gap for development.* Rome.

Finucane, M.M., Stevens, G.A., Cowan, M.J., Danaei, G., Lin, J.K., Paciorek, C.J.,

Singh, G.M., Gutierrez, H.R., Lu, Y., Bahalim, A.N., Farzadfar, F., Riley, L.M. & Ezzati, M. 2011. National, regional, and global trends in body-mass index since 1980: systematic analysis of health examination surveys and epidemiological studies with 960 country-years and 9.1 million participants. *The Lancet*, 4: 1–11.

Foresight. 2011. *The future of food and farming*. London, Government Office for Science.

Headey, D.C. 2010. *Rethinking the global food crisis: The role of trade shocks*. IFPRI Discussion Paper No. 958. Washington, DC, IFPRI.

Headey, D.C. & Fan. S. 2010. *Reflections on the global food crisis: How did it happen? How did it hurt? And how can we prevent the next one?* IFPRI Research Monograph No. 165. Washington, DC, IFPRI.

Hertel, T.W. 2010. The global supply and demand for agricultural land in 2050: A perfect storm in the making? AAEA Presidential Address 2010. Milwaukee, Wisconsin, USA, Agricultural and Applied Economics Association (AAEA).

Mazzocchi, M., Traill, W.B. & Shogren, J.F. 2009. *Fat economics. Nutrition, health, and economic policy*. Oxford, UK, Oxford University Press.

OECD. 2009. *An overview of risk-related policy measures*. TAD/CA/APM/WP(2008)24/REV1. Paris.

Patomäki, H. 2008. *The political economy of global security: War, future crises and changes in global governance*. London and New York, Routledge.

Schmidhuber, J. & Traill, W.B. 2006. The changing structure of diets in the European Union in relation to healthy eating guidelines. *Public Health Nutrition*, 9(5): 584–595.

Shetty, P. & Schmidhuber, J. 2006. Introductory lecture. The epidemiology and determinants of obesity in developed and developing countries. *Journal for Vitamin Nutrition Research*, 76(4): 157–162.

Skully, D. 2009. OECD policy and distortionary effects: A review of the evidence. *In* A. Elbehri and A. Sarris, eds. *Non-distorting farm support to enhance global food production*. Rome, FAO.

World Bank. 2010. *World Development Report 2010. Development and climate change*. Washington, DC.

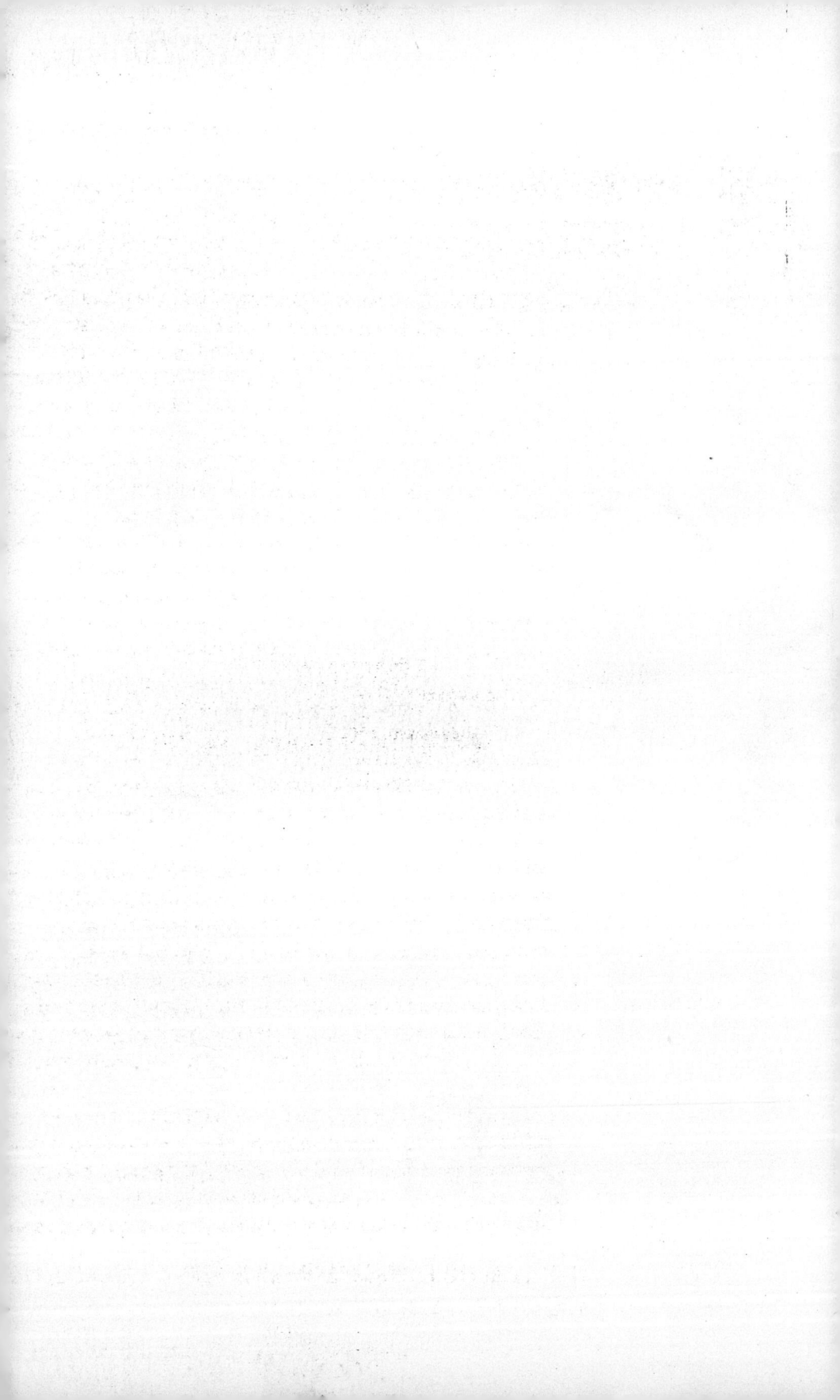